SDGs and Textiles

Series Editor

Hafeezullah Memon, International Institute of Silk, Zhejiang Sci-Tech University, Hangzhou, China

The book series "SDGs and Textiles" addresses the strategies to achieve sustainable development goals (SDGs) in the present, past, and future. It presents books about the present and future policies of textile ministries of different countries, and books related to sustainability education around different parts of the world in the textile sector. Moreover, it would welcome the conference proceeding related to SDGs and Textiles. The series would cover books comparing the sustainability and SDGs of different institutions and countries. The individual book volumes in the series are thematic. The goal of each book is to give readers a comprehensive overview of a different area of sustainability in the textile sector. As a collection, the series provides valuable resources to a broad audience in academia, the research community, industry, and anyone looking to expand their knowledge of SDGs and Textiles.

Textiles and life are together – life cannot be separated from textiles as it is the most important need for human beings after food. In 2015, the United Nations General Assembly proposed 17 interlinked global goals to be achieved by 2030. Since then, academia and industry have paid much attention to achieving these goals. Textile found its close relation with almost all of these 17 goals.

SDG 1 - No Poverty: Poverty would never be overcome by a charity only; it is essential to develop people's skills to have a better and wealthy life. Thus, the textile can be considered an excellent discipline to achieve this goal by creating jobs and small and medium businesses.

SDG 2 - Zero Hunger: Through the effective utilization of advanced application of Agrotech Textiles, it is possible to have higher crop yields and save crops from rough weather, unexpected rains, floods, insects, etc.; thus, geotextiles play an essential in achieving this goal of sustainable development.

SDG 3 - Good Health & well-being: There has been much health consciousness after Covid19, and medical textiles assist in getting good health and well-being.

SDG 4 - Learning & Education: Textile or fashion has remained a significant discipline for societies for ages, and there has always remained much to explore in this field. Textile-related universities may play a vital role by offering free access to their education resources, training and spreading information among the locals.

SDG 5 - Gender Equality: The textile sector is one of the industrial sectors that accepted gender equality long ago; in particular, the garment sector has more females than males. Thus, the textile sector has been doing gender equality. Moreover, there has been a recent trend for Gender Neutral Clothing, which need worth studying and may further assist gender equality.

SDG 6 - Clean Water & Sanitation: Textiles could be achieved through filtration, and of course, textile is one of the critical materials for filtration.

SDG 7 - Affordable & Clean Energy: With the recent advancement in material science and engineering, the textile sector has come on the front for, not only by using this clean energy during textile production but also by assisting the production of this clean energy, either in the form of wind turbines blades made of textile composites or by energy harvesting from T-Shirts, etc.

SDG 8 - Decent Work: Recently, there has been much attention that the textile workers are not paid well, labor rights are not cared about, etc.

SDG 9 - Industry and innovation: Textile Industry always follows innovation; the textile companies that do not chase innovation cannot survive in the market.

SDG 10 - Reduced Inequalities: Getting better life and well-being would help reduce inequalities in the textile industry.

SDG 11 - Sustainable Cities: Sustainable Textile Cities through Buildtech and transport textiles.

SDG 12 - Consumption and Production: Textile and garment consumption and production all come under.

SDG 13 - Climate Action: Oekotech or Ecotech Textile, waste management of textiles are upfront to achieve this goal of sustainable development.

SDG 14 - Life Below Water: Mitigating microfiber waste in rivers and oceans may come under the context of it. There has been much attention on this subject after passing the bill at the parliament level of the UK.

SDG 15 - Life on Land: Geotech or Geotextiles studies life on land.

SDG 16 - Peace, Justice, and Strong Institutions: Protective textiles are doing their best to achieve peace, justice, and strong institutions.

SDG 17 - Partnerships for the Goals: The application of textiles to achieve sustainable development goals is only an example. In all textiles sectors, combined efforts of all the goals are essential to achieve true sustainability.

Pardeep Singh
Editor

Dye Pollution from Textile Industry

Challenges and Opportunities for Sustainable Development

Springer

Editor
Pardeep Singh
Department of Environmental Studies
PGDAV College, University of Delhi
New Delhi, Delhi, India

ISSN 2948-1236 ISSN 2948-1244 (electronic)
SDGs and Textiles
ISBN 978-981-97-5340-6 ISBN 978-981-97-5341-3 (eBook)
https://doi.org/10.1007/978-981-97-5341-3

This Springer imprint is published by the registered company Springer Nature Singapore Pte Ltd.
The registered company address is: 152 Beach Road, #21-01/04 Gateway East, Singapore 189721, Singapore

Preface

The textile industry is one of the world's largest and most important industries, providing clothing, textiles, and other materials to consumers around the globe. However, the industry also contributes significantly to environmental pollution, notably dye pollution. Using dyes in textile production results in various environmental and health impacts, including water pollution, air pollution, and toxic waste. This book provides a comprehensive overview of the challenges associated with dye pollution and highlights opportunities for sustainable development in the industry. The book discusses the environmental and health impacts of textile dyeing, the regulations and standards related to dye pollution, and the available technologies and strategies for reducing dye pollution. One of the significant challenges associated with dye pollution is the contamination of water resources. Textile dyeing and printing processes require large quantities of water, often discharged into nearby water bodies without proper treatment. This can result in the contamination of water resources, affecting aquatic ecosystems and human health. The book discusses the available technologies and strategies for reducing water consumption and improving water treatment in the textile industry. Another challenge associated with dye pollution is the production of toxic waste. Many textile dyes contain toxic chemicals, which can be released into the environment during the production and disposal of textile products. It also highlights the importance of adopting sustainable production processes and waste management strategies to minimize toxic waste products in the textile industry. It will provide a valuable resource for researchers, industry professionals, policymakers, and anyone interested in the environmental impact of textile production. It emphasizes the importance of adopting sustainable development practices in the textile industry to minimize dye pollution's environmental and health effects and promote a more sustainable future.

New Delhi, India Pardeep Singh

Contents

Editor and Contributors

About the Editor

Dr. Pardeep Singh is presently an assistant professor at the Department of Environmental Science, PGDAV College, University of Delhi, New Delhi, India. He obtained his master's degree from the Department of Environmental Science, Banaras Hindu University, Varanasi, India. He obtained his doctorate from the Indian Institute of Technology, Banaras Hindu University, Varanasi. He has published more than 65 papers in international journals. He has also edited more than 45 books with various global publishers.

Contributors

Charles Oluwaseun Adetunji Applied Microbiology, Biotechnology and Nanotechnology Laboratory, Department of Microbiology, Edo State University Uzairue, Uzairue, Nigeria

Marian Onwude Agbugui Department of Biological Science, Edo State University Uzairue, Uzairue, Nigeria

Pinki Rani Agrawal Deenbandhu Chhotu Ram University of Science and Technology, Sonipat, Haryana, India

S. B. Agrawal Department of Botany, Institute of Science, Banaras Hindu University, Varanasi, India

Fatima Ahmad Sustainable Development Study Centre (SDSC), Government College University, Lahore, Pakistan

Shahnaz Ahmed Department of Chemistry, Barnagar College, Sorbhog, Assam, India

Alexander Ikechukwu Ajai Department of Chemistry, Federal University of Technology Minna, Minna, Nigeria

Jibrin Noah Akoje Department of Chemistry, Baze University Abuja, Abuja, Nigeria

Charu Arora Department of Chemistry, Bilaspur, Chhattisgarh, India

Mahesh Chandra Arya Department of Chemistry, Kumaun University, Nainital, Uttarakhand, India

Fizzah Ayub Sustainable Development Study Centre, Faculty of Physical Sciences, Government College University, Lahore, Pakistan

Bhavana Balachandran School of Civil Engineering, Vellore Institute of Technology, Chennai, India

Shiwani Berry Department of Chemistry and Chemical Sciences, Central University of Himachal Pradesh, Shahpur, Kangra, India

Manjunath Chatti Istituto Italiano di Tecnologia, Genova, Italy

Oyewale Daniel Department of Electrical and Electronic Engineering, Edo State University Uzairue, Uzairue, Nigeria

Neha Dhar Chemist/Microbiologist, M/S. Fresho Beverages, Assam, India

Rakesh Dutta Department of Chemistry, Pandit Deendayal Upadhyaya Adarsha Mahavidyalaya, Biswanath, Assam, India

Kingsley Erhons Enerijiofi Department of Microbiology, Glorious Vision University, Ogwa, Nigeria

Ohiokha Godwin Department of Accounting, Edo State University Uzairue, Uzairue, Nigeria

Aishatu Habib Idris Department of Microbiology, Edo State University Uzairue, Uzairue, Nigeria

Nazneen Hussain Department of Biosciences, School of Life Science, Assam Don Bosco University, Kamarkuchi, Assam, India

Hope Oshiolumetse Imoedemhe Department of Electrical and Electronic Engineering, Edo State University Uzairue, Uzairue, Nigeria

Abel Inobeme Department of Chemistry, Edo State University Uzairue, Uzairue, Nigeria

Jonathan Inobeme Department of Geography and Environmental Management, Ahmadu Bello University, Zaria, Nigeria

Noor Jahan Central University of Himachal Pradesh, Shahpur, Himachal Pradesh, India

Rajesh Kumar Department of Chemistry, SSJ University, Almora, Uttarakhand, India

Manjeeta Kumari Depatment of Chemistry and Chemical Sciences, Central University of Himachal Pradesh, Shahpur, Kangra, India

Roopa Kumari Department of Applied Sciences and Humanities, Indira Gandhi Delhi Technical University for Women, Delhi, India

Usha Kumari Department of Zoology, Gargi College, University of Delhi, Delhi, India

Kshitindra Kumar Singh Water Resource Management, Central Institute of Mining and Fuel Research, Dhanbad, Jharkhand, India

Divya Mahar Department of Chemistry, Kumaun University, Nainital, Uttarakhand, India

Sandeep Dutt Maindoli Department of Zoology, Kumaun University, Nainital, Uttarakhand, India

Amos Mamman Department of Chemistry, Ibrahim BadamasiBabangida University Lapai, Lapai, Nigeria

Abdullahi Mann Department of Chemistry, Federal University of Technology Minna, Minna, Nigeria;
Department of Electrical and Electronic Engineering, Edo State University Uzairue, Uzairue, Nigeria

John Tsado Mathew Department of Chemistry, Ibrahim Badamasi Babangida University Lapai, Lapai, Nigeria

Lalit Mohan Department of Chemistry, Kumaun University, Nainital, Uttarakhand, India

M. K. Mondal Department of Chemical Engineering and Technology, Indian Institute of Technology, Banaras Hindu University, Varanasi, Uttar Pradesh, India

Md. Iftakharul Muhib Department of Environmental Sciences, Jahangirnagar University, Dhaka, Bangladesh;
General Education Department, City University, Savar, Dhaka, Bangladesh

Nasreen S. Munshi Institute of Science, Nirma University, Ahmedabad, Gujarat, India

Mahnoor Nadeem Sustainable Development Study Centre (SDSC), Government College University, Lahore, Pakistan

Syeda Hamayal Zahra Naqvi Sustainable Development Study Centre, Faculty of Physical Sciences, Government College University, Lahore, Pakistan

Syeda Laiba Haider Naqvi Sustainable Development Study Centre, Faculty of Physical Sciences, Government College University, Lahore, Pakistan

Victor Oisamoje Department of Electrical and Electronic Engineering, Edo State University Uzairue, Uzairue, Nigeria

Priya Patel Institute of Science, Nirma University, Ahmedabad, Gujarat, India

Rushika Patel Varahi IO, Chicago, USA

Alka Patle Department of Chemistry, Govt. Nagarjuna Post Graduate College of Science, Raipur, Chhattisgarh, India

Tarun Kumar Patle Department of Chemistry, Pt. Sundarlal Sharma (Open) University, Chhattisgarh, Bilaspur, India;
Department of Chemistry, Guru Ghasidas Vishwavidyalaya, Bilaspur, Chhattisgarh, India

Priya Government Degree College, Nagrota Surian, Kangra, India;
Department of Chemistry and Chemical Sciences, Central University of Himachal Pradesh, Shahpur, Kangra, India

Md. Mostafizur Rahman Laboratory of Environmental Health and Ecotoxicology, Department of Environmental Sciences, Jahangirnagar University, Dhaka, Bangladesh

Sanchayita Rajkhowa Department of Chemistry, Haflong Government College, Dima Hasao, Assam, India

Rd Sabina Department of Microbiology, School of Life Science, Assam Don Bosco University, Kamarkuchi, Assam, India

P. C. Sabumon School of Civil Engineering, Vellore Institute of Technology, Chennai, India

Jyotirmoy Sarma Department of Chemistry, Assam Don Bosco University, Guwahati, Assam, India

Nirupamjit Sarmah Department of Environment and Forests, Government of Assam, Assam, India

Nitish Semwal Department of Chemistry, Kumaun University, Nainital, Uttarakhand, India

Elijah Yanda Shaba Department of Chemistry, Federal University of Technology Minna, Minna, Nigeria

Rahul Sharma Central University of Himachal Pradesh, Shahpur, Himachal Pradesh, India

Ravindra Pratap Singh Central Public Works Department, New Delhi, India

Durgesh Kumar Sinha Department of Chemistry, Bilaspur, Chhattisgarh, India

Shreya Sinha Central University of Himachal Pradesh, Shahpur, Himachal Pradesh, India

Shalini Srivastava Department of Botany, Institute of Science, Banaras Hindu University, Varanasi, India

Amtul Bari Tabinda Sustainable Development Study Centre (SDSC), Government College University, Lahore, Pakistan

Rameesha Tanveer School of Environment and Science, Australian Rivers Institute, Griffith University, West Leederville, Queensland, Australia

James Emokhai Uzualu Department of Law, Edo State University Uzairue, Uzairue, Nigeria

Iqra Wasif Sustainable Development Study Centre (SDSC), Government College University, Lahore, Pakistan

Abdullah Yasar Sustainable Development Study Centre, Faculty of Physical Sciences, Government College University, Lahore, Pakistan

Purvi Zaveri Biocare Research (I) Pvt. Ltd., Ahmedabad, India

Dye Pollution and Environmental Pollution

Understanding Dye Pollution and Its Impact on the Environment

Abel Inobeme, John Tsado Mathew, Charles Oluwaseun Adetunji, Marian Onwude Agbugui, Jonathan Inobeme, Kingsley Erhons Enerijiofi, Jibrin Noah Akoje, Elijah Yanda Shaba, Alexander Ikechukwu Ajai, Abdullahi Mann, Victor Oisamoje, Hope Oshiolumetse Imoedemhe, Aishatu Habib Idris, Ohiokha Godwin, James Emokhai Uzualu, and Oyewale Daniel

Abstract There have been rising concerns about the increasing concentrations of dyes in various environmental matrices most especially soil and wastewater. Dyes are employed for various purposes in different industries such as textile industries, food production industries, paper manufacturing, and cosmetics among others. Thus industrial activities constitute the major sources of dyes in the environment. Environmental pollution by dyes has become of serious concern due to the deleterious

A. Inobeme (✉)
Department of Chemistry, Edo State University Uzairue, Uzairue, Nigeria
e-mail: inobeme.abel@edouniversity.edu.ng

J. T. Mathew
Department of Chemistry, Ibrahim Badamasi Babangida University Lapai, Lapai, Nigeria

C. O. Adetunji · A. Habib Idris
Department of Microbiology, Edo State University Uzairue, Uzairue, Nigeria

M. O. Agbugui
Department of Biological Science, Edo State University Uzairue, Uzairue, Nigeria

J. Inobeme
Department of Geography and Environmental Management, Ahmadu Bello University, Zaria, Nigeria

K. E. Enerijiofi
Department of Microbiology, Glorious Vision University, Ogwa, Nigeria

J. N. Akoje
Department of Chemistry, Baze University Abuja, Abuja, Nigeria

E. Y. Shaba · A. I. Ajai · A. Mann
Department of Chemistry, Federal University of Technology Minna, Minna, Nigeria

A. Mann · V. Oisamoje · H. O. Imoedemhe · O. Daniel
Department of Electrical and Electronic Engineering, Edo State University Uzairue, Uzairue, Nigeria

O. Godwin
Department of Accounting, Edo State University Uzairue, Uzairue, Nigeria

P. Singh (ed.), *Dye Pollution from Textile Industry*, SDGs and Textiles,
https://doi.org/10.1007/978-981-97-5341-3_1

impact on the ecosystem and the environment at large. The removal of dye therefore becomes necessary. Various methods have been employed for the removal of dyes some of which include physical, chemical, and biological approaches some of the methods have their inherent limitations and challenges. Some of the physical approaches include filtration, flocculation/coagulation, and adsorption, while chemical methods include photocatalysis, electro-Fenton, and ozonation. Microorganisms and enzymes are employed in biological methods of removing dyes. This chapter presents a comprehensive overview on understanding dye pollution. It discusses the concept of dyes, sources of dyes, and types of dyes. It also highlights the environmental and health effects of dye pollution. Some recent reports in this regard are also presented.

Keywords Dye · Pollution · Coagulation · Natural dyes · Synthetic dyes · Ozonation

1 Introduction

One of the major issues of concern in recent times is environmental contamination from various anthropogenic activities. This has attracted the great attention of researchers. Industrialization and urbanization have been identified as the duo factors responsible for the pollution of the environment [1–4]. Among the numerous industrial activities that can result in the release of contaminants, the textile industry has been identified as one of the major sources of contamination from the dye. Dyes generated by the paper, printing, and textile industries can be of serious source of pollution in waterways and rivers. It has been estimated that more than 700,000 tons of dyestuffs are manufactured on a yearly basis. Other industries that also contribute to the pollution of the environment by dye include pulp and paper mills, distilleries, dyestuff, and tanneries [4–6]. Basically, dyes are colored substances with a remarkable affinity for the applied substrate. The colored nature of dyes occurs due to the absorption at the ultraviolet region of the electromagnetic spectrum. The colors of dye depending on the potential of the compounds to take in light within the range of the visible region of the electromagnetic ray (380–750 nm) [7]. A trace quantity of dye in an aqueous medium is capable of giving rise to a highly conspicuous color with a remarkable molar extinction coefficient. The use of dyes in pharmaceutical manufacturing, printing, processing of food, and fabric production has led to the discharge of organic and inorganic dyestuffs into the environment most especially water [8–10].

J. E. Uzualu
Department of Law, Edo State University Uzairue, Uzairue, Nigeria

Historically, William Perkin was the first to discover the first commercially available synthetic dye in 1856. Toward the end of the nineteenth century, more than ten thousand new synthetic dyes were already available. On a global scale, China is the largest producer of dye materials. There are numerous categories of dyes, which can be synthetic or natural, while synthetic dyes are more commonly employed in textile fibers, whereas naturally occurring dyes are employed as coloring agents in food [7, 11, 12].

Studies have documented the presence of various classes of dyes such as xanthenes, anthraquinone, phthalocyanine, and quinonimine in environmental matrices as a result of their structural nature. The most common dye in the environment is azo dyes, which is a result of their natural color, ease of accessibility, and less expensive nature [10, 13–15]. The elimination of dyes has become paramount due to their potential carcinogenic nature. They are capable of causing serious harm to all aquatic lives and also deter the entrance of light into the aquatic environment for plant usage. It has been reported that about 9% of the overall dye substances that are produced at the global level are released into the wastewater from textile industries [1, 2, 16].

This chapter gives a detailed overview of understanding dye pollution. It presents the concept of dyes, sources of dyes, and types of dyes. It also expatiates the environmental and health effects of dye pollution. Some recent reports in this regard are also presented (Fig. 1).

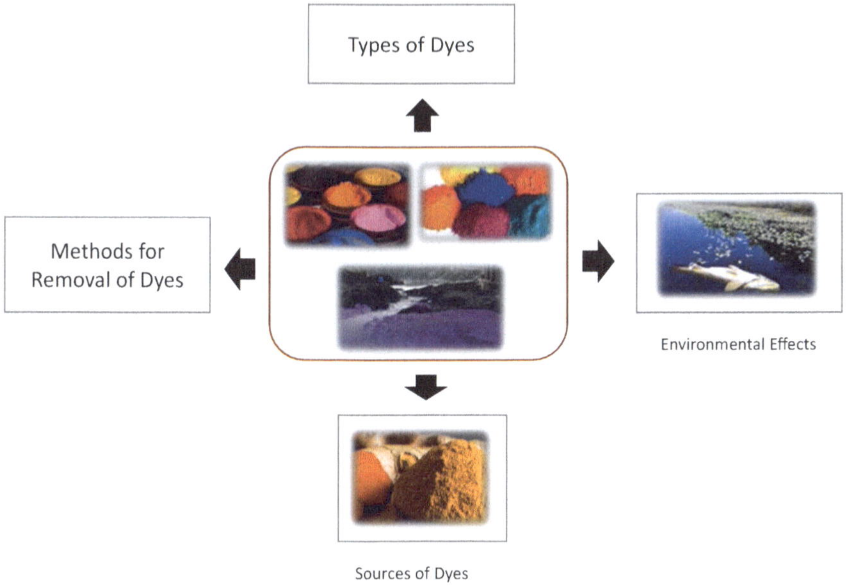

Fig. 1 Schematic representation of chapter

2 Concept of Dyes

Dyes are substances that have the potential to stick to the surface of a fabric or any other substrate where they are applied. This quality makes dye to be different from pigments in that there is no chemical binding of pigment on the materials that are colored. Dyes that take a long time to color the fabric material are not chemically stable and are less preferred. The dying of the product brings about the formation of strong covalent bonds between the dye molecules. The durability of the dye material is affected by temperature and time [17, 18].

3 Classification of Dyes

3.1 Based on Color Index (CI)

The classification of dyes is based on their color index abbreviated as CI. This development is by the Association of Dyers and Colorists and is periodically edited. In this system, dyes are first listed based on their general names by their color and application, which is then followed by an assigned 5 digit CI number. Examples of this classification include Mordant yellow 10 (1410), Reactive red 4 (18,105), and Acid blue 120 [19, 20].

3.2 Based on Functionalities and Unique Macroscopic Property

Dyes are also classified into classes which include: solvent, pigment, metallic, mordant, disperse, direct, basic, and acids, and this reveals their unique macroscopic property as well as underlying functionalities. The kind of dye employed is dependent on their unique compatibility with the substrate. The nonionic dyes include solvent, pigment, and disperse, and are also known to be soluble in water. The cationic dyes are usually basic. Disperse dyes are employed in the nonaqueous dispersion systems. Most dyes that fall into the mordant group are usually anionic, though there are some that are cationic in nature. Anionic dyes show a net charge when present in an aqueous solution which is a result of the sulphonate group that is present [21].

Some other bases that have been employed in the grouping of dyes include end-usage, chemical composition, and usage class. One of the primary ways of classifying dyes is based on the fibers on which they can be applied as well as the nature of the dye [22].

3.3 Based on Origin or Source

Dyes are further divided into two main categories natural and synthetic dyes (Fig. 2).

i. Natural dyes

They are obtained from animals and plant materials. The most common natural dyes are obtained from plant-based sources such as berries, roots, wood, leaves, and bark, as well as other non-animal sources such as lichens and fungi. The primary sources of natural dyes, sometimes referred to as natural pigments, are plant roots, stems, leaves, flowers, fruits, animals, or naturally colored ores [23]. The majority of natural dyes are vegetable dyes, which are primarily derived from different plant parts such as roots, stems, seeds, bark, leaves, and wood. Color compounds derived from natural

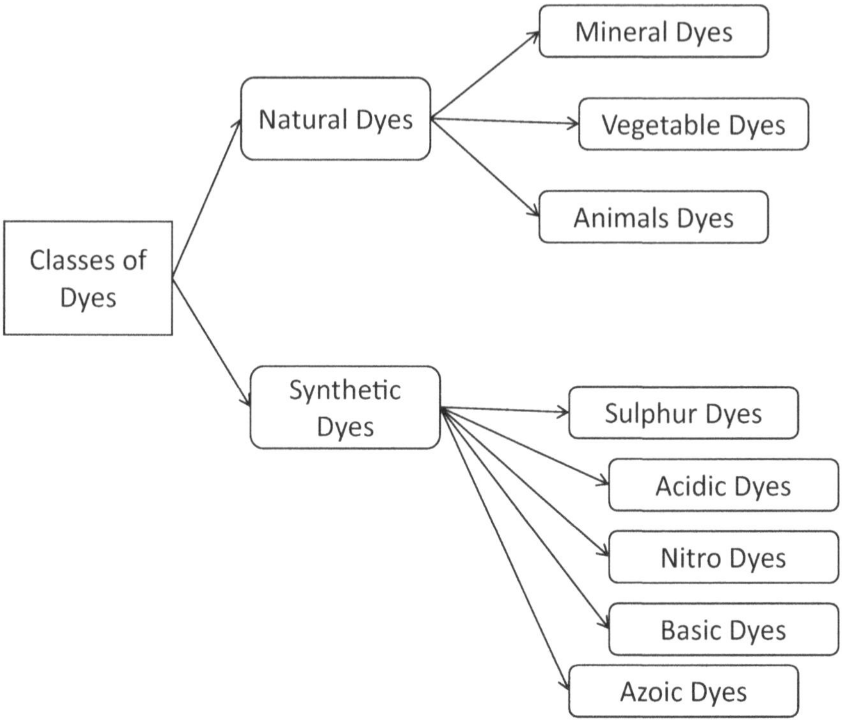

Fig. 2 Classes of dyes

sources are called natural dyes. Up to the mid-1800s, natural dyes were employed for all kinds of textile printing and dyeing. As synthetic colors were available, the use of natural dyes decreased [24].

Based on where they come from, natural dyes are divided into three categories: mineral, vegetable, and animal dyes.

Advantages of natural dyes

- Natural dyes are eco-friendly since they are derived from natural sources.
- These dyes offer superior UV protection.
- Certain natural dye has the ability to ward off mosquitoes and withstand flames.
- They are soft and soothing.
- Minimal environmental Impact: Natural dyes are quite popular since they don't hurt the environment because they are derived from natural sources. Natural colors don't pollute the environment when disposed of because they biodegrade.
- Renewable: Natural dye comes from replenishable resources that don't hurt the environment when used.
- When consumed, some natural dyes, like the carmine found in lipsticks, won't hurt you or cause any health issues.

Disadvantages of natural dyes

- The procedure of extracting dye takes a long time.
- It is challenging to replicate the same shade of color.
- Natural dyes' impurities cause the color to fade.
- Seasonal variations affect the dyes' availability.
- Standardizing the natural dye process is a challenging task.

ii Synthetic dyes

Due to the increasing demand for dyes for commercial applications, most dyes that are used today are of synthetic origin. These are dyes that are derived from inorganic and organic compounds. They include azoic dyes, sulphur dyes, acidic dyes, mordant dyes, nitro dyes, and basic dyes. Synthetic dyes are commonly prepared from petrochemicals. Most synthetic dyes are highly demanded due to their superior optical properties, high resilience, and higher market demand [10, 19].

Some of the major groups of dyes are discussed below:

i. Direct dyes

The application of this type of dye on a fabric material is done after an aqueous solution of the dye. This dyeing process occurs through the formation of hydrogen bonds between the dye and the fabric. They gained popularity as they gradually replaced the need for binders. This type of dye gives improved fastness when compared to others but however has reduced color attraction in relation to others [23].

ii. Dispersed dyes

They do not dissolve readily in water in comparison to the other types. Their introduction was for the dyeing of acetate fiber materials. Their preparation includes the grinding of the dye in fine particles, followed by their dissolution in solution with the dispersing agent. Absorption of colors occurs when this fiber is dipped inside the solution. They are commonly used in the dyeing of acetate, nylon, and polyester [11].

iii. Reactive dye

These dyes involve the formation of covalent bonds between the cellulose fibers and the dye with remarkable fastness. They show high resistance to washing and radiation. Reactive dyes have also been shown to possess numerous advantages hence their usage for some materials.

iv. Solvent dyes

They show high solubility in liquid ammonia, chlorinated hydrocarbons, and alcohols. They are widely employed in the petroleum industrial sector. They do not dissolve in water hence the solution is commonly prepared using nonpolar solvents or lipids.

v. Acid dyes

They are anionic dyes that show high solubility in water. They are commonly applied to fibers such as wool, silk, modified acrylic fibers, and nylon. The formation of salt between the cationic groups present in the fiber and the anionic components of the dye is responsible for the attachment of the fibers.

vi. Basic dyes

These are soluble in water and are mostly used on acrylic fibers though they are also used in silk and wool. The absorption of the dye on the fabric materials is aided by the addition of acetic acid to the dye bath. They are commonly employed in the paper production industry.

vii. Vat dye

They do not dissolve in water and are not used in dyeing fibers directly. When reduced in alkaline liquor, the alkali metal salt that is water soluble is then formed. This metal salt also known as leuco dye has a remarkable affinity for textile fibers.

viii. Food dyes

Other classification systems have included the role of dyes, rather than just their mechanism of action. Food dyes are used in food industries and are usually grouped as food additives. Their production is at a higher state of purity with greater standards. Such dyes can be mordant, direct, or vat dyes and their applications are strictly regulated by standards [12].

4 Structures of Some Dyes

Azo dyes

These dyes make up the majority of synthetic dyes. The auxochromes are NH_2, NR_2, and OH, and the chromophore of azo dyes is an aromatic system attached to the azo group. Azo dyes are categorized according to the number of azo groups present in a molecule, including triazole, diazo, and monoazole. Methyl orange and congo red are two significant azo dyes, and can be described as follows.

Methyl orange

N, N-dimethylaniline and diazotized sulfuric acid are coupled to yield Methyl orange. Although methyl orange is used as a coloring agent for wool and silk, it loses color when exposed to light and washed. In acid–base titrations, it is typically employed as an indicator rather than a dye. For methyl orange, the pH range is 3.1–4.4 and appears to be yellow in basic solution, while in acidic solution, it appears red (below 3.3) [25]. The shift in ion structure in an acidic and basic media causes the color change.

In an acidic medium, the ion contains p-quinonoid chromophore, while in the basic medium, the ion contains azo chromophore as shown in Fig. 2:

i. Congo red

Congo red, shown in Fig. 3, which has two azo groups is an example of a diazo dye. It is produced by combining two molecules of napthionic acid (4-aminonapthalene-1-sulphonic acid) with tetrazotized benzidine.

ii. Triphenylmethane dyes

These are the -NH_2, -NR_2, and -OH groups in the rings' derivatives. The compounds produced with this technique are known as leuco bases and are typically colorless [26]. These oxidize to produce the corresponding colorless tertiary alcohols known as carbinol bases. Due to salt production, which is known to be a dye, these carbinol bases easily convert from the colorless benzenoid form to the colored quinonoid form in the presence of acids. Two significant triphenylmethane dyes' structures and characteristics can be examined as follows [13].

iii. Malachite green

It is used to dye cotton that has been mordanted with tannin as well as wool and silk directly. When acid and base are added, its color gradually fades.

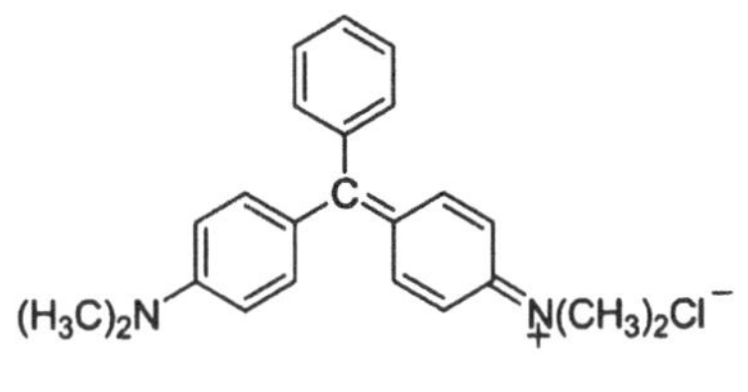

Fig. 3 Malachite green

5 Regulations on Dye Removal

International legislation concerning the elimination of dyes present in effluents from industries has become very strict recently. These policies are more pronounced in the developed countries of the world as well as the developing countries. The enforcements of this law are targeted toward ensuring that industries that produce and make use of dyes treat their waste from industries to ensure total compliance with international standards [10].

6 Environmental and Health Impacts of Dye Pollution

Dyes from various anthropogenic sources as well as other related contaminants show high toxicity and have been documented to have the potential of being carcinogenic. They are therefore linked to the degradation of the environment as well as various diseases in humans and animals [23].

The potential of being recalcitrant when present in an aerobic environment most especially in the traditional plants for treatment process has been connected to the bioaccumulation of dyes in soil and sediments as well as their transportation to public systems of water supply. Some of these dyes tend to degrade in the presence of sediments that are anoxic in nature resulting in the formation of aromatic amines which are highly dangerous. Some of their intermediate products have also been reported to be carcinogenic and mutagenic. This generally affects the overall functioning of the ecosystem. Some dyes also contain heavy metals such as cobalt, copper, and chromium. These metals have tendencies to also affect human health and the environment when present beyond particular permissible limits with deleterious effects as the outcome. Another issue associated with the recalcitrant nature is the oxidative stress that is linked to textile dyes that contain chromium [11]. Synthetic dyes are also connected with a reported incidence of cancer of the bladder especially those containing 2-napthylamine and benzidine. Other diseases due to dye pollution include central nervous system disorder, allergic reactions, rhinitis, conjunctivitis, occupational asthma, and dermatitis which are due to the substitution of cofactors or the inactivation of activities of enzymes. Allergic reactions are due to conjugate formation between the reactive dye and human albumin in human serum producing immunoglobulin E which readily combines with histamine. Dyes can also show cytotoxic effects through their actions on primary reversible inhibitors of oxidase [8].

Long-term exposure to dye can result to health issues in humans. Inhaling dye particles can cause respiratory issues, which is the most frequent hazard associated with reactive dyes. A dye can occasionally have an impact on an individual's immune system, which in severe situations may cause their body to react violently the next time they breathe the dye. This condition is known as respiratory sensitization, and symptoms include sneezing, watery eyes, itching, and wheezing and coughing which are indicative of asthma [27].

The most common health issues associated with the dyeing and finishing operations may result from exposure to irritating chemicals. These could result in irritated eyes, sneezing, itchy or blocked noses, and skin irritation [28]. A fraction of the dyes applied to the clothes always remain unfixed and wash out during the dyeing procedures, meaning that not all of the dyes are fixed on the materials. Textile effluents have been shown to contain substantial amounts of these unfixed colors [28].

The Basic Red 9 dye has a significant potential for environmental toxicity and can cause cancer in people when used in the textile, leather, paper, and ink industries. It decomposes into carcinogenic aromatic amines under anaerobic conditions, and its disposal in water bodies runs the risk of causing allergic dermatitis, skin irritation, mutations, and even cancer. Based on rat experiments, the latter could include tumors in the liver, bladder, mammary glands, hematological system, and localized sarcomas [29].

The Crystal Violet dye, which belongs to the cationic triphenylmethane group, has a very bright color and causes mitotic poisoning, which is linked to aberrant metaphase accumulation and the chromosomal damage caused by in vitro clastogenic effects in Chinese hamster ovules. This potent carcinogen induces hepatocarcinoma, reticular cell sarcoma, and fish tumors in several organs, including the uterus, ovary, bladder, and vagina. In humans, it is capable of generating chemical cystitis, irritation of the skin and digestive tract, and respiratory and renal failure, among others [30].

7 Methods for Removal of Dyes

i. Adsorption method

This is one of the cheapest approaches for the removal of dye and has therefore gained remarkable popularity since it does not give rise to byproducts and also makes it possible for the recycling and reuse of the dyes. It involves the use of various kinds of adsorbent with varying potential in the removal of dyes. The removal process is affected by factors such as the specific surface area, pore sizes, pore volume, absorption potential, and particle size distribution. Some of the most common adsorbents that have been used include alumina, zeolites, activated carbon, and silica gel among others [18].

ii. Electro Fenton Process

This is one of the most promising approaches in the removal of dyes that are difficult to remove using the conventional approaches. This has been shown to be highly efficient for organic dyes that may be difficult to process as well as some other dyes whose removal byproducts are considered carcinogenic.

iii. Biological method

This is a green approach to dye removal due to its environmentally friendly quality. It involves the use of fungi, algae, yeast, bacteria, and enzymes in the breakdown of dyes through anaerobic, aerobic, and integrated processes [21].

8 Conclusion

Dye pollution has been of increasing concern due to its environmental and health impacts. The removal of dyes has been achieved through various conventional and advanced methods. The major concern of scientists in recent times is developing highly efficient approaches to dye removal that are environmentally friendly. The need for sustainability in the dye removal approach is also paramount. Dye waste must be properly treated before being released into the environment in order to reduce environmental pollution by dyes.

Future Trend

More recently, reactive dyes have been employed commonly in the textile industry. The need for developing a reliable and efficient approach for the removal of dyes of this kind becomes necessary.

References

1. Abimbola OF, Adetunji CO, Dare BJ, Mathew JT, Inobeme A, Ghazanfar S, Olaniyan OT, Ajiboye MD, Dauda W (2023) Packaging of animal-based food products: the need for intelligent freshness sensors. In: Sensing and artificial intelligence solutions for food manufacturing. CRC Press, pp 89-102
2. Adetunji CO, Anani OA, Olaniyan OT, Inobeme A, Bodunrinde RE, Mathew JT, Adetunji JB, Hefft DI (2022) Prospect of utilizing novel nanomaterials for effective food security, safety, packaging, and preservation. In: Ameta SC, Ameta R (eds) The science of nanomaterials: basics and applications. Apple Academic Press, Inc. Co-published with CRC Press (Taylor & Francis)
3. Inobeme A, Adetunji CO, Mathew JT, Okonkwo S, Bamigboye M, Ajai AI, Afoso E, Inobeme J (2023) Advanced nanotechnology for the degradation of persistent organic pollutants. Microb Degrad Detoxif Pollut. https://doi.org/10.1515/9783110743623-003
4. Inobeme A, Adetunji CO, Ajai AI, Inobeme J, Mathew JT, Obar A, Maliki M, Nwakife N, Eziukwu C (2023) Chemical nanosensors for monitoring environmental pollution. In: Advanced application of nanotechnology to industrial wastewater. Springer Nature Singapore. https://www.springerprofessional.de/en/chemical-nanosensors-for-monitoring-environmental-pollution/25534908
5. Adetunji CO, Olaniyan OT, Singh KRB, Bodunrinde RE, Inobeme A, Mathew JT, Ogundolie AF, Akinbo O, Singh J, Nayak V, Singh RP (2023) Applications of nanofiltration for wastewater

treatment. In: Bio-nano filtration in industrial effluent treatment, vol 1. CRC-Taylor and Francis, pp 139–152
6. Ardila-Leal LD, Poutou-Piñales RA, Pedroza-Rodríguez AM, Quevedo-Hidalgo BE (2021) A brief history of colour, the environmental impact of synthetic dyes and removal by using laccases. Molecules (Basel, Switzerland) 26(13):3813
7. Berradi M, Hsissou R, Khudhair M, Assouag M, Cherkaoui O, El Bachiri A, El Harfi A (2019) Textile finishing dyes and their impact on aquatic environs. Heliyon 5(11):e02711
8. Alsukaibi AKD (2022) Various approaches for the detoxification of toxic dyes in wastewater. Processes 10(10):1968
9. Castillo-Suárez LA, Sierra-Sánchez AG, Linares-Hernández I, Martínez-Miranda V, &Teutli-Sequeira EA (2023) A critical review of textile industry wastewater: green technologies for the removal of indigo dyes. Int J Environ Sci Technol 1–38
10.]Inobeme A, Ikechukwu Ajai A, Inobeme J, Adetunji CO, Obar A (2023) Superabsorbent polymers for the development of nanofiltration. In: Properties and applications of superabsorbent polymers: smart applications. https://doi.org/10.1007/978-981-99-1102-8_7
11. Gayathiri E, Prakash P, Selvam K, Awasthi MK, Gobinath R, Karri RR, Ragunathan MG, Jayanthi J, Mani V, Poudineh MA, Chang SW, Ravindran B (2022) Plant microbe based remediation approaches in dye removal: a review. Bioengineered 13(3):7798–7828
12. Harsanto B, Primiana I, Sarasi V, Satyakti Y (2023) Sustainability innovation in the textile industry: a systematic review. Sustainability 15(2):1549
13. Hussain B, Yaseen H, Khalid-Al-Ghanim A-MF, Qasim M, Al-Mulhm N, Mahboob S (2021) A study on risk assessment of effect of hematoxylin dye on cytotoxicity and nephrotoxicity in freshwater fish: food and water security prospective research. Saudi J Biol Sci 28(4):2267–2271
14. Inobeme A, Ajai AI, Adetunji CO, Inobeme J, Maliki M, Shaba EY (2023) Wastewater treatment technologies. industrial wastewater reuse: applications, prospects and challenges, pp 201–213
15. Jamee R, Siddique R (2019) Biodegradation of synthetic dyes of textile effluent by microorganisms: an environmentally and economically sustainable approach. Eur J Microbiol Immunol 9(4):114–118
16. Inobeme A, Ajai AI, Adetunji CO, Inobeme J, Adekoya MA, Maliki M (2023) Emerging pollutants from the industries and their treatment. Mod Approaches Waste Bioremediation: Environ Microbiol 1–11:2
17. Piai L, Mei S, van Gijn K, Langenhoff A (2023) Effects of organic matter in drinking water and wastewater on micropollutant adsorption to activated carbon. Int J Environ Sci Technol 1–12
18. Rápó E, Tonk S (2021) Factors affecting synthetic dye adsorption; desorption studies: a review of results from the last five years (2017–2021). Molecules (Basel, Switzerland) 26(17):5419
19. Moyo S, Makhanya BP, Zwane PE (2022) Use of bacterial isolates in the treatment of textile dye wastewater: a review. Heliyon 8(6):e09632
20. Piccard P-J, Borges P, Cleuren B, Hooyberghs J, Buekenhoudt A (2023) Organic solvent nanofiltration and data-driven approaches. Separations 10(9):516
21. Nair VK, Selvaraju K, Samuchiwal S, Naaz F, Malik A, Ghosh P (2023) Phycoremediation of synthetic dyes laden textile wastewater and recovery of bio-based pigments from residual biomass: an approach towards sustainable wastewater management. Processes 11(6):1793
22. Nachiyar CV, Rakshi AD, Sandhya S, Jebasta NBD, Nellore J (2023) Developments in treatment technologies of dye-containing effluent: a review. Case Stud Chem Environ Eng
23. Ngo ACR, Tischler D (2022) Microbial degradation of azo dyes: approaches and prospects for a hazard-free conversion by microorganisms. Int J Environ Res Public Health 19(8):4740
24. Muruganandham M, Sivasubramanian K, Velmurugan P, Kumar SS, Arumugam N, Almansour AI, Sivakumar S (2023) An eco-friendly ultrasound approach to extracting yellow dye from Cassia alata flower petals: Characterization, dyeing, and antibacterial properties. Ultrason Sonochemistry 98:106519
25. Benkhaya S, Souad M, El A (2020) Heliyon Classi fi cations, properties, recent synthesis and applications of azo dyes. Heliyon 6:1–26. https://doi.org/10.1016/j.heliyon.2020.e03271

26. Prival M, Food US (2019) Mutagenicity of benzidine and benzidine-congener dyes and selected monoazo dyes in a modified Salmonella assay Mutagenicity of benzidine and benzidine-congener dyes and selected monoazo dyes in a modified Salmonella assay. Mutat Res 1218(1984):33–47. https://doi.org/10.1016/0165-1218(84)90132-0
27. Hassaan MA, El Nemr A (2017) Health and environmental impacts of dyes : mini review health and environmental impacts of dyes : mini review. Am J Environ Sci Eng 1:64–67. https://doi.org/10.11648/j.ajese.20170103.11
28. Lellis B, Fávaro-polonio CZ, Pamphile JA, Polonio JC (2019) Effects of textile dyes on health and the environment and bioremediation potential of living organisms. Biotechnol Res Innov 3:275–290. https://doi.org/10.1016/j.biori.2019.09.001
29. Henrique F, Bustos-obregon E, Maria D, Salvadori F (2015) Disperse Red 1 (textile dye) induces cytotoxic and genotoxic effects in mouse germ cells. Reprod Toxicol 53:75–81. https://doi.org/10.1016/j.reprotox.2015.04.002
30. Miran W, Nawaz M, Kadam A, Shin S, Heo J (2015) Microbial community structure in a dual chamber microbial fuel cell fed with brewery waste for azo dye degradation and electricity generation. Environ Sci Pollut Res 1–9. https://doi.org/10.1007/s11356-015-4582-8

Textile Dyes and Their Impact on the Natural Environment

Usha Kumari

Abstract Colours are the most important component of the textile and printing industries. These industries use different types of colouring materials to give attractive colours to clothes and prints. These colouring agents are called dyes. These dyes can be natural or synthetic. As natural dyes are present in limited colours and amounts, demand for synthetic dyes has increased many folds in recent years. Their use has increased because these are available in many colours at a reasonable price. These dyes are used in all the sectors of textile dyeing and printing industries. The natural dyes were nontoxic and harmless to human and the environment. But synthetic dyes are toxic and hazardous to the human and the environment. The pollution caused by dyes is called dye pollution. The effluents from dying and printing industries may contain toxic compounds like sulphur, nitrates, acids, heavy metals, and other materials which are used as colouring agents during the process. These dyes and colouring agents are released into water and are the major cause of high level water pollution. Printing inks contain heavy metals like titanium oxide, chromate, and iron which can cause groundwater pollution. Textile industries not only cause water pollution, but they also cause air pollution, noise, and thermal pollution. Dying and printing industries produce various gases like CO_2, NO_2, SO_2, etc. in high amounts causing air pollution. Because of high thermal and photostability, dyes remain for a longer duration in the environment and cause harmful effects on humans and other organisms. Effects of dye pollution may include cancer, skin diseases, respiratory diseases, and eye irritation. Dye pollution can be mitigated by using modern techniques which may include physical, chemical, and biological treatments using absorbents.

Keywords Dye · Synthetic · Dye pollution · Photostable · Textile · Printing

U. Kumari (✉)
Department of Zoology, Gargi College, University of Delhi, Delhi, India
e-mail: usha@gargi.du.ac.in

P. Singh (ed.), *Dye Pollution from Textile Industry*, SDGs and Textiles,
https://doi.org/10.1007/978-981-97-5341-3_2

1 Introduction

Dyes are the colouring agent that provide attractive colours to textiles and prints and play a vital role in the dying and textile industries. Initially, natural dyes produced from flowers, fruits, vegetables, etc. were used. The limitation of these dyes was that they were available in limited colours and showed low colour fastness on washing or in sunlight. Therefore, to fix the colour on fabrics, mordants (e.g. chromium) were used. These also required a high amount of water for dyeing.

In 1856, W.H. Perkins discovered synthetic dyes, which provided a wide range of colours, colour fast and various bright shades. Therefore, the use of synthetic dyes increased very rapidly. Large numbers of commercially available dyes are synthetic dyes. More than 100, 000 dyes are commercially available [39]. These are used in dyeing, textile, pulp and paper, cosmetics, leather, and plastic industries. Synthetic dyes are mainly derived from coal tar and petroleum products, having complex aromatic structures that make them highly stable and non-biodegradable in nature.

Dye pollution reduces the water quality, and poor quality of water affects all the life of the environment including mammals. Textile effluents are used for irrigation, and thus dyes present in water and accumulated in the soil enter higher organisms through the food chain [20]. Dyes enter to human body by consuming contaminated food products and contaminated drinking water. On reaching different organs in the body, these dyes can cause a number of diseases in human beings. Dyes can be degraded into the different metabolites in different tissues of human beings. These metabolites may be more toxic than parent compounds and thus lead to more toxic effects in the tissue. Harmful effects of various industrial effluents having metal dyes are reported [26]. Effluents having nickel, cobalt, copper, and chromium may accumulate in fish tissues and thus may enter human beings after their consumption. Dyes can cause dermatitis, allergy, conjunctivitis, central nervous disorder, and enzyme inactivation [31, 63]. Long-term exposure to the dyes causes genotoxic effects [23].

Dye effluents contain huge amounts of highly toxic substances such as arsenic, cadmium, lead, mercury, nickel, chromium, acids, nitrates, sulphur, and naphthol. In addition, effluents may also contain fixing agents having formaldehyde, chlorinated stain removers, softeners, and more non-biodegradable substances. Some disinfectants present in effluents may react with chlorine and other substances and may be carcinogenic in nature. These effluents when enter the fields, reduce soil fertility and crop production. The colloidal material of dye increases the turbidity of water; preventing sunlight penetration and thus photosynthesis. It also gives a bad smell to water. These dye effluents also contaminate the drinking water, and make it unfit for human consumption.

Large amounts of dyes are present in effluents because 10–15% of dyes used for the dyeing process are not adsorbed on fabric (remain unfixed to the fabrics) and get washed off, thus getting released into effluents [24].

The effluents not only contain dyes, but also contain various heavy metals like chromium, arsenic, etc. Industrial effluents having metals such as chromium, arsenic, copper, and zinc can cause various health issues such as skin ulceration, dermatitis, and nausea [46, 62]. Azo dyes can cause skin irritation, contact dermatitis, blindness, hypertension, respiratory distress, mutations, and cancer [14].

Cotton fibres are highly used fibres in various countries, cotton fibres are dyed using azo dyes, which are not easily degraded due to the presence of nitrogen-nitrogen bonds, therefore dangerous to the environment including human beings [46]. When dye particles are inhaled repeatedly, they can cause respiratory sensitization characterized by itching, coughing, and watery eyes [61].

It is not that only dyes are hazardous to environment, but the other chemicals which are used in the dying process such as fixatives, lubricants, chlorinated solvents, dust, and other by-products are also hazardous to organisms including human being.

2 Classification of Dyes

Dyes are classified as natural and synthetic dyes. Natural dyes can be obtained either from plants or animals. Synthetic dyes are divided into different groups and classes based on the source and structure [38, 43].

According to Mishra and Tripathy (1993), dyes are broadly classified into three groups based on their structure: (1) Anionic, (2) Cationic, and (3) Nonionic. Anionic dyes include direct, acid, and reactive dyes. Cationic includes all basic dyes. Nonionic includes dispersed dyes.

Synthetic dyes can be classified as azo or nonazo dyes. Azo dyes are the most used dye in industries. According to Perez and Beltran, 2000, azo dyes form the largest group, characterized by the double bond between Nitrogen, and attached to the aromatic group. These can be anionic, cationic, or nonionic depending on pH. Azo dyes again can be divided into acid dyes, basic (cationic dyes), direct dyes, disperse dyes (nonionic), reactive dyes, vat dyes, and sulphur dyes [2, 47, 49].

Acidic dyes are applied to the fabrics in an acid solution, whereas basic dyes are applied to catalytic and noncatalytic fibres. Direct dyes are applied in aquas medium containing salts and electrolytes and disperse dyes are used for polyesters. Reactive dyes are used on cellulosic fibres, protein, and nylon fibres. Sulfur dyes are used in cellulosic fibres. Vat dyes used in all areas due to excellent fastness.

3 Dye Pollution

Though synthetic dyes have wide use and are in high demand, their toxic and hazardous nature makes them harmful for life. The various processes of dye industries that produce pollutants are desizing, scouring, bleaching mercerization, dyeing, printing, and finishing. The toxic substances produced by these processes are resins, fats, waxes, caustic soda, hydrogen peroxide, sulfides, acetic acid, soap, dyes, mordants, metallic salts, etc.

The textile and printing industrial effluents contain different dyes, acids, soaps, detergents, enzymes, oxidizing agents, heavy metals, and other chemicals. These chemicals are directly poured into the lakes, rivers, etc. by the industries. As in water dyes are temperature and photostable and resistant to degradation, their life span in water is very long. They can bring about high fluctuation in water pH and strong colour change. Due to their high water solubility nature and potential to degrade into highly toxic and carcinogenic substances, synthetic dyes are hazardous to living organisms and the environment. Industrial effluents have high temperature, high biological oxygen demand (BOD) and chemical oxygen demand (COD), suspended solid particles, and bio-toxicity [36].

4 Types of Pollution Caused by Dye/textile Industrial Effluents

Industrial and printing dyes can create various types of pollution, which are as follows:

1. Water Pollution
2. Air pollution
3. Soil pollution

1. **Water pollution**:

   Wastewater-containing dyes are a significant polluter of the environment which also affects human health, as textile industries generate large amounts of highly coloured wastewater containing a diverse range of persistent pollutants [3, 4]. Printing industrial effluents mainly contain titanium oxide, chromate, and iron. Heavy metals can pollute not only surface water but groundwater also. When industrial dye effluents are present in water in high concentrations, they may cause many types of problems. Few are:

   (i) Interference with sunlight penetration- as these pollutants are in high concentration and spread on the whole surface, the penetration of sunlight reduces, due to which the photosynthetic power of phytoplankton and other photoautotrophs decreases. This ultimately leads to reduction in carbon dioxide use and low oxygen and food production.
   (ii) Low CO_2 utilization
   (iii) Low productivity
   (iv) Low oxygen
   (v) Dye pollutants also reduce the oxygen penetration in water, reducing dissolved oxygen ultimately leading to death of the aquatic organisms.
   (vi) Dyes reduce the penetration of sunlight in water, thus the duration of light-dark cycle, therefore affect the biological clocks of aquatic organisms.
   (vii) High BOD and COD

(viii) Biomagnification dyes accumulate in animal tissues and their concentration increases from the first trophic level to the second and so on.

2. **Air pollution**: During dyeing and printing process, many types of gases like CO_2, NO_2, SO_2, etc. are produced which are very harmful. Dye industries also produce various types of pollutants such as aldehydes, chlorine, dust, and dirt. These pollutants affect human health by bringing about irritation of the lungs and all other parts of the respiratory system and eyes. Increased carbon dioxide reduces the oxygen binding to haemoglobin and thus low oxygen supply to cells results and cell death. Nitrogen and sulphur oxides produced from the boiler, hydrocarbon emitted from drying oven, formaldehyde, softener, and other volatile compounds produced from textile industries cause various types of health issues including respiratory bronchitis [16].
3. **Soil pollution**: The heavy metals, harmful chemicals, dyes, and pigments released from dyeing, printing, and textile industries are the major sources of soil [36].

5 Mechanism of Action of Textile Effluents

Textile effluents contain a large number of harmful/toxic substances like sulphur, acids, heavy metals arsenic mercury, cadmium, and lead. Formaldehyde, chlorinated compounds, and hydrocarbon-based chemicals are also present in effluents. These chemicals are highly reactive, and after reaction, these form very toxic end products that are not easily removed from the body, accumulate in the body, may activate some enzymes, or act as carcinogens, allergens, mutagens, or teratogens.

6 Mechanism of Dye Affecting the Food Chain

As we already discussed, when dye enters water, its colour interferes with the sunlight penetration into water, reducing the photosynthetic potential of aquatic plants, thus also reducing the aquatic fauna [7, 53, 65]. As algae are the producers, their damage affects the whole food chain. Dyes affect the food chain by reducing the producers as well as acting as toxic agents for all other organisms of different trophic levels. As the dye bioaccumulates in different organisms of the food chain, it causes biomagnifications with highest concentration in top trophic level organisms.

7 Effects of Dye Pollution on Environment and Life

Industrial effluents affect the aesthetic quality of water bodies, increase COD and BOD, impair photosynthesis, inhibit plant growth, enter into the food chain, provide bioaccumulation, and promote carcinogenicity, mutagenicity, teratogenicity, toxicity, and allergic reactions.

Dye pollution affects both the components of the environment: biotic and abiotic.

I. Impact of dye pollution on the environment:

   (a) Impact on soil
   (b) Impact on water
   (c) Impact on air
   (d) Impact on aesthetic value
   (e) Impact on food production

II. Impact of dye pollution on aquatic organisms:

   (a) Impact on algae
   (b) Impact on fish

III. Impact of dye pollution on human health and well-being

7.1 *Impact of Dye Pollution on the Environment*

The environmental impact of industrial effluents can be broadly categorized into three levels: primary, secondary, and tertiary [36].

The primary level is seen over the surface of the ground or water in the form of a change in colour of land or water, affecting the aesthetic value of the ecosystem. When coloured effluent enters water bodies like ponds, lakes, and rivers, making them appear strong coloured, and prevents their use by animals and human beings.

At the secondary level, the influence of the primary level change is seen as effects on organisms living on the ground or in water. The colour change interferes with light penetration, inducing a reduction in photosynthesis, oxygen supply, and productivity. Thus dye pollution affects the whole food chain and food web of the ecosystem. The change colour may bring about the extinction of some colour-sensitive species, affecting the prey-predator relationship and thus unbalancing in the homeostasis of the ecosystem.

The primary and secondary level impacts bring about the degradation of the quality of the ecosystem by chemical and biological transformation called the tertiary level of impact. When industrial effluents are present in high concentrations in water, they damage the ecosystem leading to the death of living organisms. After death, degradation of these dead organisms reduces dissolved oxygen, and increases the BOD and COD. Effluents contain high amount of hydrogen sulphide and other organic matter which change the water pH (highly acidic/alkaline), and thus make water unfit for the survival of aquatic organisms.

7.1.1 Impact of Dye Pollution on Soil

Soil is a very important part of the ecosystem as it is mainly consisting of minerals important for the growth of plants. Many microorganisms are present in soil making a huge diversity of flora and fauna. In addition to providing water and nutrients to plants, soil also provides anchorage to the roots of plants. The presence of some heavy metals, chemicals such as dyes, or pigments released from chemical, printing, or textile industries makes soil unsuitable for flora and fauna; and becomes the major source of soil pollution. The presence of pollutants in soil reduces soil quality and fertility, resulting in reduced food production and thus affecting the food web [36]. Dyes have the potential to decrease the nitrogen fixation potential of bacteria by reducing ammonification and nitrification leading to the reduction in productivity. One good example of a dye that reduces the ammonifying and nitrifying bacteria is reactive black [60]. Azo dyes have the ability to change the physical and chemical properties of soil and have harmful effects on flora and fauna of soil. Dyes can kill the soil microorganisms, resulting in the loss of agricultural productivity [51].

7.1.2 Impact of Dye Pollution on Water

Dyes cause water pollution. The presence of dyes and other industrial pollutants impairs the physical, chemical, and biological properties of water. Dyes impair water qualities such as colour, odour, taste, pH, temperature, oxygen content, BOD, and COD [50]. All these affect the flora and fauna of the aquatic system. The presence of dyes makes the water unfit for human use. Dyes also reduce the photic zone and thus reduce photosynthesis [33]. The presence of high amounts of nitrite and phosphate causes the problem of eutrophication in water, affecting the life of aquatic organisms [21, 42].

7.1.3 Impact of Dye Pollution on Air

The presence of dyes and other volatile substances causes air pollution. These cause a large number of health issues in the lungs, respiratory system, allergy, and dermatitis.

7.1.4 Impact of Dye Pollution on Aesthetic Value

The presence of a variety of dyes, textiles, and other industrial pollutants in water cause the water brightly coloured and thus make it anesthetic. It may act as a warning signal for many aquatic species also, affecting flora and fauna.

7.1.5 Impact of Dye Pollution on Food Production

Dye pollution negatively affects seed germination and plant growth, and thus the flora. Toxic dyes also affect the seedling number, plant root, shoot, biomass, and pigment content in a variety of crops such as maize, beans, and soyabean [55]. As dyes prevent the entry of light, they reduce photosynthesis and thus reduce food production [29].

7.2 Impacts of Dye Pollution on Aquatic Ecosystem

Water pollution is one of the most common pollution caused by textile industries. Textile effluents contain high amounts of harmful dyes in wastewater which pollutes not only water, but also soil. These have adverse effects on plants, animals, human health, the environment, and thus the ecosystem. The industrial wastes containing synthetic dyes released due to inefficient dyeing processes cause serious effects on the flora and fauna of both aquatic and terrestrial ecosystems. Some of the effects caused by synthetic dyes are: surface water contamination, bioaccumulation/biomagnifications of toxic substances, carcinogenic, allergic [34, 48, 54].

High concentration of dyes in water dissolved oxygen concentration therefore affecting the biological activities of aquatic life [53]. Textile effluents having hydrosulfide can block light penetration through water, thus causing an imbalance in BOD and COD [30]. Oxidative stress caused by dye pollutants damages the plant's growth and reduces productivity. The dyes and their metabolites accumulate in sediments and organisms leading to increased toxicity, mutagenicity, and carcinogenicity. As these dyes are generally non-biodegradable, they persist for a longer duration and affect the food chain by accumulating in the fatty tissues of the organisms. Fish and algae are the most affected organisms by dye pollution [22, 27].

Algae being the primary producer, alteration or damage of algae population can affect the whole food web [25].

7.2.1 Impact of Dye Pollution on Fish

Fish are highly sensitive and susceptible to changes in the environment due to toxic substances, as these are in direct contact with water. The toxic substance brings about undesirable, unwanted changes in behavioural, morphological, physiological, and histological changes in fishes even at very low concentrations. Textile effluents change the physical and chemical properties of water such as colour, turbidity, and temperature; therefore affecting the fish's biological rhythm, and thus changing almost all the functions of the system by altering the hormone production. Pollutants easily accumulate in fish tissues, therefore, bringing damage to each tissue, but at different degrees. Gills are the most affected structure of fish. The availability of food for fish is affected by the change in physical and chemical properties of water, as these affect the growth of phyto and zooplankton [8, 22].

Effects of dye effluents or exposure to dyes on behavioural changes have been reported by many researchers. On *Poecila reticula,* textile effluents cause abnormal behaviour such as erratic swimming movement, hyper-excitation, fast opercular movements, and thick mucus coverings jumping out of the water [52]. Similar alterations in behavioural responses were observed in *Cirrninus mrigala* exposed to sublethal concentrations of dye effluents [28], *Oreochromis nloticus* exposed to azo acid dye like [6], *Labeo rohita* exposed to Acid orange [9]. Dye effluents also affect the rate of feeding and food absorption in fish.

Alterations in RBCs and WBCs number was also observed in fish exposed to dye effluents [6]. Studies on *Gambusia affinis*, exposed to textile dye effluents showed that there is a significant reduction in RBC count, shape, and size. It also induces cytotoxic effects on RBCs and a high rate of fish mortality [56]. Another report on *Mastacembelus armatus* showed that fish exposed to textile effluents, change ionic regulations of liver, kidney, and muscle by decreasing the concentration of sodium and chloride ions and increase in potassium, calcium, and magnesium ions [27].

Histopathological changes in fish exposed to dye include enlargement in the gill bar, the disintegration of intestine villi, and filtration of haemocytes into the lumen [52]. In liver, cell necrosis and liver degeneration were observed [58]. Textile effluents also decreased the nutritive values such as protein, carbohydrate, and lipids in *Spiralothelphusa hydrodroma* [58].

Dye pollution not only brings about behavioural, morphological, or histopathological changes, but also induce genotoxic effects on fish. Dye such as reactive red 120 (RR120) induced irreversible modification in DNA, resulting in micronuclei formation in RBCs and cell cytotoxicity in *Catla catla* [40, 41]. *Anabas testudineus* exposed to acid orange also showed genetic damage [45].

Dye pollution can induce the enzyme or influence the enzyme activity, which may alter the dye metabolism and its excretion from the body. Fish *Oreochromis niloticus and Clarias anguillaris* exposed to textile and mill effluents showed increased cytochrome monooxygenase A (CYP1A) in the liver. This enzyme is responsible for the biotransformation of parent dye into other toxic metabolites [19]. Dyes have the potential to accumulate and metabolize in fish [10]. Textile dyes such as malachite green, crystal violet, and brilliant green are reported to accumulate and metabolize in the muscle tissues of fish *Anguilla* [10].

7.2.2 Impact of Dye Pollution on Algae

The dye in water negatively affects the algae growth, protein content, pigment, and other nutrients. Algae are used as an indicator of water quality as they have high sensitivity to toxic chemicals such as dyes, pesticides, or insecticides [25]. As algae are highly sensitive to dyes, they are severely affected by dye pollution.

Indigo dye effluents can reduce the growth and biomass of *Scenedesmus quadricauda.* These effluents also affect chlorophyll and cell production in microalgae

[13]. Red, yellow, and green dyes are also reported to show similar effects on *Spirulina platensis* [17]. Ramzol, Red Brilliant dye, affects the food chain and causes an imbalance in aquatic ecosystems [16].

Green algae, *Chlorella vulgaris,* exposed to dyes drimarene blue, optilan yellow, and lanasyn brown, showed a concentration-dependent decrease in growth rate, pigments, and elements like carbon, hydrogen, and nitrogen [22].

Algae *Senedesmus obliquus* exposed to textile effluents showed an imbalance in SOD (super dismutase) activity, increased ROS production, chloroplast damage, and reduced synthesis of chlorophyll [11].

7.2.3 Impact of Dye Pollution on Human Health and Well-Being

Wastewater from textile and other industries containing different types of dye pollutants significantly affects the environments of human beings, and thus human health and well-being. Dye pollution consists of a wide range of persistent pollutants and affects almost all the systems of human beings [3, 4]. Some of these effects are listed below:

Effects on the nervous system: Dye pollutants affect the nervous system in many ways. Azure B, a metabolite of methylene blue dye, can intercalate with DNA and RNA and can inhibit the monoamine oxidase A enzyme, an intracellular enzyme of central nervous system. Therefore, dyes affect the central nervous system and human behaviour by inhibiting the intracellular enzyme of the central nervous system [32].

Effects on reproductive system: Dyes can affect the production of sperm (spermatogenesis) and ova (oogenesis) in males and females, respectively [59]. Nonylphenol and Phthalates are reported to impair reproduction and reducing fertility.

Immune system and respiratory system: Organotin affects the immune system and reproductive system. Formaldehyde affects the mucous membrane and respiratory system [36].

Chlorobenzenes are used as degreasers and are highly toxic to the liver, thyroid, and nervous system.

Effects caused by industrial effluents containing toxic dyes and other pollutants can be categorized as allergic, carcinogenic, mutagenic, or teratogenic. Some of these are listed below:

1. Allergic Effects

The azo dyes, specially dispersed dye metabolites, can cause allergic reactions in rats and mice [34, 57]. Reactive dyes cause allergic reactions including dermatitis, rhinitis, asthma, and conjunctivitis in the body of persons dealing with these dyes. The dye reacts with immunoglobulin (IgE) and forms the dye-immunoglobulin complex which binds with mast cells, which in turn secretes histamine and inflammatory mediators which induce allergic reactions [5].

2. Carcinogenic Effects

Dyes on entering the animal bodies accumulate and induce the metabolism of parent-less toxic compounds to more toxic or even carcinogenic in nature. Benzadine-based dyes are known to cause cancers as they metabolize into 4-aminobiphenyl—a cancer-causing substance [54]. Benzidine dyes are reported to cause bladder cancer [15] (Tousadi et al 2020). Sudan I dye used in textiles and food are converted to carcinogens [44]. The degree of carcinogenicity and toxicity of different dyes is different. The toxicity of red and blue dyes is greater than yellow dyes. Long-term exposure to these dyes may cause other types of cancers also such as colon and colorectal cancer, vagina [31].

3. Mutagenic Effects

Styryl and azo analogues of cyanine dyes are reported to be mutagenic by inducing chromosome damage [1]. Red 1 dye is mutagenic to human hepatoma (HepG2) and lymphocytes as it acts as a mutagen for chromosomes [64].

Disperse orange 1 dye can cause various types of mutations, cytotoxic effects, and apoptosis in HepG2 cells [12, 18]. Crystal violet (triarylmethane dye) acts as mitotic inhibitor resulting in abnormal metaphase and polyploidy. It also causes skin irritation, respiratory issues, and renal damage [35].

4. Teratogenic Effects

Dyes like 2-nitro-p-phenylenediamine and 4-nitro-o-phenylenediamine are reported to act as teratogens affecting the embryonic development of animals, leading to malformed embryos [37].

References

1. Abdou RF, Abdel-Aal SA, Waly EA, Khalil ZH (1986) Effect of cyanine dyes analogues in breaking dormancy and their mutagenic effects in garlic (*Allium sativum* L.). In: VI international symposium on medicinal and aromatic plants, XXII IHC, vol 208, pp 157–164
2. Ajmal A, Majeed I, Malik RN, Idriss H, Nadeem MA (2014) Principles and mechanisms of photocatalytic dye degradation on TiO2 based photocatalysts: a comparative overview. Rsc Adv 4(70):37003–37026
3. Ali SS, Al-Tohamy R, Sun J (2022) Performance of Meyerozyma caribbica as a novel manganese peroxidase-producing yeast inhabiting wood-feeding termite gut symbionts for azo dye decolorization and detoxification. Sci Total Environ 806:150665
4. Almroth BC, Cartine J, Jönander C, Karlsson M, Langlois J, Lindström M, Sturve J (2021) Assessing the effects of textile leachates in fish using multiple testing methods: from gene expression to behavior. Ecotoxicol Environ Saf 207:111523
5. Al-Tohamy R, Ali SS, Li F, Okasha KM, Mahmoud YAG, Elsamahy T, Sun J (2022) A critical review on the treatment of dye-containing wastewater: ecotoxicological and health concerns of textile dyes and possible remediation approaches for environmental safety. Ecotoxicol Environ Saf 231:113160

6. Amwele HR, Papirom P, Chukanhom K, Beamish FHW, Petkam R (2015) Acute and subchronic toxicity of metal complex azo acid dye and anionic surfactant oil on fish Oreochromis niloticus. J Environ Biol 36(1):199
7. Aquino JM, Rocha-Filho RC, Ruotolo LA, Bocchi N, Biaggio SR (2014) Electrochemical degradation of a real textile wastewater using β-PbO2 and DSA® anodes. Chem Eng J 251:138–145
8. Athira N, Jaya DS (2018) The use of fish biomarkers for assessing textile effluent contamination of aquatic ecosystems: a review. Nat Environ Pollut Technol 17(1)
9. Barot J, Bahadur A (2015) Toxic impacts of CI Acid Orange 7 on behavioural, haematological and some biochemical parameters of Labeo rohita fingerlings. IJSRES 3(8):0284–0290
10. Belpaire C, Reyns T, Geeraerts C, Van Loco J (2015) Toxic textile dyes accumulate in wild European eel Anguilla anguilla. Chemosphere 138:784–791
11. Cai H, Liang J, Ning XA, Lai X, Li Y (2020) Algal toxicity induced by effluents from textile-dyeing wastewater treatment plants. J Environ Sci 91:199–208
12. Chequer FMD, Angeli JPF, Ferraz ERA, Tsuboy MS, Marcarini JC, Mantovani MS, de Oliveira DP (2009) The azo dyes Disperse Red 1 and Disperse Orange 1 increase the micronuclei frequencies in human lymphocytes and in HepG2 cells. Mutat Res/Genet Toxicol Environ Mutagen 676(1–2):83–86
13. Chia MA, Musa RI (2014) Effect of indigo dye effluent on the growth, biomass production and phenotypic plasticity of Scenedesmus quadricauda (Chlorococcales). An Acad Bras Ciências 86:419–428
14. Chung KT (1983) The significance of azo-reduction in the mutagenesis and carcinogenesis of azo dyes. Mutat Res/Rev Genet Toxicol 114(3):269–281
15. Clark M (ed) (2011) Handbook of textile and industrial dyeing: principles, processes and types of dyes. Elsevier.
16. de Sousa ML, de Moraes PB, Lopes PRM, Montagnolli RN, de Angelis DDF, Bidoia ED (2012) Contamination by remazol red brilliant dye and its impact in aquatic photosynthetic microbiota. Environ Manag Sustain Dev 1(2):129
17. Dwivedi S (2013) Effect of textile dyes on Spirulina platensis. J Chem Pharm Res 5(4):66–80
18. Ferraz ER, Grando MD, Oliveira DP (2011) The azo dye Disperse Orange 1 induces DNA damage and cytotoxic effects but does not cause ecotoxic effects in Daphnia similis and Vibrio fischeri. J Hazard Mater 192(2):628–633
19. Gadagbui BKM, Goksrøyr A (1996) CYP1A andother biomarker responses to effluents from a textile mill in the Volta River (Ghana) using caged tilapia (Oreochromis niloticus) andsediment-exposed mudfish (Clarias anguillaris). Biomarkers 1(4):252–261
20. Garg VK, Kaushik P (2008) Influence of textile mill wastewater irrigation on the growth of sorghum cultivars. Appl Ecol Environ Res 6(1):1–12
21. Ghaedi M, Hajjati S, Mahmudi Z, Tyagi I, Agarwal S, Maity A, Gupta VK (2015) Modeling of competitive ultrasonic assisted removal of the dyes—Methylene blue and Safranin-O using Fe_3O_4 nanoparticles. Chem Eng J 268:28–37
22. Gita S, Hussan A, Choudhury TG (2017) Impact of textile dyes waste on aquatic environments and its treatment. Environ Ecol 35(3C):2349–2353
23. Haq I, Raj A (2018) Biodegradation of Azure-B dye by Serratia liquefaciens and its validation by phytotoxicity, genotoxicity and cytotoxicity studies. Chemosphere 196:58–68
24. Hassan M, Hassan R, Mahmud MA, Pia HI, Hassan MA, Uddin MJ (2017) Sewage waste water characteristics and its management in urban areas-A case study at pagla sewage treatment plant, dhaka. Urban Regional Plan 2(3):13
25. Hernández-Zamora M, Martínez-Jerónimo F (2019) Exposure to the azo dye Direct blue 15 produces toxic effects on microalgae, cladocerans, and zebrafish embryos. Ecotoxicology 28(8):890–902
26. Ito T, Adachi Y, Yamanashi Y, Shimada Y (2016) Long–term natural remediation process in textile dye–polluted river sediment driven by bacterial community changes. Water Res 100:458–465

27. Karthikeyan S, Jambulingam M, Sivakumar P, Shekhar AP, Krithika J (2006) Impact of textile effluents on fresh water fish Mastacembelus armatus (Cuv. & Val). J Chem 3:303–306
28. Kaur HARBHAJAN, Kalotra ROHIT, Walia GK, Handa DIANA (2013) Dyeing industry effluent induced behavioural and morphological changes in the fish, Cirrhinus mrigala. IJZR 3:13–20
29. Khan MD, Abdulateif H, Ismail IM, Sabir S, Khan MZ (2015) Bioelectricity generation and bioremediation of an azo-dye in a microbial fuel cell coupled activated sludge process. PLoS One 10(10):e0138448
30. Khan S, Malik A (2018) Toxicity evaluation of textile effluents and role of native soil bacterium in biodegradation of a textile dye. Environ Sci Pollut Res 25:4446–4458
31. Lellis B, Fávaro-Polonio CZ, Pamphile JA, Polonio JC (2019) Effects of textile dyes on health and the environment and bioremediation potential of living organisms. Biotechnol Res Innov 3(2):275–290
32. Li H, Zhang R, Tang L, Zhang J, Mao Z (2014) Evaluation of Bacillus sp. MZS10 for decolorizing Azure B dye and its decolorization mechanism. J Environ Sci 26(5):1125–1134
33. Liang J, Ning XA, Kong M, Liu D, Wang G, Cai H, Yuan Y (2017) Elimination and ecotoxicity evaluation of phthalic acid esters from textile-dyeing wastewater. Environ Pollut 231:115–122
34. Malinauskiene L, Bruze M, Ryberg K, Zimerson E, Isaksson M (2013) Contact allergy from disperse dyes in textiles–a review. Contact Dermat 68(2):65–75
35. Mani S, Bharagava RN (2016) Exposure to crystal violet, its toxic, genotoxic and carcinogenic effects on environment and its degradation and detoxification for environmental safety. Rev Environ Contam Toxicol 237:71–104
36. Mohan MD, Shukla SP (2022) Hazardous consequences of textile mill effluents on soil and their remediation approaches. Clean Eng Technol 7:100434
37. Marks TA, Gupta BN, Ledoux TA, Staples RE (1981) Teratogenic evaluation of 2-nitro-p-phenylenediamine, 4-nitro-o-phenylenediamine, and 2, 5-toluenediamine sulfate in the mouse. Teratology 24(3):253–265
38. Mishra G, Tripathy M (1993) A critical review of the treatments for decolourization of textile effluent. Colourage 40:35–35
39. Olakunle MO, Inyinbor AA, Dada AO, Bello OS (2018) Combating dye pollution using cocoa pod husks: a sustainable approach. Int J Sustain Eng 11(1):4–15
40. Parmar AVNI, Jagruti B (2016) Determination of genotoxic effect of azo dye CI RR 120 on fish Catla catla. Biotechnol Res 2(2):77–80
41. Parmar A, Shah A (2019) Cytogenotoxicity of azo dye Reactive Red 120 (RR120) on fish Catla catla. Environ Exp Biol 17(3)
42. Patel H (2018) Charcoal as an adsorbent for textile wastewater treatment. Sep Sci Technol 53(17):2797–2812
43. Perez-Urquiza M, Beltran JL (2000) Determination of dyes in foodstuffs by capillary zone electrophoresis. J Chromatogr A 898(2):271–275
44. Piątkowska M, Jedziniak P, Olejnik M, Żmudzki J, Posyniak A (2018) Absence of evidence or evidence of absence? A transfer and depletion study of Sudan I in eggs. Food Chem 239:598–602
45. Rishin A, Priyatha CV, Chitra KC (2019) Induction of genetic damage in peripheral erythrocytes of the fish, Anabas testudineus exposed to sublethal concentration of Acid Orange 7. Res Rev AJ Life Sci 9(2):1–10
46. Robert L, Joseph F, Alexander A (2008) Fisher's contact dermatitis in textiles and shoes. BC Decker Inc., Ontario, pp 339–401
47. Robinson T, McMullan G, Marchant R, Nigam P (2001) Remediation of dyes in textile effluent: a critical review on current treatment technologies with a proposed alternative. Bioresour Technol 77(3):247–255
48. Sani ZM, Abdullahi IL, Sani A (2018) Toxicity evaluation of selected dyes commonly used for clothing materials in urban Kano, Nigeria. Eur Exp Biol 8(4):26
49. Saquib M, Muneer M (2003) TiO2-mediated photocatalytic degradation of a triphenylmethane dye (gentian violet), in aqueous suspensions. Dye Pigment 56(1):37–49

50. Sarvajith M, Reddy GKK, Nancharaiah YV (2018) Textile dye biodecolourization and ammonium removal over nitrite in aerobic granular sludge sequencing batch reactors. J Hazard Mater 342:536–543
51. Savin II, Butnaru R (2008) Wastewater characteristics in textile finishing mills. Environ Eng Manag J (EEMJ) 7(6)
52. Selvaraj D, Leena R, Kamal DC (2015) Toxicological and histopathological impacts of textile dyeing industry effluent on a selected teleost fish Poecilia reticulata. Asian J Pharmacol Toxicol 3(10):26–30
53. Setiadi T, Andriani Y, Erlania M (2006) Treatment of textile wastewater by a combination of anaerobic and aerobic processes: a denim processing plant case. In: Proceedings of the Southeast Asian water environment, vol 1
54. Singh GK, Singh NB, Shukla SP, Markandeya (2019) Remediation of COD and color from textile wastewater using dual stage electrocoagulation process. SN Appl Sci 1(9):1000
55. Singh KARUNA, Kumar PANKAJ, Srivastava RAJANI (2017) An overview of textile dyes and their removal techniques: Indian perspective. Pollut Res 36:790–797
56. Soni P, Sharma S, Sharma S, Kumar S, Sharma KP (2006) A comparative study on the toxic effects of textile dye wastewaters(untreated and treated) on mortality and RBC of a freshwater fish Gambusia affinis(Baird and Gerard). J Environ Biol 27(4):623–628
57. Sontag JM (1981) Carcinogenicity of substituted-benzenediamines (phenylenediamines) in rats and mice. J Natl Cancer Inst 66(3):591–602
58. Sripriya L, Vijayalakshmi M, Sumathy R, Sharmila J (2014) The impact of textile dyes on the biochemistry and histology of liver, a freshwater fish, tilapia, *Oreochromis mossambicus* (Peters.). Int J Pharma Bio Sci 5(3)
59. Suryavathi V, Sharma S, Sharma S, Saxena P, Pandey S, Grover R, Sharma KP (2005) Acute toxicity of textile dye wastewaters (untreated and treated) of Sanganer on male reproductive systems of albino rats and mice. Reprod Toxicol 19(4):547–556
60. Topaç FO, Dindar E, Uçaroğlu S, Başkaya HS (2009) Effect of a sulfonated azo dye and sulfanilic acid on nitrogen transformation processes in soil. J Hazard Mater 170(2–3):1006–1013
61. Tounsadi H, Metarfi Y, Taleb M, El Rhazi K, Rais Z (2020) Impact of chemical substances used in textile industry on the employee's health: epidemiological study. Ecotoxicol Environ Saf 197:110594
62. Tüfekci N, Sivri N, Toroz İ (2007) Pollutants of textile industry wastewater and assessment of its discharge limits by water quality standards. Turk J Fish Aquat Sci 7(2)
63. Vikrant K, Giri BS, Raza N, Roy K, Kim KH, Rai BN, Singh RS (2018) Recent advancements in bioremediation of dye: current status and challenges. Bioresour Technol 253:355–367
64. Will Y, McDuffie JE, Olaharski AJ, Jeffy BD (eds) (2016) Drug discovery toxicology: from target assessment to translational biomarkers. Wiley
65. Zaharia C, Suteu D, Muresan A, Muresan R, Popescu A (2009) Textile wastewater treatment by homogenous oxidation with hydrogen peroxide. Environ Eng Manag J 8(6):1359–1369

Dyes and Unveiling the Environmental Story of Dye Pollution

Shreya Sinha, Noor Jahan, and Pinki Rani Agrawal

Abstract Dyes, being natural or synthetic, are renowned for their vibrant colours, which come from a group of atoms known as chromophores. There are around 14 different types of dye. The majority of dyes are used by textile industries for fabric manufacturing, and many of these dyes are toxic in nature and possess complex structures. During dyeing and finishing operations, nearly 2 lakh tonnes of dye are introduced into water bodies. Due to their exceptional stability against heat, temperature, light, chemicals, water, and other parameters, these dyes are sustained in the environment for a long period of time, causing severe harm to marine and aquatic bodies. In order to have minimal effects of dyes on the environment, we must look for alternatives or find out effective methods to degrade these hazardous chemicals. This chapter aims to delve into the creation of dyes, their widespread usage, the detrimental effects on our environment, and the techniques available to identify dye pollution.

Keywords Dye · Chromophores · Hazardous · Environment · Pollution

1 Introduction

The increasing demand for colourful products across various industries has led to the widespread production of dyes. These dyes are coloured substances extensively used in many industries such as textiles, papers, pharmaceuticals, foods, cosmetics, construction, and plastics for colouring their products [1]. In spite of significant contributions to dye production and dye usage and successfully meeting global demands, these industries also bear responsibility for environmental pollution [2–4].

S. Sinha · N. Jahan (✉)
Central University of Himachal Pradesh, Shahpur, Himachal Pradesh, India
e-mail: noor.jahan@hpcu.ac.in

P. R. Agrawal
Deenbandhu Chhotu Ram University of Science and Technology, Murthal, Sonipat, Haryana, India

P. Singh (ed.), *Dye Pollution from Textile Industry*, SDGs and Textiles,
https://doi.org/10.1007/978-981-97-5341-3_3

Primarily, textile industries are major contributors to chemical waste production, generating 200,000 tonnes of wastewater every year [5–7]. Industries like these produce large amounts of coloured waste that get dumped into the environment. Textile manufacturing processes such as dyeing and printing, washing and finishing, de-sizing, wet processing, cooling, cleaning, and maintenance consume substantial amounts of water, leading to the generation of large volumes of wastewater containing dyes. The environment faces considerable challenges in dealing with these harmful chemicals, particularly dyes, due to their strong thermal stability, complex structures, high pH value, and high chemical oxygen demand [5–7].

For textiles only, around 8×10^5 tonnes per year of synthetic dyes are being manufactured in the world at present [8]. Although dyeing has been an integral part of human culture and history, it has much evidence. Ignorance of waste effluents during the dyeing process poses a significant risk to the ecosystem. Many health-attributable diseases are the consequences of allowing solid waste dyestuff, along with dye-containing wastewater, to enter our environment through lakes, rivers, soil, and air.

Dyes, either natural or synthetic, are made in such a manner that they can effortlessly bind with the substrates permanently. Dyeing includes the synthesis of coloured organic substances, which not only get adsorbed onto the surfaces of the substrate's molecule but also get diffused within those substrates [9]. Numerous dyes have been developed since the discovery of the first synthetic dye. These dyes are heavily utilized by industries, and waste generated from them has severe impacts on plants, agriculture, aquatic bodies, and human health, including some serious health issues such as kidney failure, disorders in reproductive systems, and damage to the central nervous system.

From ground to air, water to soil, almost every facet of the environment is directly or indirectly affected by the advancement of industries and high demand for dyes. A study by ETAD (Ecological and Toxicological Association of the Dyestuff Manufacturing Industry) identified that severe health risks are attributed to dyes and their toxic by-products. In this survey, more than 90% of 4000 dyes were found to have greater LD50 (Lethal Dose 50) values than 2×10^3 mg/kg, indicating their potential harm [10].

This chapter focuses on the in-depth knowledge of dyes. It comprehensively explores their source of origin and classifications based on their applications. In addition to this, it examines the harmful consequences of dyes on the ecosystem, providing concrete, real-life examples of dye production and its worldwide consequences. Furthermore, the chapter provides a comprehensive understanding of various analytical techniques for the detection of dyes in wastewater, soil, and the atmosphere.

2 Dyes

Dyes are organic-coloured compounds which contain chromophores and auxochromes. This was proposed by Witt in 1876 to explain the origin of colours in dye molecules. Basically, the chromophores are the group of atoms for example $-C = O-$, $-C = N-$, $-C = C-$, $-N = O-$, $-NO_2$, and $-N = N-$ groups [11] which impart colour to dyes, commonly known as electron-acceptor groups whereas auxochromes are the group of atoms for example $-NH_2$, $-COOH$, $-NR_2$, $-OH$, and $-SO_3H$ groups [11] responsible for enhancing the colour of a dye [12] while reducing water-solubility [13]. These auxochromes are often referred to as electron-donors. Chromophores and auxochromes are linked together through a conjugated system [12]. Additionally, the colour which a dye can absorb depends on its wavelength and ultimately on the structure. So, changes in structure will directly affect the colour of a dye. These groups absorb radiations in the visible region from 350 to 750 nm [14]. Further, the absorption band can be shifted to a longer wavelength by increasing the electron-accepting power of chromophores, electron-donating power of auxochromes, and length of conjugation. All this will help achieve a bathochromic shift, i.e., an absorption band shift to longer wavelengths [12]. The level to which a certain dye is toxic depends on its structure rather than the dyeing process [12].

3 Classification of Dyes

In general, there are two categories of dyes based on the origin.

- Natural dyes.
- Synthetic dyes.

3.1 Natural Dyes

Natural dyes are organic-coloured substances that are obtained from natural sources such as plants, animals, minerals, microorganisms, and fungi [15, 16]. These sources are both cost-effective as well as readily bio-degradable since they are obtained from nature [17]. Natural dyes have a history dating back to prehistoric times, with the earliest known use dating back to around 2600 BC. In Spain, the paintings inside 'El Castillo cave' were found to be painted around 40,000 years ago through scientific techniques like radiocarbon dating and archaeological evidence. Also, textile dyeing practices trace their origins to the Neolithic period or New Stone Age [18]. Natural dyes have minimal impact on the environment as their disposal does not pollute the ecosystem. These sources are safe to use with less or no chemical treatment [16].

3.2 *Classification of Natural Dyes*

On the basis of their source origin, natural dyes are classified as

- Plant origin-based dyes.
- Animal origin-based dyes.
- Mineral origin-based dyes.
- Microbial origin-based dyes.

3.2.1 Plant Origin-Based Dyes

Dyes which are acquired from the various parts of plants—roots, wood, leaves, seeds, flowers, bark, and berries—are known as plant-based dyes. Due to their abundance and easy extraction methods, these sources are commonly in use. Almost all the parts of plants are utilized for extracting colours and used as natural dyes known for their antifungal, antimicrobial, antioxidant, insect-repellent, and disinfectant properties. They are biodegradable in nature, free from toxic substances, and even possess certain medicinal qualities [19]. Some of the commonly employed plant-based dyes are madder, safflower, and morinda for red dyes; saffron, turmeric, and marigold for yellow dyes; and indigo for producing blue colour dye (Table 1).

3.2.2 Animal Origin-Based Dyes

Another type of natural dyes originated from animals includes insects, lichens, and shellfish. The dye obtained from marine mollusc secretions, Murex, is the oldest animal origin dye known as Tyrian purple. Animal dyes mostly give red colour except for the Tyrian purple which provided purple colour. Tyrian purple imparts deep violet colour on the textile fabrics. This dye was considered as one of the most expensive dyes among others and was used to dye clothes of the royal family. Lichens produce yellow dye which is obtained by boiling lichens in water [16] (Table 2).

3.2.3 Mineral Origin-Based Dyes

These dyes are obtained from abundant minerals present on earth. Minerals like red ochre, yellow ochre, malachite, and cinnabar, among others, were used for colouration purposes. These dyes are not as commonly used as plant-based dyes and synthetic dyes, but they are used in traditional dyeing processes. The surface of the rock was scratched to get powder form which was further mixed with water and oil for usage [16]. A few examples of mineral-based dyes are mentioned in Table 3.

Table 1 Examples of plant origin-based dyes

S. no	Botanical name	Common name	Part of the plant	Colour	Image	References
1	*Indigofera Tinctoria*	Indigo	Leaves	Indigo blue		[20]
2	*Morinda Citrifolia*	Morinda	Roots and bark	Red		[21]
3	*Tagetus spp.*	Marigold	Flower	Yellow		[22]
4	*Punica granatum*	Pomegranate	Fruit	Yellow		[23, 24]

Table 2 Examples of animal origin-based dyes

S.No	Animal species	Colour	Dye stuff	Image	References
1	*Sea Mollusc Murex*	Purple	Tyrian purple		[25, 26]
2	*Dactylopius coccus*	Crimson red	Cochineal		[27]
3	*Kermes Ilicis*	Red	Kermes		[28]

(continued)

Table 2 (continued)

S.No	Animal species	Colour	Dye stuff	Image	References
4	*Kerria lacca*	Deep red	Lac		[29, 30]
5	*Lichen*	Red and purple	Orchal		[31]

3.2.4 Microbial Origin-Based Dyes

Some microbes or bacteria produce coloured substances. The dyes or pigments that are produced from microbes are known as microbial-based dyes. Common examples of these microbes are Bacillus, Pseudomonas, Rhodococcus spp., and Serratia [16]. The main advantage of microbial origin-based dyes is that their growth requires cheap substrates under normal growing conditions. According to the study, stability to heat and dyeing characteristics of the Serratia strain were discouraged due to its poor fastness to light ability, and its future use in textile industries was avoided. Also, pigments produced from the fungus 'Monascus purpureus' are employed to colour food items. The table below contains some of the examples of microbes-based dyes or pigments (Table 4).

4 Why Natural Dyes Were Replaced?

With the advancement in human civilization, the demand for dyes also expanded. Natural dyes were the primary choice until the mid-nineteenth century. However, the business of natural dyes experienced a downfall following the creation of the first artificial dye in 1856. After this year, synthetic dyes were preferred over natural dyes because natural dyes exhibited limitations such as poor ability to bind with

Table 3 Examples of mineral origin-based dyes

S.No	Mineral origin	Colour	Image	References
1	Prussian Blue	Blue		[32]
2	Iron	Black		[33]
3	Manganese	Brown		[34]

(continued)

Table 3 (continued)

S.No	Mineral origin	Colour	Image	References
4	Malachite	Green		[35]
5	Chrome	Yellow		[36]

textile materials and extensively long and complex dyeing processes. Additionally, the range of dyes that could be obtained from natural sources is restricted. This led to the quick replacement of natural dyes [40].

4.1 Synthetic Dyes

Synthetic dyes are man-made or artificial colouring substances designed to be used in numerous applications, particularly in industries such as textiles, papers, and plastics. William Henry Perkin accidentally discovered the first synthetic dye, named Mauveine, in 1856. With this accidental invention, the world saw the emergence of the synthetic dye industry [41]. Unlike natural dyes, synthetic dyes are chemically synthesized. These dyes offer advantages like enhanced colour fastness, a variety of shades, and the ability to adhere effectively to various materials. They have played an important role in meeting the growing global demand for vibrant and long-lasting colours.

Table 4 Examples of microbial origin-based dyes

S.No	Microbial origin	Colour	Image	References
1	Serratia	Red	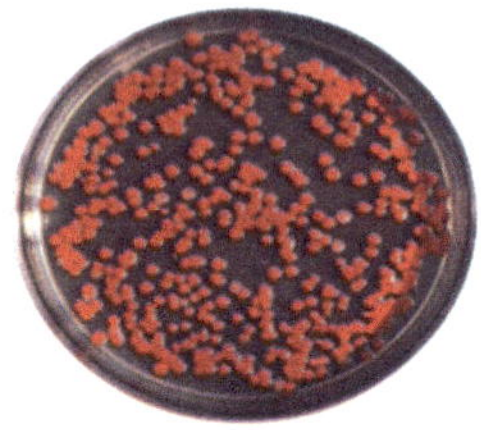	[37]
2	Pseudomonas	Green	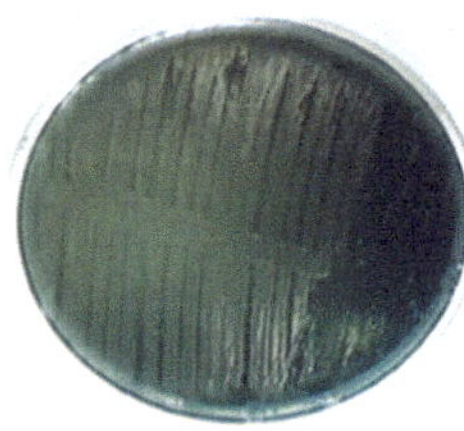	[38]
3	Staphylococcus sp.	Golden	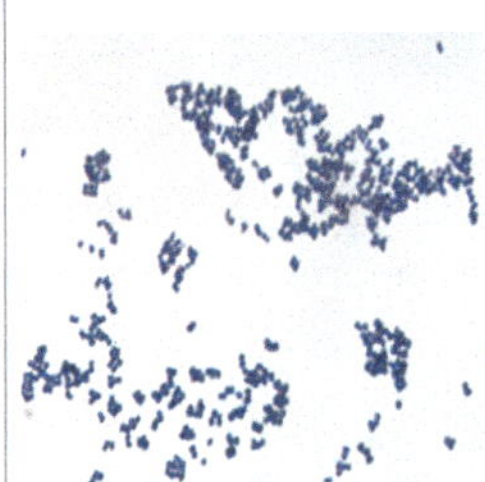	[39]

4.2 *Classification of Synthetic Dyes*

These dyes can further be categorized based on their chemical structure, dye chemistry, colour, and application method. Here, we have chosen the simplest method to remember their types, which is the classification of synthetic dyes based on their application method.

- Direct dyes.
- Reactive dyes.
- Acid or anionic dyes.
- Basic or cationic dyes.
- Mordent dyes.
- Disperse dyes.
- Vat dyes.
- Sulphur dyes.

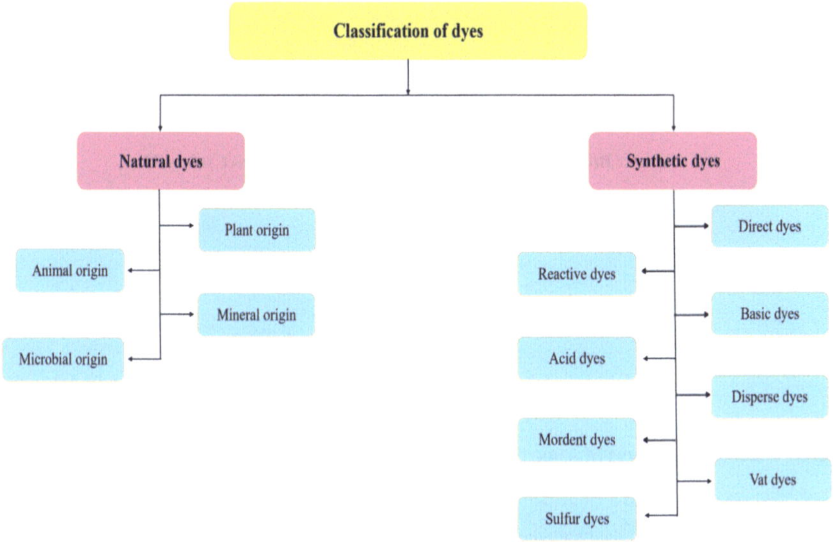

Fig. 1 Classification of dyes

4.2.1 Direct Dyes

The discovery of the first direct dye (Congo Red) in 1884 by P. Bottiger led to the beginning of direct cotton dyestuffs [42]. Direct dyes are water-soluble anionic dyes that constitute the largest group of dyes [43]. As the name implies, direct dyes do not require pre-treatment with a metal salt (or mordant) [44] turning the complex dyeing process into a simpler one. Most of the direct dyes belong to the azo class. These dyes are the cheapest when compared with other dyes [45]. The main advantage of direct dyes include a variety of colours and easy application. Even though these dyes have a variety of hues and ease of application, they have an inadequate capacity to endure washing. This has been the reason behind the substitution of reactive dyes that have a far stronger opposition to washing and to the humidity qualities of fabrics [46, 47]. They are directly applied to the salt solution made up of sodium chloride or sodium sulphate, and the wettability of fibre is facilitated by auxiliary products [48].

Examples of direct dyes are Direct Blue 1, Direct Blue 86, Direct Blue 106, Direct Yellow 59, Direct Red 227, etc. Figure 2 represents the structures of a few direct dyes.

4.2.2 Reactive Dyes

These are the second largest group of dyes [46] containing azo or anthraquinone-based chromophore groups. First, these dyes get adsorbed on the surface of the fibre and then start bonding covalently with the textile fibres [49]. In an attempt to overcome the drawback of direct dye's weak wet-fastness, direct dye provided

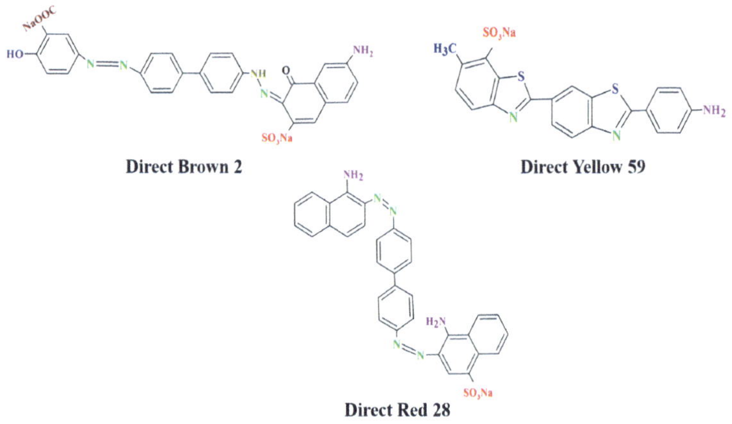

Fig. 2 Chemical structures of some direct dyes

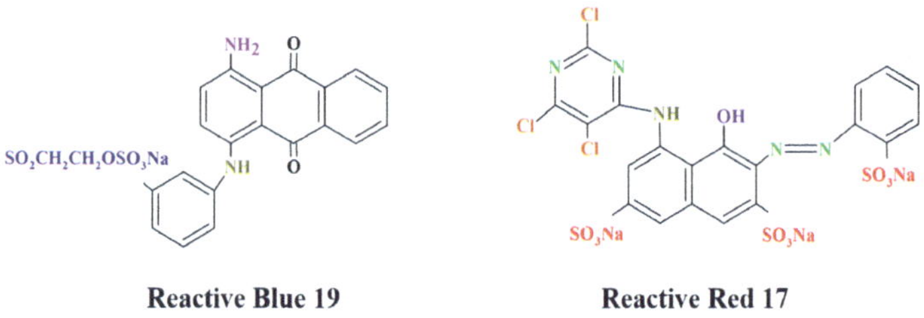

Fig. 3 Chemical structures of some reactive dyes

the foundation for the development of a class of reactive dyes. The fundamental distinction between these two classes of dyes lies in the strong covalent bonding between reactive groups [46]. Like direct dyes, reactive dyes are also water-solvable anionic dyes. A group which has the ability to react with the fibre sites, for example, -OH groups in cellulose fibre and -SH, -OH, and -NH_2 groups in wool, is a potential active group that can be utilized in a reactive dye. The diffusion of reactive dye in the innermost layers of fibres improves the ability of the dye to stick to fibres for a longer period of time [50, 51], giving stronger and permanent attachment of dye to the fibre. Reactive dyes have characteristics such as high long-term wet-fastness, variety of colours, high photolytic stability, and brilliance [46, 52]. Few examples of reactive dyes with their chemical structures are shown in Fig. 3.

4.2.3 Basic Dyes

Basic dyes, also known as cationic dyes, contain colourless acid with a coloured base [53]. These dyes are known as basic or cationic dyes because they are formed from

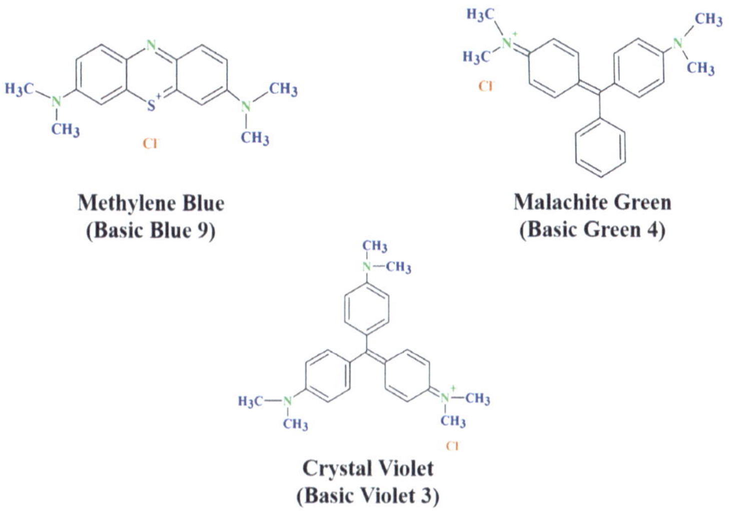

Fig. 4 Chemical structures of some basic dyes

organic substances and undergo ionization to produce coloured cation. These dyes offer some outstanding properties, such as intensity and brightness [13]. Modern nylons, modified polyesters, acrylic, and paper are usually dyed using basic dyes [54, 55].

Earlier silk, cotton, and wool were dyed using these dyes. Coloured cations are produced when dissolved in solution [54]. These cations are drawn to negatively charged substrates electrostatically. The positive charge on basic or cationic dye can either be distributed as a delocalized charge over the dye cation or localized on an ammonium group [55]. Common examples of basic dyes are depicted in Fig. 4.

4.2.4 Acid Dyes

Acid dyes are anionic dyes with the presence of sulphonated groups making them soluble in water [46]. Here, the acid component is coloured with a colourless base [53]. The molecules of dye have one or more functional groups such as SO_3H^- and COOH which are acidic in nature and hence were named acid dyes. They are used to dye cotton, wool, silk, nylon, and leather. Since acid dyes are often more stable in an acidic environment, the majority of applications were performed in a weakly acidic bath with a pH ranging from 4.5 to 5 [46]. The interaction and extent of colour fastness vary with the size of dye molecules. Azo chromophoric systems, anthraquinone, triphenylmethane, or copper phthalocyanine are examples of acid dyes [55]. The examples of acid dyes are as follows (see Fig. 5).

Fig. 5 Chemical structures of some acid dyes

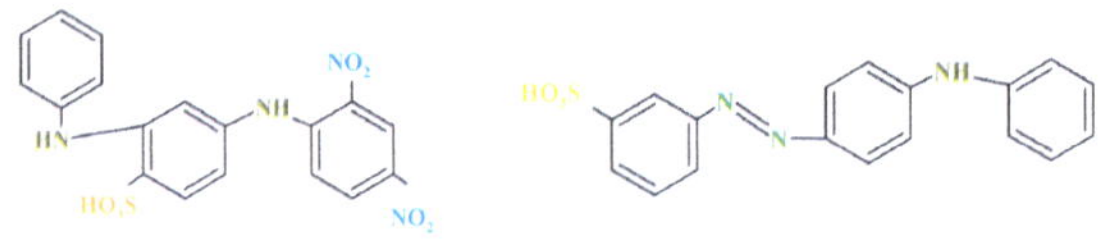

Acid Orange 3

Acid Yellow 36

Fig. 6 Chemical structures of some mordant dyes

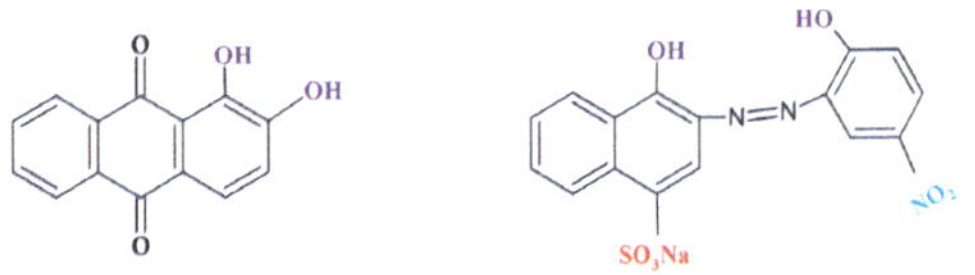

Mordant Red 11

Mordant Brown 25

4.2.5 Mordant Dyes

Mordant dyes are a type of dye that requires the use of chemicals or substances known as mordant, with the help of which dyes easily and effectively attach to the fibre. These dyes are known for their vibrant colour fastness and light fastness results. The term 'mordant' comes from the Latin word 'mordere', which means 'to bite' [56]. This signifies the role of the mordant in fixing the dyes to the material by forming a coordination complex with the fibre [57]. Some of the examples of mordant dyes are shown in Fig. 6. Mordant dyes are often used to achieve a large variety of shades and colour variations in textiles and other materials.

4.2.6 Disperse Dyes

Disperse dyes are also known as plastosolubles, exceedingly hydrophobic, and used as a fine powder scattered in the dye bath. They contain one or more fibres that are hydrophobic like polyester, nylon, cellulose, or cellulose acetate [13, 54]. These dyes are non-ionic and coloured aromatic compounds with azo and nitro groups that are frequently employed to colour synthetic textiles [46]. The examples of disperse dyes are shown in Fig. 7. Disperse dye is non-ionic in nature and thus gets quickly volatile [13]. These dyes are extensively employed in dyeing fibres, particularly polyester and polyamide, when the dispersing agent is present [48].

Fig. 7 Chemical structures of some disperse dyes

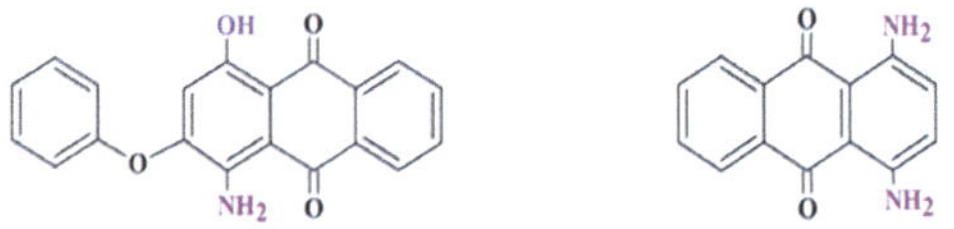

Disperse Red 60

Disperse Violet 1

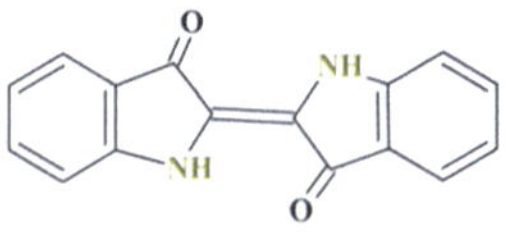

Vat Blue 1

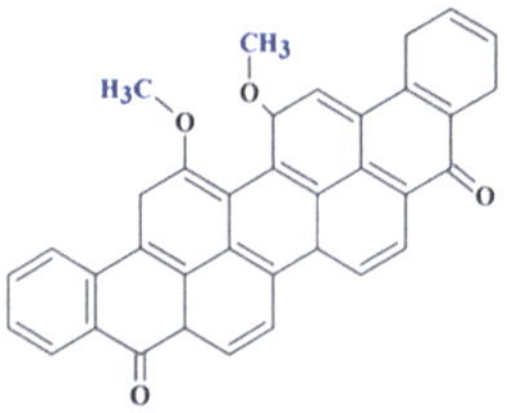

Vat Green 1

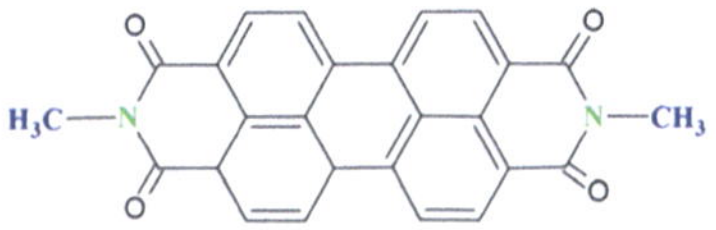

Vat Red 23

Fig. 8 Chemical structures of some vat dyes

4.2.7 Vat Dyes

The oldest known vat dye is indigo, and it has been in use for 4000 years now. These dyes were fermented in wooden vessels known as vats, hence named vat dyes [58]. These dyes are originally water-insoluble pigments but are referred to as dyes due to chemical breakdown in an alkaline solution, which changes pigment into the leuco form, which is water soluble [59, 60]. Once the chemically reduced dye has permeated the fibre, it oxidizes and turns into an insoluble state of water. This leads to excellent wash fastness. Despite being costly, vat dyes are extensively employed in various applications. The availability of oxygen atom double-bonded to an atom of carbon is the main feature of these dyes [54]. The reproducibility of colour, excellent fastness properties [61], and colour value are the important advantages of vat dyes over other dyes and it is easier to wash off dyeing [46]. Examples of vat dyes are shown in Fig. 8.

4.2.8 Sulphur Dyes

Sulphur dyes differ from vat dyes in terms of structure as they contain Sulphur-containing molecules with high molecular weight, but they are alike in terms of method of use [48]. Croissant and Bretonniere prepared the first Sulphur dye in 1873 [46] by heating cellulose materials, which contain wood, cotton waste, and paper waste, along with alkali and polysulphide [61]. Since then, they are frequently used on cellulose fibres and are one of the famous classes of dye. These dyes became quite a class and were employed extensively for cellulose with an inexpensive dyeing process [46]. This class of dyes is water-insoluble, similar to VAT dyes, but can be made soluble by adding reducing agents and alkalis. The chemical structure of Sulphur

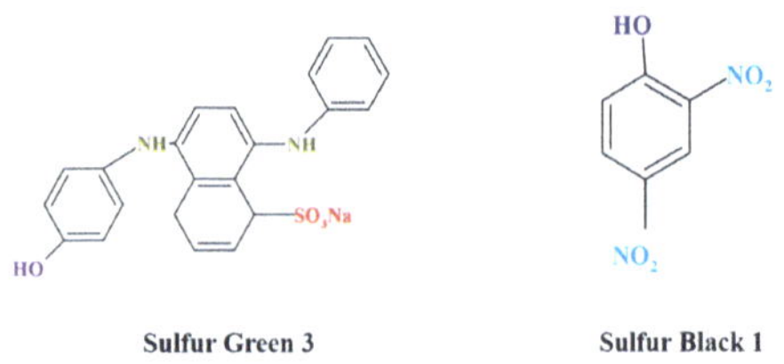

Fig. 9 Chemical structures of some sulphur dyes

dyes contains disulphide linkage (R–S–S–R) where R represents the chromophores [61]. A few examples are shown in Fig. 9.

5 Dye Pollution Sources

There are numerous industries responsible for causing dye pollution. Some include paper industries, textile and apparel industries, pharmaceuticals, dye industries, food industries, cosmetics and paint industries, etc. The pie chart (see Fig. 10) shows the comparison between a few industries on the basis of their contribution to dye pollution [62]. With 54% contributors to dye pollution, textile industries are leading and hence the matter of concern for our environment. Below, we have discussed some of the industries and their role in causing dye pollution.

5.1 Paper Industry

Paper industries require large amounts of dyes and pigments to produce papers with various designs and functions, and a part of it is released into water streams, causing

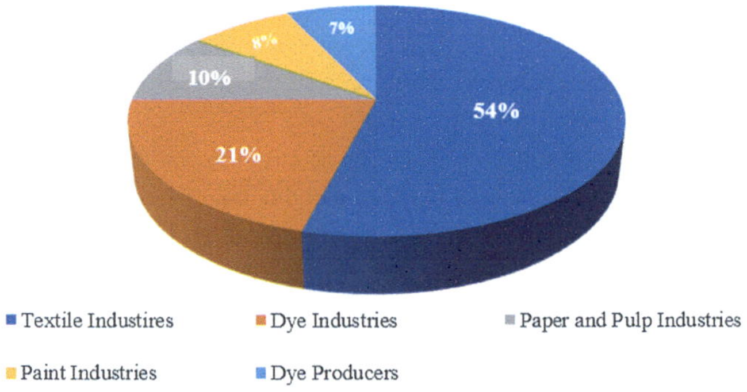

Fig. 10 Percentage of dye pollution caused by various industries. (Redrawn from Ref. [62])

aquatic pollution. A Paper mill located in Uttarakhand, India (M/S Century Pulp and Paper Mill) releases 302,832.8 L of wastewater every day, and there are over 10,000 paper and pulp mills worldwide [63]. The amount of contaminated water which is used in the paper industry released worldwide is above six hundred million metre cubes, and the use of pollutants in the process results in high oxygen demand and toxicity. No doubt that industrialization has harmed the environment from the very beginning. So alternative eco-friendly methods should be adopted for the production of papers with minimal waste.

5.2 *Textile Industry*

Nearly 280,000 tonnes of dyes are annually discharged by textile industries into the environment [64]. Dyeing process in itself affects the environment to some extent whether used for the production of natural dyes or synthetic dyes. It has been reported that during the production of dyes approximately 10–15% dyes are discarded into the wastewater. With the dazzling of colourful clothes and fashion, most of the world is still unaware of the harmful effects behind it. Since the dyes are created in a manner that they have high reliability in light and water, it becomes very hard to remove them from the contaminated water by conventional degradation technique. The effect of the textile dyes on aquatic life is more concerning to the ecosystem, flora, and fauna, and the algae present in the water is more reactive towards the pollutants [65]. With the advancement in industrialization, proportionally there is an enlargement in the consumption of dyes, and similarly, a large amount of fresh water is required some of which is released into the rivers, water stream impacts the water for drinking and water bodies and causes harm to humans and the environment. There are certainly about 8000 different chemicals that are used in textile production processes [66]. So we can assume the damage it must be causing to our surroundings.

5.3 *Pharmaceuticals*

Dyes used in pharmaceutical industries are generally used as colourants in capsules, syrups, tablets, and ointments to make them look more appealing and eye-catching [67]. This in return increases the demand of the product and overall business of the industry. The colours used in these medicines are also used for the identification of medicines [68]. Also, the use of dyes in pharmaceuticals can be regulated on the application as biological stains, the process became an important diagnostic of fungi, protozoa, and virus infection [69]. Drugs play a huge role in polluting water and its constituents include various organic pollutants which are responsible for water contamination [70].

5.4 Food Industry

The food dyes functioning in the product have various aims including commercial, as an identifier, and some dyes also elevate the appetite [71]. For many years, a large number of synthetic dyes have been used to design food to make it look appealing and attractive and also make people happy. Mostly the dyes used in these industries are azo dye, erythrosine, quinone, etc. The accepted levels of dye content in food products are regulated by different international organizations, and the use of many dyes is prohibited by these organizations and still few dyes are detected in food products [72]. Malachite green, a common basic dye which is hardly soluble in water, was banned by the EU in the year 2000 due to its usage in the treatment of fishes [73]. The dyes used in food colourants are mostly synthetic and are organic complex compounds that are produced from coal tar and petroleum. The choice of dyes used by the industries stuck to the synthetic because they are cheaper, stable, and brighter than the natural ones [74].

5.5 Tannery Industry

The tannery sector is the place where the animal's skin gets converted into leather, and the worldwide production of leather is 1.85 billion metres square annually [75]. This produced leather gets converted into various products like footwear, bags, and textiles which increases the demand of dyeing these products to make them more appealing; this ultimately results in wastewater from these industries. More than 90% of leather is tanned using Cr (III) salts. With the conventional chrome tanning process, about 70% of all the chrome agent is assimilated by the leather, resulting in the concentration of Cr (III) in tannery wastewater surpassing 3000 ppm [76].

5.6 Cosmetics

Cosmetic industry, driven by consumer attraction to vibrant colours, is currently witnessing a trend in the use of cosmetic colourants. The colourants used are mainly dyes and pigments. Hair dyes are known to be the largest hit since the discovery of first hair dye; also in ancient times, Egyptians used henna to dye their grey colour [77]. The experimental confirmation by Bruce et al. showed that 89% of the hair dyes are mutagenic and carcinogenic [78]. The percentage of synthetic to natural dye is very high because of the low cost of the synthetic dye and its long-lasting nature. The major cause of waste coming from cosmetic industries is the packaging which is less practical and more fashionable.

6 Adverse Effects of Dyes on Environment

One of the intense issues faced by humanity is the inaccessibility and scarcity of superior quality water for drinking [79]. Though most of the Earth's surface is predominantly covered by water, approximately 98% of it consists of seawater which is not suitable for drinking, only about 2% of Earth's surface is freshwater with approximately 1.5% frozen solid in glaciers. In subterranean wells and aquifers, around 0.36% of water is present beneath the Earth. About 0.036% of the world's total water supply is contained in lakes and rivers. However, the rapid pace of industrialization and modernization has led to an exponential increase in contamination of these freshwater resources [80]. Bar graph shown below represents the world's most water-stressed cities [81] (Fig. 11).

Every year around ten thousand tonnes of distinct synthetic dyes are produced globally. Majorly highest number of dyes are consumed by textile industries [82]. During dyeing and washing process, dyes do not get attached to the fibre completely and this unattached dye is washed away with water and released into the environment [83–85]. These textile dyes are highly toxic and carcinogenic [86] even when used in low concentrations [87–92]. The wastewater generated by textile industries contains toxic dye solution with high concentration [93].

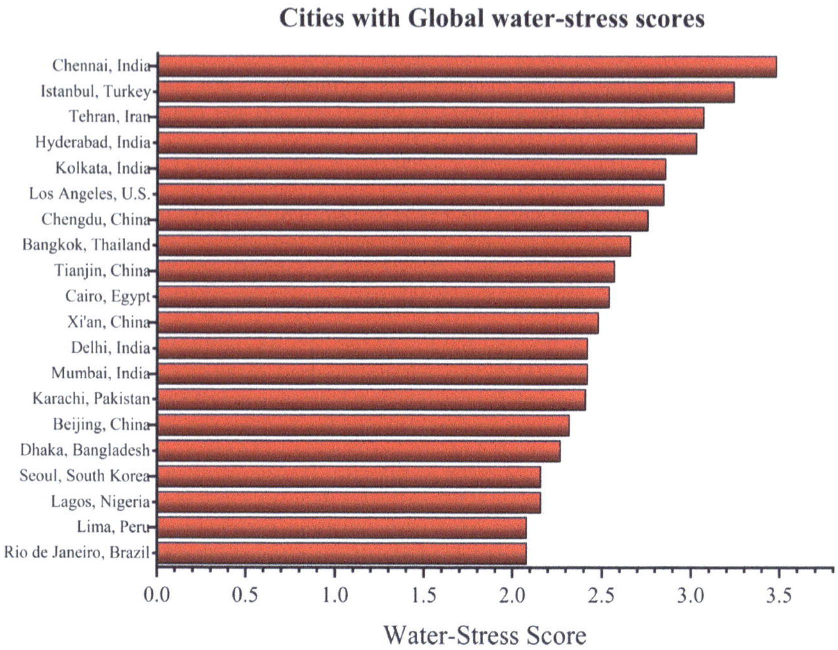

Fig. 11 Bar graph of global cities suffering from water scarcity. (Redrawn from Ref. [81])

The recalcitrant behaviour of dyes affects the functioning of ecosystems; once dyes are discarded into the aquatic ecosystem without proper investigation, it gets accumulated in certain tissues of fish and ultimately reaches human organs through the food chain. From dermatitis to disorders in the central nervous system, human health is severely affected by inhalation and ingestion of dyes [94, 95] Although there is no proof that every dye stuff employed in finishing and dyeing processes is hazardous to human beings' health but with long term exposure it does causes severe irritation in skin, respiratory problems, irritation in eyes and nose, etc.

Textile dyeing effluents contain not just dyes but moreover include harmful heavy metals such as copper, arsenic, zinc, and above everything, chromium, causing many health problems including haemorrhage, nausea, etc. [96]. The presence of metals, dyes, and aromatic compounds in aquatic bodies affects the photosynthesis process by obstructing the penetration of light. In addition to this, dyes are also mutagenic and can last in the environment for longer period of time [87–90, 92, 97–99]. Despite prohibiting the use of more than 4000 dyes, around 100 of them are still being used and available in the market [100].

Synthetic dyes are not biodegradable because of their complex chemical structure, stability, and properties. The amount of harm these dyes can cause on health and environment depends on individual dye and the amount of exposure. The most frequently used dyes are azo dyes with low cost, colour fastness, and high intensity. The increase in cotton usage subsequently leads to an increase in the use of azo dyes as cotton fibres are mainly dyed with azo dyes. These dyes alter the chemical and physical properties of soil, causing deterioration of aquatic life and ultimately harming the ecosystem. They have complex structures, and are known to be mutagenic and carcinogenic in nature. Most of these azo dyes are highly poisonous to the environment [59, 96, 101–103]. In addition, a recent study reported that ~15% discharge of azo dye effluent comes from textile industries every year [104, 105]. Along with this, Reactive dyes are known to be problematic, with poor dye-to-fibre fixation fibre, leading to 5–50% of dye waste production [84].

Several dyes have mutagenic potential, and most of the dyes are soluble in water, posing serious threats to aquatic bodies 106. Crystal violet dye is responsible for causing tumours in fish [5], skin irritation, and respiratory failure in humans. These dyes impart a very intense colour [106]. Methylene blue produces aromatic amines such as benzidine and is one of the widely used dyes in textile industries. These aromatic amines are known as potential carcinogens, trigger allergic reactions, and cause dermal effects. The toxicity level of this dye was investigated on microalgae named S. platensis and C. vulgaris by exposing them to different doses of methylene blue. The results revealed a decrease in concentration with a specific growth rate [107]. Azo dyes enter the human body through the digestive system and damage the DNA, and human exposure to these dyes results in gastrointestinal, circulatory, respiratory, and lung diseases [108].

Moreover, various biological and physicochemical parameters, such as total oxygen demand, chemical oxygen demand, total dissolved solids, and biochemical oxygen demand, reach elevated levels in dye-polluted environment [109, 110]. The major health risks from dyes and intermediates include allergic sensitization, acute irritations, and carcinogenicity [10]. Many dyes contain cancer-producing metals and their use in industries is questionable. The side-effects of dyes pose serious risks to our ecosystem and also present challenges to our lifestyle. There is a need to maintain an ecological balance between growing textile demand and proper treatment of effluents.

7 National and Global Dye Effluent Emission/National and Global Statistics of Dye Effluent Emission?

The sources of dye pollutants and their adverse effects have been out in open for a long time but the extent to which these industries and states contribute to this pollution is widely known. These dye-contaminants are severely impacting our environment and polluting the surroundings with each passing day. Here, we have discussed some national and global dye effluent emissions.

7.1 National Level

According to a report by the Central Pollution Central Board (CPCB) for the year 2019–20, there are 23 wastewater drains in the National Capital Territory (NCT) of Delhi, out of which 20 drains are connected to Yamuna River and rest of the 3 drains join Agra/ Gurugram canal. This has been the root cause of river pollution in India.

Rajasthan, being the largest state in India covering 10.4% of the country's land area, ranks second among producers of textile fibres. Bhilwara, also known as the Textile city of India, is famous for textile production. However, water-deficient regions like Rajasthan are unable to cope with the critical problems of widespread water contamination [111].

Due to the huge amount of discharge and wastewater generation, textile plants are considered as the major polluting industries. Tirupur, Tamil Nadu has around 760 textile dyeing units and untreated dyes were discarded into the Noyyal River till year 2006. This led to the increased levels of total dissolved solids (TDS) and chlorides in the river as well as in the Orathupalayam dam located across the river [112]. The increased value of TDS also negatively impacts the underground water further impacting the plant's growth. Furthermore, excessive chlorine concentrations were detected at the sites of Kasipalayam due to dyeing and bleaching processes adopted in textile dyeing units in the year 2008–2009 [113].

Kanpur is known for its prosperous leather industries which is located on the west bank of the Ganga River. But leather industries are contaminating the river with organic and inorganic compounds which are hard to degrade [114]. The Ganga River is facing a huge discharge of 450 million litres of industrial effluents and municipal sewage daily. The majority of drains, sewage, and effluents were managed through common effluent treatment plants before the launch of the Namami Gange Programme. Still, many industries' untreated effluents are directly entering into the river [115]. This underscores the major need for effective wastewater management techniques and methods to facilitate degradation.

7.2 International Level

The ecological imbalance is not confined to India but also extends globally, affecting many countries. As much as textile industries are leading to economic growth, they also contribute to ecological disruptions through the release of dye pollutants into the field, water, and air. Below are some real-life examples of the countries which face pollution due to dyes and textile industries.

China, being the largest textile fibre producing country in 2009, accounted for yearly 90,100-tonne production of dyes and pigments [84]. The fibre production processes generate both solid and liquid waste, seriously impacting the ecosystem.

With a total of 662 licensed and 1000 small-scale unlicensed textiles and apparel industries, Malaysia ranked as the 15th largest producer of textile fibre in 2008. This substantial production contributes to a large volume of waste, which has a notable impact on the environmental system. The famous industry of Malaysia is a man-made textile fibre. While the textile industry was reported to contribute only about 0.1% of waste as per a report given in 2009, this figure increased exponentially from 744 to 1559 tonnes within just two years span. Notably, the majority of wastewater pollution from the textile sector is concentrated in Peninsular Malaysia, with Pulau Pinang (28.2%), Selangor (15.6%), and Johor (28.6%) having the largest concentrations [84].

The sudden growth in population and industrialization has also affected Mexico City by polluting its rivers severely. The most polluted rivers of Mexico include the Lerma-Santiago River, the Tula River, and the Atoyac River. Numerous efforts have been taken to maintain permissible levels of BOD, pH, TDS, and salinity of the Atoyac River, but the insufficiency in addressing the water quality and sediment quality fails to govern the efforts. The Atoyac River basin, covering an area of approximately 4130 km^2, is situated in Central Mexico. The increase in river contamination arises from 14 industrial factories which comprise textile, pharmaceutical, food sectors, and others upon entering Puebla city [116].

In Bangladesh, surface water is exposed to pollution from untreated effluents, wastewaters, runoff from chemical fertilizers, etc. Research on the water quality of Karnafully River showed seasonal variation in temperature, pH, BOD, salinity, and transparency. The discharge from the industries mainly consists of organic and inorganic wastes according to the survey. The organic wastes include degradable wood chips, pulps, and tannery effluents whereas inorganic wastes include dyes, pigments, acids, heavy metals, and salts. Major polluting industries in Bangladesh comprise 26 industries of textiles, 19 tanneries, 5 fish processing units, 4 paint and dye manufacturing plants, 1 steel mill, and 1 oil refinery. Among these, the major contributing industries in the economy as well as in pollution are textile industries. All the effluents ultimately enter into the Karnafully River polluting the source to a large extent [117]. A study revealed the amount of time taken by dyes to degrade was more than three years [118].

Considering these findings, there is an urgent need to address the steps required to mitigate dye pollution. This also demands the development of various strategies, policies, regulatory frameworks, and techniques to effectively control dye pollution so that our environment can thrive.

8 Analytical Methods

To detect the presence of dyes and concentration of dyes, several techniques are available which help us in investigating the quantity of dyes and separate them from mixtures. These techniques involve chromatographic and spectroscopic techniques.

8.1 Chromatographic Techniques

Chromatography is a technique which is used to separate the mixture into its individual components. Not only does it separate them but also identify and analyse their characteristics. It is a special technique used to separate dyes. There are two different types of phases that are present in chromatography—stationary phase and mobile phase. Stationary phase is composed of solid or liquid whereas mobile phase is composed of liquid or gaseous component. This technology helps in identifying the dye content. Various types of chromatography techniques are gas chromatography (GC), liquid chromatography (LC), thin layer chromatography (TLC), ion exchange chromatography (IC), high pressure/performance liquid chromatography (HPLC), etc. Out of these, only TLC, HPLC, IC, and GC are preferred over others for wastewater treatment [119], and a combination of these with other techniques like mass spectrometry results in LC–MS, HPLC–MS, etc. Here, we will discuss a few important techniques which are used in wastewater treatment or in removing dyes from the wastewater.

8.2 Gas Chromatography (GC)

In this technique, inert gas is used in the mobile phase typically He, Ar, N_2, or CO_2. An inert gas is passed through the stationary phase (a solid support) with a coat of a thin layer of non-volatile liquid (column) under high pressure. The sample is vapourized and transported through the column into the mobile phase which is a gas. The components of a sample are dispersed between the mobile and stationary phase on the solid support. A technique where gases are used to separate the individual components of a sample is known as gas chromatography. This is a very simple, extremely sensitive, and highly employed technique used to separate very minute molecules [120]. This technique is applied for gases, volatile liquids, and solid materials. Because of low thermal stability and less volatile nature of dyes, gas chromatography is not best employed for them. Although combining gas chromatography with mass spectrometry, i.e., GC–MS can easily identify intermediates and reaction pathways. GC–MS technique enables the identification of a large number of organic compounds present in water or air and is being commonly used [121].

8.3 High-Performance Liquid Chromatography (HPLC)

High-performance liquid chromatography or High-pressure liquid chromatography is a technique which involves the use of high pressure in distinguishing non-volatile and thermally unstable species within a short span of time. The application of high pressure (10 to 400 atm pressure) on small molecules with a high flow solvent rate (0.1–5 cm/s) enables fast separation and completion of structural and functional analysis in a short time [122]. The mobile phase is liquid hence the name liquid chromatography. This technique helps in purifying a number of organic and inorganic compounds. HPLC is a technique used for the detection of dyes like Rhodamine B, Methylene Blue, Crystal Violet, Malachite Green, etc. [123]. For separating historical dyes, reverse phase-high performance liquid chromatography (RP-HPLC) is most commonly employed [10]. RP-HPLC also made it possible to separate water-soluble substances [119]. HPLC when used in conjunction with mass spectrometry (HPLC–MS) is the most effective method of identifying dyes and their consecutive intermediates. It also helps in monitoring their byproducts and aromatic intermediates thereby establishing a reaction mechanism 126. In 2013, Xian et al. first reported the combination of liquid chromatography with mass spectrometry for the analysis of dyes in cosmetics [124].

8.4 Ion Chromatography

Ion-exchange chromatography is the electrostatic interaction between charged protein groups and solid support material [120]. During dye degradation, the formation of organic and inorganic ions can be identified using this technique. The most commonly identified ions in dyes are organic ions such as carboxylic acids, nitrate, nitrite, and ammonium ions and inorganic ions such as sulphate or chlorine. Ion chromatography when combined with other techniques like TOC helps in mineralizing the dye completely [125].

8.5 *Thin Layer Chromatography (TLC)*

Another very common technique that is employed for the detection, separation, and analysis of dyes is thin layer chromatography. Thin layer chromatography consists of a stationary phase in which a thin layer of finely ground matrix made up of alumina, silica gel, or similar materials is coated on a glass plate, metal plate, or a plastic film. To increase the visualization, the stationary phase is combined with the fluorescent powder. Once the plate is set, choosing a solvent is the next crucial step in TLC and it requires a lot of trial and error. Upon completion of separation, the individual molecules show up as spots [126].

Each spot has its own retention factor which is the ratio of distance travelled by the sample and the solvent. TLC is a comparatively older technique yet it is relevant and used for the analysis of many dyes like amaranth, rhodamine B, etc.

8.6 *Spectroscopic Techniques*

The study of the interaction between matter and electromagnetic radiation is known as spectroscopy. The molecules absorb a particular wavelength of light and transmit the remaining light. Spectroscopy is one of the widely used techniques [123]. Many spectroscopic methods such as UV–VIS spectroscopy, Fourier transform Infrared spectroscopy (FTIR), Near-infrared spectroscopy (NIR), and nuclear magnetic resonance (NMR) are suitable for dye detection and wastewater treatment [119].

8.7 *UV–VIS Spectroscopy*

A very well-known technique for the determination of dyes and their degradation rate is the UV–VIS spectroscopic technique. This is the easiest technique based on the determination of absorbance peaks of different dyes. Methylene Blue, rhodamine B, malachite green, methyl orange, acid red, and Crystal red are examples of very few dyes among others which have been detected using UV–VIS spectroscopy. Different dyes show absorbance peaks at different wavelengths which makes it easier to detect the presence of these dyes in wastewater or powdered sample dissolved in a solvent. The wavelength at which the absorbance value is highest is called λ_{max} and it denotes the colour of that dye. Due to its simplicity and versatile nature, UV–VIS spectroscopy is a widely used method.

8.8 Fourier Transform Infrared Spectroscopy (FTIR)

Among other spectroscopic techniques, the most commonly used technique is Fourier Transform infrared spectroscopy. Fourier Transform Infrared Spectroscopy identifies the alterations in dye molecules' structures. Depending on the initial dye, the presence of certain ions can be observed with peaks indicating the creation of chloride ions, HCO_3^-, NO_3^-, OCl^-, etc. The existence of C–N symmetric stretching suggests the breakage of dyes into their byproducts and guides us in understanding the degradation mechanisms. Peaks arising from = C–H bending and C–O–H stretching are most often detected due to the aromatic nature of organic dyes [125]. FTIR spectroscopy can identify dyes as well as their concentration.

8.9 Nuclear Magnetic Resonance (NMR)

Nuclear magnetic resonance is a technique which provides valuable information about the chemical structure of dyes. When a sample containing dye molecules is analysed using NMR spectroscopy, the protons present in the dye molecules give off signals which give an overall NMR spectrum.

In spite of having the ability to determine intermediates and analyse the degradation mechanism, this technique is less used. Nuclear magnetic resonance only shows the presence of aromatic compounds in the dyes and how much particular amount of intermediates are generated during the dye degradation process [125].

8.10 Future Directions and Conclusion

The ever-evolving domain of dye pollution and expanding modernization demands the exploration of emerging technologies for wastewater treatment. Adopting green and sustainable dyeing practices in order to minimize the generation of dye wastewater which includes low-impact dyeing methods, natural and eco-friendly dyes will be helpful in developing a clean and green planet. There is a need to make time-to-time improvements and potential changes in regulatory frameworks and compliance standards to address the growing dye pollution. Additionally, the environmental impacts of dyeing processes can be understood with the help of life cycle assessment methodologies. Establishing sustainable economy in the apparel and dyeing sector should be a priority with the strategy to recover, reuse, and recycle dyes. Along with this, public awareness campaigns equally play an influential role in promoting sustainable practices. Organizing training programmes and educating and inspiring students, workers, and the public will promote sustainable dyeing practices. The importance of these programmes cannot be overstated as they will inculcate new ideas and innovations. Further, the practices that have been successfully implemented worldwide will provide valuable insight, and this will motivate industries and researchers to look for better and fast alternatives in mitigating dye pollution. These prospective directions collectively shape a roadmap towards more environmentally aware and sustainable textile and dyeing industries [8, 127, 128].

References

1. Grand View Research (2015). http://www.grandviewresearch.. Google Scholar. Accessed 26 Septe 2023. https://scholar.google.com/scholar?hl=en&as_sdt=0%2C5&q=Grand+View+Research+%282015%29.+http%3A%2F%2Fwww.grandviewresearch.com%2Findustry-analysis%2Fdyes-andpigments-+market&btnG=
2. Singh H, Kumar D, Soni V (2020) Copper and mercury induced oxidative stresses and antioxidant responses of Spirodela polyrhiza (L.) Schleid. Biochem Biophys Rep 23:100781. https://doi.org/10.1016/j.bbrep.2020.100781
3. Dil EA, Ghaedi M, Ghaedi AM et al (2016) Modeling of quaternary dyes adsorption onto ZnO–NR–AC artificial neural network: analysis by derivative spectrophotometry. J Ind Eng Chem 34:186–197. https://doi.org/10.1016/j.jiec.2015.11.010
4. Saleh TA, Gupta VK (2014) Processing methods, characteristics and adsorption behavior of tire derived carbons: a review. Adv Colloid Interface Sci 211:93–101. https://doi.org/10.1016/j.cis.2014.06.006
5. Mani S, Bharagava RN (2016) Exposure to crystal violet, its toxic, genotoxic and carcinogenic effects on environment and its degradation and detoxification for environmental safety. In: De Voogt WP (ed) Reviews of environmental contamination and toxicology, vol 237. Reviews of environmental contamination and toxicology. Springer International Publishing, pp 71–104. https://doi.org/10.1007/978-3-319-23573-8_4
6. Senthilkumaar S, Kalaamani P, Subburaam C (2006) Liquid phase adsorption of crystal violet onto activated carbons derived from male flowers of coconut tree. J Hazard Mater 136(3):800–808. https://doi.org/10.1016/j.jhazmat.2006.01.045
7. Li S (2010) Removal of crystal violet from aqueous solution by sorption into semi-interpenetrated networks hydrogels constituted of poly(acrylic acid-acrylamide-methacrylate) and amylose. Bioresour Technol 101(7):2197–2202. https://doi.org/10.1016/j.biortech.2009.11.044
8. Slama HB, Chenari Bouket A, Pourhassan Z et al (2021) Diversity of synthetic dyes from textile industries, discharge impacts and treatment methods. Appl Sci 11(14):6255. https://doi.org/10.3390/app11146255
9. Rattee ID (1972) The chemistry of dyeing. Chem Soc Rev 1(2):145–162. https://doi.org/10.1039/CS9720100145
10. Shahid M, Wertz J, Degano I, Aceto M, Khan MI, Quye A (2019) Analytical methods for determination of anthraquinone dyes in historical textiles: a review. Anal Chim Acta 1083:58–87. https://doi.org/10.1016/j.aca.2019.07.009
11. Verma AK, Dash RR, Bhunia P (2012) A review on chemical coagulation/flocculation technologies for removal of colour from textile wastewaters. J Environ Manage 93(1):154–168. https://doi.org/10.1016/j.jenvman.2011.09.012
12. Affat SS Classifications, advantages, disadvantages, toxicity effects of natural and synthetic dyes: a review
13. Sharma J, Sharma S, Soni V (2021) Classification and impact of synthetic textile dyes on Aquatic Flora: a review. Reg Stud Mar Sci. 45:101802. https://doi.org/10.1016/j.rsma.2021.101802
14. Kishor R, Bharagava RN, Saxena G (2018) Industrial wastewaters. In: Bharagava RN (ed) Recent advances in environmental management, 1st ed. CRC Press, pp 1–25. https://doi.org/10.1201/9781351011259-1

15. Natural dyes: a traditional craft experiencing new attention - ProQuest. Accessed 28 Sept 2023. https://www.proquest.com/openview/ca293db4748d2714af1e46c4c3ceb789/1?cbl=47978&pq-origsite=gscholar&parentSessionId=9VyQ9c2uKAu7Kkn5g3YYF6M%2BBo6oq1DM%2Bzvh6%2Fhe91g%3D
16. Saxena S, Raja ASM (2014) Natural dyes: sources, chemistry, application and sustainability issues. In: Muthu SS (ed). Roadmap to sustainable textiles and clothing. Textile science and clothing technology. Springer, Singapore, pp 37–80. https://doi.org/10.1007/978-981-287-065-0_2
17. Yusuf M (ed) (2018) Handbook of renewable materials for coloration and finishing, 1st edn. Wiley. https://doi.org/10.1002/9781119407850
18. Focus T (2021) Sources and applications of natural dyes textile focus. Textile focus. Published February 26, 2021. Accessed 28 Sept 2023. https://textilefocus.com/sources-applications-natural-dyes/
19. Verma S, Gupta G Natural dyes and its applications: a brief review 4(4)
20. Indigo. Natural Dyes. Accessed 5 Oct 2023. https://naturaldyes.ca/indigo
21. Colours of Nature—Sustainable Dyeing with Aal (Morinda citrifolia). Sahapedia. Accessed 5 Oct 2023. https://www.sahapedia.org/colours-of-nature-sustainable-dyeing-aal-morinda-citrifolia
22. Dried Marigold Blossoms - KnittyVet. Accessed 21 Oct 2023. https://knittyvet.com/product/dried-marigold-blossoms/
23. How to prepare a pomegranate. BBC Good Food. Accessed 5 Oct 2023. https://www.bbcgoodfood.com/howto/guide/how-eat-pomegranate
24. Dyeing with Pomegranate | Wild Colours natural dyes. Accessed 5 Oct 2023. http://www.wildcolours.co.uk/html/pomegranate.html
25. murex dye. ClothRoads. Accessed 5 Oct 2023. https://www.clothroads.com/tag/murex-dye/
26. Dekel S (2023) Imperial purple and holy blue - Mollusk Dyes. DekelDyes. Published August 2, 2021. Accessed 5 Oct 2023. https://www.suzannedekel.com/post/imperial-purple-and-holy-blue-mollusk-dyes
27. The Law Of Food Coloring Agents From Insects | LPPOM MUI. Published August 14, 2014. Accessed 5 Oct 2023. https://halalmui.org/en/law-dye-food-from-insects/
28. Cardon D Natural dyes, our global heritage of colors
29. Lac Extract - Botanical Colors. Accessed 5 Oct 2023. https://botanicalcolors.com/shop/natural-dyes/natural-dye-extracts/lac-extract/
30. Lac. Natural Dyes. Accessed 5 Oct 2023. https://naturaldyes.ca/lac
31. Lichen Dyes (In-Person) – The ArtsCenter. Accessed 5 Oct , 2023. https://artscenterlive.org/performance/lichen-dyes-in-person/
32. Discover the Versatile Use Cases of Permatex Prussian Blue. SkyGeek. Accessed 5 Oct 2023. https://skygeek.com/blog/discover-the-versatile-use-cases-of-permatex-prussian-blue/
33. Iron. Natural Dyes. Accessed 5 Oct 2023. https://naturaldyes.ca/iron
34. KG KPG& C. Manganese Brown Intense Pigments. Kremer Pigmente Online Shop. Accessed 21 Oct 2023. https://www.kremer-pigmente.com/en/shop/pigments/40623-manganese-brown-intense.html
35. gd-admin. manufacturering in stock Wool fabric malachite green dye price basic green 4 high quality Paper dyes Malachite Green for paper textile silk wool and Acetate fiber Basic Green 4/Magenta Green/Malachite Green CAS 14426-28-9. https://www.mit-ivy.com/. Accessed 5 Oct 2023. https://www.mit-ivy.com/manufacturer-in-stock-discount-30-basic-green-4magenta-greenmalachite-green-cas-14426-28-9-product/
36. Potassium chromate. In: Wikipedia. 2023. Accessed 5 Oct 2023. https://en.wikipedia.org/w/index.php?title=Potassium_chromate&oldid=1163637588
37. Dahal P (2023) Serratia marcescens-an overview. Published May 13, 2023. Accessed 5 Oct 2023. https://microbenotes.com/serratia-marcescens-an-overview/
38. Zr AH, Ss A (2016) Antimicrobial effect of pyocyanin extracted from pseudomonas aeroginosa. Eur J Exp Biol. Published online 2016

39. Staphylococcus | Description, Characteristics, Diseases, & Antibiotic Resistance | Britannica. Published September 14, 2023. Accessed 19 Oct 2023. https://www.britannica.com/science/Staphylococcus
40. Yadav S, Tiwari KS, Gupta C, Tiwari MK, Khan A, Sonkar SP (2023) A brief review on natural dyes, pigments: recent advances and future perspectives. Results Chem 5:100733. https://doi.org/10.1016/j.rechem.2022.100733
41. Making Color. Published 2015. Accessed 21 Oct 2023. https://library.si.edu/exhibition/color-in-a-new-light/making
42. Standing HA (1945) The dyeing of cellulose with direct dyes. Part I. A review of the literature. Trans Faraday Soc ;41:410. https://doi.org/10.1039/tf9454100410
43. Considine DM, Considine GD (2013) Van Nostrand's scientific encyclopedia. Springer, Science & Business Media
44. King D (2007) Dyeing of cotton and cotton products. In: Cotton, Elsevier, pp 353–377. https://doi.org/10.1533/9781845692483.2.353
45. Ghaly AE, Ananthashankar R (2013) Production, characterization and treatment of textile effluents: a critical review. J Chem Eng Process Technol 05(01). https://doi.org/10.4172/2157-7048.1000182
46. Muthu SS, Khadir A (eds) (2022) Dye Biodegradation, mechanisms and techniques: recent advances. Springer, Singapore. https://doi.org/10.1007/978-981-16-5932-4
47. Burkinshaw SM (1995) Chemical principles of synthetic fibre dyeing. Springer Science & Business Media
48. Berradi M, Hsissou R, Khudhair M et al (2019) Textile finishing dyes and their impact on aquatic environs. Heliyon 5(11):e02711. https://doi.org/10.1016/j.heliyon.2019.e02711
49. Asgher M (2012) Biosorption of reactive dyes: a review. Water Air Soil Pollut 223(5):2417–2435. https://doi.org/10.1007/s11270-011-1034-z
50. Khatri Z, Ahmed F, Jhatial AK, Abro MI, Mayakrishnan G, Kim IS (2014) Cold pad-batch dyeing of cellulose nanofibers with reactive dyes. Cellulose 21(4):3089–3095. https://doi.org/10.1007/s10570-014-0320-5
51. Khatri A, White M, Padhye R, Momin NH (2014) The use of reflectance measurements in the determination of diffusion of reactive dyes into cellulosic fiber. Color Res Appl 39(1):63–69. https://doi.org/10.1002/col.21764
52. Pereira L, Alves M (2012) Dyes—environmental impact and remediation. In: Malik A, Grohmann E (eds) Environmental protection strategies for sustainable development. Springer, Netherlands, pp 111–162. https://doi.org/10.1007/978-94-007-1591-2_4
53. Puchtler H, Meloan SN, Spencer M (1985) Current chemical concepts of acids and bases and their application to anionic ("acid") and cationic ("basic") dyes. Histochemistry 82(4):301–306. https://doi.org/10.1007/BF00494057
54. Gürses A, Açıkyıldız M, Güneş K, Gürses MS (2016) Classification of dye and pigments. In: Dyes and pigments. SpringerBriefs in Molecular Science. Springer International Publishing, pp 20+16:31–45. https://doi.org/10.1007/978-3-319-33892-7_3
55. Benkhaya S, Harfi SE, Harfi AE (2017) Classifications, properties and applications of textile dyes: a review
56. Mordant (2023). In: Wikipedia. Accessed 21 Oct 2023. https://en.wikipedia.org/w/index.php?title=Mordant&oldid=1179853027
57. Annika. What is a Mordant? (Mordants for Natural Dyes). Rosemary and pines fiber arts. Published January 22, 2021. Accessed 21 Oct 2023. https://www.rosemaryandpinesfiberarts.de/what-is-a-mordant/
58. Vat Dyes and Their Application
59. Hassaan MA, El Nemr A, Madkour FF (2017) Testing the advanced oxidation processes on the degradation of direct blue 86 dye in wastewater. Egypt J Aquat Res 43(1):11–19. https://doi.org/10.1016/j.ejar.2016.09.006
60. Arun Prasad AS, Satyanarayana VSV, Bhaskara Rao KV (2013) Biotransformation of direct blue 1 by a moderately halophilic bacterium Marinobacter sp. strain HBRA and toxicity assessment of degraded metabolites. J Hazard Mater 262:674–684. https://doi.org/10.1016/j.jhazmat.2013.09.011

61. Teli MD, Paul R, Landage SM, Aich A. Ecofriendly processing of sulphur and vat dyes-An overview.
62. Velusamy S, Roy A, Sundaram S, Kumar MT (2021) A review on heavy metal ions and containing dyes removal through graphene oxide-based adsorption strategies for textile wastewater treatment. Chem Rec 21(7):1570–1610. https://doi.org/10.1002/tcr.202000153
63. Kumar A, Chandra R (2021) Biodegradation and toxicity reduction of pulp paper mill wastewater by isolated laccase producing Bacillus cereus AKRC03. Clean Eng Technol 4:100193. https://doi.org/10.1016/j.clet.2021.100193
64. Zainith S, Saxena G, Bharagava RN Microbes: an eco-friendly tools for the treatment of industrial wastewaters
65. Gita S, Hussan A, Choudhury TG Impact of textile dyes waste on aquatic environments and its treatment
66. Kant R (2012) Textile dyeing industry an environmental hazard. Nat Sci 04(01):22–26. https://doi.org/10.4236/ns.2012.41004
67. Wainwright M (2011) Dyes for the medical industry. In: Handbook of textile and industrial dyeing. Elsevier, pp 204–230. https://doi.org/10.1533/9780857094919.2.204
68. Šuleková M, Smrčová M, Hudák A, Heželová M, Fedorová M (2017) Organic colouring agents in the pharmaceutical industry. Folia Vet 61(3):32–46. https://doi.org/10.1515/fv-2017-0025
69. Tkaczyk A, Mitrowska K, Posyniak A (2020) Synthetic organic dyes as contaminants of the aquatic environment and their implications for ecosystems: a review. Sci Total Environ 717:137222. https://doi.org/10.1016/j.scitotenv.2020.137222
70. Liu L, Chen Z, Zhang J et al (2021) Treatment of industrial dye wastewater and pharmaceutical residue wastewater by advanced oxidation processes and its combination with nanocatalysts: a review. J Water Process Eng 42:102122. https://doi.org/10.1016/j.jwpe.2021.102122
71. Hussain S, Khan N, Gul S, et al (2019) Contamination of water resources by food dyes and its removal technologies. In: Water chemistry. IntechOpen. https://doi.org/10.5772/intechopen.90331
72. Tripathi M, Khanna SK, Das M (2007) Surveillance on use of synthetic colours in eatables vis a vis prevention of food adulteration act of India. Food Control 18(3):211–219. https://doi.org/10.1016/j.foodcont.2005.09.016
73. Sudova E, Machova J, Svobodova Z, Vesely T (2007) Negative effects of malachite green and possibilities of its replacement in the treatment of fish eggs and fish: a review. Veterinární Medicína 52(12):527–539. https://doi.org/10.17221/2027-VETMED
74. Kobylewski S, Jacobson MF (2012) Toxicology of food dyes. Int J Occup Environ Health 18(3):220–246. https://doi.org/10.1179/1077352512Z.00000000034
75. Cao S, Wang K, Zhou S et al (2018) Mechanism and effect of high-basicity chromium agent acting on Cr-wastewater-reuse system of leather industry. ACS Sustain Chem Eng 6(3):3957–3963. https://doi.org/10.1021/acssuschemeng.7b04282
76. Liu G, Yu R, Lan T, Liu Z, Zhang P, Liang R (2019) Gallic acid-functionalized graphene hydrogel as adsorbent for removal of chromium (III) and organic dye pollutants from tannery wastewater. RSC Adv 9(46):27060–27068. https://doi.org/10.1039/C9RA05664E
77. Final Report on the Safety Assessment of Disperse Blue 71 (2007). Int J Toxicol 26(2_suppl):65–77. https://doi.org/10.1080/10915810701351210
78. Ames BN, Kammen HO, Yamasaki E (1975) Hair dyes are mutagenic: identification of a variety of mutagenic ingredients. Proc Natl Acad Sci 72(6):2423–2427. https://doi.org/10.1073/pnas.72.6.2423
79. Gupta VK, Suhas (2009) Application of low-cost adsorbents for dye removal—a review. J Environ Manage 90(8):2313–2342. https://doi.org/10.1016/j.jenvman.2008.11.017
80. Progress on Sanitation and Drinking Water : 2015 Update and MDG Assessment. World Health Organization (2015)
81. Most water-stressed cities globally 2019. Statista. Accessed 2 Oct 2023. https://www.statista.com/statistics/1040184/water-stress-score-city-globally/
82. Chen H, Wang RF, Cerniglia CE (2004) Molecular cloning, overexpression, purification, and characterization of an aerobic FMN-dependent azoreductase from Enterococcus faecalis. Protein Expr Purif 34(2):302–310. https://doi.org/10.1016/j.pep.2003.12.016

83. Wantala K, Sriprom P, Pojananukij N, Neramittagapong A, Neramittagapong S, Kasemsiri P (2013) Optimal decolorization efficiency of reactive red 3 by Fe-RH-MCM-41 catalytic wet oxidation coupled with Box-Behnken Design. Key Eng Mater 545:109–114. https://doi.org/10.4028/www.scientific.net/KEM.545.109
84. Pang YL, Abdullah AZ (2013) Current status of textile industry wastewater management and research progress in Malaysia: a review. CLEAN - Soil Air Water 41(8):751–764. https://doi.org/10.1002/clen.201000318
85. Al-Degs Y (2000) Effect of carbon surface chemistry on the removal of reactive dyes from textile effluent. Water Res 34(3):927–935. https://doi.org/10.1016/S0043-1354(99)00200-6
86. Sharma B, Dangi AK, Shukla P (2018) Contemporary enzyme based technologies for bioremediation: a review. J Environ Manag 210:10–22. https://doi.org/10.1016/j.jenvman.2017.12.075
87. Anliker R, Butler GC, Clarke EA et al (1980) Anthropogenic compounds, vol 3/3A. Springer, Berlin. https://doi.org/10.1007/978-3-540-38522-6
88. Banat IM, Nigam P, Singh D, Marchant R (1996) Microbial decolorization of textile-dyecontaining effluents: a review. Bioresour Technol 58(3):217–227. https://doi.org/10.1016/S0960-8524(96)00113-7
89. Fu Y, Viraraghavan T (2001) Fungal decolorization of dye wastewaters: a review. Bioresour Technol 79(3):251–262. https://doi.org/10.1016/S0960-8524(01)00028-1
90. Chu HC, Chen KM (2002) Reuse of activated sludge biomass: I. Removal of basic dyes from wastewater by biomass. Process Biochem 37(6):595–600. https://doi.org/10.1016/S0032-9592(01)00234-5
91. Fu Y, Viraraghavan T (2002) Removal of Congo Red from an aqueous solution by fungus Aspergillus niger. Adv Environ Res 7(1):239–247. https://doi.org/10.1016/S1093-0191(01)00123-X
92. Mittal AK, Gupta SK. Biosorption of cationic dyes by dead macro fungus Fomitopsis carnea: batch studies
93. Saravanan R, Gupta VK, Narayanan V, Stephen A (2014) Visible light degradation of textile effluent using novel catalyst ZnO/γ-Mn2O3. J Taiwan Inst Chem Eng 45(4):1910–1917. https://doi.org/10.1016/j.jtice.2013.12.021
94. Khan S, Malik A (2018) Toxicity evaluation of textile effluents and role of native soil bacterium in biodegradation of a textile dye. Environ Sci Pollut Res 25(5):4446–4458. https://doi.org/10.1007/s11356-017-0783-7
95. Rawat D, Mishra V, Sharma RS (2016) Detoxification of azo dyes in the context of environmental processes. Chemosphere 155:591–605. https://doi.org/10.1016/j.chemosphere.2016.04.068
96. Hassaan M, El Nemr A (2017) Health and environmental impacts of dyes: mini review. Am J Environ Sci Eng. 1:64–67. https://doi.org/10.11648/j.ajese.20170103.11
97. Gupta GS, Prasad G, Singh VN (1990) Removal of chrome dye from aqueous solutions by mixed adsorbents: fly ash and coal. Water Res 24(1):45–50. https://doi.org/10.1016/0043-1354(90)90063-C
98. Lazar T (2005) Color chemistry: synthesis, properties, and applications of organic dyes and pigments, 3rd revised edn. Color Res Appl 30(4):313–314. https://doi.org/10.1002/col.20132
99. Kadirvelu K, Kavipriya M, Karthika C, Radhika M, Vennilamani N, Pattabhi S (2003) Utilization of various agricultural wastes for activated carbon preparation and application for the removal of dyes and metal ions from aqueous solutions. Bioresour Technol 87(1):129–132. https://doi.org/10.1016/S0960-8524(02)00201-8
100. Lacasse K, Baumann W (2012) Textile chemicals: environmental data and facts. Springer, Science & Business Media
101. Venkata Mohan S, Chandrasekhar Rao N, Karthikeyan J (2002) Adsorptive removal of direct azo dye from aqueous phase onto coal based sorbents: a kinetic and mechanistic study. J Hazard Mater 90(2):189–204. https://doi.org/10.1016/S0304-3894(01)00348-X
102. Manu B, Chaudhari S (2003) Decolorization of indigo and azo dyes in semicontinuous reactors with long hydraulic retention time. Process Biochem 38(8):1213–1221. https://doi.org/10.1016/S0032-9592(02)00291-1

103. Hassaan M, El Nemr A (2017) Advanced oxidation processes for textile wastewater treatment. Int J Photochem Photobiol. 2:85–93. https://doi.org/10.11648/j.ijpp.20170203.13
104. Stolz A (2001) Basic and applied aspects in the microbial degradation of azo dyes. Appl Microbiol Biotechnol 56(1–2):69–80. https://doi.org/10.1007/s002530100686
105. Chen BY, Chang JS (2007) Assessment upon species evolution of mixed consortia for azo dye decolorization. J Chin Inst Chem Eng 38(3–4):259–266. https://doi.org/10.1016/j.jcice.2007.04.002
106. Ali HM, Shehata SF, Ramadan KMA (2016) Microbial decolorization and degradation of crystal violet dye by Aspergillus niger. Int J Environ Sci Technol 13(12):2917–2926. https://doi.org/10.1007/s13762-016-1117-x
107. Spirulina-an over view 4(3)
108. Islam MdA, Ali I, Karim SMA et al (2019) Removal of dye from polluted water using novel nano manganese oxide-based materials. J Water Process Eng 32:100911. https://doi.org/10.1016/j.jwpe.2019.100911
109. Kagalkar AN, Jagtap UB, Jadhav JP, Govindwar SP, Bapat VA (2010) Studies on phytoremediation potentiality of Typhonium flagelliforme for the degradation of Brilliant Blue R. Planta 232(1):271–285. https://doi.org/10.1007/s00425-010-1157-2
110. Mijin DŽ, Avramov Ivić ML, Onjia AE, Grgur BN (2012) Decolorization of textile dye CI basic yellow 28 with electrochemically generated active chlorine. Chem Eng J 204–206:151–157. https://doi.org/10.1016/j.cej.2012.07.091
111. Rajput RS, Pandey S, Bhadauria S (2017) Status of water pollution in relation to industrialization in Rajasthan. Rev Environ Health 32(3). https://doi.org/10.1515/reveh-2016-0069
112. Report on assessment of pollution from textile dyeing units in Tirupur, Tamil Nadu and measures taken to achieve zero liquid discharge
113. Prabha S, Kumar M, Kumar A, Das P, Ramanathan Al (2013) Impact assessment of textile effluent on groundwater quality in the vicinity of Tirupur industrial area, southern India. Environ Earth Sci. 70(7):3015–3022. https://doi.org/10.1007/s12665-013-2361-8
114. Nagpure N, Srivastava R, Kumar R, Dabas A, Kushwaha B, Kumar P (2015) Assessment of pollution of river Ganges by tannery effluents using genotoxicity biomarkers in murrel fish, Channa punctatus (Bloch). Published online, Indian J Exp Biol
115. Rout AK, Dehury B, Parida PK et al (2022) Taxonomic profiling and functional gene annotation of microbial communities in sediment of river Ganga at Kanpur, India: insights from whole-genome metagenomics study. Environ Sci Pollut Res 29(54):82309–82323. https://doi.org/10.1007/s11356-022-21644-6
116. Mora A, García-Gamboa M, Sánchez-Luna MS, Gloria-García L, Cervantes-Avilés P, Mahlknecht J (2021) A review of the current environmental status and human health implications of one of the most polluted rivers of Mexico: the Atoyac River, Puebla. Sci Total Environ 782:146788. https://doi.org/10.1016/j.scitotenv.2021.146788
117. Bhuyan MdS, Islam MdS (2017) Status and impacts of industrial pollution on the karnafully river in Bangladesh: a review. Int J Mar Sci. https://doi.org/10.5376/ijms.2017.07.0016
118. Ito T, Shimada Y, Suto T (2018) Potential use of bacteria collected from human hands for textile dye decolorization. Water Resour Ind 20:46–53. https://doi.org/10.1016/j.wri.2018.09.001
119. Renita AA, Senthil Kumar P, Srinivas S, Priyadharshini S, Karthika M (2017) A review on analytical methods and treatment techniques of pharmaceutical wastewater. Desalination Water Treat 87:160–178. https://doi.org/10.5004/dwt.2017.21311
120. Coskun O (2016) Separation tecniques: chromatography. North Clin Istanb. https://doi.org/10.14744/nci.2016.32757
121. Muralikrishna IV, Manickam V (2017) Analytical methods for monitoring environmental pollution. In: Environmental management. Elsevier; pp 495–570. https://doi.org/10.1016/B978-0-12-811989-1.00018-X
122. Regnier FE (1983) High-performance liquid chromatography of biopolymers. Science 222(4621):245–252. https://doi.org/10.1126/science.6353575

123. Hakami AAH, Wabaidur SM, Khan MA, AlOthman ZA, Siddiqui MR (2021) Extraction procedures and analytical methods for the determination of methylene blue, rhodamine B and crystal violet - an overview. Curr Anal Chem 17(5):708–728. https://doi.org/10.2174/1573411017999201125203536
124. Guerra E, Llompart M, Garcia-Jares C (2018) Analysis of dyes in cosmetics: challenges and recent developments. Cosmetics 5(3):47. https://doi.org/10.3390/cosmetics5030047
125. Fernández C, Larrechi MS, Callao MP (2010) An analytical overview of processes for removing organic dyes from wastewater effluents. TrAC Trends Anal Chem 29(10):1202–1211. https://doi.org/10.1016/j.trac.2010.07.011
126. Chen J, Wang W, Liu H, Xu X, Xia J (2021) A review on the occurrence, distribution, characteristics, and analysis methods of microplastic pollution in ecosystem s. Environ Pollut Bioavailab. 33(1):227–246. https://doi.org/10.1080/26395940.2021.1960198
127. Yagub MT, Sen TK, Afroze S, Ang HM (2014) Dye and its removal from aqueous solution by adsorption: a review. Adv Colloid Interface Sci 209:172–184. https://doi.org/10.1016/j.cis.2014.04.002
128. Ardila-Leal LD, Poutou-Piñales RA, Pedroza-Rodríguez AM, Quevedo-Hidalgo BE (2021) A brief history of colour, the environmental impact of synthetic dyes and removal by using laccases. Molecules 26(13):3813. https://doi.org/10.3390/molecules26133813

Technological Solution for Remediation of Dye from the Environment

Technological Solutions for Dye Removal from the Natural Environment

John Tsado Mathew, Abel Inobeme, Charles Oluwaseun Adetunji, and Amos Mamman

Abstract Dyes are extensively used in various industries, including textile, paper, and leather, to enhance the appearance of products. However, the discharge of dye-containing effluents into water bodies poses significant environmental challenges due to their high color intensity and potential toxicity. Consequently, the development of effective and sustainable technological solutions for dye removal has become a critical area of research. Several innovative approaches, including adsorption, photocatalysis, membrane filtration, and advanced oxidation processes, have been investigated for their effectiveness in removing dyes from wastewater. Adsorption-based methods utilizing materials such as activated carbon, zeolites, and bioadsorbents have demonstrated promising results in dye removal, owing to their high surface area and adsorptive capacity. Photocatalytic degradation techniques, utilizing semiconductor-based catalysts like titanium dioxide and zinc oxide, have gained attention due to their ability to degrade dyes under light irradiation, leading to complete mineralization. Membrane filtration, including processes like ultrafiltration and nanofiltration, effectively remove dyes by size exclusion and electrostatic interactions. The study highlights the growing emphasis on eco-friendly and cost-effective methods that minimize energy consumption and waste generation.

Keywords Dyes · Photocatalysis · Membrane filtration · Nanofiltration · Wastewater

J. T. Mathew (✉) · A. Mamman
Department of Chemistry, Ibrahim BadamasiBabangida University Lapai, Lapai, Nigeria
e-mail: johntsadom@gmail.com; jmathew@ibbu.edu.ng

A. Inobeme
Department of Chemistry, Edo State University Uzairue, Uzairue, Nigeria

C. O. Adetunji
Applied Microbiology, Biotechnology and Nanotechnology Laboratory, Department of Microbiology, Edo University Uzairue, PMB 04, Auchi, Edo State, Uzairue, Nigeria

P. Singh (ed.), *Dye Pollution from Textile Industry*, SDGs and Textiles,
https://doi.org/10.1007/978-981-97-5341-3_4

1 Introduction

The textile industry is a cornerstone of global manufacturing, providing clothing, furnishings, and myriad other products that are integral to our daily lives. However, the textile dyeing process poses significant environmental challenges, particularly in the discharge of wastewater containing residual dyes. The release of these dyes into water bodies can result in severe ecological harm and compromise water quality for both human and aquatic life. Addressing this issue is critical to achieving sustainable industrial practices and safeguarding our environment [63].

In response to these challenges, researchers, engineers, and environmentalists have been actively exploring and developing technological solutions for dye removal from wastewater. These innovative technologies aim to reduce the environmental footprint of the textile industry while ensuring compliance with stringent regulatory requirements and promoting responsible resource management [16].

This report delves into some of the cutting-edge technological solutions available for dye removal in the textile industry. We will explore various methods, both established and emerging, that offer effective and sustainable means of treating dye-contaminated wastewater. These solutions span physical, chemical, and biological processes, each with its unique advantages and applications [27].

As we navigate the dynamic landscape of environmental sustainability, the adoption of these advanced dye removal technologies promises to play a crucial role in mitigating the environmental impact of textile dyeing. By implementing these solutions, industries can not only reduce their ecological footprint but also contribute to a cleaner and healthier world for current and future generations. This report will provide insights into the latest advancements in dye removal technologies, shedding light on the potential for a more sustainable future for the textile industry (Fig. 3.1) [33].

In recent years, nanotechnology has shown promise in enhancing the performance of dye removal technologies, offering improved adsorption capacities and catalytic

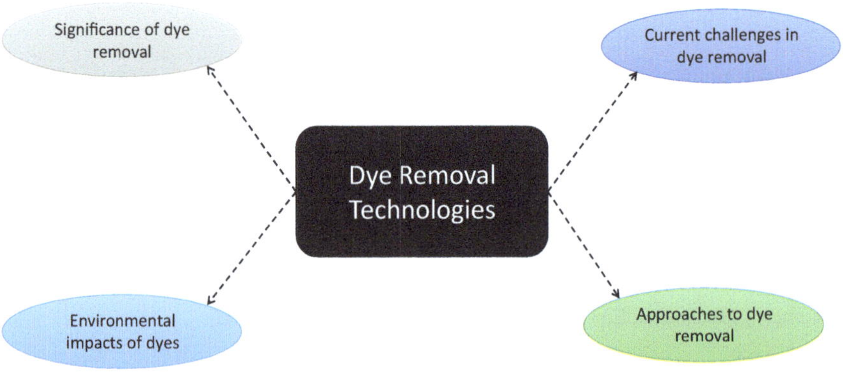

Fig. 3.1 Schematic representation of chapter

Table 3.1 The significance of dye removal in the textile industry

Aspect	Significance of dye removal	References
Environmental	• Protects water quality by removing toxic dyes • Preserves aquatic biodiversity and ecosystems • Prevents harm to human health from contaminated water	[59]
Regulatory	• Ensures compliance with strict environmental laws • Helps avoid legal penalties and maintains reputation	[10]
Economic	• Increases resource efficiency, reducing costs • Facilitates market access, ensuring business growth	[15]
Social	• Demonstrates corporate responsibility and sustainability • Meets consumer demand for eco-friendly products	[81]

properties. Additionally, the development of green and sustainable materials for adsorption and photocatalysis contributes to more environmentally friendly solutions [69].

The choice of the most suitable dye removal technology depends on various factors, including the type of dye, wastewater characteristics, scale of operation, and economic considerations. Often, a combination of multiple methods may provide the most effective and cost-efficient approach [7].

2 Significance of Dye Removal

The significance of dye removal from wastewater in the textile industry cannot be overstated. It holds immense importance for various reasons, encompassing environmental, social, and economic aspects. Dye removal is not merely a technical challenge but a vital component of responsible industrial practices (Table 3.1). It serves as a bridge between industry, environmental stewardship, and societal expectations, playing a pivotal role in shaping a more sustainable future for the textile sector and beyond. The continued development and adoption of dye removal technologies are essential steps toward achieving a harmonious balance between industrial growth and environmental preservation [28].

3 Dye Contamination and Environmental Impacts

Dyes have diverse environmental effects (Fig. 3.2). Dye contamination refers to the release of synthetic dyes into the environment, typically as a result of industrial processes such as textile manufacturing, paper production, and dyeing of various materials. These dyes can have significant environmental impacts, as they often contain chemicals that are harmful to ecosystems and human health [39]. Here are

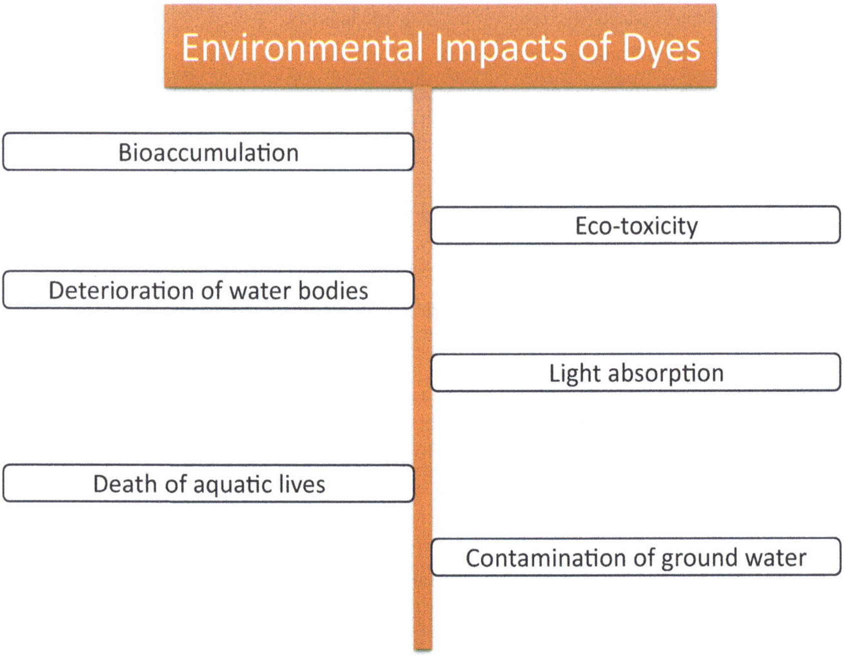

Fig. 3.2 Environmental effects of dyes

some key points to consider regarding dye contamination and its environmental impacts:

Chemical Composition: Synthetic dyes are composed of a wide range of chemicals, some of which can be toxic or hazardous. These chemicals may include heavy metals, aromatic hydrocarbons, and various organic compounds. Some of the commonly used dyes, such as azo dyes, can break down into aromatic amines, which are known carcinogens [65].

Water Pollution: Dye contamination primarily affects water bodies. When dyes are discharged into rivers, lakes, or oceans, they can disrupt aquatic ecosystems. Dyes can reduce light penetration into the water, hindering photosynthesis and affecting aquatic plant and algae growth. This, in turn, can impact the entire food chain [12].

Toxicity: Many dyes are toxic to aquatic life. They can harm fish, insects, and other aquatic organisms. Some dyes can interfere with the ability of fish to respire properly, leading to reduced survival rates and reproductive success [35].

Bioaccumulation: Dye chemicals can accumulate in the tissues of aquatic organisms, a process known as bioaccumulation. This can lead to the transfer of toxins up the food chain when predators consume contaminated prey. Ultimately, this can affect human health if people consume contaminated seafood [12].

Color and Light Absorption: Dye contamination can alter the color of water bodies, making them visually unattractive and reducing their aesthetic and recreational value.

Dyes can also absorb sunlight and heat, contributing to elevated water temperatures, which can harm aquatic ecosystems [83].

Groundwater Contamination: Dye pollutants can leach into groundwater, potentially contaminating drinking water sources. This poses a direct risk to human health if the dye chemicals are not effectively removed during water treatment processes [13].

Regulation and Mitigation: Many countries have regulations in place to limit the discharge of dyes into water bodies. Industries are encouraged to adopt cleaner production techniques, such as recycling dye baths and using more environmentally friendly dyes. Advanced wastewater treatment technologies can also help remove dye contaminants from industrial effluents [27].

Alternative Dyes: Researchers are developing alternative dyeing methods that are less harmful to the environment. These include natural dyes derived from plants and microbes, as well as innovative dyeing processes that require less water and energy [77].

Furthermore, dye contamination can have severe environmental impacts, primarily on water bodies and aquatic ecosystems. Efforts to mitigate these impacts include stricter regulations, cleaner production processes, and the development of more eco-friendly dyeing methods. It is essential for industries to be responsible and take measures to minimize their environmental footprint in the production and disposal of synthetic dyes [14].

4 Current Challenges in Dye Removal

The removal of synthetic dyes from wastewater and the environment presents several ongoing challenges due to the complex nature of these pollutants. These challenges arise from the diversity of dye types, their persistence, and the need for effective, eco-friendly removal methods. There are some current challenges in dye removal;

Diversity of Dye Types: There are thousands of synthetic dye compounds, each with its unique chemical structure and properties. This diversity makes it challenging to develop universal removal methods that can effectively treat all types of dyes [43].

Persistence and Stability: Many synthetic dyes are highly stable and persistent in the environment, making them difficult to break down or degrade naturally. Some dyes can remain in the environment for extended periods, leading to long-term contamination issues [8].

Toxicity and Health Concerns: Dyes often contain toxic or carcinogenic compounds, such as heavy metals or aromatic amines. Removing these toxic components while ensuring the complete removal of dyes is a significant challenge [62].

Complex Wastewater Streams: Dye-contaminated wastewater is often mixed with other pollutants, making the treatment process more complex. Effective dye removal methods must be compatible with treating diverse wastewater compositions [76].

Energy and Resource Intensity: Some conventional dye removal methods, like activated carbon adsorption or advanced oxidation processes, can be energy-intensive and require substantial resources. Developing more energy-efficient and cost-effective techniques is a challenge [11].

Environmental Impact of Treatment: The methods used to remove dyes can generate secondary pollutants or waste by-products. Ensuring that the treatment itself does not lead to additional environmental problems is crucial.

Scale-Up and Implementation: Laboratory-scale research often needs to be scaled up for practical implementation in wastewater treatment plants or industrial processes. Scaling up and integrating effective dye removal methods can be challenging [41].

Regulatory Compliance: Meeting stringent environmental regulations and standards for dye removal is a challenge for industries and municipalities. Staying up-to-date with evolving regulations and ensuring compliance can be a significant hurdle [86].

Natural and Eco-Friendly Solutions: While synthetic dyes pose environmental challenges, the search for natural and eco-friendly alternatives that can replace synthetic dyes in various applications is ongoing [61].

Monitoring and Detection: Developing sensitive and accurate methods for detecting and monitoring dye concentrations in wastewater is essential for effective removal and regulatory compliance.

Research and Innovation: Keeping pace with emerging dye compounds and developing innovative removal technologies is an ongoing challenge. Research efforts must continue to evolve to address new dye types and environmental concerns [46].

To address these challenges, ongoing research and collaboration between scientists, engineers, and environmental regulators are crucial. Developing more sustainable and efficient dye removal methods, improving monitoring techniques, and promoting responsible dye use and disposal practices are essential steps toward mitigating the environmental impact of synthetic dyes [9].

5 Technological Approaches for Dye Removal

There are various technological approaches for dye removal from wastewater and contaminated environments. These methods range from traditional to advanced techniques, each with its advantages and limitations. The choice of method depends on factors such as the type of dye, the concentration of contamination, available resources, and regulatory requirements [60] (Table 3.2).

The selection of the appropriate dye removal technology depends on factors like the nature of the dye, the volume of wastewater, the required treatment efficiency, and economic considerations. In many cases, a combination of methods may be necessary to achieve effective dye removal and treatment. Additionally, ongoing research is focused on developing more efficient, cost-effective, and sustainable technologies for addressing dye contamination challenges [58].

Table 3.2 Some technological approaches for dye removal

Method	Description	Advantages	Limitations	References
Adsorption	Utilizes adsorbents to bind dye molecules from wastewater	• Effective for various dye types • Relatively low-cost • Can be combined with other treatments	• Adsorbent regeneration may be required • Saturation of adsorbent limits capacity	[84]
Chemical precipitation	Chemicals are added to induce dye particle precipitation	• Effective for coagulation of dye particles • Widely used in water treatment	• Requires careful dosing and pH control • Produces sludge that requires disposal	[25]
Biological treatment	Microorganisms metabolize and degrade dye molecules in activated sludge or biofilters	• Eco-friendly and sustainable • Reduces chemical usage	• Slower treatment rates may be a limitation • May not work for all dye types	[51]
Advanced Oxidation Processes (AOPs)	Oxidative methods break down dye molecules (e.g., ozone, UV photocatalysis)	• Effective for persistent dyes • Generates non-toxic by-products	• High energy consumption • Complex system design and maintenance	[78]
Membrane filtration	Uses membranes to physically separate dye particles	• Effective for fine particle removal	• Membrane fouling can decrease efficiency	[48]
Ion exchange	Resins selectively remove dye ions by ion exchange	-High selectivity for specific ions • Potential for ion recovery and reuse	• Limited capacity and regeneration needed • Not suitable for all dye types	[24]
Electrochemical methods	Use electrical current to destabilize or oxidize dyes (e.g., electrocoagulation, electrooxidation)	• Effective for certain dye degradation • May reduce chemical usage	• Energy-intensive and may require skilled operation • Electrode maintenance and replacement	[54]

(continued)

Table 3.2 (continued)

Method	Description	Advantages	Limitations	References
Green and sustainable technologies	Utilizes environmentally friendly methods and materials (e.g., green adsorbents, natural coagulants)	• Reduces environmental impact—Promotes eco-friendly practices	• Limited availability of suitable green materials • Variable performance depending on materials	[46]
Sorption and desorption	Focuses on dye recovery and reuse through sorption and desorption processes	• Sustainable approach for dye recycling • Potential cost savings and reduced waste	• Additional steps needed for desorption and recovery • May not be suitable for all dyes	[74]
Hybrid processes	Combines multiple treatment methods (e.g., adsorption followed by photocatalysis) to enhance removal	• Can improve treatment efficiency • Tailored solutions for specific challenges	• Complex system design and operation • Increased operational complexity	[57]
Microfiltration, nanofiltration, ultrafiltration, reverse osmosis	(microfiltration, ultrafiltration)	• Suitable for smaller dye molecules • Provides consistent effluent quality	• High operational and maintenance costs	[26]

6 Adsorption-Based Methods

Adsorption-based methods are a group of techniques used in chemistry, chemical engineering, and various other fields to separate, purify, or analyze substances by exploiting the adsorption properties of materials. Adsorption refers to the process by which molecules or ions adhere to the surface of a solid or liquid material, known as the adsorbent (Table 3.3). This process is distinct from absorption, where substances are taken up and distributed within the bulk of the material. Adsorption-based methods play a crucial role in various industries, from analytical chemistry to environmental protection, and they continue to evolve with advancements in materials science and technology [74].

Table 3.3 Certain common adsorption-based methods, their applications, adsorbent materials used, and examples

Adsorption-based method	Application	Adsorbent materials	Example(s)	References
Liquid Chromatography (LC)	Separation and analysis of liquid samples	Stationary phase (e.g., C18)	High-performance liquid Chromatography	[50]
Gas Chromatography (GC)	Separation and analysis of gas samples	Stationary phase (e.g., silica)	Gas analysis, drug testing	[42]
Thin-Layer Chromatography (TLC)	Separation and analysis of compounds	Thin layer of adsorbent	Drug screening, compound identification	[32]
Pressure Swing Adsorption (PSA)	Gas separation and purification	Adsorbents (e.g., zeolites)	Natural gas processing, air separation	[80]
Activated carbon filtration	Removal of impurities from air and water	Activated carbon	Water purification, air pollution control	[71]
Ion exchange resins	Water softening and purification	Resin beads (e.g., ion exchange)	Water softeners, demineralization	[45]
Adsorption heat pumps	Efficient cooling and heating	Adsorbents (e.g., silica gel)	Eco-friendly refrigeration, heating	[22]
Gas sensors	Detection of specific gases	Sensing materials (e.g., metal oxides)	Environmental monitoring, safety	[49]
Heterogeneous catalysis	Catalyzing chemical reactions	Catalysts (e.g., platinum)	Petrochemical processes, fuel cells	[6]
Soil and water remediation	Removal of contaminants from soil and water	Sorbent materials (e.g., activated carbon)	Environmental cleanup, groundwater treatment	[34]
DNA and protein purification	Isolation and purification of biomolecules	Affinity resins, chromatography columns	DNA sequencing, protein research	[47]

7 Photocatalysis

Photocatalysis is a fascinating and rapidly advancing field of chemistry that involves the use of light to initiate and accelerate chemical reactions on the surface of a catalyst. This process harnesses the energy of photons to drive chemical transformations, making it a sustainable and environmentally friendly technology with a wide range of applications. In this discussion, we will explore the principles, mechanisms, and applications of photocatalysis [82].

Principles of Photocatalysis: At the heart of photocatalysis is the use of a photocatalyst, typically a semiconductor material like titanium dioxide (TiO_2) or zinc oxide (ZnO). These materials possess a unique property: when exposed to light with energy greater than their bandgap, they generate electron–hole pairs. This excited state allows them to participate in redox reactions, where electrons are transferred between molecules, facilitating various chemical transformations [85].

Mechanisms of Photocatalysis

Photocatalysts absorb photons from a light source, promoting electrons from the valence band to the conduction band, creating electron–hole pairs. The photogenerated electrons and holes can participate in redox reactions. Electrons reduce molecules, while holes oxidize them. This enables a wide range of chemical reactions to occur on the catalyst's surface. One challenge in photocatalysis is the recombination of electron–hole pairs, which can limit the efficiency of the process. To combat this, researchers often introduce co-catalysts or modify the catalyst's surface to trap and utilize these charge carriers more effectively [64].

Applications of Photocatalysis

Photocatalysis can degrade organic pollutants, such as pesticides, dyes, and volatile organic compounds, in water and air. This has significant potential for water purification and air quality improvement. Photocatalysts are used in solar cells and photovoltaic devices to harness sunlight and convert it into electricity or fuel, such as hydrogen production through water splitting [70].

Photocatalytic coatings on surfaces like glass and concrete can break down organic contaminants, making them self-cleaning and reducing maintenance costs. Photocatalysis has applications in disinfection and sterilization processes, potentially reducing the use of harsh chemicals and antibiotics. Researchers are exploring photocatalysis as a green and sustainable approach to various organic syntheses, including the production of pharmaceuticals and fine chemicals [31].

Challenges and Future Prospects: While photocatalysis holds immense promise, it also faces challenges such as catalyst stability, selectivity, and the need for optimal light sources. Researchers are actively working on improving the efficiency and expanding the scope of photocatalytic reactions. The field is also moving towards the development of novel materials with enhanced photocatalytic properties [70].

Furthermore, photocatalysis is a dynamic field with diverse applications that have the potential to address pressing environmental and energy-related challenges. As our understanding of photocatalytic mechanisms deepens and new materials are developed, this technology is likely to play an increasingly vital role in shaping a sustainable and cleaner future [29].

8 Membrane Filtration

Membrane filtration is a separation process that plays a critical role in various industries and applications by selectively allowing the passage of specific substances while retaining others based on their size, shape, and chemical properties. This technique relies on semi-permeable membranes to separate particles, molecules, or ions from a fluid mixture. Membrane filtration has a wide range of applications, from water purification to pharmaceutical manufacturing, and it is a crucial tool for achieving high-purity products and ensuring the safety of various processes [68].

Principles of Membrane Filtration: Membrane filtration operates on the principle of size exclusion, where a porous membrane acts as a physical barrier that permits the passage of particles or molecules smaller than its pore size while blocking larger entities. The selection of membrane materials and pore sizes depends on the specific application and the target species to be separated. Common membrane materials include polymeric materials like polyethersulfone (PES), cellulose acetate, and inorganic materials such as ceramics or metals [19].

Types of Membrane Filtration

Microfiltration (MF): This process uses membranes with relatively large pore sizes (0.1 to 10 μm) to remove suspended solids, bacteria, and larger microorganisms. It is commonly employed in water treatment for turbidity reduction and in the food and beverage industry for clarification [30].

Ultrafiltration (UF): UF membranes have smaller pores (0.001–0.1 μm) and are used to separate macromolecules such as proteins and colloids from solutions. Applications include dairy processing, wastewater treatment, and the pharmaceutical industry [2].

Nanofiltration (NF): NF membranes have even smaller pores (0.001–0.01 μm) and are capable of selectively removing ions and small organic molecules. They are used in water softening, desalination, and the pharmaceutical sector [52].

Reverse Osmosis (RO): RO membranes have the smallest pores (typically less than 0.001 μm) and are used for the removal of salts, minerals, and other solutes from water. This technology is vital in desalination, water purification, and some industrial processes [2].

Applications of Membrane Filtration

Water Treatment: Membrane filtration is widely used for producing potable water, treating wastewater, and desalinating seawater. It helps remove contaminants, pathogens, and impurities, ensuring safe and clean water supplies [17].

Food and Beverage Industry: It plays a crucial role in clarifying juices, concentrating dairy products, and removing unwanted components in beer and wine production [1].

Pharmaceuticals: Membrane filtration is used for sterilizing drugs, separating and purifying proteins, and ensuring the quality of pharmaceutical products [4].

Biotechnology: It is employed in cell harvesting, protein purification, and the production of biopharmaceuticals [56].

Chemical Processing: Membrane filtration is used in various chemical processes for separating and purifying chemicals, solvents, and intermediates [4].

Oil and Gas Industry: It assists in the separation of oil–water emulsions, the removal of contaminants from produced water, and the recovery of valuable hydrocarbons.

Challenges and Advancements

Challenges in membrane filtration include fouling (accumulation of particles or substances on the membrane surface), membrane durability, and energy consumption. Researchers continually work on developing more efficient and durable membranes, improving fouling resistance, and optimizing operating conditions to enhance the effectiveness of this separation technique [2].

In addition, membrane filtration is a versatile and indispensable technology that finds applications in a multitude of industries, addressing various separation and purification needs. As advancements continue, membrane filtration is likely to play an increasingly crucial role in ensuring the quality, safety, and sustainability of many processes and products [38].

9 Ultrafiltration

Ultrafiltration (UF) is a membrane filtration process used for the separation and purification of fluids by selectively allowing the passage of solvents and small solute particles while retaining larger molecules, particles, and macromolecules. It operates on the principles of size exclusion and is commonly used in various industries for applications ranging from water treatment to biopharmaceutical manufacturing [40].

Key Principles of Ultrafiltration

Membrane Selection: UF membranes are semi-permeable and typically made from materials like polyethersulfone (PES), polypropylene, or polysulfone. These membranes have defined pore sizes that determine the range of particles and molecules they can separate [44].

Size Exclusion: UF membranes have pore sizes typically in the range of 0.001–0.1 μm. This allows them to selectively pass small solute molecules such as ions, sugars, and proteins while retaining larger molecules like macromolecules, colloids, and microorganisms [20].

Hydrostatic Pressure: The separation process is driven by a hydrostatic pressure gradient applied across the membrane. The pressure forces the solvent and solutes through the membrane while leaving larger particles behind [72].

Applications of Ultrafiltration

Water Treatment: UF is widely used for drinking water and wastewater treatment. It effectively removes particulate matter, bacteria, viruses, and colloidal impurities, producing high-quality potable water and safe effluent discharge [37].

Food and Beverage Industry: UF is used for concentrating and purifying liquids such as fruit juices, dairy products, and whey proteins. It aids in the removal of bacteria and spores, extending the shelf life of products [67].

Biopharmaceuticals: UF plays a critical role in the biopharmaceutical industry by concentrating and purifying biomolecules like proteins, enzymes, and antibodies. It is essential for the production of monoclonal antibodies and other biopharmaceutical products [75].

Dairy Processing: UF is used to produce cheese, remove lactose from milk, and concentrate milk proteins for the manufacturing of dairy products [75].

Wastewater Reuse: In industrial applications, UF can treat wastewater to recover valuable materials and reduce disposal costs.

Oil and Gas: UF is employed for water treatment in the oil and gas industry to remove oil and solids from produced water, enabling safe disposal or reuse [36].

Challenges and Advances: Despite its numerous advantages, UF faces challenges such as fouling (accumulation of particles on the membrane surface), membrane integrity, and energy consumption. Researchers and engineers are continually working on improving membrane materials and designs to mitigate fouling, enhance durability, and reduce energy requirements. Additionally, innovations like backwashing and air scouring have been employed to maintain membrane performance [68].

Ultrafiltration is a versatile and efficient separation technique with widespread applications in various industries. Its ability to selectively separate molecules and particles based on size makes it an indispensable tool for ensuring product quality, safety, and sustainability in numerous processes and applications. Advances in membrane technology and process optimization continue to expand the reach and effectiveness of ultrafiltration [18].

10 Nanofiltration

Nanofiltration (NF) is a membrane-based separation process that falls between ultrafiltration (UF) and reverse osmosis (RO) in terms of membrane pore size and selectivity. NF membranes have pores with a size range typically between 0.001 and 0.01 μm, making them capable of selectively separating molecules, ions, and particles based on both size and charge. This versatile technology finds applications in various industries due to its ability to remove specific contaminants while allowing the passage of desirable components [79].

Major Principles of Nanofiltration

Membrane Properties: NF membranes are semi-permeable and are typically made from materials such as polymeric materials like polyamide (PA) or thin-film composite (TFC) membranes. These materials have specific surface chemistry and charge properties that influence the selectivity of the membrane [53].

Size and Charge Exclusion: NF membranes can separate solute molecules based on size and charge. They effectively reject particles and molecules that are larger than the membrane's pore size, as well as ions and molecules with high charge density [21].

Operating Pressure: The separation process in NF is driven by an applied pressure differential across the membrane. The pressure forces the feed solution through the membrane while retaining the solute particles and molecules that do not pass through the membrane [66].

Applications of Nanofiltration

Water Treatment: NF is used in drinking water and wastewater treatment to selectively remove divalent ions, such as calcium and magnesium, responsible for water hardness. It is also employed for the removal of color, organic matter, and contaminants like pesticides and pharmaceuticals [23].

Desalination: NF can be used for brackish water desalination, selectively removing ions while allowing water molecules to pass through, making it more energy-efficient compared to RO in certain applications.

Food and Beverage Industry: NF is used for concentrating and fractionating solutions, as well as selectively removing undesirable components such as color, bitter-tasting compounds, and ions [3].

Pharmaceuticals: NF is employed for the concentration and purification of pharmaceutical products, as well as the removal of impurities and ions.

Dairy Processing: NF is used for concentrating milk and whey proteins, fractionating milk components, and improving the shelf life of dairy products [73].

Challenges and Advances: NF faces challenges such as membrane fouling, which can reduce its efficiency over time. Researchers are working on developing anti-fouling membranes and optimizing operating conditions to mitigate fouling. Additionally, advancements in membrane technology aim to improve the selectivity and durability of NF membranes [5].

Nanofiltration is a valuable separation technique with applications in various industries where selective separation of ions, molecules, and particles is essential. Its ability to remove specific contaminants while allowing the passage of desirable components makes it a versatile and efficient solution for water treatment, desalination, food processing, pharmaceuticals, and more. Continued research and development efforts are expected to expand the capabilities and enhance the performance of nanofiltration in the coming years [55].

11 Conclusion

The quest for effective technological solutions for dye removal from wastewater and polluted water bodies is of paramount importance to address environmental concerns and ensure water quality. Various methods, including adsorption, chemical coagulation, advanced oxidation processes (AOPs), and membrane filtration, offer viable approaches to tackle this challenge.

Adsorption techniques, especially activated carbon, remain one of the most widely used methods for dye removal due to their versatility and effectiveness. However, their cost and regenerability may limit their application in some cases. Chemical coagulation and flocculation techniques can be effective, particularly in treating large volumes of wastewater, but they often require the addition of chemicals and generate sludge, posing secondary pollution concerns.

Advanced oxidation processes, such as photocatalysis and ozonation, exhibit high efficiency in dye degradation but may require specific conditions, costly equipment, and energy consumption. Membrane filtration, including ultrafiltration and nanofiltration, is a promising and sustainable solution for dye removal, especially when combined with other methods to address fouling issues.

As the global community faces increasing challenges related to water pollution and the textile industry's environmental impact, continued research and innovation in dye removal technologies are essential. Collaborative efforts among scientists, engineers, and policymakers are necessary to develop and implement efficient, affordable, and environmentally friendly solutions to ensure clean water resources and a more sustainable future. By integrating these advanced technologies and practices into wastewater treatment processes, we can mitigate the environmental impact of dye pollution and safeguard the health of our ecosystems and communities.

References

1. Abimbola OF, Adetunji CO, Dare BJ, Mathew JT, Inobeme A, Ghazanfar S, Olaniyan OT, Ajiboye MD, Dauda W (2023) Packaging of animal-based food products: the need for intelligent freshness sensors. Sensing and artificial intelligence solutions for food manufacturing, CRC Press, pp 89–102
2. Abuwatfa WH, AlSawaftah N, Darwish N, Pitt WG, Husseini GA (2023) A review on membrane fouling prediction using Artificial Neural Networks (ANNs). Membranes 13(7):685
3. Adetunji CO, Anani OA, Olaniyan OT, Inobeme A, Bodunrinde RE, Mathew JT, Adetunji JB, Hefft DI (2022) Prospect of utilizing novel nanomaterials for effective food security, safety, packaging, and preservation. In: Ameta SC, Ameta R (eds) The science of nanomaterials: basics and applications. Apple Academic Press, Inc. Co-published with CRC Press (Taylor & Francis)
4. Adetunji CO, Olaniyan OT, Singh KRB, Bodunrinde RE, Inobeme A, Mathew JT, Ogundolie AF, Akinbo O, Singh J, Nayak V, Singh RP (2023) Applications of nanofiltration for wastewater treatment. In: Bio-nano filtration in industrial effluent treatment, vol 1. CRC-Taylor and Francis, pp 139–152

5. Ahmad NNR, Mohammad AW, Mahmoudi E, Ang WL, Leo CP, Teow YH (2022) An overview of the modification strategies in developing antifouling nanofiltration membranes. Membranes 12(12):1276
6. Akhter M, Alam MM (2023) Catalysis. In: Physical pharmacy and instrumental methods of analysis. Springer Nature Switzerland, Cham, pp 97–107
7. Alsukaibi AKD (2022) Various approaches for the detoxification of toxic dyes in wastewater. Processes 10(10):1968
8. Ardila-Leal LD, Poutou-Piñales RA, Pedroza-Rodríguez AM, Quevedo-Hidalgo BE (2021) A brief history of colour, the environmental impact of synthetic dyes and removal by using laccases. Molecules (Basel, Switzerland) 26(13):3813
9. Ayilara M, Olanrewaju O, Babalola O, Odeyemi O (2020) Waste management through composting: challenges and potentials. Sustainability 12(11):4456
10. Bandari V (2023) Enterprise data security measures: a comparative review of effectiveness and risks across different industries and organization types. Int J Bus Intell Big Data Anal 6(1):1–11
11. Bao J, Guo S, Fan D, Cheng J, Zhang Y, Pang X (2023) Sonoactivated nanomaterials: a potent armament for wastewater treatment. Ultrason Sonochemistry 106569
12. Bashir I, Lone FA, Bhat RA, Mir SA, Dar ZA, Dar SA (2020) Concerns and threats of contamination on aquatic ecosystems. Bioremediation Biotechnol: Sustain Approaches Pollut Degrad 1–26
13. Benjelloun M, Miyah Y, Evrendilek GA, Zerrouq F, Lairini S (2021) Recent advances in adsorption kinetic models: their application to dye types. Arab J Chem 14(4):103031
14. Berradi M, Hsissou R, Khudhair M, Assouag M, Cherkaoui O, El Bachiri A, El Harfi A (2019) Textile finishing dyes and their impact on aquatic environs. Heliyon 5(11):e02711
15. Brockmann-Hosseini N, Jell-Ojobor M, Windsperger J (2023) Market entry through multilateral networks in developing countries: the case of public-private development partnership in Zambia. Networks in international business: managing cooperatives, franchises and alliances. Springer International Publishing, Cham, pp 279–307
16. Castillo-Suárez LA, Sierra-Sánchez AG, Linares-Hernández I, Martínez-Miranda V, Teutli-Sequeira EA (2023) A critical review of textile industry wastewater: green technologies for the removal of indigo dyes. Int J Environ Sci Technol 1–38
17. Cescon A, Jiang J-Q (2020) Filtration process and alternative filter media material in water treatment. Water 12(12):3377
18. Cevallos-Mendoza J, Amorim CG, Rodríguez-Díaz JM, Montenegro MDCBSM (2022) Removal of contaminants from water by membrane filtration: a review. Membranes 12(6):570
19. Chen C, Guo L, Yang Y, Oguma K, Hou LA (2021) Comparative effectiveness of membrane technologies and disinfection methods for virus elimination in water: a review. Sci Total Environ 801:149678
20. Conidi C, Castro-Muñoz R, Cassano A (2020) Membrane-based operations in the fruit juice processing industry: a review. Beverages 6(1):18
21. Covaliu-Mierlă CI, Păunescu O, Iovu H (2023) Recent advances in membranes used for nanofiltration to remove heavy metals from wastewater: a review. Membranes 13(7):643
22. Deshmukh SA, Barmavatu P, Das MK, Naik BK, Sikarwar VS, Shivakrishna A, Subash R (2023) Advances in solar-driven air-conditioning systems for buildings. In: Pragmatic engineering and lifestyle. Emerald Publishing Limited, pp 39–60
23. Devaisy S, Kandasamy J, Nguyen TV, Ratnaweera H, Vigneswaran S (2023) Membranes in water reclamation: treatment, reuse and concentrate management. Membranes 13(6):605
24. El-Moselhy MM, Hakami OM (2015) Selective removal of chromate using hybrid anion exchanger. Desalin Water Treat 56(11):2917–2924
25. El-taweel RM, Mohamed N, Alrefaey KA, Husien S, Abdel-Aziz AB, Salim AI, Radwan AG (2023) A review of coagulation explaining its definition, mechanism, coagulant types, and optimization models; RSM, and ANN. Curr Res Green Sustain Chem 100358
26. Gaioto FC, Matheus MC, de Souza-Chaves BM, Campos JC, Barra TA, de Almeida Azevedo D, Dezotti M (2023) Treatment of a pesticide-containing wastewater by biological and physicochemical processes: seeking the best conditions towards reuse. Braz J Chem Eng 1–17

27. Gayathiri E, Prakash P, Selvam K, Awasthi MK, Gobinath R, Karri RR, Ragunathan MG, Jayanthi J, Mani V, Poudineh MA, Chang SW, Ravindran B (2022) Plant microbe based remediation approaches in dye removal: a review. Bioengineered 13(3):7798–7828
28. Gazzola P, Pavione E, Pezzetti R, Grechi D (2020) Trends in the fashion industry. the perception of sustainability and circular economy: a gender/generation quantitative approach. Sustainability 12(7):2809
29. Goodarzi N, Ashrafi-Peyman Z, Khani E, Moshfegh AZ (2023) Recent progress on semiconductor heterogeneous photocatalysts in clean energy production and environmental remediation. Catalysts 13(7):1102
30. Gul A, Hruza J, Yalcinkaya F (2021) Fouling and chemical cleaning of microfiltration membranes: a mini-review. Polymers 13(6):846
31. Hamdany AH, Satyanaga A, Zhang D, Kim Y, Kim JR (2022) Photocatalytic cementitious material for eco-efficient construction—a systematic literature review. Appl Sci 12(17):8741
32. Han C, Wang Q, Yao Y, Zhang Q, Huang J, Zhang H, Qu L (2023) Thin layer chromatography coupled with surface enhanced Raman scattering for rapid separation and on-site detection of multi-components. J Chromatogr A 464217
33. Harsanto B, Primiana I, Sarasi V, Satyakti Y (2023) Sustainability innovation in the textile industry: a systematic review. Sustainability 15(2):1549
34. Hassanisaadi M, Riseh RS, Rabiei A, Varma RS, Kennedy JF (2023) Nano/micro-cellulose-based materials as remarkable sorbents for the remediation of agricultural resources from chemical pollutants. Int J Biol Macromol 125763
35. Hussain B, Yaseen H, Khalid-Al-Ghanim, Al-Misned F, Qasim M, Al-Mulhm N, Mahboob S (2021) A study on risk assessment of effect of hematoxylin dye on cytotoxicity and nephrotoxicity in freshwater fish: food and water security prospective research. Saudi J Biol Sci 28(4):2267–2271
36. Ibrahim M, Nawaz MH, Rout PR, Lim J-W, Mainali B, Shahid MK (2023) Advances in produced water treatment technologies: an in-depth exploration with an emphasis on membrane-based systems and future perspectives. Water 15(16):2980
37. Inobeme A, Ajai AI, Adetunji CO, Inobeme J, Maliki M, Shaba EY, Osarenre EJ, Mathew JT, Eziukwu CA, Kelani T, Bamigboye O, Okonkwo S (2023) Wastewater treatment technologies. In: Industrial wastewater reuse: applications, prospects and challenges. Springer Nature Singapore, pp 201–213
38. Inobeme A, Nayak V, Mathew JT, Okonkwo S, Ekwoba L, Ajai AI, Bernard E, Inobeme J, Agbugui MM, Singh KRB (2022) Chemometric approach in environmental pollution analysis: a critical review. J Environ Manage 309(114653):1–18
39. Jamee R, Siddique R (2019) Biodegradation of synthetic dyes of textile effluent by microorganisms: an environmentally and economically sustainable approach. Eur J Microbiol Immunol 9(4):114–118
40. Ji K, Liu C, He H, Mao X, Wei L, Wang H, Zhang M, Shen Y, Sun R, Zhou F (2023) Research progress of water treatment technology based on nanofiber membranes. Polymers 15(3):741
41. Jia M, Farid MU, Kharraz JA, Kumar NM, Chopra SS, Jang A, An AK (2023) Nanobubbles in water and wastewater treatment systems: small bubbles making a big difference. Water Res 120613
42. Jian W, Yu Q, He H, Zhao A, Wang Y, Dong N (2023) A new organic soluble cucurbit [7] uriltruxene derivative as stationary phase for gas chromatographic separation of some challenging isomers. Talanta 125197
43. Josephy PD, Allen-Vercoe E (2023) Reductive metabolism of azo dyes and drugs: toxicological implications. Food Chem Toxicol 113932
44. Joshi R, Sebat N, Chi K, Khan M, Johnson KI, Alhamzani AG, Habib MA, Lindstrom T, Hsiao BS (2023) Low fouling nanostructured cellulose membranes for ultrafiltration in wastewater treatment. Membranes 13(2):147
45. Karunanithi MK, Qanungo K (2023) A review of fluoride removal from ground water by various adsorption techniques. J Chil Chem Soc 68(2):5839–5846

46. Khanzada AK, Al-Hazmi HE, Śniatała B, Joseph TM, Majtacz J, Abdulrahman SA, Mąkinia J (2023) Hydrochar-nanoparticle integration for arsenic removal from wastewater: challenges, possible solutions, and future horizon. Environ Res 117164
47. Kilgore R, Minzoni A, Shastry S, Smith W, Barbieri E, Wu Y, Menegatti S (2023) The downstream bioprocess toolbox for therapeutic viral vectors. J Chromatogr A 464337
48. Kim SY, Curko J, Matosic M, Herrera A, Lopez-Vazquez CM, Brdjanovic D, Garcia HA (2023) Effects of a sidestream concentrated oxygen supply system on the membrane filtration performance of a high-loaded membrane bioreactor. Environ Res 237:116914
49. Kong L, Yuan Z, Gao H, Meng F (2023) Recent progress of gas sensors based on metal oxide composites derived from bimetallic metal-organic frameworks. TrAC Trends Anal Chem 117199
50. Kongasseri A, Madhesan T, Mitra S, Rao CB, Nagarajan S, Chinaraga PK, Mohan AM (2023) Fast and selective reversed-phase high performance liquid chromatographic separation of UO 2 2+ and Th 4+ ions using surface modified C 18 silica monolithic supports with target specific ionophoric ligands. RSC Adv 13(5):3317–3328
51. Koul B, Yadav D, Singh S, Kumar M, Song M (2022) Insights into the Domestic Wastewater Treatment (DWWT) regimes: a review. Water 14(21):3542
52. Lazarenko NS, Golovakhin VV, Shestakov AA, Lapekin NI, Bannov AG (2022) Recent advances on membranes for water purification based on carbon nanomaterials. Membranes 12(10):915
53. Liu M, Zhang L, Geng N (2023) Effect of interlayer construction on TFC nanofiltration membrane performance: a review from materials perspective. Membranes 13(5):497
54. Łuba M, Mikołajczyk T, Pierożyński B, Smoczyński L, Wojtacha P, Kuczyński M (2020) Electrochemical degradation of industrial dyes in wastewater through the dissolution of aluminum sacrificial anode of cu/Al macro-corrosion galvanic cell. Molecules (Basel, Switzerland) 25(18):4108
55. Mahmoud AED, Mostafa E (2023) Nanofiltration membranes for the removal of heavy metals from aqueous solutions: preparations and applications. Membranes 13(9):789
56. Mathew JT, Dauda BEN, Mann A, Ndamitso MM, Fadipe AL, Shaba EY (2017) Nutrient and anti-nutrient compositions of fermented and unfermented seed of *Cissuspopulnea* from Niger State, Nigeria. Biotechnology as a change agent for national development book of proceedings, 30th August, 2017, pp 467–473
57. Mei J, Gao X, Zou J, Pang F (2023) Research on photocatalytic wastewater treatment reactors: design, optimization, and evaluation criteria. Catalysts 13(6):974
58. Moosavi S, Lai CW, Gan S, Zamiri G, Akbarzadeh Pivehzhani O, Johan MR (2020) Application of efficient magnetic particles and activated carbon for dye removal from wastewater. ACS Omega 5(33):20684–20697
59. Mousavi SH, Kavianpour MR, Alcaraz JLG, Yamini OA (2023) System dynamics modeling for effective strategies in water pollution control: insights and applications. Appl Sci 13(15):9024
60. Moyo S, Makhanya BP, Zwane PE (2022) Use of bacterial isolates in the treatment of textile dye wastewater: a review. Heliyon 8(6):e09632
61. Muruganandham M, Sivasubramanian K, Velmurugan P, Kumar SS, Arumugam N, Almansour AI, Sivakumar S (2023) An eco-friendly ultrasound approach to extracting yellow dye from Cassia alata flower petals: characterization, dyeing, and antibacterial properties. Ultrasonics Sonochemistry 98:106519
62. Nachiyar CV, Rakshi AD, Sandhya S, Jebasta NB, Nellore J (2023) Developments in treatment technologies of dye-containing effluent: a review. Case Stud Chem Environ Eng 100339
63. Nair VK, Selvaraju K, Samuchiwal S, Naaz F, Malik A, Ghosh P (2023) Phycoremediation of synthetic dyes laden textile wastewater and recovery of bio-based pigments from residual biomass: an approach towards sustainable wastewater management. Processes 11(6):1793
64. Navidpour AH, Abbasi S, Li D, Mojiri A, Zhou JL (2023) Investigation of advanced oxidation process in the presence of TiO2 semiconductor as photocatalyst: property, principle, kinetic analysis, and photocatalytic activity. Catalysts 13(2):232

65. Ngo ACR, Tischler D (2022) Microbial degradation of azo dyes: approaches and prospects for a hazard-free conversion by microorganisms. Int J Environ Res Public Health 19(8):4740
66. Obotey-Ezugbe E, Rathilal S (2020) Membrane technologies in wastewater treatment: a review. Membranes 10(5):89
67. Ogundolie FA, Adetunji CO, Mathew JT, Inobeme A, Ghazanfar S, Olaniyan OT, Ajiboye MD, Dauda WP, Adetunji JB, Akinbo O, Adetuyi BO (2023) Chapter twenty-two - intelligent freshness indicators in packaging of plant-based food products. In: Adetunji CO, Hefft DI, Mbuge DO, Workneh TS (eds) Evaporative coolers for the postharvest management of fruits and vegetables. Academic Press, pp 323–331
68. Othman NH, Alias NH, Fuzil NS, Marpani F, Shahruddin MZ, Chew CM, David Ng KM, Lau WJ, Ismail AF (2021) A review on the use of membrane technology systems in developing countries. Membranes 12(1):30
69. Patra R, Dash P, Panda PK, Yang P-C (2023) A breakthrough in photocatalytic wastewater treatment: the incredible potential of g-C3N4/titanateperovskite-based nanocomposites. Nanomaterials 13(15):2173
70. Pavel M, Anastasescu C, State R-N, Vasile A, Papa F, Balint I (2023) Photocatalytic degradation of organic and inorganic pollutants to harmless end products: assessment of practical application potential for water and air cleaning. Catalysts 13(2):380
71. Piai L, Mei S, van Gijn K, Langenhoff A (2023) Effects of organic matter in drinking water and wastewater on micropollutant adsorption to activated carbon. Int J Environ Sci Technol 1–12
72. Piccard P-J, Borges P, Cleuren B, Hooyberghs J, Buekenhoudt A (2023) Organic solvent nanofiltration and data-driven approaches. Separations 10(9):516
73. Pires AF, Marnotes NG, Rubio OD, Garcia AC, Pereira CD (2021) Dairy by-products: a review on the valorization of whey and second cheese whey. Foods 10(5):1067
74. Rápó E, Tonk S (2021) Factors affecting synthetic dye adsorption; desorption studies: a review of results from the last five years (2017–2021). Molecules (Basel, Switzerland) 26(17):5419
75. Reig M, Vecino X, Cortina JL (2021) Use of membrane technologies in dairy industry: an overview. Foods 10(11):2768
76. Sartaj KM, Prasad R, Matsakas L, Patel A (2023) Transforming recalcitrant wastes into biodiesel by oleaginous yeast: an insight into the metabolic pathways and multi-omics landscape. Chem Eng J 145625
77. Shinde RM, Ingle PU, Trivedi HR, Wasule D, Gaharwar A, Gade AK, Biswas JK (2023) Bioremediation of industrial dye waste effluents aided by GIS applications: a comprehensive review. Environ, Dev Sustain 1–27
78. Solayman HM, Hossen MA, Abd Aziz A, Yahya NY, Hon LK, Ching SL, Zoh KD (2023) Performance evaluation of dye wastewater treatment technologies: a review. J Environ Chem Eng 109610
79. Suhalim NS, Kasim N, Mahmoudi E, Shamsudin IJ, Mohammad AW, Mohamed Zuki F, Jamari NL (2022) Rejection mechanism of ionic solute removal by nanofiltration membranes: an overview. Nanomaterials (Basel, Switzerland) 12(3):437
80. Sun Y, Zhang C, Zhu X, Dong L, Sun X (2023) Mass and heat transfer of pressure swing adsorption oxygen production process with small adsorbent particles. Processes 11(8):2485
81. Suttikun C, Mahasuweerachai P (2023) The influence of status consumption and social media marketing strategies on consumers' perceptions of green and CSR strategies: How the effects link to emotional attachment to restaurants. J Hosp Tour Manag 56:546–557
82. Tahir MB, Sohaib M, Sagir M, Rafique M (2022) Role of nanotechnology in photocatalysis. Encycl Smart Mater 578–589
83. Tkaczyk A, Mitrowska K, Posyniak A (2020) Synthetic organic dyes as contaminants of the aquatic environment and their implications for ecosystems: a review. Sci Total Environ 717:137222
84. Xiao W, Sun R, Hu S, Meng C, Xie B, Yi M, Wu Y (2023) Recent advances and future perspective on lignocellulose-based materials as adsorbents in diverse water treatment applications. Int J Biol Macromol 126984

85. Yang H, Yang B, Chen W, Yang J (2022) Preparation and photocatalytic activities of TiO2-based composite catalysts. Catalysts 12(10):1263
86. Yaro NSA, Sutanto MH, Baloo L, Habib NZ, Usman A, Yousafzai AK, Noor A (2023) A comprehensive overview of the utilization of recycled waste materials and technologies in asphalt pavements: towards environmental and sustainable low-carbon roads. Processes 11(7):2095

Advances in Dye-Containing Wastewater Treatment Technologies

Tarun Kumar Patle, Charu Arora, Durgesh Kumar Sinha, and Alka Patle

Abstract The global community is very concerned about the ongoing contamination of the drinking water body by both organic and inorganic pollutants. The textile industry discharged a significant volume of effluent that was poisonous to humans, animals, and the environment because it included xenobiotics and other dangerous organic substances. Hazardous colors found in textile wastewater block sunlight and impair aquatic life. Thus, before wastewater containing dyes is released into the environment, a treatment method needs to be developed. Textile effluent can be treated to remove dyes using a variety of previously discovered physical, chemical, and biological approaches. The technological solution for dye removal is condensed within this chapter. Both the benefits and drawbacks of the strategies are thoroughly covered in the chapter.

Keywords Dye · Removal · Technology · Remediation · Physical · Chemical · Biological

1 Introduction

The population of the world continuously increases which leads to rapid socioeconomic development, urbanization, and industrialization, these all are major sources of water pollution [1]. They used natural resources of colorant obtained from flora and fauna. But in 1856 a scientist named Perkin invented the first synthetic dye Mauve was a bloom for the chemical industry and at present the synthetic dyes are

T. K. Patle
Department of Chemistry, Pt. Sundarlal Sharma (Open) University, Chhattisgarh, Bilaspur 495009, India

T. K. Patle · C. Arora (✉) · D. K. Sinha
Department of Chemistry, Guru Ghasidas Vishwavidyalaya, Bilaspur, Chhattisgarh 495009, India
e-mail: charuarora77@gmail.com

A. Patle
Department of Chemistry, Govt. Nagarjuna Post Graduate College of Science, Raipur, Chhattisgarh 492010, India

P. Singh (ed.), *Dye Pollution from Textile Industry*, SDGs and Textiles,
https://doi.org/10.1007/978-981-97-5341-3_5

available in more than six digit numbers [2]. Tones of dyes are produced every year, among, more than a half percentage are associated with the textile industry which uses huge amounts of water as well as chemicals. The dyes contain inorganic as well as organic chemicals that possess various structural variations either the cationic, non-ionic, or anionic type [3]. The anionic dyes are very reactive acidic dyes that are easily soluble in water and bright in color. These two properties of anionic dyes are very problematic because they can easily pass through the treatment system. The non-ionic dyes are also called disperse dyes due to their non-ionization in solution. The chromophoric groups present in dyes are called Azo dyes, Nitro dyes, and Nitroso dyes [4].

A large amount of water is used by the textile industry during the wet process and generates a huge amount of waste containing color and other chemicals. Dyes are visible in water even at lower concentrations of ppm level [5]. The improper dye uptake, degree of dying on the substrate, substrate-to-solution ratio, dying method, pH, depth of shade, and many other factors are responsible for large amounts of wastewater production from textile industries that contain toxic color and chemicals [6]. The aquatic life of plants and animals is affected by dye pollution, the sunlight is inhibited by dye pollutants causing a lower photosynthesis process and affecting the growth and development of plants and animals [7]. The carcinogenic benzidine has been detected in various colorants [8]. Nowadays sliver cancer is caused by packaged beverages because they may contain tartrazine, erythrosine, and carmoisine which are toxic [9]. Natural as well as synthetic dyes cause serious health-related problems to humans as well as other living organisms of environment. That is why the removal of dyes before entering into water and other environmental components is necessary.

Ultrafiltration and Nanofiltration techniques are unique and advanced techniques for the removal of many classes of dyes. These techniques are effective but require attention to the membrane fouling problem [10]. The biodegradation of dye is poor; hence, conventional biological techniques are less effective for the treatment of dye-containing wastewater [11]. In that situation, chemical and physical techniques are employed. The introduction of dye-containing wastewater into natural water bodies causes mutagenicity, lethal effects, carcinogenicity, and genotoxicity to all living organisms [12]. Textile industries used digestors for degradation of dyes before entering into environmental water bodies. It can decrease the color and chemicals as well as sludge and reduce the toxicity of dyes [13]. Several conventional technologies are available for the removal of dye such as oxidation, degradation, coagulation, froth floatation, photocatalysis, advanced oxidation, and adsorption processes [4]. In this chapter, the technological and scientific advancements regarding treatment and removal of dye are presented and discussed. Physical, chemical, and biological technologies for removal of dyes from textile and other wastewaters are discussed.

2 Dye Removal Technologies

2.1 Physical Techniques

In this technique, some physical forces are involved in the separation of dyes from water. The mechanism associated with the physical removal of color from wastewater is mass transfer which is simple and very useful in dye treatment [14]. This method is nonchemical and undependable as compared to other methods, these two significant properties of the physical method lead to frequent use in dye removal. Physical processes including adsorption, coagulation-flocculation, filtration, etc., are widely used technologies for color removal [14].

2.1.1 Adsorption

In this technique, various inorganic and organic materials may be employed as adsorbent materials in dye removal from wastewater [15]. This technique is low-cost, more effective, and has been tested the efficiency in removing days. The adsorption capability depends upon the organic and inorganic materials, further, it is observed that inorganic materials have greater stability, larger area, and are not affected by microbial growth [16]. Whereas, the renewability of organic materials is a characteristic property, that enhances the capability of adsorbent in the removal of dyes. Many types of dyes can be removed with the help of carbon-based inorganic materials [17, 18]. The factors that can affect the removal of dyes are their molecular size, solubility, polarity of dye as well as inorganic framework. Using carbon material, it was shown that basic and direct dyes had higher adsorption rates than acid and reactive dyes, which have lower molecular weights [19]. Some other dyes have colloidal nature in solution and lower adsorption occurs at room temperature for removal of that kind of dyes using carbon-based adsorbent. A large number of research articles have been published for the removal of dye using activated carbon material [20–22].

The surface diffusion method has been significantly used to know the various aspects of the adsorption of dye [23]. Several types of acid and basic dyes have been removed by using lignite-based carbon and wood charcoal [24]. There was some literature suggesting that sulfonated carbon has significant potential for the removal of dyes [25]. It is well discovered the efficiency of carbon-based adsorbents for successfully removing dyes from textile effluent; however, the development of this kind of material is so expensive and requires large amounts of carbon sources are major drawback. Several other inorganic supports have already been invented to overcome this problem of carbon-based adsorbent [26]. Inorganic materials such as silica, dolomite, mud, sawdust, groundnut shell, bauxite, activated aluminum, and sugarcane dust were applied for the treatment of color with remarkable results [27].

Organic materials such as a slurry of biogas waste, chitosan fiber, orange peel, banana peel, and other fruit waste, polymers, metal organic frameworks, Layered double hydroxides were frequently tested for removal of dyes [28]. Basic violet 10 dye was successfully removed from wastewater with the help of modified orange peel and biogas waste [29]. The direct dyes can also be removed by chitosan-linked fibers as an organic framework. The first-order kinetic was observed after plotting Langmuir and Freundlich adsorption isotherm in those kinds of organic adsorbents [30]. A major factor that could affect the capability of sorbent material is the pH of the solution. Usually, it has been observed that an alkaline condition does not promote adsorption but it can help in desorption, whereas an acidic condition is good for adsorption of dyes [31]. The adsorption of acidic dyes in acidic media and desorption in basic media suggested that the ion exchange mechanism is associated with acidic dye removal, but some literature on the removal of basic dyes also suggested that the variation of pH from acidic to basic did not affect the adsorption or desorption. Hence, there is a possibility of a different mechanism in dye removal called the chemosorption process.

Several complexometric agents have been employed as dye-complexing agents for the removal of dyes [32–34]. These agents are usually metal–organic frameworks [73–76], layerd double hydroxides [77], and polymer that is capable of forming a host–guest complex in the cavity of its aromatic organic framework with dyes. Using this complexometric technique, almost 100% removal of dye can be achieved [35]. When curcurbituril polymer is used as a complexometric agent only dye is bound with the cavity of the polymer and other organic compounds present in wastewater of the textile industry cannot interfere with the mechanism of binding [36].

2.1.2 Ion Exchange

Removal of dyes from textile wastewater has been demonstrated by several ion exchange resins. The ion exchange resins can be prepared with the help of various agricultural wastes such as banana peel, maiz, cob, coir pith, sugar beet fiber, sunflower stalks, soybean pulp [37, 38]. The ion exchange resins have tremendous properties and are capable of removing dyes from textile wastewater as well as other organic toxins. It was observed that ion exchange resins modified with the help of sugarcane waste, paper waste, and chitin have been capable of removal of color as well as other organic species from wastewater samples generated from textile industries [39]. When the wastewater containing acidic or basic dyes is treated with ion exchange resins it forms a complex that can be easily separated. It has been reported that cellulose-modified ion exchange resins have the property to bind with reactive and hydrolyzed dyes for their removal from water samples [37]. It was well demonstrated that the van der Waals force and hydrogen bonding were involved during the interaction between ion exchange resin and dyes. Hence, these two types of forces increase the capability of that resin for the removal of dyes [40]. Somewhere, this interaction cannot be enough efficient in the removal of dispersed dyes having hydrophobic character. There are several disadvantages of the ion exchange resins that make this technique less effective as compared to carbon-modified adsorbents. The ion exchange techniques require high pressure and flow rate of water, costly, and sludge production is its major disadvantage.

2.1.3 Coagulation-Flocculation

Coagulation is one more physical method that has also been extensively used for the treatment of dyes. The zeta potential of dye particles decreases while using coagulation agents like anionic, cationic, or non-ionic polymers [41]. There are some major drawbacks of the method such as required costly chemicals, the resistance of coagulation by high water-soluble dyes, generation of solid waste, and toxic sludge production [42]. However, some research also suggested that the sludge can be reused as an adsorbent for example poly aluminum chloride, carbon, and bentonite clay were combined and used for the removal of dyes and the coagulated sludge produced after the removal of dye can be reused as an adsorbent again [43]. The combination of some chemicals like polyelectrolyte with coagulating agents has the potential to remove highly colored dyes like remazole from the sample, additionally, the quality of sludge was also enhanced by using bentonite and polyelectrolyte with sodium alginate and alum [44]. The water-insoluble and dispersed dyes can be easily removed by this method whereas the removal of water-soluble dyes is a little difficult by using this method [45]. This method produces a large amount of toxic sludge containing an excess amount of aluminum and iron. The salt of iron in the form of ferrous or ferric possesses low toxicity due to oxidation moreover it could be toxic to marine animals. Hence, the treatment of sludge is required before entering into the environment.

2.1.4 Membrane Filtration

The membrane is a layer and works as a blockade to dye in textile effluents. This filtration technique allows the dye to separate in its original form without any structural changes and can be reused. The use of this technique was successfully demonstrated for the removal of dyes from wastewater by several researchers [46]. Various types of membranes can be used depending on the process of dying, the type of dye, and other chemicals associated with the dyeing process. The membranes that are pH sensitive, temperature sensitive, and high thermal and chemical durability can be used for the removal of dyes. Several membrane filtration techniques like reverse osmosis, ultrafiltration, nanofiltration, etc., are used in textile industries for successful treatments of dye [47]. The pores of micro filters are big and are applied in the removal of large particle-size dyes and other organic chemicals. But ultrafiltration is used for the removal of high molecular weight dye [48]. Carbonate, amide, styrene, acrylonitrile, and sulfone-based polymers are associated with ultrafiltration and reverse osmosis [32]. It was well described that nanofiltration was capable of removal of several reactive dyes from wastewater [49]. Nanofiltration and reverse osmosis have the potential to recover hydrolyzed reactive dyes. The recycling of water can be achieved to some extent using the filtration method depending upon the low concentration of dyes and other chemicals. However, the higher concentration of dyes and other chemicals cannot allow for the recycling of water using the membrane filtration technique. This technique is economical costlier, power power-consuming, also life of the membrane is short, and requires subsequent disposal, these are major disadvantages of the method [50].

2.1.5 Irradiation

In this technique, gamma radiation is applied to irradiate wastewater containing dyes. Most of the dyes have been degraded by the irradiation process; hence, irradiation is the most effective technique for color removal [51]. Radiation power and oxygen are two main factors concerned in the degradation. The irradiation process was successfully demonstrated for degradation and removal of reactive, acid, and disperses dyes. The breakdown of basic, reactive as well as acidic dyes has been reported [52].

2.2 *Chemical Method*

2.2.1 Oxidation

Chemical oxidation is a simple and widely used method for the removal of dyes from textile wastewater. This method involved the breaking of dye into small fragments of colorless chemicals based on carbon, hydrogen, nitrogen, oxygen, sulfur, etc. [53]. The breaking of dye into small molecules truly depends upon the types of oxidants used in the reaction. Conventional oxidants like chlorine and its dioxide, ozone, hydrogen peroxide, and permanganate can be used in the treatment of dye [54]. The chemical bleaching process starts with hydroxyl radical which further reacts with the chromophoric group and leads to the formation of organic radicals, further, this radical breakdown of aromatic rings and other rings into small molecules. Ozone is a strong and effective oxidant because of reactivity of ozone can be possible with many classes of dyes as well as the higher efficiency of removal of color has been achieved using ozone as an oxidant [55]. The fast removal of color from wastewater was associated with the breaking down into organic acids, aldehydes, and ketones of a chromophoric group of dye by ozone as an oxidant. In ozonation, no sludge formation makes this oxidant a promising agent for the removal of dye using, additionally, the color waste can be removed in a single-step process. Noteworthy that ease of installation, no other chemical use, easy decomposition of ozone into oxygen, and easy operational, and economical properties of the ozonation process make this a more effective tool in dye removal techniques. Several articles suggested the capability of the ozonation process for efficient removal of CI reactive yellow 145 and CI reactive red 194 [56].

The Fenton's reagent was combined with aniron catalyst and hydrogen peroxide for the removal of color effluent. This reagent has tremendous potential to remove soluble or insoluble dyes as well as pigments as compared to other oxidants such as peroxide and ozone [57]. A few dyes like vat and disperse show resistance to Fanton's reagent, hence cannot be removed using this reagent. However, this reagent has the potential to remove helizarin blue BGT, remazol brilliant blue B, irgalan blue FGL, and indanthrene blue GCD from the sample [4]. The major disadvantage of the method is sludge production which is very harmful to living beings. The chemical sludge requires a treatment process that can be achieved by a burning process which produces power. Acidic, direct, vat, soluble, as well as reactive dyes have been applicably treated by this reagent.

2.2.2 Electrochemical

The electrochemical technique is an employed technique for the treatment of dyes and other chemicals from textile effluents. This method is simple, effective, and economical for wastewater treatment, additionally, it requires fewer chemicals and no excess sludge production occurs in this process. The technique required two electrodes to act as coagulant (anode) and hydrogen gas production (cathode) to peak up flocculated matters [58]. The oxidation–reduction occurred at the electrochemical reactor for on-the-spot production of coagulants. The treatment of dye depends upon several factors of the electrochemical process like the density of current, concentration of dye, material used in anode, pH, etc. Some article shows that disperse dye removal using an electrochemical process are more efficient as compared to reactive dyes. The electrocoagulation process was successfully demonstrated for the treatment of CI Reactive Blue 19 [59]. Recently a new glow discharge electrolytic technique was discovered for the treatment of wastewater [60]. The color was converted rapidly into CO_2 and minerals by electron and hydroxyl radicals. Several works on electrochemical techniques for dye removal suggested that the technique is quite simple, rapid, and eco-friendly.

2.3 *Biological Method*

Biological techniques are cheap, less sludge produced, and eco-friendly as compared to physicochemical techniques of dye removal. The wastewater from textiles contains color as well as other organic chemicals that can be removed biologically by aerobic and anaerobic processes. The microorganism interacts with organic molecules and turn into its minerals.

2.3.1 Bacterial Treatment

Bacterial treatment of dyestuff is either aerobic or anaerobic or both modes could take place during the treatment of textile wastewater. The charge bacterial cell makes effective and strong links with azo dye resulting in higher degradation for the removal of color. The linked dye has been broken down in an anaerobic manner and further, the broken molecules are treated either by aerobic or anaerobic process [61]. Several studies have been already done in this field including bacteria such as *Bacillus subtilis* [62]. The removal of azo days through adsorption on the biomass of isolated microbes was studied, perhaps, this process is tough for long-term use, and disposal of biomass is required. However, the breakdown of nitrogen bonds by the redox process can be an alternative for the use of bacterial removal of dye. It has been mentioned that the mixed strain of bacteria was more effective in decolorizing textile wastewater as compared to single-strain bacterial treatment [63]. This method is more appropriate for the treatment of several dyes. The cheap, eco-friendly approach and production of the lower amount of sludge make this method more attractive.

2.3.2 Fungal Treatment

Several kinds of phenolic, non-phenolic, soluble, insoluble, etc., dyes can be decolorized by fungal treatment, and degradation occurs with laccases, reductases, and peroxidises [64]. The physicochemical interaction between fungi and dye occurred followed by biosorption for catalytic degradation and removal from textile wastewater. A wide range of fungal treatment of textile wastewater has been already demonstrated for the degradation of azo dye [65]. The efficiency of degradation of azo dye can be affected by the substituent like hydroxyl, amino, nitro, etc., group enhance the degradation as compared to unsubstituted dye. Some conditions that occurred for lower efficiency in the fungal treatment of dye like lower pH, the activity of enzymes, and degradation time [66].

2.3.3 Algal Treatment

Photosynthetic microorganisms like algae can be used in the treatment of dye. The chromospheres of dye go through a biosorption process followed by the transformation of dye into colorless carbon, nitrogen, and hydrogen base molecules. The algal-based treatment of azo dye is initiated with the degradation of azo dye by azoreductase and transforms into amines [67]. It was reported that several azo dyes have been removed *Chlorella* and *Oscillatoria* sp. by degradation and conversion into aromatic amines [68]. *Caulerpa lentillifera* was used in the removal of several kinds of basic dyes such as CI Basic Blue 3, 9, 159, and CI Basic Red 18. Similarly, *Hypnea valentiae* and *Ulva fasciate* were tested for efficient removal of simulated dye [32]. Some factors like pH, concentration of dye, temperature, and biomass of algae are the major factors in the algal treatment of dye. A similar result was obtained with *Spirogyra* sp. algae for the treatment of CI reactive Red 8 and CI Direct Red 28 [69].

2.3.4 Phytoremediation

Plants play a crucial role in the removal of different kinds of environmental pollutants from soil, water, and other components [70]. Plant-based remediation of dye is a successful technique for the removal of dye from wastewater of textile industries. The bioaccumulation of dye in plants can be an alternate natural dye removal, but taking out the accumulated dyes from a particular plant is a tough task. Several kinds of plants including agricultural waste materials, ferns, graces, and weeds are used for the treatment of dye [71]. The combination of plant and microorganism can enhance the efficiency of removal of dye from wastewater, in this circumstance; *Medicago sativa, Sesbania cannabina* with *Gracili bacillus* were tested for removal of acid red B and acid scarlet GR from wastewater [72]. Plant-based remediation is cheap, easy to handle, eco-friendly, and long-term use making this method far better as compared to other remediation techniques.

3 Conclusion

Recently textile industries generated millions of gallons of effluents containing harmful chemicals including dyes. Dyes are colored organic chemicals that are toxic to all living organisms. Several technological solutions are available for the removal of dye from textile wastewater. These techniques include physical methods, chemical and biological. Physical methods such as adsorption, ion exchange, irradiation, and filtration are promising techniques with several limits. Some physicochemical methods such as adsorption, chemical treatment, and ion pair extractions were costly and less effective. Biological treatment methods are relatively inexpensive and measured to be an effective substitute with appropriate study and environmental control. In the chemical method advanced oxidation process by ozon, peroxide, and Fanton's reagent worked as promising agents for chemical treatment methods for decolorization and degradation of textile dyes. Biological treatment is better in terms of cost, simplicity, easy operation, and eco-friendly nature as compared to physical as well as chemical technologies. At present, biological treatments are known to be competent in removing color through conventional anaerobic, aerobic, and combined anaerobic–aerobic biodegradation techniques. The major concern about aerobic treatment is its ineffectiveness in color removal from wastewater. The utilization of the salt- and thermotolerant microorganisms in the biotreatment method would be a huge enhancement of conventional biological treatment and the biotreatment idea. The use of dispersed dyes in textiles is growing day by day and given the shortage of efficient methods, the need to develop protocols for the removal of these dyes is fuelled. Recently combined technology can enhance the efficiency of the dye removal process like a combination of physical adsorption with biological techniques using bacteria, fungi, or algal biomass to gain more attention. Shortly, researchers and scientists should work on complex dye treatment as well as single dye treatment in an easy way and develop a cheaper, simple, rapid, and environmentally friendly technology.

References

1. Singh J, Yadav P, Pal AK, Mishra V (2020) Water pollutants: origin and status. In: Sensors in water pollutants monitoring: role of material, pp 5–20
2. Hagan E, Poulin J (2021) Statistics of the early synthetic dye industry. Herit Sci 9(1):33
3. Benkhaya S, M'rabet S, El Harfi A (2020) A review on classifications, recent synthesis and applications of textile dyes. Inorg Chem Commun 115:107891
4. Nachiyar CV, Rakshi AD, Sandhya S, Jebasta NBD, Nellore J (2023) Developments in treatment technologies of dye-containing effluent: a review. Case Stud Chem Environ Eng 100339
5. Abid MZ, Rafiq K, Rauf A, Shah SSA, Jin R, Hussain E (2023) Synergism of Co/Na in BiVO 4 microstructures for visible-light driven degradation of toxic dyes in water. Nanoscale Adv
6. Khalil E, Sarkar J, Rahman MM, Shamsuzzaman M, Das D (2023) Advanced technology in textile dyeing. In: Advanced technology in textiles: fibre to apparel. Springer Nature Singapore, Singapore, pp 97–138
7. Tkaczyk A, Mitrowska K, Posyniak A (2020) Synthetic organic dyes as contaminants of the aquatic environment and their implications for ecosystems: a review. Sci Total Environ 717:137222
8. Tomioka K, Obayashi K, Saeki K, Okamoto N, Kurumatani N (2015) Increased risk of lung cancer associated with occupational exposure to benzidine and/or beta-naphthylamine. Int Arch Occup Environ Health 88:455–465
9. Yang WS, Zeng XF, Liu ZN, Zhao QH, Tan YT, Gao J, Li HL, Xiang YB (2020) Diet and liver cancer risk: a narrative review of epidemiological evidence. Br J Nutr 124(3):330–340
10. Ramachandran SK, Sathishkumar P (2023) Membrane-based techniques for pollutants removal: an outlook on recent advancements. Curr Opin Environ Sci Health 100513
11. Katheresan V, Kansedo J, Lau SY (2018) Efficiency of various recent wastewater dye removal methods: a review. J Environ Chem Eng 6(4):4676–4697
12. Josephy PD, Allen-Vercoe E (2023) Reductive metabolism of azo dyes and drugs: toxicological implications. Food Chem Toxicol 113932
13. Amin MSA, Mozumder MSI, Stüber F, Giralt J, Fortuny A, Fabregat A, Font J (2023) Biofilm model development and process analysis of anaerobic bio-digestion of azo dyes. Environ Technol Innov 29:102962
14. Piaskowski K, Świderska-Dąbrowska R, Zarzycki PK (2018) Dye removal from water and wastewater using various physical, chemical, and biological processes. J AOAC Int 101(5):1371–1384
15. Lambert SD, Graham NJD, Sollars CJ, Fowler GD (1997) Evaluation of inorganic adsorbents for the removal of problematic textile dyes and pesticides. Water Sci Technol 36(2–3):173–180
16. Sabzehmeidani MM, Mahnaee S, Ghaedi M, Heidari H, Roy VA (2021) Carbon based materials: a review of adsorbents for inorganic and organic compounds. Mater Adv 2(2):598–627
17. Madima N, Mishra SB, Inamuddin I, Mishra AK (2020) Carbon-based nanomaterials for remediation of organic and inorganic pollutants from wastewater. a review. Environ Chem Lett 18:1169–1191
18. Smith SC, Rodrigues DF (2015) Carbon-based nanomaterials for removal of chemical and biological contaminants from water: a review of mechanisms and applications. Carbon 91:122–143
19. Gusain R, Kumar N, Ray SS (2020) Recent advances in carbon nanomaterial-based adsorbents for water purification. Coord Chem Rev 405:213111
20. Arora C, Kumar P, Soni S, Mittal J, Mittal A, Singh B (2020) Efficient removal of malachite green dye from aqueous solution using *Curcuma caesia* based activated carbon. Desalin Water Treat 195:341–352. https://doi.org/10.5004/dwt.2020.25897
21. Arora C, Sahu D, Bharti D, Tamrakar V, Soni S, Sharma S (2019) Adsorption of hazardous dye crystal violet from industrial waste using low-cost adsorbent *Chenopodium album*. Desalin Water Treat 167:324–332. https://doi.org/10.5004/dwt.2019.24595

22. Patel A, Soni S, Mittal J, Mittal A, Arora C (2021) Sequestration of crystal violet from aqueous solution using ash of black turmeric rhizome. Desalin Water Treat 220:342–352. https://doi.org/10.5004/dwt.2021.26911
23. Muthukkumaran A, Aravamudan K (2017) Combined homogeneous surface diffusion model–design of experiments approach to optimize dye adsorption considering both equilibrium and kinetic aspects. J Environ Manag 204:424–435
24. Tsang DC, Hu J, Liu MY, Zhang W, Lai KC, Lo IM (2007) Activated carbon produced from waste wood pallets: adsorption of three classes of dyes. Water Air Soil Pollut 184:141–155
25. Zhu S, Xu J, Kuang Y, Cheng Z, Wu Q, Xie J, Wang B, Gao W, Zeng J, Li J, Chen K (2021) Lignin-derived sulfonated porous carbon from cornstalk for efficient and selective removal of cationic dyes. Ind Crops Prod 159:113071
26. Forgacs E, Cserháti T, Oros G (2004) Removal of synthetic dyes from wastewaters: a review. Environ Int 30(7):953–971
27. NagaBabu A (2019) Development and characterization of non-conventional low cost adsorbents for the removal of diverse pollutants from wastewater
28. Sahoo JK, Hota A, Khuntia AK, Sahu S, Sahoo SK, Sabar AK, Rath J (2022) Removal of dyes using various organic peel-based materials: a systematic review. Appl NanoBioSci 11:3714–3727
29. Namasivayam C, Yamuna R, Arasi DJSE (2001) Removal of acid violet from wastewater by adsorption on waste red mud. Environ Geol 41:269–273
30. Anirudhan TS, Ramachandran M (2015) Adsorptive removal of basic dyes from aqueous solutions by surfactant modified bentonite clay (organoclay): kinetic and competitive adsorption isotherm. Process Saf Environ Prot 95:215–225
31. Mittal A, Mittal J, Malviya A, Gupta VK (2009) Adsorptive removal of hazardous anionic dye "Congo red" from wastewater using waste materials and recovery by desorption. J Colloid Interface Sci 340(1):16–26
32. Singh K, Arora S (2011) Removal of synthetic textile dyes from wastewaters: a critical review on present treatment technologies. Crit Rev Environ Sci Technol 41(9):807–878
33. Xu Y, Ren B, Wang R, Zhang L, Jiao T, Liu Z (2018) Facile preparation of rod-like MnO nanomixtures via hydrothermal approach and highly efficient removal of methylene blue for wastewater treatment. Nanomaterials 9(1):10
34. Van der Schueren L, De Meyer T, Steyaert I, Ceylan Ö, Hemelsoet K, Van Speybroeck V, De Clerck K (2013) Polycaprolactone and polycaprolactone/chitosan nanofibres functionalised with the pH-sensitive dye Nitrazine Yellow. Carbohyd Polym 91(1):284–293
35. Rai HS, Bhattacharyya MS, Singh J, Bansal TK, Vats P, Banerjee UC (2005) Removal of dyes from the effluent of textile and dyestuff manufacturing industry: a review of emerging techniques with reference to biological treatment. Crit Rev Environ Sci Technol 35(3):219–238
36. Kanth PC, Trivedi MU, Patel K, Misra NM, Pandey MK (2021) Cucurbituril-functionalized nanocomposite as a promising industrial adsorbent for rapid cationic dye removal. ACS Omega 6(4):3024–3036
37. Hassan MM, Carr CM (2018) A critical review on recent advancements of the removal of reactive dyes from dyehouse effluent by ion-exchange adsorbents. Chemosphere 209:201–219
38. Alalwan HA, Kadhom MA, Alminshid AH (2020) Removal of heavy metals from wastewater using agricultural byproducts. J Water Supply: Res Technology—AQUA 69(2):99–112
39. Swami D, Buddhi D (2006) Removal of contaminants from industrial wastewater through various non-conventional technologies: a review. Int J Environ Pollut 27(4):324–346
40. Zhang H, Li Y, Cheng B, Ding C, Zhang Y (2020) Synthesis of a starch-based sulfonic ion exchange resin and adsorption of dyestuffs to the resin. Int J Biol Macromol 161:561–572
41. Lau YY, Wong YS, Teng TT, Morad N, Rafatullah M, Ong SA (2015) Degradation of cationic and anionic dyes in coagulation–flocculation process using bi-functionalized silica hybrid with aluminum-ferric as auxiliary agent. RSC Adv 5(43):34206–34215
42. Verma AK, Dash RR, Bhunia P (2012) A review on chemical coagulation/flocculation technologies for removal of colour from textile wastewaters. J Environ Manag 93(1):154–168

43. Fosso-Kankeu E, Webster A, Ntwampe IO, Waanders FB (2017) Coagulation/flocculation potential of polyaluminium chloride and bentonite clay tested in the removal of methyl red and crystal violet. Arab J Sci Eng 42:1389–1397
44. Gürses A, Yalçin M, Dogar C (2003) Removal of Remazol Red RB by using Al (III) as coagulant-flocculant: effect of some variables on settling velocity. Water Air Soil Pollut 146:297–318
45. Iwuozor KO (2019) Prospects and challenges of using coagulation-flocculation method in the treatment of effluents. Adv J Chem-Sect A 2(2):105–127
46. Rashidi HR, Sulaiman NMN, Hashim NA, Hassan CRC, Ramli MR (2015) Synthetic reactive dye wastewater treatment by using nano-membrane filtration. Desalin Water Treat 55(1):86–95
47. Cinperi NC, Ozturk E, Yigit NO, Kitis M (2019) Treatment of woolen textile wastewater using membrane bioreactor, nanofiltration and reverse osmosis for reuse in production processes. J Clean Prod 223:837–848
48. Purkait MK, DasGupta S, De S (2004) Removal of dye from wastewater using micellar-enhanced ultrafiltration and recovery of surfactant. Sep Purif Technol 37(1):81–92
49. Vishnu G, Joseph K (2008) Decolorization of reactive dyebath containing high exhaust and low salt dyes by nanofiltration and its reuse. Environ Prog 27(1):30–39
50. Ahmad A, Mohd-Setapar SH, Chuong CS, Khatoon A, Wani WA, Kumar R, Rafatullah M (2015) Recent advances in new generation dye removal technologies: novel search for approaches to reprocess wastewater. RSC Adv 5(39):30801–30818
51. Rauf MA, Ashraf SS (2009) Radiation induced degradation of dyes—an overview. J Hazard Mater 166(1):6–16
52. Ghodbane H, Hamdaoui O (2010) Decolorization of antraquinonic dye, CI Acid Blue 25, in aqueous solution by direct UV irradiation, UV/H2O2 and UV/Fe (II) processes. Chem Eng J 160(1):226–231
53. Mascolo G, Lopez A, Bozzi A, Tiravanti G (2002) By-products formation during the ozonation of the reactive dye Uniblu-A. Ozone: Sci Eng 24(6):439–446
54. Ramseier MK, Peter A, Traber J, von Gunten U (2011) Formation of assimilable organic carbon during oxidation of natural waters with ozone, chlorine dioxide, chlorine, permanganate, and ferrate. Water Res 45(5):2002–2010
55. Eren HA, Yiğit İ, Eren S, Avinc O (2020) Ozone: an alternative oxidant for textile applications. Sustain Text Appar Ind: Prod Process Sustain 81–98
56. Gül Ş, Özcan Ö, Erbatur O (2007) Ozonation of CI Reactive Red 194 and CI Reactive Yellow 145 in aqueous solution in the presence of granular activated carbon. Dyes Pigm 75(2):426–431
57. Ghasemi H, Aghabarari B, Alizadeh M, Khanlarkhani A, Abu-Zahra N (2020) High efficiency decolorization of wastewater by Fenton catalyst: magnetic iron-copper hybrid oxides. J Water Process Eng 37:101540
58. Sharma L, Prabhakar S, Tiwari V, Dhar A, Halder A (2021) Optimization of EC parameters using Fe and Al electrodes for hydrogen production and wastewater treatment. Environ Adv 3:100029
59. Khataee AR, Vatanpour V, Ghadim AA (2009) Decolorization of CI Acid Blue 9 solution by UV/Nano-TiO2, Fenton, Fenton-like, electro-Fenton and electrocoagulation processes: a comparative study. J Hazard Mater 161(2–3):1225–1233
60. Sen Gupta SK (2017) Contact glow discharge electrolysis: a novel tool for manifold applications. Plasma Chem Plasma Process 37:897–945
61. Mishra S, Maiti A (2018) The efficacy of bacterial species to decolourise reactive azo, anthroquinone and triphenylmethane dyes from wastewater: a review. Environ Sci Pollut Res 25:8286–8314
62. Ikram M, Naeem M, Zahoor M, Hanafiah MM, Oyekanmi AA, Islam NU, Ullah M, Mahnashi MH, Ali AA, Jalal NA, Bantun F (2022) Bacillus subtilis: as an efficient bacterial strain for the reclamation of water loaded with textile azo dye, orange II. Int J Mol Sci 23(18):10637
63. Karim ME, Dhar K, Hossain MT (2018) Decolorization of textile reactive dyes by bacterial monoculture and consortium screened from textile dyeing effluent. J Genet Eng Biotechnol 16(2):375–380

64. Cagide C, Castro-Sowinski S (2020) Technological and biochemical features of lignin-degrading enzymes: a brief review. Environ Sustain 3:371–389
65. Mani A, Hameed SAS (2019) Improved bacterial-fungal consortium as an alternative approach for enhanced decolourisation and degradation of azo dyes: a review. Nat Environ Pollut Technol 18(1):49–64
66. Kaushik P, Malik A (2009) Fungal dye decolourization: recent advances and future potential. Environ Int 35(1):127–141
67. Shi Y, Yang Z, Xing L, Zhang X, Li X, Zhang D (2021) Recent advances in the biodegradation of azo dyes. World J Microbiol Biotechnol 37:1–18
68. Touliabah HES, El-Sheekh MM, Ismail MM, El-Kassas H (2022) A review of microalgae-and cyanobacteria-based biodegradation of organic pollutants. Molecules 27(3):1141
69. Khataee AR, Vafaei F, Jannatkhah M (2013) Biosorption of three textile dyes from contaminated water by filamentous green algal Spirogyra sp.: Kinetic, isotherm and thermodynamic studies. Int Biodeterior Biodegrad 83:33–40
70. Sharma S, Singh B, Manchanda VK (2015) Phytoremediation: role of terrestrial plants and aquatic macrophytes in the remediation of radionuclides and heavy metal contaminated soil and water. Environ Sci Pollut Res 22:946–962
71. Bharathiraja B, Jayamuthunagai J, Praveenkumar R, Iyyappan J (2018) Phytoremediation techniques for the removal of dye in wastewater. Bioremediation: Appl Environ Prot Manag 243–252
72. Zhou X, Xiang X (2013) Effect of different plants on azo-dye wastewater bio-decolorization. Procedia Environ Sci 18:540–546
73. Soni S, Bajpai PK, Bharti D, Mittal J, Arora C (2020) Removal of crystal violet from aqueous solution using iron-based metal-organic framework. Desalin Water Treat 205:386–399. https://doi.org/10.5004/dwt.2020.26387
74. Soni S, Bajpai PK, Mittal J, Arora C (2020) Utilisation of cobalt doped Iron based MOF for enhanced removal and recovery of methylene blue dye from wastewater. J Mol Liq 314. https://doi.org/10.1016/j.molliq.2020.113642
75. Arora C, Soni S, Sahu S, Mittal J, Kumar P, Bajpai PK (2019) Iron-based metal organic framework for efficient removal of methylene blue dye from industrial waste. J Mol Liq 284:343–352. https://doi.org/doi.org/10.1016/j.molliq.2019.04.012
76. Soni S, Bajpai PK, Mittal J, Arora C (2022) Enhanced sequestration of an acidic dye on novel bimetallic metal-organic framework. J Dispers Sci Technol. https://doi.org/10.1080/01932691.2022.2135521
77. Kameliya J, Verma A, Dutta P, Arora C, Vyas S, Varma RS (2023) A review on recent advances in synthesis, characterization and applications of layered double hydroxides. Inorganics 11(3):121. https://doi.org/10.3390/inorganics11030121

Nanomaterials Across the Periodic Table as the Promising Agents for the Remediation of Dye-Contaminated Water: A Review

Divya Mahar, Nitish Semwal, Lalit Mohan, Rajesh Kumar, Sandeep Dutt Maindoli, Kshitindra Kumar Singh, Manjunath Chatti, and Mahesh Chandra Arya

Abstract Availability and supply of drinking water have become a challenge for the Government and the people alike. Contamination of water sources adds to the misery. The dyes present in the effluents contribute to water pollution to a large extent. Some of the methods employed to eliminate these dyes from polluted water are such as oxidative removal with oxidizing agents, application of UV and visible radiation, adsorption, distillation, membrane filtration, chemical precipitation, solvent extraction ion exchange, reverse osmosis, and bioremediation. The nanomaterials have not only helped to overcome the limitations of these methods but also strengthened our war with pollutants to a great extent. The nanomaterials have found their applications either as adsorbents or catalysts (or photocatalysts) for the treatment of wastewater to eliminate dyes. This exhaustive literature review explores the synthesis, properties, applications, and efficiency of different nanomaterials and highlights the superiority of nanotechnological approaches for the remediation of dyes over conventional materials. There is a big list of metal-based nanomaterials engaged in the elimination of dyes from wastewater. These can be classified on the basis of various metals belonging to *s*-, *p*-, *d*-, and *f*-blocks of the modern periodic table.

D. Mahar · N. Semwal · L. Mohan · M. C. Arya (✉)
Department of Chemistry, Kumaun University, DSB Campus, Nainital, Uttarakhand, India
e-mail: mcarya181@gmail.com

R. Kumar
Department of Chemistry, SSJ University, SSJ Campus, Almora, Uttarakhand, India

S. D. Maindoli
Department of Zoology, Kumaun University, DSB Campus, Nainital, Uttarakhand, India

K. Kumar Singh
Water Resource Management, Central Institute of Mining and Fuel Research, Dhanbad, Jharkhand, India

M. Chatti
Istituto Italiano di Tecnologia, Via Morego 30, 16163 Genova, Italy

P. Singh (ed.), *Dye Pollution from Textile Industry*, SDGs and Textiles,
https://doi.org/10.1007/978-981-97-5341-3_6

Keywords Dye contamination · Dye degradation · Nanomaterials · Photocatalyst · Water treatment

1 Introduction

Colors have fascinated man since time immemorial. With the advent of science, dyes and other coloring agents gave wings to his imagination.

> Colour is the touch of the eye, music to the deaf, a word out of the darkness. .... I'm so fortunate to be red! I'm fiery. I'm strong. I know men take notice of me and that I cannot be resisted..... Wherever I'm spread, I see eyes shine, passions increase, eyebrows rise and heartbeats quicken. Behold how wonderful it is to live! Behold how wonderful to see. I am everywhere. Life begins with and returns to me. Have faith in what I tell you.
>
> Orhan Pamuk, *My Name is Red,* (Page 152, [1])

Dyes are the chemical stuff that transmits their color to an object, thing, commodity article, etc., and are soluble in the aqueous medium. The dyes are of various categories and include natural and synthetic dyes. The chief source of natural dyes is biota, which includes both Flora and Fauna while synthetic dyes are artificially made products that are used to add pigment to a wide range of media. They are commonly known as "coal tar dyes". It was under control till things started at a large scale and natural pigments were replaced with synthetic dyes. Applications of synthetic dyes in textile, printing, leather, light, pharmaceutical, and chemical industries have raised alarm because of environmental pollution. Due to various reasons, the untreated and under-treated effluents of these industries fall into water bodies such as rivers, ponds, and lakes and contaminate them. About 300,000 tons of synthetic dyes go annually into effluent, which consists of a hefty 50% of the dyes used in various dyeing processes [2, 3]. Dyes present in the effluent can cause environmental pollution in more than one way. Dyes such as anthraquinone dyes show their recalcitrant nature and remain in aquatic systems for long periods. Their stay in water bodies is highly detrimental to aquatic flora and fauna as it causes less light penetration and low photosynthesis and affects the solubility of gases [4]. The organic substances present in these effluents show biological toxicity because of their carcinogenic, mutagenic, and teratogenic properties [5]. Though the dyes used across the industries vary in their chemical nature, approximately two-thirds of these dyes belong to the azo dye category [6]. Due to their excellent fasting and spectra of colors, these dyes are extensively used for dyeing in textiles, leather, paints, plastics, papers, pharmaceuticals, and food industries [7].

The harmful effects of dyes demand their effective removal from the industrial discharge. This means simple decolorization of discharge is not sufficient but this must include effective degradation of dyes into harmless substance. Because of the considerable variation in the chemical composition of dyes and fair variability of effluent concoction, it is often impossible to achieve the desired results of dye removal using single treatment technology. So, a broad spectrum of techniques has been used for dye removal and to minimize their influence on the environment. These methods encompass both physical and chemical treatments. Physical methods such as distillation, chemical precipitation, membrane filtration, solvent extraction, ion exchange, and reverse osmosis have been used as the classical treatment of effluents [8]. Chemical methods named as coagulation, precipitation, oxidation, etc. have also been proved useful. Also, biological methods viz. enzyme treatment and microbial degradation have also been explored as the remedy to the problem. In the recent literature, the removal of vat dyes was done by cost-effective, facile, and nature-friendly adsorbent from the leftover material after the oil extraction of seeds of *Moringa oleifera*. The obtained green adsorbents were then effectively studied and resulted in 94% removal of Indigo Carmine dye [9]. The hybrid nanoparticles of nickel and copper are obtained through the rhizome extract of *Zingiber officinale*, which provides the stabilizing and reducing agent for the synthesis of nanoparticles. These nanoparticles show superior antimicrobial activity and excellent catalytic removal efficiency for the elimination of Crystal Violet dye [10]. Similarly, a metal-containing organic framework named poly-o-anisidine/ZnO@ZIF-8 nanomaterial shows effective potential in the elimination of toxic dyes such as Malachite Green from wastewater [11].

As understood, each of these methods has its own merits and limitations. With the progress in the field of relevant research, there have been efforts to invoke nanotechnology to get rid of the problem. Several researchers have tried to synthesize nanoadsorbents, nanophotocatalysts, and nanocatalysts which can be handy in degrading the residual dyes present in the effluents. The present literature review begins with the harmful effect of dyes on aquatic flora and fauna. It explores the vast literature available on the removal of the dye using various methods. For the sake of study, the chapter has been divided into sections such as traditional removal methods and biological methods. The latter half of the chapter encompasses the various methods based on nanomaterials which have been prepared using different metals. As there is a big list of metal-based materials engaged for dye removal from polluted water, the review has been classified on the basis of different metals belonging to *s*-, *p*-, *d*-, and *f*-blocks of the modern periodic table.

2 Harmful Effects of Dyes on Aquatic Biota and Human Beings

Freshwater is a crucial natural resource necessary for human consumption and the survival of aquatic ecosystems. However, the increasing human population and advancements in civilization have led to the expanding use and misuse of these limited freshwater resources. This situation calls for concerted efforts to enhance our understanding of the structural and functional aspects of freshwater systems. Various dye industries discharge textile dyes, leading to pollution in rivers, seas, and oceans. These dyes are resistant to natural decomposition in standard environmental conditions. Aquatic flora and fauna are at the receiving end, significantly disrupting aquatic ecology by lowering the rate of photosynthesis in comparison to native aquatic plants. These dyes raise concerns regarding eutrophication by releasing excessive phosphate and nitrate ions. Fluctuating pH and a high degree of COD (Chemical oxygen demand), BOD (biochemical oxygen demand), TSS (total suspended solids), and high level of coloration indicate the existence of textile pigments. Metal elements obtained by pigment and metalliferous impose toxicity which causes an imbalance in aquatic life; the high concentration of these elements absorbed by aquatic plants such as algae, pteridophytes, bryophytes, and vascular plants accelerates the uncontrolled growth which results in oxygen lacking by limiting the photosynthesis [12].

Many terms such as aquatic ecology and hydrobiology have been coined for the study of freshwater. Like any other ecosystem, the existence of freshwater ecosystems depends on the interaction between abiotic and biotic components. The abiotic components include various types of gaseous and micro- and macronutrients present in the systems. Various types of biota in aquatic water bodies include plankton, benthos, nekton, neuston, periphyton, etc. This section embodies and aims to assess the impact of chemical dyes on the biota of the freshwater that includes both lentic (standing) and lotic (running) water bodies. The grazing food chain in these aquatic bodies consists of various tropic levels which include phytoplankton, zooplankton, macrozoobenthos, and fish. The presence of residual dyes in water bodies adversely affects each level directly or indirectly.

2.1 *Harmful Effects of Dyes on Aquatic Life*

2.1.1 Effect of Chemical Dyes on Phytoplankton

Phytoplankton are microscopic algae, which capture and transform most of the solar energy into food energy in aquatic ecosystems. They are also chief producers of oxygen in water bodies. Being the first tropic level in wetlands, they respond more quickly to any interventions than any other biota of the system. Aquatic vegetation encompasses a diverse range of algae and macrophytic plants. Algal plants are found in various habitats and play crucial roles in sediment stabilization, nutrient cycling, and water quality regulation. These organisms act as primary producers and are highly responsive indicators of environmental changes within aquatic systems. Microalgae, in particular, have short lifecycles and are highly sensitive to toxic pollutants. Dye contamination hinders the energy transfer process by restraining the growth and development of microalgae. Dyes show diverse effects on aquatic flora. One of the frequently observed effects is mutations, often induced by azo dyes and anionic dyes, among other impacts. The presence of dyes in aquatic systems affects various growth parameters of algae, including protein and pigment content.

Dyes like Drimarene blue, Optilan yellow, and Lanasyn Brown exhibit toxicity toward green algae. Cyanobacteria, which are phototrophic and diazotrophic, produce oxygen through photosynthesis. The growth of cyanobacteria such as *Anabaena* is inhibited by both azo dyes and anionic dyes [13].

Furthermore, textile pigment pollution significantly affects the rate of photosynthesis by obstructing the penetration of sunlight into the water. This reduction in light availability decreases the rate of photosynthesis in algae and other aquatic plants. Algae occupy the producer level in ecological pyramids, impacting herbivorous organisms in the food chain. The persistence of these textile pigments as pollutants is particularly concerning due to biomagnification, wherein their contamination levels increase as they move up the trophic levels, starting from lower to higher levels [14].

The excessive use of synthetic textile dyes, such as Azo, Disperse, and Vat dyes, reduces phytoplankton growth. The inappropriate discarding of azo dyes in freshwater bodies cause great concern to phytoplanktons due to their toxicity and carcinogenicity. Due to bio-decolorization in producer organisms, it directly stimulates the proliferation of microorganisms in contaminated water. The excess concentration of dye increases the contact of microorganisms.

Rhodamine B, a synthetic dye, is discharged in significant amounts into freshwater bodies, posing a hazard due to its vibrant color and disrupting the metabolic and physiological functions of aquatic vegetation [15, 16]. These nitrogen-containing molecules, upon dissociation, exhibit a highly stable and vivid coloration [17], 18. The intense coloration of Rhodamine B dye proves lethal to aquatic organisms, impacting both their health and human well-being, leading to mutagenic and oncogenic changes in living organisms [19], 18.

2.1.2 Effect of Chemical Dyes on Macrozoobenthos

Macrozoobenthos are organisms that spend most of their time living, crawling, burrowing, or attaching themselves to the bottom sediments. Because they typically remain in one location, they are excellent indicators of pollution and eutrophication. Additionally, through bioturbation, they significantly impact nutrient cycling.

2.1.3 Harmful Impacts of Dyes on Fish

Fish occupy the top position in the aquatic ecosystem's food chain and are referred to as nekton. They play a crucial role in balancing ecosystems in various ways. However, fish are highly affected by textile dyes, which enter their bodies through gill respiration. These contaminants damage biochemical and hematological processes, harming the kidneys, liver, nephrons, and gills [20]. Additionally, certain fish vendors introduce colorants into the fish diet to alter their color. In a similar vein, the crucian carp (*Carassius auratus*), an omnivorous freshwater fish, is utilized as a biological marker to assess the status of aquatic water. Tartrazine, an azo dye, demonstrates toxicity in crucian carp even at lower concentrations upon consumption and adversely affects human health upon consumption [21].

The acute toxicity of Reactive Red dye (RR239) varies based on the species it interacts with. For instance, abnormal growth in the length of *Daphnia similis* (a species of sawfly) has been observed, while malformation was noted in the embryos of *Biomphalaria glabrata* (snail) species [22]. Auramine O and Auramine dyes exhibit both chronic and acute toxicity toward aquatic organisms. Various organisms across different tropic levels, including *Daphnia similis, Danio rerio, Raphidocelis subcapitata*, and *Hydra attenuata*, are impacted by the toxic effects of these dyes. Both types of dyes caused a reduction in the growth rate of the exposed algae, leading to adverse effects on the reproduction of *D. similis* and causing deformities in *H. attenuate* [23].

2.2 Adverse Impacts of Dyes on Human Beings

Dyes affect the water quality which influences a range of organisms from the simplest to the most complex. Human beings are also highly susceptible to the lethal effects of textile colorants. In many countries, dye-containing wastewater is commonly used for irrigation, leading to significant dye penetration into the soil. From there, plants come into contact with these dyes. Although the dyes settle into the soil quickly, they remain for a long period. Human beings get affected by these dyes directly and indirectly: directly when consuming crops, vegetables, and other plants grown in contaminated soil, and indirectly when consuming meat from grazing animals. Dye-contaminated dust particles, when inhaled or ingested, cause acute toxicity and can induce skin and eye irritation. Moreover, workers involved in the handling and production of textile dyes can be vulnerable to various health conditions like asthma, dermatitis, and allergic reactions because of their exposure to the substances used in the dyeing processes [14].

Efforts to reduce the release of these textile colorants and improve wastewater treatment are essential to minimize the health and environmental hazards associated with their presence in water.

3 Classical Methods of Dye Degradation

Industries like textile, plastic, food, leather, and cosmetics are producing synthetic dyes as byproducts, in which the textile sector plays a key role in the consumption of dyes. Synthetic dyes are composed of aromatic structures, making them highly stable and relatively easy to detect but difficult to eliminate from wastewater systems. There are various technologies for dye removal involving chemical, physiochemical, and biological processes, each with its advantages and disadvantages. Generally, dye removal methods are classified into two types: non-destructive and destructive methods (Fig. 1).

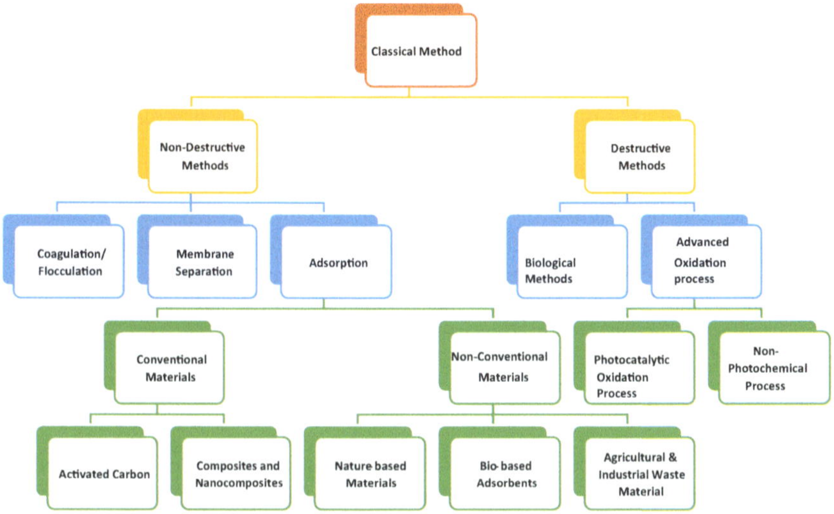

Fig. 1 Schematic illustration of classical approaches for the removal of dye

3.1 Non-destructive Methods

Non-destructive methods are classified into three subparts: 1. Coagulation/ Flocculation, 2. Membrane-based separation, and 3. Adsorption.

3.1.1 Coagulation/Flocculation

Coagulation or Flocculation methods are used when the contaminants are present on the water surface and are non-settleable; in this process various coagulating chemicals are used to agglomerate small particles into larger ones, known as flocculation process. Coagulation methods have proven effective for the removal of phosphate, often referred to as a single treatment process, as the small particles are too lightweight to settle and are not large enough to be trapped during filtration. Multivalent metal inorganic salts have traditionally served as coagulants due to their cost-effectiveness and ease of handling compared to chemical flocculants. However, the use of alum can increase the presence of total solids. Moreover, the neutralization process increases the chemical cost of the coagulation process. Another limiting factor of the use of inorganic coagulants is that they produce a large quantity of toxic (hydroxide) sludge, and the requirement of large quantity, sensitivity toward pH, and ineffectiveness toward fine particles are significant drawback [24].

3.1.2 Membrane Separation

Membrane separation is widely acknowledged for its high efficiency and excellent selectivity, offering manageable operating conditions that make it a prominent technique for eliminating dyes from polluted water. The advanced techniques of membrane separation, reverse osmosis, and nanofiltration show effective rejection of pollutants but face limitations such as high operating pressure and significant energy consumption. On the other hand, microfiltration and ultrafiltration operate at lower pressures but exhibit lower removal efficiency due to the presence of large-sized pores [25].

3.1.3 Adsorption

Adsorption methods are classified into two methods named as conventional materials and non-conventional materials.

Conventional Materials

Activated carbon obtained from waste materials has long been utilized as a promising adsorbent due to its high surface area and reactivity, highly porous structure, and high adsorption capabilities. However, a major drawback of the activated carbon adsorbents is the formation of secondary pollution as separating these adsorbents from water bodies can be challenging, limiting their widespread use. To address these challenges, magnetic separation emerged with activated charcoal, as it provides fast and easy recovery of adsorbents and high adsorption efficiency. This method includes an adsorption phase based on physicochemical interactions with a filtration phase employing magnetic properties [26].

Composites and Nanocomposites—In recent years, various composites and nanocomposites have been employed for dye adsorption from polluted water [27], introduced an economically viable adsorbent called activated carbon based on graphene oxide, which proved effective in removing cationic dyes and was designed for reusability. Additionally, [28] synthesized a copper-containing organic-metal framework on graphene oxide using the ultrasonication method, demonstrating excellent performance in removing methylene blue dye from wastewater [28].

Non-conventional Materials

Non-conventional materials are classified into three main types; natural materials (zeolites, clay, chitosan), biosorbents (biomass, algal waste, fungi), and agricultural and industrial waste (orange peel, ash, lignin, coconut shell).

1. Nature-based Materials—Nature-based materials, a subset of non-conventional materials, exhibit high dye removal efficiency, cost-effectiveness, and non-toxicity in nature. Natural clay, zeolites, fly ash, and chitosan are some common examples of nature-based adsorbent materials. Natural zeolites have been identified as a cost-effective adsorption material for removing Crystal Violet dye, outperforming activated carbon with a reported removal capacity of 177.75 mg/g compared to activated carbon's 84.11 mg/g [29]. Similarly, naturally occurring biopolymer chitosan is widely used for the treatment of wastewater. For instance, carbon/Fe_3O_4 incorporated chitosan was synthesized by a modified hydrothermal process for the removal of anionic dyes named Methyl Orange [30].

2. Bio-based adsorbents—Biosorbents obtained by algal waste, bacterial, living or non-living fungi, and waste biomass are recognized as effective adsorbents for the elimination of anionic and cationic dyes from wastewater. Algal adsorbents possess the ability to decolorize dyes through enzymatic mechanisms, a process also known as phytoremediation. Biological-based treatments have been seen as effective alternative rather than the classical physicochemical methods for the elimination of dyes [31]. Moreover, the removal of azo dyes was performed by Bayramoglu and Yilmaz [32]. They utilized unfixed fungal biomass and a composite of fungal biomass from *Lentinus concinnus* embedded in polyvinyl alcohol for adsorbing Reactive Yellow 86 dye. Reactive Blue 19 was successfully decolored using a biosorbent called white-rot fungus (*Panus tigrinus*). Deniz and Ersanli presented experimental results on the biosorption capabilities of a phyco-composite made up of filamentous green algae *Spirogyra sp.* and *Rhizoclonium sp.*, which effectively removed Basic Red 46 dye [33].

3. Agricultural and Industrial Waste Material—Agricultural and industrial waste garnered significant attention for dye decolorization processes. Agricultural waste, often sourced from the food industry, is inexpensive and readily available. Furthermore, it can be converted into nanomaterials for high efficiency in the treatment of wastewater [34]. Ma et al. [35] reported the utilization of grape peels as an adsorbent for the adsorption of MB dye. The grape peel was treated with the microwave-assisted hydrothermal method to attain greater dye removal efficiency. Similarly, an iron-based organic framework derived from industrial waste exhibited a 94.74% dye removal efficiency for methylene Blue [36]. Recent research by Samimi and Shahriari-Moghadam [37] highlighted the effectiveness of Lantana camara L. stem biomass as a cost-effective biosorbent for removing Malachite Green from aquatic environments through adsorption. They also conducted comprehensive kinetics, equilibrium, and thermodynamic analyses. Ali and a colleague investigated the isotherm, kinetics, and thermodynamic aspects of the Methyl Violet dye's adsorption onto a produced bio-adsorbent made from seeds of date [38].

3.2 *Destructive Methods*

These methods are of mainly two types Bio-degradation (bacteria and enzymes, fungi., yeast, green algae) and oxidation processes.

3.2.1 Biological Methods

An intriguing alternative to chemical methods for color removal from wastewater is the use of biological approaches. These methods often involve activated sludge that has been inoculated with a variety of aerobic and anaerobic microorganisms, such as fungus, algae, and bacteria. Applications for biological material including algae, bacteria, fungi, and yeasts, which can decompose and absorb various kinds of synthetic colors, have been recommended by researchers [39]. Compared to other treatments, biological degradation (also known as bioremediation) is more environmentally benign, economically viable, and generates less sludge overall. This process involves breaking the chromophoric groups in synthetic dyes, leading to their decomposition into less hazardous organic products, ultimately aiding in color removal [40]. The breakdown of azo dyes occurs in two stages: first, the colors are broken down into amines by the breaking of the azo bonds, and then the aromatic amines break down into small, non-toxic molecules in an oxygen-rich environment [41]. The methods under investigation leverage bacteria's capacity to thrive in both oxygen-rich (aerobic) and oxygen-limited (anaerobic) conditions to fully break down the azo linkages present within the dyes. The capacities of the *Streptomyces microflavus* CKS6 microbe in the remediation of textile wastewater containing Crystal Violet (CV) and Safranin T dyes have been demonstrated by research by Bunti and co-authors [42]. This environmentally friendly bioremediation approach involved two processes: the adsorption of dye on the bacterium's surface and complete biodegradation through the microorganism's enzyme system. Using fungal strains, Mustafa and co-authors explored various approaches to dye biodegradation using fungal strains [43]. Hortaea sp. was isolated from soil polluted with petroleum by Al Farraj et al. [44] and its capacity to break down solvent green was investigated. The research discovered that at the lowest beginning dosage of 10 mg/L, the dye was degraded by the bacteria in just 24 h. Malachite Green, Crystal Violet, Methyl Violet, and Cotton Blue are four regularly used TPM dyes, and [45] demonstrated that different actinobacteria extracted from soil displayed exceptional degradation efficiency in breaking down these dyes. *Among the bacterial strains identified, N. alba showed the highest degradation rates* over 14 days for Malachite Green (97.0%), Methyl Violet (95.1%), Crystal Violet (95.8%), and Cotton Blue (83.8%).

3.2.2 Advanced Oxidation Process

The oxidation method is further divided into two groups one is a non-photochemical method which includes an electrochemical method, Fenton process, Electro-Fenton process, and ozonation, whereas the second sub-group is a photochemical method which includes photocatalytic oxidation and photo-Fenton process.

Photocatalytic Oxidation Process

The basic principle of photocatalysis involves the redox reactions of a catalyst in the presence of solar energy, transforming organic contaminants into byproducts such as CO_2 and H_2O. The Fenton oxidation process utilizes the formation of $\dot{O}H$ radicals by the chain reaction on H_2O_2 and ferrous (Fe^{2+}) ions. These $\dot{O}H$ radicals are highly effective at oxidizing organic pollutants, whereas the photocatalysis-assisted Fenton process includes the irradiation of light with a Fenton catalyst [46]. Wang et al. [47] reported a reusable heterogeneous g-C_3N_4/$MnFe_2O_4$ catalyst, which provides a solar photo-Fenton system for the removal of organic pollution by self-supplying of H_2O_2.

Non-photochemical Method

In this method, the dye degradation process operates without utilizing radiation. Ozone is a strong oxidant and can effectively oxidize various inorganic and organic pollutants. The ozone oxidation method can be done by the two pathways, one involves the direct reaction of ozone with dyes, which reduces the multiple bonds present in organic pollutants and thus are highly selective for the unsaturated aromatic and aliphatic organic pollutants, whereas the other involves an indirect method that produces the $\dot{O}H$ radical, which then acts as a highly reactive species for the oxidation process. However, the high consumption of energy and comparatively high cost make this method limited. Moreover, the basic principle of electrochemical reactions is the use of electrode reactions for oxidation purposes. This can be further obtained by two methods; direct oxidation, in which $\dot{O}H$ radicals are produced on the anode, followed by the oxidation of pollutants by $\dot{O}H$ radical, whereas indirect oxidation uses the reactive intermediate species ($\dot{O}H$ and $^{\cdot}O_2$) obtained through anodic reactions for the removal of organic pollutants [48]. The fundamental concept behind wet air oxidation relies on the generation of $\dot{O}H$ radicals under high pressure and elevated temperatures, making it suitable for treating small-sized pollutants. The downside of this method is that it is unable to provide one-step degradation of pollutants, which can be overcome by the catalyst wet air oxidation process [46].

4 Application of Different Metal-Based Nanomaterials Toward the Removal of Dyes

A significant environmental concern is the rising number of synthetic dyes that are released into natural water basins. Traditional color removal techniques have been shown to be ineffective and frequently harmful to the environment. This has prompted researchers to look into alternate and environmentally friendly wastewater treatment materials, such as *f*-block-based nanomaterials.

4.1 s-Block (Alkali Metals and Alkaline Earth Metals) Metals

The *s*-block metals are vital in nanomaterial synthesis for efficient dye removal applications. The lattice structure or doping materials of nanoparticles used for effective dye adsorption and removal from diverse solutions include these metals as crucial components. These elements are found in Groups 1 and 2 of the periodic table, and they offer a flexible lattice framework on which different nanomaterials may be precisely grown or put together.

4.1.1 Alkali Metals (Group 1)

Due to their extreme reactivity with oxygen and moisture, alkali metals like lithium (Li), sodium (Na), and potassium (K) are rarely used as a lattice material for nanomaterial production. Their reactivity makes them challenging to handle and limits control over particle characteristics, compatibility with common solvents, and often leads to unwanted reactions that can damage nanomaterials and reduce yield.

In one study, the hydrothermal approach was employed to synthesize Caesium lead iodide ($CsPbI_3$) perovskite nanostructures, which were then examined for their photocatalytic properties for the removal of organic dyes under natural light exposure. The maximum degrading efficiency was obtained when Methyl Violet was used as the target dye, with a decolorization rate of 81.7% [49].

Various metal-based nanoparticles were synthesized by waste lithium-ion batteries. For instance, cobalt oxide nanoparticles were produced from discarded Li-ion batteries to create a photocatalyst for degrading Methyl Blue dye [50], whereas, cobalt ferrite nanoparticles were successfully generated using a co-precipitation method that involved combining recycled iron and cobalt derived from waste iron ore tailings and used lithium-ion batteries. These nanoparticles efficiently facilitated the degradation of CR dye in the presence of visible light [51]. In some cases, alkali metals are employed as a stabilizing reagent for the nanoparticle's production. Silver nanoparticles (Ag-NPs) were synthesized by employing sodium dodecyl sulfate (SDS) as a stabilizer. These Sd@Ag-NPs particles proved to be effective catalysts for the removal of Methyl Orange (MO) and Rhodamine B (Rh-B) dyes

[52]. A photocatalyst with g-C_3N_4 doped with sodium ion, combined with ultrafine silver nanoparticles, was synthesized by an efficient one-step method. This catalyst proved highly effective in degrading Rhodamine B and tetracycline hydrochloride when exposed to natural light [53].

Rubidium tin chloride (Rb_2SnCl_6) and caesium-doped rubidium perovskites [$(RbXCs_{1-}X)_2SnCl_6$] were synthesized with the intention of using them for photocatalytic applications. These materials were employed to break down Methylene Blue dye in the UV light. Remarkably, $(Rb_{0.5}Cs_{0.5})_2SnCl_6$ showed higher photocatalytic efficiency (85%) compared to Rb_2SnCl_6 (78%) after 120 min of exposure [54].

4.1.2 Alkaline Earth Metals (Group 2)

MgO nanoparticles have been widely researched due to their potential application in the photocatalytic elimination and adsorption-based removal of dyes. Synthesized through various methods such as reflux condensation and thermal breakdown, MgO nanoparticles have demonstrated strong photocatalytic activity in the deterioration of various dyes, such as Methyl Orange and Methylene Blue [55]. Furthermore, MgO nanoparticles have shown their photocatalytic effectiveness in degrading Methylene Blue and Rhodamine B dyes under natural light exposure [56]. Incorporating erbium into MgO nanoparticles has been found to enhance their photocatalytic activity for the removal of Methylene Blue dye [57].

For improved Reactive Yellow 84 dye adsorption, Calcium Carbonate nanoparticles were modified with EDDS (ethylenediamine-N, N′-disuccinic acid) and EDTA (ethylenediaminetetraacetic acid). In comparison to CC-EDDS and plain CC nanoparticles, CC-EDTA nanoparticles had the best dye removal effectiveness of 99% [58]. Polyvinyl alcohol/Calcium Carbonate nanocomposites were employed as the adsorbent for Safranin and Methylene Blue. The nanocomposite successfully removes cationic dyes from wastewater with a removal efficiency of 87% for Safranin (SF) and 82% for Methylene Blue (MB) [59].

A successful synthesis of oxide of mixed calcium and magnesium ($CaMgO_2$) nanoflakes was achieved using ultrasound-facilitated co-precipitation. Even after undergoing five cycles, the catalyst made of mixed $CaMgO_2$ nanoflakes exhibited outstanding photocatalytic performance, degrading 4-nitrophenol (4-NP) and Methylene Blue (MB) by approximately 88% and 96%, respectively [60].

The sol–gel process was effectively utilized to produce nanoparticles of strontium titanate that were doped with lanthanum and had a grain size of around 20 nm. These nanoparticles were then evaluated for their photocatalytic activity, demonstrating strong effectiveness, particularly in the degradation of the Rhodamine B Blue process under UV light exposure [61]. In another research, the hydrothermal technique was used to successfully synthesize strontium titanate nanocubes (STNCs). When subjected to UV light exposure, these synthesized STNCs showed excellent photocatalytic effectiveness in the removal of Tartrazine (TZ) and Methylene Blue (MB) dyes. When subjected to UV light exposure, the synthesized STNCs degraded the MB and TZ dyes by around 78.9% and 76.4%, respectively [62].

However, MgAlNi barium-ferrite (MgAlNi-BaFe) nanocomposites were also used for the degradation of Methyl Orange from wastewater with high regeneration performance [63]. Cu-doped Strontium Titanate (Cu/$SrTiO_3$) nanoparticles were synthesized using a sol–gel technique. These nanoparticles displayed effective photocatalytic activity when degrading Methylene Blue dye in visible light exposure. The findings suggest that within 2 h of visible light exposure, Cu/$SrTiO_3$ nanoparticles achieved a 66% elimination of Methylene Blue dye [64].

4.2 p-*Block Metals for Nanomaterial Synthesis and Dye Removal Applications*

The synthesis of zinc-aluminum oxide nanoparticles doped with lead (LD-ZAO-NPs) was accomplished using a facile procedure with sodium hydroxide (NaOH) as the precipitating agent. Within just 30 min, the Methyl Orange elimination efficiency reached a remarkable rate of 99.60% [65]. An electrochemical method was employed to synthesize aluminum oxide (Al_2O_3) nanoparticles. These nanoparticles were assessed for their effectiveness in degrading Malachite Green dye in the presence of sunlight. Two distinct methodologies were examined: adsorption accompanied by photocatalysis and the combined approach of adsorption and photocatalysis. The combined process exhibited significantly higher photodegradation performance, achieving a removal rate of 45% for Malachite Green, compared to the sequential approach, which achieved a removal rate of 32% [66].

Selenium nanoparticles coated with Brij-58 denoted as Brij-58@Se NPs were successfully synthesized and demonstrated a remarkable adsorption efficiency of 98% within a remarkably short time frame of just 6 min for the elimination of Bromophenol Blue dye [67]. A photocatalyst composed of a nanocomposite of ferric oxide (Fe_3O_4), zinc oxide (ZnO), and silicon nitride (Si_3N_4) was synthesized. The photocatalyst's performance was examined in the context of degrading two dyes, Sunset Yellow and Methyl Orange, within aqueous solutions. The findings revealed an exceptionally high removal rate, achieving 90% degradation for Sunset Yellow and an impressive 96% degradation for Methyl Orange [13]. While, by utilizing the sol–gel technique, spherical TiO_2/SiO_2 nanoparticles were successfully synthesized and assessed for their photocatalytic capabilities in degrading Methyl Orange and Methylene Blue dyes. Notably, they exhibited superior photodegradation performance, with Methylene Blue degradation reaching 80–85% in comparison to Methyl Orange's 70–75% in the presence of mercury light when exposed for a 2-h duration [68]. A nanocomposite consisting of SiO_2 nanoparticles derived from biomass, combined with TiO_2 semiconductors, was utilized to break down Rhodamine B dye molecules under UV–visible light [69].

Germanium oxide nanorods (GeO_2Nrs) were successfully synthesized at room temperature using the co-precipitation method. These GeO_2Nrs demonstrated an impressive ability to achieve over 50% removal of Methylene Blue dye within an irradiation time of 60 min [70]. A wet chemical synthesis process was employed to synthesize germanium dioxide doped with magnesium (Mg-GeO_2) for the investigation of photocatalytic decolorization and elimination of Methyl Violet dye [71]. Through an ultrasonic/hydrothermal process, nickel sulfide (NiS)-indium oxide (In_2O_3) nanostructures, as well as NiS-In_2O_3 composites anchored on graphene oxide (GO) were synthesized. An evaluation was conducted to assess the photocatalytic efficiency of these nanomaterials under UV lamp illumination, for the removal of Methylene Blue dye [72]. A photocatalyst composed of Ga_2O_3 and TiO_2 was synthesized using a mechanomixing process followed by sonication. This Ga_2O_3-TiO_2 photocatalyst displayed exceptional performance in degrading Rhodamine B dye in the presence of UV radiation [73]. Tin oxide (SnO_2) nanoparticles were synthesized in an environmentally friendly and cost-efficient manner, utilizing a green synthesis technique that involved utilizing an extract derived from guava leaves (*Psidium Guajava*). Subsequently, an evaluation of the photocatalytic efficiency of these nanoparticles was conducted under sunlight exposure to degrade Reactive Yellow 186 dye [74]. Through a wet chemical co-precipitation technique, bismuth-doped tin dioxide (Bi-SnO_2) nanoparticles were synthesized and employed as a photocatalyst to facilitate the removal of Congo Red dye [75].

For the photocatalytic degradation of anthraquinone-derived nuclear Fast Red dye, orthorhombic and tetragonal lead oxide nanoparticles were used [76]. Bismuth oxychloride, synthesized via a hydrothermal process, demonstrates photocatalytic activity for the removal of Rhodamine B [77]. Utilizing a bismuth-based MOF (Bi-BTC) as both the substrate and precursor, a hierarchical composite of Co_3O_4-Bi_2O_3 was synthesized through an adsorption-calcination approach which was used for the photocatalytic removal of Rhodamine B [78]. Thallium-doped Tl_4CdI_6 nanostructures were produced using a sonochemical approach, and the results demonstrated the highest degradation achieved for Acid Black 1, reaching 85.7% within 110 min [79].

4.3 *Application of Nanomaterials of* d-*Block Transition Metals for Dye Removal*

4.3.1 *3d*-Series Metals

Nanomaterials derived from *3d*-metals are extensively utilized for the purpose of dye degradation in wastewater treatment. TiO_2 nanoparticles have found applications in photocatalytic, antiparasitic, and antimicrobial purposes. While these nanoparticles can be synthesized by various physicochemical approaches, their applicability in the medical field is constrained due to the necessity of elevated temperatures, increased pressures, and the use of hazardous chemicals. Consequently, there is a growing trend toward eco-friendly production of titanium oxide using diverse sources such as plant extracts, biological substances, and microbes [80]. A similar method of eco-friendly production of mesoporous TiO_2 using *Azadirachta indica* leaves was reported by Dash et al. [81]. These mesoporous TiO_2 obtained by the calcination of *Azadirachta indica* leaves show 98% photodegradation of Rhodamine 6 G dye. Moreover, Vanadium pentoxide shows excellent photocatalytic properties due to its suitable band gap, high surface area, and appropriate stability with low-temperature needs. These properties increase their applicability in various fields such as photocatalysis, supercapacitors, electrochromic devices, and gas sensors. To illustrate its photocatalysis properties for the removal of Methylene Blue dye [82], prepared vanadium pentoxide (V_2O_5) nanoparticles by using the hydrothermal technique, which shows 68.58% degradation within 80 min using a very small catalyst quantity. Moreover, a sol–gel-synthesized V_2O_5 material doped with nickel was reported for wastewater remediation purposes. The incorporation of nickel dopants was employed to control the electron–hole pairs recombination rate. This resulted in the material exhibiting a remarkable 100% efficiency in photodegrading Rhodamine dye [83]. Sustainable and eco-friendly advanced oxidation processes (AOPs) are the primary techniques for eradicating organic pollutants from wastewater. In a recent study, a deposition precipitation technique was utilized to fabricate chromium-doped cerium oxide nanostructures. The photocatalytic characteristics of the catalyst, particularly in the degradation of Methylene Blue (MB) dye, were enhanced with increasing dopant concentration, as it helps in decreasing the band gap of CeO_2 [84]. In 2020, Qamar and colleagues developed a hybrid catalyst known as chromium-doped ZnO dispersed onto g-C_3N_4 nanosheets, designed for the photocatalytic decomposition of Methylene Blue dye [85]. The photocatalyst, when exposed to visible light, achieves a remarkable 93% degradation capacity within 90 min. The utilization of medicinal plants in the synthesis of various metal nanoparticles is considered eco-friendliness, facile, and cost-efficient. Similarly, manganese nanoparticles were biosynthesized using *Cinnamomum verum's* bark and manganese acetate as precursors. These nanoparticles show 87.5% photocatalytic degradation toward Congo Red dye and also show good antimicrobial activity toward bacteria [86]. Moreover, manganese dioxide (MnO_2) nanoparticles fabricated on activated carbon obtained through peanut shells were synthesized by a chemical reduction process. These nanoparticles show improved dye removal efficiency in comparison to manganese oxide nanoparticles toward Congo Red. The Congo Red degradation efficiency was reported 98.53% within 5 min under UV–visible radiation, whereas manganese oxide nanoparticles show 56.78% dye removal capacity [87]. Bhuiyan et al. [88] adopted a greener approach for the removal of azo dyes. They synthesize iron oxide (Fe_2O_3) nanoparticles using ferric chloride as a precursor and *Carica papaya* leaf extract as a stabilizing agent in the nanoparticle synthesis process. The photocatalytic removal of the textile dye (Reactive Blue) was also shown by the Fe_2O_3 nanoparticles obtained through the seeds extract of pomegranate. The seed extract functions exceptionally well as a stabilizing agent and redox mediator; this is assigned to the presence of secondary metabolites within the plants [89]. The photocatalytic removal of Malachite Green dye was obtained by oleic and citric acid-designed Cobalt oxide photocatalyst. The sol–gel synthesized cobalt nanoparticles altered with citric acid show a greater chelating effect thus showing 91.2% photocatalytic degradation, while oleic acid-altered cobalt nanoparticles show 66.6% degradation toward Malachite Green dye [90]. Moreover, a nanoadsorbent known as fulvic acid-modified cobalt ferrite ($CoFe_2O_4$) was synthesized by co-precipitation process, to eliminate Remazol Red dye. Under optimized conditions, the fulvic acid-altered cobalt ferrite shows 99.5% dye removal capacity. The fulvic acid route of synthesis enhances both the adsorption potential of the adsorbent and its ability for uptake during recycling [91]. Nickel oxide nanoparticles are employed for the photocatalytic degradation of azo dyes, because of their semiconductor and non-poisonous nature, excellent chemical stability, and strong catalytic efficiency [92]. The nickel-incorporated reduced graphene oxide (Ni-RGO) nanocomposites were reported by [93], for adsorbing of Rhodamine B dye. The prepared adsorbent shows good regeneration and reusability with efficiency maintained for more than 6 cycles. However, a decrease in efficiency to 85% was observed after the first recycling, attributed to the clustering of the nanoparticles. Moreover, biosynthesized copper nanoparticles from bacterial sources show an environment-friendly nature, cost-effectiveness, and excellent biological and physical–chemical properties which is attributed to its photocatalytic removal capacity toward the removal of azo dyes. The prepared Copper nanoparticles showed 97.07% removal efficiency for Congo Red, 90.55% for Malachite Green, and 88.42% and 83.61% for Direct Blue-1 and Reactive Black, respectively [94]. The copper sulfide (CuS) nanoparticles show distinctive characteristics such as low cost, strong chemical stability, great biocompatibility, and impressive electrical and optical properties, which makes Copper sulfide an exceptional photocatalyst for the removal of several dyes. The decolorization of a trinary combination of Crystal Violet, Rhodamine B, and Methylene Blue dye was done by copper sulfide (CuS_3) nanoparticles synthesized by a single precursor source (Cu (II) dithiocarbamate). The smallest nanoparticles of CuS_3 show 100% dye removal efficiency toward Methylene Blue, 85% for Crystal Violet, and 81% for Rhodamine B [95]. The nanoparticles of ZnO show high degradation ability and are inexpensive and eco-friendly in nature. There are several approaches and methods available for the production of ZnO nanoparticles, but the biosynthesis approach restricts the use of harmful surfactants and chemicals and is thus preferable for the synthesis of nanoparticles. Rupa et al. [96] reported a biosynthesis approach that employed buckthorn fruit for generating ZnO nanoparticles through the co-precipitation method. It shows 99% removal capacity toward Malachite Green, Congo Red, Methylene Blue, and Eosin Y dyes.

4.3.2 *4d*-Series Metals

The incorporation of different metals offers distinct advantages, such as enhancing porosity, increasing surface area, and improving adsorption and photocatalytic performance. As a result, a number of doped-metal oxides belonging to *4d*-series elements have been synthesized and explored for their dye removal efficiency. The nanoparticles of yttrium vanadate (YVO_4) were investigated for photocatalytic purposes toward Methylene Blue dye and phenol. It displays 98.5% decolorization of Methylene Blue dye and 93.2% photocatalytic decomposition of phenol [97]. Yttrium oxide-doped zinc oxide (Y_2O_3-doped ZnO) was fabricated using the ball milling technique and investigated for the removal of Basic Fuchsin dye. It shows 75.33 mg/g removal efficiency and is also reusable [98]. Studies have highlighted a significant enhancement in the adsorption capacity of biochar, which can be effectively increased when it is modified with different metal-based nanomaterials.

Biochar has high thermal and chemical stability with diverse possibilities of structure. The biochar-supported $ZrFe_2O_5$ nanocomposites show significant adsorption potential for the elimination of Tartrazine dye, showcasing notable reusability for up to five cycles, with only a 26% reduction in efficiency from 89 to 63% [99]. Furthermore, a novel adsorbent named zirconium-organic framework nanomaterial was reported for the elimination of both cationic (Methylene Blue) and anionic (Methyl Orange) dyes. The high stability is provided by the Zr_6 clusters in the Zr-based metal–organic framework and also contains defects in their structure, which can be altered by several methods for better removal efficiency toward organic contaminants [100]. The combination of different semiconductors has been used for photocatalytic purposes, therefore, the Ni-modified mesoporous niobium was synthesized by using the oxides of niobium (n-type semiconductor with the nickel oxide (p-type semiconductor by using evaporation induced self-assembly method. These nanostructures decolorize the Indigo Carmine dye effectively degraded through photocatalysis and photo-Fenton-like reaction mechanisms [101]. Molybdenum has been found in various forms of oxide stoichiometries and has poly morphology. Molybdenum is recognized to exist in several prominent phases, including α-MoO_3 (orthorhombic), β-MoO_3 (monoclinic), h-MoO_3 (hexagonal), and MoO_3 (II) (monoclinic under high pressure), among all of these α-MoO_3 exhibits excellent catalytic and electrochemical activities. The α-MoO_3 (molybdenum trioxide) nanosorbent was reported by Rakass et al. [102], for the 100% efficient degradation at pH 11 of Methylene Blue dye from wastewater. Subsequently, in 2022, Wang and colleagues explored the adsorption properties of molybdenum disulfide (MoS_2) adsorbents synthesized through a hydrothermal method in the context of Methylene Blue dye [103]. The MoS_2 nanostructures possess 93.47% removal capacity and show a change of 99.5% to 89.34% when recycled 5 times, demonstrating good regeneration and recyclability. The assessment of Methylene Blue dye degradation was performed using RuO_2 nanoparticles that were biosynthesized through the leaf extracts of *Moringa oleifera* and *Catharanthus roseus*. These nanoparticles are highly stable at high temperatures. *Catharanthus roseus*-derived nanoparticles show 79.02% photocatalytic activity with greater antibacterial and antioxidant activities than the *Moringa oleifera*-derived nanoparticles [104]. In another approach, ruthenium nanoparticles anchored onto porous silicon (Si@RuNP) were synthesized using an acid etching method followed by a galvanic displacement process for efficient removal of azo dyes [105].

Moreover, the palladium nanoparticles obtained from agricultural waste (peels of cotton bolls) were used for the removal of azo dyes including Methyl Orange, Congo Red, Sunset Yellow, and Tartrazine dye by using $NaBH_4$, functioning as an electron donor. These peel extract-mediated palladium nanoparticles show a 95–97% reduction of azo dyes in 14 min [106]. Later on [107], synthesized green nanoparticles of palladium by using glycerol as an environment-friendly reducing agent and maghemite (γ-Fe_2O_3) as a supporting material. The Pd/γ-Fe_2O_3 shows 90% efficient elimination of Methyl Orange dye through both the Fenton process and the adsorption method. Silver nanoparticles (Ag-NPs) have also been used for the removal of organic dyes. The flower extract of *Thunbergia grandiflora* provides a low-cost and simple synthesis route for the production of silver nanoparticles. These Ag-NPs show catalytic degradation toward Congo Red dye. Similarly, Silver nanoparticles were produced via a green synthesis method employing an extract derived from *Viburnum opulus* fruit, as this fruit extract serves as a bioactive component source, which acts as a reducing and stabilizing agent throughout the synthesis. These Ag-NPs efficiently remove Carmoisine, Tartrazine, and Brilliant Blue FCF dye, from wastewater, and thus can be very useful for environmental remediation [108].

The catalyst's specific structure significantly influences the catalytic reaction, making it sensitive to morphology variations. The crystal plane and morphology of hydrothermally synthesized CdS nanostructures play a crucial role in amplifying their catalytic reactivity. A straightforward and environmentally friendly hydrothermal process was employed to create CdS nanostructures with petal and flower-like morphologies. These structures exhibited substantial photocatalytic effectiveness against Rhodamine B dye. Specifically, 78% photocatalytic degradation was shown by nanoflower morphology and 63% dye removal capacity by nanometal CdS catalyst for Rhodamine B dye [109]. In the context of doping, a variety of nonmetals and metals have been employed. Among these, cadmium demonstrates notable synergistic effects when doped with WO_3 and g-C_3N_4, particularly in the degradation of various organic dyes. Doping cadmium leads to a reduction in the bandgap, resulting in significantly improved visible light catalytic activity [110].

4.3.3 *5d*-Series Metals

Traditionally, titanium and zirconium oxides have been recognized as conventional semiconductors within the Group IV elements for photocatalytic activity. However, recent research has highlighted the exceptional semiconducting properties of hafnium oxide (HfO_2), positioning it alongside conventional metal oxide semiconductors like titanium and zirconium.

A basic co-precipitation process was used for the synthesis of HfO_2-graphene nanocomposites, using different (10, 30, 50) wt% loading of graphene. Among 3 wt%, 50 wt% HfO_2-graphene exhibits 99% photocatalytic dye removal activity toward Methylene Blue. Additionally, it demonstrated a notable 81% efficiency in removing Tetracycline dye within 180 min [111].

Tantalum oxide (Ta_2O_5) acts as an n-type semiconductor due to its substantial band gap energy and elevated refractive index. Furthermore, these properties can be tailored using diverse optical adsorption techniques. Nanoparticles of tantalum oxide (Ta_2O_5) synthesized through a sol–gel process using tantalum alkoxy carboxylate exhibit exceptional photocatalytic efficacy in eliminating Rhodamine dye under UV–Vis irradiation [112]. Extensive research has been conducted on semiconductors for photocatalysis, revealing metal tungstates as promising candidates due to their structural versatility. Metal tungstates, generally represented by the formula MWO4, can be categorized into wolframite and scheelite types, depending on the bivalent cation "M" (e.g., Cu, Zn, Co, Mn, Cd, etc.). Among these, cobalt tungstate ($CoWO_4$) stands out as a notable p-type semiconductor, showcasing wide-ranging applicability across electrical, optical, magnetic, and catalytic properties. The desirable phase composition and notable chemical stability of $CoWO_4$ make it an excellent photocatalyst for eliminating Brilliant Green dye, as it is capable of removing 94% of dye within 120 min [113].

Osmium in a zero-oxidation state is highly unstable in its aqueous solution due to its reactivity upon exposure to atmospheric oxygen. Consequently, to prevent oxidation of osmium (Os^0), different stabilizing agents like surfactants and polymers are utilized. Various materials, such as zeolite, reduced graphene oxide (rGO), silica, and carbon nanomaterials, have been employed as supportive matrices for osmium (Os^0). Alloyed nanoparticles like tellurium-osmium and osmium-platinum exhibit improved charge separation capacity and strong interaction between metals, making them excellent catalysts for organic contaminant breakdown [114]. Tungsten oxide is recognized for its narrow band gap, expansive surface area, and strong thermal stability, making it highly versatile in applications such as photocatalysis, biological activities, and gas sensors. However, the electron–hole pairs separation is challenging due to strong Coulombic forces in a single semiconductor catalyst. Therefore, the introduction of different metals through doping has emerged as a viable approach to narrow the band gap, leading to improved photocatalytic performance. Iridium, in particular, has gained significant interest as a dopant metal due to its ability to enhance photocatalytic activities. In a study by Dhanalakshmi et al. [115], iridium-loaded tungsten oxide (Ir/WO_3) was used for the photocatalytic elimination of Methylene Blue and Crystal Violet dyes from wastewater.

Iridium nanoparticles were utilized alongside hexacyanoferrate ions for the oxidative degradation of the azo dye Orange G. The addition of iridium nanocatalyst resulted in a proportional enhancement in the degradation of the dye. This catalytic process effectively transformed the Orange G dye into less harmful and simpler compounds, showcasing the catalytic potential of iridium. Moreover, the catalyst's presence accelerated the reaction rate by four to five times in comparison to the non-catalyzed reaction [116].

Platinum is known for its ability to reduce the recombination rate of electron–hole pairs, thereby enhancing photocatalysis rates. Hence, nanoparticles composed of cupric oxide silicon dioxide doped with platinum ($Pt\text{-}CuO\text{-}SiO_2$), synthesized through a sol–gel process, were employed to degrade the Acridine Orange dye. The photocatalyst, doped with 3 wt% platinum, demonstrated complete removal capacity for the Acridine and exhibited stability in photocatalysis throughout five cycles [117]. The process of synthesizing gold nanoparticles (Au NPs) through plant mediation has gained immense popularity due to its simplicity and highly effective results in degrading Methylene Blue dye in wastewater. In the phytogenic production of gold nanoparticles, the reducing, stabilizing, and capping factors are sourced from natural origins, making the process non-toxic and cost-effective. Phytochemical constituents like alkaloids, flavonoids, terpenoids, carotenoids, polyphenols, amino acids, and more found within the plant show vital importance in the synthesis of these gold (Au) nanoparticles [118].

4.4 f-*Block-Based Nanomaterials Employed for the Elimination of Dyes from Wastewater*

The rare earth metals, which belong to the *f*-block, possess unique electrical structures that render them interesting candidates for a range of uses, including water filtration [119, 120]. In recent years, there has been growing attention toward rare earth metal oxide nanostructures, largely driven by their exceptional physicochemical attributes. Their remarkable optical, magnetic, and electrical attributes arise from the partially filled 4f orbitals [121] prompting detailed research into their utilization in diverse applications such as memory, logic device switching mechanisms, and optoelectronic devices. Rare earth oxides also exhibit exciting catalytic activity, high surface basicity, and quick oxygen ion mobility, which makes sense given that metal catalysts encompass a wide range of transition metals and other elements that exhibit catalytic activity [122, 123]. Moreover, additional strategies such as doping, incorporating structural guiding agents, and adjusting pH have been employed to enhance the photo-response properties. For instance, samarium (Sm), a rare earth metal, can be doped to modify the electrical, optical, magnetic, and photoresponsive properties of the host material, potentially leading to the formation of oxygen vacancies [124]. Several studies have been conducted utilizing metal-doped Lanthanides nanoparticles and Lanthanides to remove dye from wastewater. The photocatalytic effectiveness of CeO_2 nanocrystals for degrading Methylene Blue has been reported by Pouretedal and Kadkhodaie [125]. The enhanced catalytic performance of smaller particles was evident in the photocatalytic decomposition of an Acridine Orange solution in water, using CeO_2 nanoparticles under UV light [126]. According to research, the increased calcination temperature caused the CeO_2 nanoparticles' absorption maxima to shift toward the blue and their band gap to drop as a result of the larger particle size. In turn, this improved their capacity for Congo Red photocatalytic degradation [127]. The utilization of CeO_2 nanoparticles for the photocatalytic decomposition of Rhodamine B dye utilizing natural light was recently described by Akbari-Fakhrabadi et al. [128]. Shahin Ahmadi and their research team conducted a study focusing on the elimination of AB92 dye from aqueous solutions through the adsorption process using Nd_2O_3 nanoparticles. Ni-doped TiO_2, synthesized at the nanoscale through hydrothermal methods, exhibited a 63% increase in photocatalytic efficiency for the degradation of Methylene Blue under visible light [129]. According to a research study, the utilization of an alternative impregnation technique in the production of nickel-doped titanium dioxide nanorods resulted in enhanced photocatalytic performance for Methylene Blue under visible light exposure [130]. Due to the substantial oxygen mobility within the doped CeO_2 lattice, researchers have investigated Ni-doped CeO_2 nanoparticles and observed that they display a modest level of ferromagnetism at room temperature, as reported by Ranjith et al. [131] and Bear et al. [132]. In a recent study conducted by Ali and colleagues in 2018, we explored the photocatalytic performance of CeO_2 doped with Ni (Ni-doped CeO_2) in the context of decomposing the azo dye Tartrazine under both visible and UV light [133]. We have previously reported on the photocatalytic capabilities of both pristine and Mn-doped CeO_2 nanoparticles in the context of degrading the anthraquinone dye Remazol Brilliant Blue R when exposed to visible and UV irradiation [134]. Sm-Zn ferrite was synthesized by the co-precipitation process and demonstrated efficient reduction of Methyl Orange dye when exposed to sunlight [135]. Samarium (Sm)-doped zinc ferrite ($ZnFe_2O_4$) was reported by El-Kholy et al. [136], for increased photocatalytic removal of hazardous dye from aquatic systems. El-Kholy et al. [137], investigated the impact of the synthesized pure TiO_2 and (Er^{3+}/TiO_2) nanocomposite catalysts on the processes of adsorption, mineralization, and degradation of Methylene Blue dye. The photocatalytic effectiveness decreases in the following order: Er-doped ZnO (94.32%) is at the top, followed by Ce-doped ZnO (93.81%), La-doped ZnO (72.16%), and Pr-doped ZnO (67.67%). The Ce and Er-doped ZnO photocatalysts have demonstrated effectiveness in inducing organic dye oxidation in wastewater, making them interesting candidates for future solar-driven photocatalysts [138].

Based on the aforementioned studies, it can be inferred that f-block-based nanomaterials derived from rare earth metals offer a promising solution to the pressing issue of dye pollution in industrial effluents. Due to their unique properties and tunability, they represent a valuable resource in the quest for effective and sustainable water treatment methods. With further research and development, these nanoparticles hold significant potential to revolutionize the wastewater treatment industry.

5 Conclusion

Dyes are extensively employed in various industries in large quantities. However, a considerable portion of these dyes find their way into the environment through industrial effluents, raising concerns about their proper removal. In response, numerous approaches and methodologies have been embraced for their effective disposal and removal from wastewater and water bodies. Non-destructive conventional methods such as coagulation, membrane separation, and adsorption are commonly employed, alongside destructive methods like biological treatments and oxidation processes. Among these, adsorption and catalytic oxidation methods have been widely utilized in water treatment. Several adsorbents have been used for dye removal and water treatment which include nature-based materials such as zeolites, chitosan, and agricultural waste. Conventional methods have inherent limitations such as low efficiency, high energy input, non-economic, and high specificity, which have prompted researchers to explore nanomaterials. Several nanomaterials and nanocomposites have been developed and applied as adsorbents, catalysts, and photocatalysts for the degradation of dyes. Metals from all over the periodic tables have been used for these synthesizes with varying degrees. The *s*-block metals are explored as dopants and matrix providers for doping due to their flexible lattice framework. However, alkali metals exhibit high reactivity toward oxygen and moisture, making them seldom used as a lattice material for nanomaterial production. Rb and Cs are the main metals used for nanomaterial synthesis among alkalis. Alkaline earth metals are used for the synthesis of nanomaterials more often than their alkali counterparts. Nanomaterials viz. MgO and CaO synthesized by different routes have been employed for dye removal. Also, several metals have been used as dopants to improve the adsorptive and photocatalytic potential of nanomaterials. Al_2O_3, Si_3N_4, SiO_2, GeO_2, Ga_2O_3, and SnO_2 are some of the *p*-block element compounds which have been explored for their dye removal potential as adsorbent or photocatalysts independently or as mixed metal compounds. The *d*- and *f*-blocks elements have been extensively used as single, doped, or mixed metal oxides. TiO_2 nanoparticles are one of the most popular materials from the *3d*-series, extensively explored for their photocatalytic potential. MnO_2, Fe_2O_3, $CoFe_2O_4$, and NiO are some other compounds which have been used as such or incorporated with other materials to enhance their potential. YVO_4, Y_2O_3-doped ZnO, zirconium-organic framework nanomaterial, MoO_3, palladium nanoparticles, and CdS nanostructures are *4d*-series compounds which have been reported as efficient nanomaterials either as adsorbent or photocatalysts toward the removal of dyes from water. HfO_2, Ta_2O_5, Metal tungstate, iridium nanoparticles, platinum and doped Pt, and gold nanoparticles show excellent photocatalytic potential and have been found as efficient materials for the removal of dyes from wastewaters. Apart from these materials, rare earth metals have attracted the interest of researchers due to their distinct optical, magnetic, and electrical properties arising out of the partially filled 4 *f*-orbitals. These metals have been explored as dopants and found to alter the electrical, optical, magnetic, and photoresponsive properties of the host. CeO_2 nanoparticles are the most explored materials used as photocatalysts. A number of dopants from all over the periodic table have been studied to enhance the photocatalytic potentials of these nanomaterials. These substances have demonstrated varying levels of effectiveness in eliminating different dyes from wastewater, under diverse reaction conditions.

The nanomaterials investigated in this chapter have been found effective against many stable, recalcitrant dyes including azo and anthraquinone dyes under different conditions. Still, there are opportunities in the field of nanocatalysis/photocatalysis considering the size and shape-dependent properties of the nanomaterials. For example, the literature review reveals that the nanomaterials derived from precious *4d*-metals like rhenium (Rh), osmium (Os), iridium (Ir), and platinum (Pt) have not received comprehensive research attention for their potential in polluted water treatment processes, especially concerning the removal of dyes.

References

1. Pamuk O (2010) My name is red: written and introduced by Orhan Pamuk. Everyman's Library
2. Hassan MM, Alam MZ, Anwar MN (2013) Biodegradation of textile Azo dyes by bacteria isolated from dyeing industry effluent. Int Res J Biol Sci 2:27–31
3. Saini RD (2017) Textile organic dyes: polluting effects and elimination methods from textile waste water. Int J Chem Eng Res 9:121–136
4. Ali SR, Kumar R, Kadabinakatti SK, Arya MC (2019) Enhanced UV and visible light—driven photocatalytic degradation of tartrazine by nickel-doped cerium oxide nanoparticles. Mater Res Express 6:025513. https://doi.org/10.1088/2053-1591/aaee44
5. Liu Q (2020) Pollution and treatment of dye waste water. IOP Conf Ser.: Earth Environ Sci 514:052001
6. Benkhaya S, M'rabet S, El Harfi A (2020) Classifications, properties, recent synthesis and applications of azo dyes. Heliyon 6(1):e03271. https://doi.org/10.1016/j.heliyon.2020.e03271
7. Kunz A, Peralta-Zamora P, Moraes SG, Durán N (2002) Degradation of reactive dyes by the system metallic iron/hydrogen peroxide. Quim Nova 25:78–81
8. Gupta VK, Ali I, Saleh TA, Nayak A, Agarwal S (2012) Chemical treatment technologies for waste-water recycling—an overview. RSC Adv 2(16):6380–6388
9. El-Kammah M, Elkhatib E, Gouveia S, Cameselle C, Aboukila E (2022) Enhanced removal of Indigo Carmine dye from textile effluent using green cost-efficient nanomaterial: adsorption, kinetics, thermodynamics and mechanisms. Sustain Chem Pharm 29:100753
10. Hussain T, Faisal S, Rizwan M, Zaman N, Iqbal M, Iqbal A, Ali Z (2022) Green synthesis and characterization of copper and nickel hybrid nanomaterials: investigation of their biological and photocatalytic potential for the removal of organic crystal violet dye. J Saudi Chem Soc 26(4):101486
11. Mohd S, Wani AA, Khan AM (2022) ZnO/POA functionalized metal-organic framework ZIF-8 nanomaterial for dye removal. Clean Chem Eng 3:100047
12. Berradi M, Hsissou R, Khudhair M, Assouag M, Cherkaoui O, El Bachiri A, El Harfi A (2019) Textile finishing dyes and their impact on aquatic environs. Heliyon 5(11)
13. Sharma G, Kumar A, Sharma S, Naushad M, Dhiman P, Vo DVN, Stadler FJ (2020) Fe3O4/ZnO/Si3N4 nanocomposite based photocatalyst for the degradation of dyes from aqueous solution. Mater Lett 278:128359. https://doi.org/10.1016/j.matlet.2020.128359
14. Mehra S, Singh M, Chadha P (2021) Adverse impact of textile dyes on the aquatic environment as well as on human beings. Toxicol Int 28(2):165
15. Hamdaoui O (2011) Intensification of the sorption of Rhodamine B from aqueous phase by loquat seeds using ultrasound. Desalination 271:279–286
16. Imam SS, Babamale HF (2020) A short review on the removal of rhodamine B dye using agricultural waste-based adsorbents. Asian J Chem Sci 7:25–37
17. Benbrika C, Menasra H, Kularkar A, Smaili L, Sbaihi A (2024) Enhanced photocatalytic degradation of rhodamine B using Fe/Ba-doped Bi4Ti3O12 nanostructures: mechanism and performance evaluation. J Phys Chem Solids 184:111702. https://doi.org/10.1016/j.jpcs.2023.111702
18. Vuˇcurovi´c VM, Razmovski RN, Milji´c UD, Puˇskaˇs VS (2014) Removal of cationic and anionic azo dyes from aqueous solutions by adsorption on maize stem tissue. J Taiwan Inst Chem Eng 45:1700–1708
19. Goscianska J, Ptaszkowska M, Pietrzak R (2015) Equilibrium and kinetic studies of chromotrope 2R adsorption onto ordered mesoporous carbons modified with lanthanum. Chem Eng J 270:140–149
20. Hussain B, Yaseen H, Khalid-Al-Ghanim, Al-Misned F, Qasim M, Al-Mulhm N, Mahboob S (2021) A study on risk assessment of the effect of hematoxylin dye on cytotoxicity and nephrotoxicity in freshwater fish: food and water security prospective research. Saudi J Biol Sci 28(4):2267–2271. https://doi.org/10.1016/j.sjbs.2021.01.019

21. Wu L, Xu Y, Lv X, Chang X, Ma X, Tian X, ... Kong X (2021) Impacts of an azo food dye tartrazine uptake on intestinal barrier, oxidative stress, inflammatory response and intestinal microbiome in crucian carp (Carassius auratus). Ecotoxicol Environ Saf 223:112551
22. Garcia VSG, de Freitas Tallarico L, Rosa JM, Suzuki CF, Roubicek DA, Nakano E, Borrely SI (2021) Multiple adverse effects of textile effluents and reactive Red 239 dye to aquatic organisms. Environ Sci Pollut Res 28:63202–63214
23. de Jesus Azevedo CC, de Oliveira R, Suares-Rocha P, Sousa-Moura D, Li AT, Grisolia CK, ... Montagner CC (2021) Auramine dyes induce toxic effects to aquatic organisms from different trophic levels: an application of predicted non-effect concentration (PNEC). Environ Sci Pollut Res 28:1866–1877
24. Iwuozor KO (2019) Prospects and challenges of using the coagulation-flocculation method in the treatment of effluents. Adv J Chem-Section A 2(2):105–127
25. Chen W, Mo J, Du X, Zhang Z, Zhang W (2019) Biomimetic dynamic membrane for aquatic dye removal. Water Res 151:243–251
26. Moosavi S, Lai CW, Gan S, Zamiri G, Akbarzadeh Pivehzhani O, Johan MR (2020) Application of efficient magnetic particles and activated carbon for dye removal from wastewater. ACS Omega 5(33):20684–20697
27. Abd-Elhamid AI, Kamoun EA, El-Shanshory AA, Soliman HM, Aly HF (2019) Evaluation of graphene oxide-activated carbon as effective composite adsorbent toward the removal of cationic dyes: composite preparation, characterization and adsorption parameters. J Mol Liq 279:530–539
28. Firouzjaei MD, Afkhami FA, Esfahani MR, Turner CH, Nejati S (2020) Experimental and molecular dynamics study on dye removal from water by a graphene oxide-copper-metal organic framework nanocomposite. J Water Process Eng 34:101180
29. Sarabadan M, Bashiri H, Mousavi SM (2019) Removal of crystal violet dye by an efficient and low-cost adsorbent: modeling, kinetic, equilibrium and thermodynamic studies. Korean J Chem Eng 36:1575–1586
30. Tanhaei B, Ayati A, Iakovleva E, Sillanpää M (2020) Efficient carbon interlayed magnetic chitosan adsorbent for anionic dye removal: synthesis, characterization and adsorption study. Int J Biol Macromol 164:3621–3631
31. Ayele A, Getachew D, Kamaraj M, Suresh A (2021) Phycoremediation of synthetic dyes: an effective and eco-friendly algal technology for the dye abatement. J Chem 2021:1–14
32. Bayramoglu G, Yilmaz M (2018) Azo dye removal using free and immobilized fungal biomasses: isotherms, kinetics, and thermodynamic studies. Fibers Polym 19:877–886
33. Deniz F, Ersanli ET (2016) Removal of colorant from simulated wastewater by phyco-composite material: equilibrium, kinetic and mechanism studies in a lab-scale application. J Mol Liq 220:120–128
34. Kadhom M, Albayati N, Alalwan H, Al-Furaiji M (2020) Removal of dyes by agricultural waste. Sustain Chem Pharm 16:100259
35. Ma L, Jiang C, Lin Z, Zou Z (2018) Microwave-hydrothermal treated grape peel as an efficient biosorbent for methylene blue removal. Int J Environ Res Public Health 15(2):239
36. Arora C, Soni S, Sahu S, Mittal J, Kumar P, Bajpai PK (2019) Iron based metal organic framework for efficient removal of methylene blue dye from industrial waste. J Mol Liq 284:343–352
37. Samimi M, Shahriari-Moghadam M (2023) The Lantana camara L. stem biomass as an inexpensive and efficient biosorbent for the adsorptive removal of malachite green from aquatic environments: kinetics, equilibrium and thermodynamic studies. Int J Phytoremediat 1–9
38. Ali NS, Jabbar NM, Alardhi SM, Majdi HS, Albayati TM (2022) Adsorption of methyl violet dye onto a prepared bio-adsorbent from date seeds: isotherm, kinetics, and thermodynamic studies. Heliyon 8(8)
39. Ali H (2010) Biodegradation of synthetic dyes—a review. Water Air Soil Pollut 213:251–273
40. Babu SS, Mohandass C, Vijayaraj AS, Dhale MA (2015) Detoxification and color removal of Congo Red by a novel Dietzia sp. (DTS26)–a microcosm approach. Ecotoxicol Environ Saf 114:52–60

41. Chequer FMD, Dorta DJ, de Oliveira DP (2011) Azo dyes and their metabolites: does the discharge of the azo dye into water bodies represent human and ecological risks. Adv Treat Text Effl 48:28–48
42. Buntić AV, Pavlović MD, Antonović DG, Šiler-Marinković SS, Dimitrijević-Branković SI (2017) A treatment of wastewater containing basic dyes by the use of new strain Streptomyces microflavus CKS6. J Clean Prod 148:347–354
43. Mustafa MM, Jamal P, Mahmod SS, Jimat DN, Ilyas NN (2017) Panus tigrinus as a potential biomass source for reactive blue decolorization: isotherm and kinetic study. Electron J Biotechnol 26:7–11
44. Al Farraj DA, Elshikh MS, Al Khulaifi MM, Hadibarata T, Yuniarto A, Syafiuddin A (2019) Biotransformation and detoxification of antraquione dye green 3 using halophilic Hortaea sp. Int Biodeter Biodegr 140:72–77
45. Adenan NH, Lim YY, Ting ASY (2020) Discovering decolorization potential of triphenylmethane dyes by actinobacteria from soil. Water Air Soil Pollut 231(12):560
46. Ma D, Yi H, Lai C, Liu X, Huo X, An Z, ... Yang L (2021) Critical review of advanced oxidation processes in organic wastewater treatment. Chemosphere 275:130104
47. Wang J, Yue M, Han Y, Xu X, Yue Q, Xu S (2020) Highly-efficient degradation of triclosan attributed to peroxymonosulfate activation by heterogeneous catalyst g-C3N4/$MnFe_2O_4$. Chem Eng J 391:123554
48. Rodríguez-Narváez OM, Picos AR, Bravo-Yumi N, Pacheco-Alvarez M, Martínez-Huitle CA, Peralta-Hernández JM (2021) Electrochemical oxidation technology to treat textile wastewaters. Curr Opin Electrochem 29:100806
49. Karami M, Ghanbari M, Amiri O, Salavati-Niasari M (2020) Enhanced antibacterial activity and photocatalytic degradation of organic dyes under visible light using cesium lead iodide perovskite nanostructures prepared by hydrothermal method. Sep Purif Technol 253:117526. https://doi.org/10.1016/j.seppur.2020.117526
50. Dhiman S, Gupta B (2021) Co3O4 nanoparticles synthesized from waste Li-ion batteries as photocatalyst for degradation of methyl blue dye. Environ Technol Innov 23:101765. https://doi.org/10.1016/j.eti.2021.101765
51. Behura R, Sakthivel R, Das N (2021) Synthesis of cobalt ferrite nanoparticles from waste iron ore tailings and spent lithium-ion batteries for photo/sono-catalytic degradation of Congo red. Powder Technol 386:519–527. https://doi.org/10.1016/j.powtec.2021.03.066
52. Naseem K, Ali F, Tahir MH, Afaq M, Yasir HM, Ahmed K, Aljuwayid AM, Habila MA (2022) Investigation of catalytic potential of sodium dodecyl sulfate stabilized silver nanoparticles for the degradation of methyl orange dye. J Mol Struct 1262:132996. https://doi.org/10.1016/j.molstruc.2022.132996
53. Wang KL, Li Y, Sun T, Mao F, Wu JK, Xue B (2019) Ultrafine silver nanoparticles deposited on sodium-doped graphitic carbon nitride towards enhanced photocatalytic degradation of dyes and antibiotics under visible light irradiation. Appl Surf Sci 476:741–748. https://doi.org/10.1016/j.apsusc.2019.01.168
54. Ganesan R, Muralidharan R, Parthipan G, Vinodhini SP, Balasubramani V, Sridhar TM, Arulmozhi R, Leelavathi H (2021) Investigations on caesium-incorporated rubidium tin chloride-defect perovskite nanomaterial as highly efficient ultraviolet photocatalysts. J Mater Sci: Mater Electron 32(20):25409–25424. https://doi.org/10.1007/s10854-021-07001-2
55. Zheng Y, Cao L, Xing G, Bai Z, Huang J, Zhang Z (2019) Microscale flower-like magnesium oxide for highly efficient photocatalytic degradation of organic dyes in aqueous solution. RSC Adv 9(13):7338–7348. https://doi.org/10.1039/c8ra10385b
56. Pachiyappan J, Gnanansundaram N, Sivamani S, Sankari NPBP, Senthilnathan N, Kerga GA (2022) Preparation and characterization of magnesium oxide nanoparticles and its application for photocatalytic removal of Rhodamine B and methylene blue dyes. J Nanomater 2022:1–6. https://doi.org/10.1155/2022/6484573
57. Ameur I, Khantoul AR, Boudine B, Brien V, Horwat D, Sebais M, Halimi O (2022) Effects of erbium incorporation on structural, surface morphology, and degradation of methylene blue dye of magnesium oxide nanoparticles. J Inorg Organomet Polym Mater 33(1):30–46. https://doi.org/10.1007/s10904-022-02482-y

58. Brinza L, Maftei AE, Tascu S, Brinza F, Neamtu M (2022) Advanced removal of Reactive Yellow 84 azo dye using functionalised amorphous calcium carbonates as adsorbent. Sci Reports 12(1). https://doi.org/10.1038/s41598-022-07134-2
59. Jahani D, Nazari A, Ghourbanpour J, Ameli A (2020) Polyvinyl alcohol/calcium carbonate nanocomposites as efficient and cost-effective cationic dye adsorbents. Polymers 12(10):2179. https://doi.org/10.3390/polym12102179
60. Karuppusamy I, Samuel MS, Ethiraj S, Shanmugam S, Kumar PSM, Brindhadevi K, Pugazhendhi A (2021) Ultrasound-assisted synthesis of mixed calcium magnesium oxide (CaMgO2) nanoflakes for photocatalytic degradation of methylene blue. J Colloid Interface Sci 584:770–778. https://doi.org/10.1016/j.jcis.2020.09.112
61. Deshmukh V, Nagaswarupa H, Ravikumar C, Anil Kumar M, Shashi Shekhar T, Ananda Murthy H (2020) Lanthanum doped strontium titanate nanomaterial for photocatalytic and supercapacitor applications. Asian J Chem 32(8):2013–2020. https://doi.org/10.14233/ajchem.2020.22725
62. Kiran KS, Shashanka R, Lokesh SV (2022) Enhanced photocatalytic activity of hydrothermally synthesized perovskite strontium titanate nanocubes. Topics Catal. https://doi.org/10.1007/s11244-021-01558-2
63. Mu'azu ND, Jarrah N, Zubair M, Manzar MS, Kazeem TS, Qureshi A, Haladu SA, Blaisi NI, Essa MH, Al-Harthi MA (2020) Mechanistic aspects of magnetic MgAlNi barium-ferrite nanocomposites enhanced adsorptive removal of an anionic dye from aqueous phase. J Saudi Chem Soc 24(10):715–732. https://doi.org/10.1016/j.jscs.2020.08.001
64. Rahman QI, Ahmad M, Misra SK, Lohani M (2012) Efficient degradation of methylene blue dye over highly reactive Cu Doped Strontium Titanate (SrTiO3) nanoparticles photocatalyst under visible light. J Nanosci Nanotechnol 12(9):7181–7186. https://doi.org/10.1166/jnn.2012.6494
65. Jethave G, Fegade U, Attarde S, Ingle S (2017) Facile synthesis of Lead Doped Zinc-Aluminum Oxide Nanoparticles (LD-ZAO-NPs) for efficient adsorption of anionic dye: kinetic, isotherm and thermodynamic behaviors. J Ind Eng Chem 53:294–306. https://doi.org/10.1016/j.jiec.2017.04.038
66. Pathania D, Katwal R, Kaur H (2016) Enhanced photocatalytic activity of electrochemically synthesized aluminum oxide nanoparticles. Int J Miner Metall Mater 23(3):358–371. https://doi.org/10.1007/s12613-016-1245-9
67. Chauhan P, Bhasin KK, Chaudhary S (2021) High selectivity and adsorption proficiency of surfactant-coated selenium nanoparticles for dye removal application. Environ Sci Pollut Res 28(43):61344–61359. https://doi.org/10.1007/s11356-021-15024-9
68. Nandanwar R, Bamne J, Singh N, Sharma PK, Singh P, Umar A, Haque FZ (2022) Synthesis of Titania/Silica nanocomposite for enhanced photodegradation of methylene blue and methyl orange dyes under UV and mercury lights. ES Mater Manuf. https://doi.org/10.30919/esmm5f628
69. Easwaran G, Packialakshmi JS, Syed A, Elgorban AM, Vijayan M, Sivakumar K, Bhuvaneswari K, Palanisamy G, Lee J (2022) Silica nanoparticles derived from Arundo donax L. ash composite with Titanium dioxide nanoparticles as an efficient nanocomposite for photocatalytic degradation dye. Chemosphere 307:135951. https://doi.org/10.1016/j.chemosphere.2022.135951
70. Kim WJ, Soshnikova V, Markus J, Oh KH, Anandapadmanaban G, Mathiyalagan R, Perez ZEJ, Kim YJ, Yang DC (2019) Room temperature synthesis of germanium dioxide nanorods and their in vitro photocatalytic application. Optik 178:664–668. https://doi.org/10.1016/j.ijleo.2018.09.182
71. Abdullah MM, Albargi HB, Ahmad MZ, Ahmad J (2023) Synthesis and characterization of pure and magnesium doped germanium dioxide (Mg–GeO2) nano crystals for photocatalytic applications. J Nanoelectron Optoelectron 18(2):168–175. https://doi.org/10.1166/jno.2023.3385
72. Ashraf MA, Li C, Zhang D, Fakhri A (2019) Graphene oxides as support for the synthesis of nickel sulfide–indium oxide nanocomposites for photocatalytic, antibacterial and antioxidant performances. Appl Organomet Chem 34(3). https://doi.org/10.1002/aoc.5354

73. Amdeha E, El-Salamony RA, Al-Sabagh AM (2019) Enhancing the photocatalytic activity of Ga2O3–TiO2 nanocomposites using sonication amplitudes for the degradation of Rhodamine B dye. Appl Organometallic Chem 34(2). https://doi.org/10.1002/aoc.5336
74. Kumar M, Mehta A, Mishra A, Singh J, Rawat M, Basu S (2018) Biosynthesis of tin oxide nanoparticles using Psidium Guajava leave extract for photocatalytic dye degradation under sunlight. Mater Lett 215:121–124. https://doi.org/10.1016/j.matlet.2017.12.074
75. Nauman Ali NA, Adnan khan AK, Aaisha Riaz AR, Asiri AM, Tahseen Kamal AMAATK (2020) Photocatalytic performance evaluation of bismuth doped tin-dioxide under UV and direct sunlight irradiation for congo red dye degradation. J Chem Soc Pakistan 42(5):687–687. https://doi.org/10.52568/000689
76. Netzahual-Lopantzi A, García-Nieto E, Juárez-Santacruz L, Romero-Ibarra IC (2023) Photocatalytic evaluation of tetragonal and orthorhombic lead oxide nanoparticles in dye degradation treatment. Appl Phys A 129(8). https://doi.org/10.1007/s00339-023-06871-z
77. Ismail M, Wu Z, Zhang L, Ma J, Jia Y, Hu Y, Wang Y (2019) High-efficient synergy of piezo catalysis and photocatalysis in bismuth oxychloride nanomaterial for dye decomposition. Chemosphere 228:212–218
78. Li Z, Tang X, Huang G, Luo X, He D, Peng Q, Huang J, Ao M, Liu K (2020) Bismuth MOFs based hierarchical Co3O4-Bi2O3 composite: an efficient heterogeneous peroxymonosulfate activator for azo dyes degradation. Separ Purif Technol 242:116825. https://doi.org/10.1016/j.seppur.2020.116825
79. Ghanbari M, Salavati-Niasari M (2018) Tl4CdI6 nanostructures: facile sonochemical synthesis and photocatalytic activity for removal of organic dyes. Inorg Chem 57(18):11443–11455. https://doi.org/10.1021/acs.inorgchem.8b01293
80. Nadeem M, Tungmunnithum D, Hano C, Abbasi BH, Hashmi SS, Ahmad W, Zahir A (2018) The current trends in the green syntheses of titanium oxide nanoparticles and their applications. Green Chem Lett Rev 11(4):492–502
81. Dash L, Biswas R, Ghosh R, Kaur V, Banerjee B, Sen T, ... Haldar KK (2020) Fabrication of mesoporous titanium dioxide using azadirachta indica leaves extract towards visible-light-driven photocatalytic dye degradation. J Photochem Photobiol A: Chem 400:112682
82. Babar BM, Pisal KB, Sutar SH, Mujawar SH, Kadam LD, Pathan HM, ... Patil PS (2022) Hydrothermally prepared vanadium oxide nanostructures for photocatalytic application. ES Energy Environ 15:82–91
83. Rafique M, Hamza M, Shakil M, Irshad M, Tahir MB, Kabli MR (2020) Highly efficient and visible light–driven nickel–doped vanadium oxide photocatalyst for degradation of Rhodamine B Dye. Appl Nanosci 10:2365–2374
84. Habib IY, Burhan J, Jaladi F, Lim CM, Usman A, Kumara NTRN, ... Mahadi AH (2021) Effect of Cr doping in CeO2 nanostructures on photocatalysis and H2O2 assisted methylene blue dye degradation. Catalysis Today 375:506–513
85. Qamar MA, Shahid S, Javed M, Iqbal S, Sher M, Akbar MB (2020) Highly efficient g-C3N4/Cr-ZnO nanocomposites with superior photocatalytic and antibacterial activity. J Photochem Photobiol, A 401:112776
86. Kamran U, Bhatti HN, Iqbal M, Jamil S, Zahid M (2019) Biogenic synthesis, characterization and investigation of photocatalytic and antimicrobial activity of manganese nanoparticles synthesized from Cinnamomum verum bark extract. J Mol Struct 1179:532–539
87. Khan I, Sadiq M, Khan I, Saeed K (2019) Manganese dioxide nanoparticles/activated carbon composite as efficient UV and visible-light photocatalyst. Environ Sci Pollut Res 26:5140–5154
88. Bhuiyan MSH, Miah MY, Paul SC, Aka TD, Saha O, Rahaman MM, ... Ashaduzzaman M (2020) Green synthesis of iron oxide nanoparticle using Carica papaya leaf extract: application for photocatalytic degradation of remazol yellow RR dye and antibacterial activity. Heliyon 6(8)
89. Bibi I, Nazar N, Ata S, Sultan M, Ali A, Abbas A, ... Iqbal M (2019) Green synthesis of iron oxide nanoparticles using pomegranate seeds extract and photocatalytic activity evaluation for the degradation of textile dye. J Mater Res Technol 8(6):6115–6124

90. Verma M, Mitan M, Kim H, Vaya D (2021) Efficient photocatalytic degradation of Malachite green dye using facilely synthesized cobalt oxide nanomaterials using citric acid and oleic acid. J Phys Chem Solids 155:110125
91. Badawy AA, Ibrahim SM, Essawy HA (2020) Enhancing the textile dye removal from aqueous solution using cobalt ferrite nanoparticles prepared in presence of fulvic acid. J Inorg Organomet Polym Mater 30:1798–1813
92. Akbari A, Sabouri Z, Hosseini HA, Hashemzadeh A, Khatami M, Darroudi M (2020) Effect of nickel oxide nanoparticles as a photocatalyst in dyes degradation and evaluation of effective parameters in their removal from aqueous environments. Inorg Chem Commun 115:107867
93. Jinendra U, Bilehal D, Nagabhushana BM, Kumar AP (2021) Adsorptive removal of Rhodamine B dye from aqueous solution by using graphene–based nickel nanocomposite. Heliyon 7(4)
94. Noman M, Shahid M, Ahmed T, Niazi MBK, Hussain S, Song F, Manzoor I (2020) Use of biogenic copper nanoparticles synthesized from a native Escherichia sp. as photocatalysts for azo dye degradation and treatment of textile effluents. Environ Pollut 257:113514
95. Ajibade PA, Oluwalana AE (2021) Enhanced photocatalytic degradation of ternary dyes by copper sulfide nanoparticles. Nanomaterials +11(8), 2000
96. Rupa EJ, Kaliraj L, Abid S, Yang DC, Jung SK (2019) Synthesis of a zinc oxide nanoflower photocatalyst from sea buckthorn fruit for degradation of industrial dyes in wastewater treatment. Nanomaterials 9(12):1692
97. Mohamed RM, Harraz FA (2020) Mechanistic investigation and photocatalytic activity of yttrium vanadate (YVO4) nanoparticles for organic pollutants mineralization. J Market Res 9(3):5666–5675
98. Ben Aissa MA, Khezami L, Taha K, Elamin N, Mustafa B, Al-Ayed AS, Modwi A (2021) Yttrium oxide-doped ZnO for effective adsorption of basic fuchsin dye: equilibrium, kinetics, and mechanism studies. Int J Environ Sci Technol 1–14
99. Perveen S, Nadeem R, Nosheen F, Asjad MI, Awrejcewicz J, Anwar T (2022) Biochar-mediated zirconium ferrite nanocomposites for tartrazine dye removal from textile wastewater. Nanomaterials 12(16):2828
100. Nguyen LHT, Nguyen HTT, Le BQG, Dang MHD, Nguyen TTT, Mai NXD, Doan TLH (2022) Microwave-assisted solvothermal synthesis of defective zirconium-organic framework as a recyclable nano-adsorbent with superior adsorption capacity for efficient removal of toxic organic dyes. Colloid Interface Sci Commun 46:100511
101. Silva IF, Martins AR, Krambrock K, Rosmaninho MG, Binatti I, Moura FC (2020) Understanding photocatalytic activity and mechanism of nickel-modified niobium mesoporous nanomaterials. J Photochem Photobiol, A 388:112168
102. Rakass S, Oudghiri Hassani H, Abboudi M, Kooli F, Mohmoud A, Aljuhani A, Al Wadaani F (2018) Molybdenum trioxide: efficient nanosorbent for removal of methylene blue dye from aqueous solutions. Molecules 23(9):2295
103. Wang X, Zhang P, Xu F, Sun B, Hong G, Bao L (2022) Adsorption of methylene blue on azo dye wastewater by molybdenum disulfide nanomaterials. Sustainability 14(13):7585
104. Anjum SM, Riazunnisa K (2022) Fine ultra-small ruthenium oxide nanoparticle synthesis by using Catharanthus roseus and Moringa oleifera leaf extracts and their efficacy towards in vitro assays, antimicrobial activity and catalytic: adsorption kinetic studies using methylene blue dye. J Cluster Sci 1–15
105. Sahoo A, Patra S (2018) A combined process for the degradation of azo-dyes and efficient removal of aromatic amines using porous silicon supported porous ruthenium nanocatalyst. ACS Appl Nano Mater 1(9):5169–5178
106. Narasaiah BP, Mandal BK (2020) Remediation of azo-dyes based toxicity by agro-waste cotton boll peels mediated palladium nanoparticles. J Saudi Chem Soc 24(2):267–281
107. Freitas NS, Alzamora M, Sanchez DR, Licea YE, Senra JD, Carvalho NM (2021) Green palladium nanoparticles prepared with glycerol and supported on maghemite for dye removal application. J Environ Chem Eng 9(1):104856

108. David L, Moldovan B (2020) Green synthesis of biogenic silver nanoparticles for efficient catalytic removal of harmful organic dyes. Nanomaterials 10(2):202
109. Iqbal M, Ali A, Nahyoon NA, Majeed A, Pothu R, Phulpoto S, Thebo KH (2019) Photocatalytic degradation of organic pollutant with nanosized cadmium sulfide. Mater Sci Energy Technol 2(1):41–45
110. Hanif MA, Akter J, Islam MA, Sapkota KP, Hahn JR (2021) Visible-light-driven enhanced photocatalytic performance using cadmium-doping of tungsten (VI) oxide and nanocomposite formation with graphitic carbon nitride disks. Appl Surf Sci 565:150541
111. Jayaraman V, Mahalingam S, Chinnathambi KKS, Pandian GN, Prakasarao A, Ganesan S, ... Ahn YH (2022) Facile synthesis of hafnium oxide nanoparticle decorated on graphene nanosheet and its photocatalytic degradation of organic pollutants under UV-Light irradiation. Appl Sci 12(21):11222
112. Ambreen S, Danish M, Pandey ND, Pandey A (2017) Investigation of the photocatalytic efficiency of tantalum alkoxy carboxylate-derived Ta2O5 nanoparticles in rhodamine B removal. Beilstein J Nanotechnol 8(1):604–613
113. Taneja P, Sharma S, Umar A, Mehta SK, Ibhadon AO, Kansal SK (2018) Visible-light driven photocatalytic degradation of brilliant green dye based on cobalt tungstate (CoWO4) nanoparticles. Mater Chem Phys 211:335–342
114. Hejazi SA, Zaheer Z, Kosa SA (2023) Silver-osmium core-shell nanoparticles: synthesis and heterogeneous persulfate activator. Mater Chem Phys 305:127927
115. Dhanalakshmi M, Prabavathi SL, Saravanakumar K, Jones BF, Muthuraj V (2020) Iridium nanoparticles anchored WO3 nanocubes as an efficient photocatalyst for removal of refractory contaminants (crystal violet and methylene blue). Chem Phys Lett 745:137285
116. Goel A, Lasyal R (2016) Degradation of orange G dye by hexacyanoferrate (III) ions in the presence of Iridium nanoparticles: effect of system parameters and kinetic study. Desalin Water Treat 57(37):17547–17556
117. Gadah RH, Basaleh AS (2020) Influence of doped platinum nanoparticles on photocatalytic performance of CuO–SiO2 for degradation of Acridine orange dye. Ceram Int 46(2):1690–1696
118. Dabhane H, Chatur S, Jadhav G, Tambade P, Medhane V (2021) Phytogenic synthesis of gold nanoparticles and applications for removal of methylene blue dye: a review. Environ Chem Ecotoxicol 3:160–171
119. Liyanage AD, Perera SD, Tan K, Chabal Y, Balkus KJ Jr (2014) Synthesis, characterization, and photocatalytic activity of Y-doped CeO2 nanorods. ACS Catal 4(2):577–584
120. Sifontes AB, Rosales M, Méndez FJ, Oviedo O, Zoltan T (2013) Effect of calcination temperature on structural properties and photocatalytic activity of ceria nanoparticles synthesized employing chitosan as template. J Nanomater 2013:1–1
121. Yue L, Zhang XM (2009) Structural characterization and photocatalytic behaviors of doped CeO2 nanoparticles. J Alloy Compd 475(1–2):702–705
122. Buddee S, Suwanchawalit C, Wongnawa S (2017) Nickel doped nanorod titanium dioxide photocatalyst with enhanced visible light photocatalytic performance. Dig J Nanomater Biostruct 12(3):829–839
123. Channei D, Inceesungvorn B, Wetchakun N, Ukritnukun S, Nattestad A, Chen J, Phanichphant SJSR (2014) Photocatalytic degradation of methyl orange by CeO2 and Fe–doped CeO2 films under visible light irradiation. Sci Rep 4(1):5757
124. Chahal S, Rani N, Kumar A, Kumar P (2020) Electronic structure and photocatalytic activity of samarium doped cerium oxide nanoparticles for hazardous rose bengal dye degradation. Vacuum 172:109075
125. Pouretedal HR, Kadkhodaie A (2010) Synthetic CeO2 nanoparticle catalysis of methylene blue photodegradation: kinetics and mechanism. Chin J Catal 31(11–12):1328–1334
126. Herrling T, Seifert M, Jung K (2013) Cerium dioxide: future UV-filter in sunscreen. SOFW J 139(5):10–14
127. Mohamed RM, Aazam ES (2011) Characterization and catalytic properties of nano-sized Au metal catalyst on titanium containing high mesoporous silica (Ti-HMS) synthesized by photo-assisted deposition and impregnation methods. Int J Photoenergy 91–97

128. Akbari-Fakhrabadi A, Saravanan R, Jamshidijam M, Mangalaraja RV, Gracia MA (2015) Preparation of nanosized yttrium doped CeO2 catalyst used for photocatalytic application. J Saudi Chem Soc 19(5):505–510
129. Ahmadi S, Mohammadi L, Rahdar A, Rahdar S, Dehghani R, Adaobi Igwegbe C, Kyzas GZ (2020) Acid dye removal from aqueous solution by using neodymium (III) oxide nanoadsorbents. Nanomaterials 10(3):556
130. Nakhate GG, Nikam VS, Kanade KG, Arbuj S, Kale BB, Baeg JO (2010) Hydrothermally derived nanosized Ni-doped TiO2: a visible light driven photocatalyst for methylene blue degradation. Mater Chem Phys 124(2–3):976–981
131. Ranjith KS, Saravanan P, Chen SH, Dong CL, Chen CL, Chen SY, Asokan K, Rajendra Kumar RT (2014) Enhanced room-temperature ferromagnetism on Co-doped CeO2 nanoparticles: mechanism and electronic and optical properties. J Phys Chem C 118(46):27039–27047
132. Bear JC, McNaughter PD, Southern P, O'Brien P, Dunnill CW (2015) Nickel-doped ceria nanoparticles: the effect of annealing on room temperature ferromagnetism. Crystals 5(3):312–326
133. Ali SR, Kumar R, Kadabinakatti SK, Arya MC (2018) Enhanced UV and visible light—driven photocatalytic degradation of tartrazine by nickel-doped cerium oxide nanoparticles. Mater Res Express 6(2):025513
134. Ali SR, Kumar R, Arya MC (2018) Enhanced photocatalytic activity of manganese-doped cerium oxide nanoparticles under visible and UV irradiation. Asian J Chem 30(11):2544–2550
135. Rashmi SK, Naik HB, Jayadevappa H, Viswanath R, Patil SB, Naik MM (2017) Solar light responsive Sm-Zn ferrite nanoparticle as efficient photocatalyst. Mater Sci Eng, B 225:86–97
136. Keerthana SP, Yuvakkumar R, Kumar PS, Ravi G, Velauthapillai D (2021) Rare earth metal (Sm) doped zinc ferrite (ZnFe2O4) for improved photocatalytic elimination of toxic dye from aquatic system. Environ Res 197:111047
137. El-Kholy RA, Isawi H, Zaghlool E, Soliman EA, Khalil MM, Said MM, El-Aassar AEM (2023) Preparation and characterization of rare earth element nanoparticles for enhanced photocatalytic degradation. Environ Sci Pollut Res 30(26):69514–69532
138. Rooshde MS, Abdullah WRW, Ibrahim NF, Ghazali MSM, Wan WMN (2018) Photocatalytic efficiency of lanthanide-doped zinc oxide for degradation of methylene blue dye. J Sustain Sci Manag 13(2):153–161

Carbon Nanotube Applications in the Bioremediation of Heavy Metals and Dyes from Wastewater

Shalini Srivastava, S. B. Agrawal, and M. K. Mondal

Abstract Heavy metals and dyes have become one of the major environmental concerns because of their intractable and obstinate nature and thrilling level of toxicity on living beings. Therefore, it is of high importance to remove both heavy metals and dyes from the environment to reduce their toxic effects. Among many existing techniques of pollutant remediation from aqueous solution, adsorption is one of the most simple, effective, and efficient one used by number of researchers. Recently, carbon nanotubes-based adsorbents gained too much attention in metal ions and dyes remediation process due to their unique properties in the form of large surface area, cylindrical hollow structure, hydrophobic wall, high aspect ratios, more adsorbate binding sites, and easily modified surfaces. This paper critically reviewed the application of unmodified and modified CNTs in removal of both heavy metals and dyes from water and wastewater.

Keywords Heavy metal · Dyes · Environment · Remediation · Carbon nanotubes

1 Introduction

The increasing and unplanned pace of industrialization, urbanization, and very exploitative habit of agricultural land utilization for more and more crop production during the last few decades created the serious issues of water pollution and global water scarcity in our modern societies [1]. Contamination of water bodies by toxic heavy metals and dyes offers great environmental concerns.

However, in the environment, heavy metals are present and these naturally present heavy metals are found to be tightly bound to the soil and cannot impose any kind of harmful effects on the living beings [2]. On the other hand, industrial activities

S. Srivastava · S. B. Agrawal
Department of Botany, Institute of Science, Banaras Hindu University, Varanasi, India

M. K. Mondal (✉)
Department of Chemical Engineering and Technology, Indian Institute of Technology BHU Varanasi, Varanasi, Uttar Pradesh 221005, India
e-mail: mkmondal13@yahoo.com

P. Singh (ed.), *Dye Pollution from Textile Industry*, SDGs and Textiles,
https://doi.org/10.1007/978-981-97-5341-3_7

such as mining and smelting of metalliferous compounds, energy and fuel production units, surface finishing industries, fertilizer and pesticide industry and their application, metallurgy, electroplating, electrolysis, electro-osmosis, energy and fuel production, iron and steel industry, photography, textiles and tanneries, electric appliances manufacturing, aerospace, and atomic energy installation units produce and discharge heavy metal contaminated wastewater into nearby waterbodies and result in a tremendous level of heavy metal pollution [3]. Because of the toxic effects of these heavy metals on plants, animals, and food, they are kept under the environmental pollutant category. Arsenic (As), chromium (Cr), cadmium (Cd), lead (Pb), mercury (Hg), nickel (Ni), etc., are some of the listed heavy metals in the most concern category because of their cumulative toxic effects. In the environment, these heavy metals show persistent behaviors and have accumulative and biomagnification property. These heavy metals do not metabolize in other intermediated compound and not easily breakdown into less harmful or beneficial metallic forms. Their toxicity results in reduction of mental and central nervous system functioning, damage to hematological activity, lungs, kidneys, liver, and other vital organ malfunctioning and lower energy levels are some significant symptoms of metal toxicity. Also in children, the heavy metal exposure leads to dangerous health issues that leads to severe adverse health effects and deformities like slower growth rate, learning and behavioral disorders, and low level of intelligence quotient (IQ) [4]. In the aquatic ecosystem, heavy metal toxicity results in noteworthy threat to surrounding flora and fauna and the gill damage of the fish are identified as the most common one [5].

When dealing with the problem of increased toxicity of dyes as a matter of special concern then the wastewater and byproducts generated from textile, leather, paper and pulp industries, cosmetic, printing, plastics, and other related industrial units play a crucial role because of the use of synthetic dyes to color their products [6]. The chief source of dyes as pollutant are the textile industries with a share of approximately two third of the total dyestuff market [7, 8]. Dyes are used in different forms by the textile industries and these dyes do not bind tightly with the fabrics. As a result, dyes are easily washed out in the effluents of textile industries [9]. Dyes are mainly categorized into ionic and nonionic forms and, among these two, ionic dyes have gained more attention from the researchers due to their very high reactivity and cancer-causing properties [10]. Azo dyes characterized by the presence of azo group (–N=N–) linked to aromatic species in their chemical structure consist more than 50% of total dyes used in the textile industries [11]. The complex chemical confirmation of azo dyes makes them very toxic and leads to the problem of their difficult disposal and degradation in the environment.

As the dyes and their derived forms enter the environment, their toxicity starts negatively affecting organisms with potent mutagenic and carcinogenic properties. They also possess toxic influence on nutrient and water cycles and may results in skin irritation, increased sensitivity, and even dermatitis [12].

Bearing in mind all these harmful effects of heavy metals and dyes on environment and human beings, it is of utmost important to develop the most sustainable and appropriate treatment technologies with high level of sensitivity and selectivity to remove heavy metal ions and dyes from water bodies. The existing conventional technologies such as ion exchange, electrochemical oxidation, membrane filtration, osmosis, coagulation and flocculation, ion exchange, solvent extraction, etc., are not all the time found to be effective in targeting the stern regulatory limits [13–15]. Many of the above-mentioned techniques have complicated issues that includes higher capital cost involvement process operation, lesser efficiency, high maintenance cost, and excess sludge generation. These drawbacks in terms of cost-effectiveness make them unsuitable for their economical use [16]. However, adsorption process is recognized as one of the most appealing techniques due to its simplicity, smooth operation, high pollutant removal efficiency, cost-effectiveness, and sustainability [17]. Furthermore, most of the adsorption processes are reversible, and therefore, the used adsorbents may be easily regenerated and reused repeatedly in another cycle of adsorption making them more cost-effective.

A wide range of classical and modified adsorbents, both of biological and chemical origin, have been discovered and successfully applied in the field of heavy metals and dyes remediation since the last few decades [3, 18]. Nowadays, scientists pay more attention on advanced material to be utilized as adsorbent for more efficient results. Nanostructured materials, viz. nanoparticles, nanotubes, and nanowires are among the most effective ones due to their fascinating properties like large surface area to volume ratio, high thermal and chemical stability, numerous adsorption sites presence in their structure, easy functional group attachment, and also range of fabrication opportunity [19, 20].

In recent years, carbon nanotubes have been applied as a novel adsorbent for confiscation of variety of contaminants form wastewater owing to their large surface area, hollow structure, high porosity, and strong pollutant binding affinity [21]. The present review article aims to cover the latest widespread application of carbon nanotubes in the adsorption of heavy metals and dyes from wastewater.

2 Morphology of Carbon Nanotubes

Carbon nanotubes were discovered by a Japanese physicist, Sumio Iijima in 1991. He used arc evaporation method for CNT synthesis. The CNT structures are actually carbon allotropes with an aromatic surface rolled up to form a cylindrical structure [22]. The detailed morphology of nanotubes shows that they are made up of graphene or graphite sheets rolled up into tubular form. These tubes have diameter in nanometer range and length in micrometer range. Both ends of the tube are covered with a fullerene-like structure in a hemisphere form. Carbon nanotubes are highly hydrophobic in nature, consist of different sidewall curvature and have pi-conjugative structure [23]. The higher surface area and more porosity in CNT structure comes from its hollow and layered structure and ultimately these farfetched morphology of CNT offers more and more electronic and thermal stability with a range of enthralling mechanical strength. CNTs may have metallic or semiconductor behavior due to different bond positions (chirality) [24].

3 Types of Carbon Nanotubes

Based on their structure carbon nanotubes are broadly categories into to two forms, i.e., single-walled carbon nanotubes and multiwalled carbon nanotubes. The single-walled carbon nanotubes (SWCNTs) are made up of single graphene sheet rolled up into a cylindrical structure. The concentric stacking of two or more graphene sheet in cylindrical form constitute the multiwalled carbon nanotubes (MWCNTs). In MWCNTs, the adjacent sheets of graphene have an interspace of approximately 0.34 nm and held together through van der Waals interactions [25]. For both SWCNTs and MWSNTs, their total length may vary from 10 s of nm to 10 s of mm. SWCNTs have diameter in the range of 0.7–2.5 nm and for MWCNTs the diameter may be in the range of 4–150 nm [26].

In MWCNTs, the nature of the pores can be categorized into two groups: the small diameter-inner hollow pores in the range of 3–6 nm and aggregated pores formed by the interaction of isolated MWCNTs in the range of 20–40 nm [27]. The unique structure of both types of CNTs and the presence of various functional groups on their surfaces facilitates strong interaction with both organic and inorganic molecules via different intermolecular bonding such as van der Waals attraction, hydrogen bonding and π-π stacking (especially in aromatic compounds) [28, 29].

4 Methods Used for Synthesis of Carbon Nanotubes

The extensively used methods for CNT synthesis are arc discharge method, laser ablation method, and Chemical Vapor Deposition (CVD). In all the three mentioned methods, energy and carbon sources are used for CNT synthesis.

4.1 *Arc Discharge Method for CNT Synthesis*

The first method used for the synthesis of CNTs is arc discharge method [30]. Briefly, in the presence of either helium or argon at a pressure of 500 Torr, direct current of 50–100 A and nearly 20 V potential difference is established in between a pair of graphite electrodes. Because of the supply of electric current in the system at low pressure, in the presence of inert gas and catalyst, very high temperature has been generated in the system and results in the carbon electrode surface evaporation that results in the formation of cylindrical-shaped carbon nanotubes [31, 32]. The presence of mixed metal catalysts like iron, cobalt, and nickel, is essential for SWCNT synthesis, while MWCNTs can be fabricated using arc discharged method even in the absence of these metallic catalyst. Structure and size of CNTs synthesized via arc discharged method may be affected by various reaction conditions such as chamber temperature, type of catalyst and their concentration in the medium, presence of hydrogen atoms, etc. A detailed study shows higher levels of structural accuracy in CNTs synthesized through above-mentioned method [33].

4.2 Laser Ablation Method for CNT Synthesis

In 1995, Guo and his team used the laser ablation technique to synthesize CNTs for the first time [34]. In this process, at very high temperature of range 800–1000 °C and 500 Torr pressure in the presence of inert gas argon, the graphite object is embattled by a laser beam with high energy [35]. In this process, the required energy is provided by laser pulse and source of carbon used is a graphite object. In the first step after laser beam application, larger size CNT particles are synthesized which break down into smaller ones after consecutive beam application. In most of the cases, the used catalysts are transition metals. Nearly 100 mm long CNTs that have diameter in the range of 10–20 nm can be formed with rope-shaped structure during the process.

4.3 Chemical Vapor Deposition (CVD) Method for CNT Synthesis

CVD is the most popular method for bulk CNT fabrication. Briefly, in this method, gases like carbon monoxide or hydrocarbons are used as carbon source and the very high temperature of 500–1000 °C provide the required energy for CNT fabrication. During the process, because of the high temperature inside reaction chamber, the decomposition of the gas, i.e., carbon source occurs in the presence of carrier gas. This results in generation of a substrate containing carbon atoms on the catalyst surface and ultimately CNTs form [36]. This method of CNT synthesis is most attractive since for a highly advanced field of electronics and optoelectronics application it allows precise level of tailored control on the CNT architecture which can be easily scalable. For the nucleation and growth of CNTs, transition metal nanoparticles, mainly cobalt, iron, nickel, and yttrium, either singly or in combination with some suitable supporting materials (alumina, silica, zeolite, and other metal oxides) are preferred substrates.

5 Application of Carbon Nanotubes for Heavy Metal Removal

So many adsorbents having either of industrial or chemical or biological origin in their original or modified forms are used by various researchers for the removal of a variety of noxious heavy metals. In the recent time, CNTs gain much more attention in the same field. Some of the key work done for metal ions removal in single-metallic system and multi-metallic system present in waste water have been described below.

The study is conducted by Kosa et al. [37] with the MWCNTs in its original form and after chemical modification with 8-hydroxyquinoline. The modified MWCNTs show increased removal efficiency for Zn^{+2} (96%) than unmodified MWCNTs (79%). Lu and Chiu [38] used SWCNTs and MWCNTs for the removal of Zn^{+2} from aqueous solution. Both the adsorbents were purified with NaOCl before application. The non-purified SWCNTs and MWCNTs have a maximum adsorption capacity of 10.21 and 11.23 mg/g, respectively. On the other side, after purification their adsorption capacity was increased by 43.66 and 32.68 mg/g for SWCNTs and MWCNTs, respectively. The lower concentration of Zn^{+2} ions in the solution favors more removal efficiency by the used adsorbent compared to higher concentration. Also, pH 8 is most suitable for the maximum metal ion removal because of the fact that at this pH the dominant zinc species is Zn^{2+}.

Recently, Sobhanardakani et al. [39] synthesized oxidized MWCNTs and used it for the remediation of Cu^{2+} from aqueous medium. At a concentration of 200 mg/l of Cu^{2+} ions, 25 °C temperature and optimum pH of 6, the maximum removal percentage was achieved to be 99.5%.

Al Hamouz et al. [40] synthesized composites of CNTs with pure polymer of Melamine–Formaldehyde–Diaminohexane (MFDH) and used it for the removal of Pb^{2+} ions from wastewater. The results showed that a composite containing 0.1% of chlorinated CNTs removed 99% of total Pb^{2+} ions. In another study conducted in the same year by Jiang et al. [41], 95.5% of Pb^{2+} were removed from wastewater through the application of iron oxide-modified CNTs. Hydrazine, ammonium ferrous sulfate, and ferric sulfate functionalized MWCNTs were used by Jiang et al. [42] for the removal of Pb^{2+} and Zn^{2+}. The modified MWCNTs give very high adsorption capacity with a value of 195.81 mg/g and 169.89 mg/g for Pb^{2+} and Zn^{2+}, respectively, while, in another study [43] when they applied magnetic MWCNTs for the removal of same set of metal ions, it was only 67.25 mg/g and 3.76 mg/g only for Pb^{2+} and Zn^{2+}, respectively. Wang et al. (2007) and Zhao et al. [44] used TiO_2 and MnO_2 functionalized hybrid CNTs, respectively for remediation of Pb^{2+}. Their study showed that these modified CNTs can be used as very efficient adsorbent in metal removal. In case of TiO_2 functionalized CTNs, the Pb^{2+} adsorption capacity was calculated to be 137 mg/g, while 78.74 mg/g was shown by MnO_2-altered CNTs.

Removal of both the forms of arsenic (arsenate As(V) and arsenite (As(III)) were examined by Velickovic et al. [45] using Fe_3O_4 modified ethylenediamine MWCNTs (e-MWCNTs/Fe^{2+} and w-MWCNTs/Fe^{3+}). Both the adsorbents showed higher removal capacity for As(V) compared to As(III). The maximum adsopriton capacity was found to be 24.69 mg/g and 14.45 mg/g for As(V) using e-MWCNTs/Fe^{2+} and w-MWCNTs/Fe^{3+}, respectively, at a fixed temperature of 45 °C. In the research conducted by Chen et al. [150], authors synthesized MWCNTs incorporated with cerium-iron mixed oxide along with the mixer of alkylbenzene sulfonate surfactant (NASDBS) for the adsorption of As in its both oxidation state, i.e., As(III) and As(V). The modified MWCNTs shows 30.96 and 28.74 mg/g adsorption capacity for As(V) and As(III), respectively, at 25 °C.

CNTs and modified CNTs application in removal of Hg have also been investigated by many researchers. A study by Shadbad and Mohebbi (2011) reports Hg removal by MWCNTs synthesized through chemical vapor deposition technique. Neutral pH range (6.5–7.5) seems to be most suitable condition for maximum Hg(II) removal and maximum adsorption capacity was noticed to be 93.45 mg/g at 37 °C. Chen et al. [46] used MWCNTs modified with hydroxyl (MWCNTs-OH) and carboxyl group (MWCNTs-COOH). At room temperature, the maximum adsorption capacity of 120.1 mg/g and 127.6 mg/g were shown by MWCNTs-OH and MWCNTs-COOH, respectively. In another study, , Pillay et al. [47] modified MWCNTs with sulfur and the results obtained were quite attractive in terms of great enhancement in the Hg(II) removal percentage. Almost 100% removal of Hg(II) was reported by the authors. This result was for the maximum ratio of sulfur which can be associated with MWCNTs and it also have a high reactivity rate in wide pH range of up to 14.

Chromium is also very toxic heavy metal in its hexavalent form and exist in most of the industrial wastewater. The CNTs and their modified forms were also applied by many researchers for the removal of Cr(VI) form contaminated environment. Cr(VI) removal by polydiallyldimethyl ammonium chloride (PDDA) modified MWCNTs through non-covalent wrapping method was used by Zhang et al. [48]. Almost 100% removal of Cr(VI) was achieved by MWCNTs-PDDA. The Cr(VI) adsorption on the surface of MWCNTs-PDDA was highly pH-dependent and maximum removal occurred in acidic range of pH (2.0–5.1). On the other hand, pristine modified MWCNTs shows on 65% of Cr(VI) removal at 5.1 pH value [49]. Ansari et al. [50] developed MWCNT composite with graphene oxide after modification with polyaniline (MWCNTs-GO-PANI) and applied it for Cr(VI) removal from wastewater. The results show a very high adsorption capacity of 142.85 mg/g in acidic pH range. MnO_2/Fe_2O_3/acid oxidized MWCNT nanocomposites were prepared by Luo et al. [51] for the removal of Cr^{6+}. At very low pH condition (pH 2.0), the maximum adsorption capacity was calculated to be 186.9 mg/g and highest removal efficiency was 85%.

The MWCNTs were surface functionalized through chemical modification technique by ethylenediamine using O-(7-azabenzotriazol-1-yl)-N,N,N′,N′-tetramethyluronium hexafluorophosphate (e-MWCNTs) for the removal of Cd^{2+} from aqueous solution by Aleksandar et al. (2009). The oxidized MWCNTs (o-MWCNTs) were also be tested parallelly in the experiment. The results concluded that both types of adsorbents removed Cd^{2+} ions strongly depending on pH range (best in the range of 8–9). The maximum adsorption capacity was calculated for e-MWCNTs with the value of 25.7 mg/g at 45 °C temperature. Li et al. [153] carried out a comparative study using H_2O_2, HNO_3, and $KMnO_3$ oxidized CNTs for Cd^{2+} remediation. The results show greater Cd^{2+} removal by HNO_3 oxidized CNTs because of its very high reactivity that can help in opening of the CNT chains in smaller form and simultaneously its tip opening that ultimately provide more and more functional groups for Cd^{2+} binding with high pore volume value. Maximum adsorption capacity by HNO_3 functionalized CNTs was 5.1 mg/g, while 1.1 mg/g and 2.6 mg/g were the calculated adsorption capacity by $KMnO_4$ and H_2O_2 oxidized CNTs, respectively.

The details of various studies carried out in previous years for remediation of heavy metals using CNTs and modified CNTs are extensively reviewed and summarized in Tables 1 and 2.

Table 1 Adsorption of heavy metal ions on to unmodified CNTs

Adsorbent	Adsorbate	Maximum adsorption capacity (mg/g)	References
Raw MWCNTs	Cr(VI)	1.02	[64]
Raw SWCNTs		3.12	[64]
SWCNTs		20.3	[65]
MWCNTs		2.48	[65]
MWCNTs		13.2	[66]
MWCNTs	Hg(II)	40.16	[67]
MWCNTs		5.48	[68]
Raw MWCNTs		32.4	[46]
MWCNTs		25.64	[69]
MWCNTs	Cd(II)	43.10	[70]
Raw CNTs		1.66	[147]
CNTs		1.1	[153]
CNTs	Cu(II)	8.25	[71]
MWCNTs		45.0	[72]
CNTs		55.25	[104]
MWCNTs	Pb(II)	162.16	[159]
CNTs		2.96	[73]
MWCNTs		48.0	[72]
MWCNTs	Ni(II)	35.0	[72]
MWCNTs		6.09	[74]
MWCNTs		14.2	Ahmadi et al. (2019)
MWCNTs	Zn(II)	10.21	[75]
MWCNTs		156.25	[76]
Raw CNTs		250.0	[77]
MWCNTs		90.9	[155]
MWCNTs		10.21	[38]
MWCNTs	As(V)	2.99	[78]
CNTs		0.373	[79]
MWCNTs		32.68	[80]

Table 2 Adsorption of heavy metal ions on to modified CNTs

Adsorbent	Adsorbate	Maximum adsorption capacity (mg/g)	References
Oxidized MWCNTs	Cr(VI)	2.46	[81]
CeO_2/CNTs		30.2	[82]
AC-CNTs		9.0	[83]
Oxidized-MWCNTs		67.0	[72]
IL-oxi-MWCNTs		85.83	[66]
Acid modified MWCNTs		1.31	[64]
Ptsa-pani@GO-CNTs		142.85	[50]
Magnetic MWCNTs		16.23	[84]
Magnetic nanoparticle-MWCNTs		17.09	[85]
Chitin/magnetite/MWCNTs		11.3	[86]
Activated carbon/MWCNTs		113.29	[87]
FeOOH/AlOOH/MWCNTs		60.6	[88]
MWCNTs-COOH	Hg(II)	127.6	[46]
MWCNTs-OH		120.1	[46]
MWCNTs-Amidoamine		45.05	[89]
MWCNTs-Thiolated		84.66	[90]
MWCNTs-Iodide		123.5	[68]
MWCNTs-Sulfur		151.5	[68]
MWCNTs-COOH		134.0	[91]
MWCNTs-NH_2		205.0	[91]
MWCNTs-OH		78.9	[91]
MnO_2-CNTs		58.82	[92]
Thiol-derived SWCNTs		131.58	[67]
Ag-MWCNTs	Cd(II)	51.97	[93]
Oxidized-MWCNTs		66.0	[72]
Al_2O_3/MWCNTs		27.21	[94]
Acid treated CNTs4.35		4.35	[147]
PPy/O-MWCNTs		392.0	[95]
MWCNTs-COOH		212.77	[70]
MWCNTs-f		400.0	[70]
Microwave-assisted MWCNTs		88.62	[96]
Oxidized CNTs		34.36	[97]
Oxidized CNTs		10.86	Ding et al.(2003)
Sulfonated MWCNTs	Cu(II)	7.9	[98]
Sulfur-MWCNTs		59.6	[98]

(continued)

Table 2 (continued)

Adsorbent	Adsorbate	Maximum adsorption capacity (mg/g)	References
SWCNTs-OH-Reduced Graphene Oxide		256	[99]
Oxidized-CNTs		28.49	[154]
Microwave-induced MWCNTs		99.0	[100]
NN-mSiO@MWCNTs		66.58	[101]
Ag-MWCNTs		53.29	[93]
CS/MWCNTs		454.55	[102]
Oxidized MWCNTs		70.4	[72]
Oxidized MWCNTs		200.0	[39]
MWCNTs-TA		30.85	[103]
Acid-treated CNTs		82.64	[104]
Chitosan-treated CNTs		120.48	[104]
SWCNs/WSh	Pb(II)	185.18	[105]
Oxidized MWCNTs		75.0	[72]
O-MWCNTs		76.7	[106]
MWCNTs-NCO		196.1	[107]
Fe_3O_4/O-MWCNTs		67.25	[42]
N_2H_4-SH-Fe_3O_4/O-MWCNTs		40.0	[43]
NiO/CNT		24.63	[108]
Acid-functionalized MWCNTs		166.0	[109]
Fe_3O_4/CNT		21.55	[41]
Oxidized CNTs		97.08	[154]
Microwave assisted MWCNTs		104.2	[156]
MnO_2/CNTs		78.74	[160]
Oxidized MWCNTs		91.0	[161]
Oxidized with CNTs-DES		288.4	[110]
Oxidized MWCNTs	Ni(II)	2.71	[111]
$NaClO_3$ oxidized SWCNTs		51.28	[112]
$NaClO_3$ oxidized MWCNTs		40.00	[112]
Oxidized CNTs		49.26	[113]
MWCNT/Fe_2O_4		9.18	[114]
Oxidized MWCNTs		38.46	[115]
Oxidized MWCNTs		59.2	[72]
PPy/O-MWCNTs		409.4	[95]
Acid functionalized MWCNTs		95.0	[109]
Functionalized MWCNTs	Zn(II)	1.05	[116]

(continued)

Table 2 (continued)

Adsorbent	Adsorbate	Maximum adsorption capacity (mg/g)	References
Fe_3O_4/O-MWCNTs		3.76	[42]
N_2H_4-SH-Fe_3O_4/O-MWCNTs		21.28	[43]
O-MWCNTs (HNO_3)		27.2	[117]
O-MWCNTs ($KMnO_4$)		28.01	[117]
O-MWCNTs (NaClO)		32.68	[117]
O-MWCNTs (HNO_3 + H_2SO_4)(3:1)		18.14	[117]
Oxidized MWCNTs	As(V)	3.61	[78]
e-MWCNTs		12.18	[78]
CF-CNTs-A		16.84	[151]
CF-CNTs		30.96	[151]
Fe_3O_4-MWCNTs		39.09	[132]
MWCNTs-zirconia nanohybrid		5.0	[118]
Fe^{2+}-coated ethylenediamine functionalized MWCNTs		24.69	[119]
Fe^{3+}-coated ethylenediamine functionalized MWCNTs		14.45	[119]
CeO_2-CNTs		78.8	Peng et al. (2005)

6 Application of Carbon Nanotubes for Dyes Removal

Although synthetic dyes production and industrial application on large scale are important but, during the process, so many poorly biodegradable unwanted effluents are generated that possess serious threat to the society. So many adsorbents were used by the researchers for the removal of these hazardous dyes from wastewater, but, nowadays, CNTs and modified CNTs have been widely exploited for the repossession of the same and also these newly developed adsorbents show a strong affinity towards dye removal.

Methylene blue removal on CNTs was reported by Yao et al. [52]. The maximum adsorption capacity of 41.63 mg/g was calculated at 333 k temperature in batch experiment. Methylene blue (MB) and acid red 183 (AR183) adsorption studies onto MWCNTs was reported by Wang et al. [53]. Authors used synergistic approach for the simultaneous removal of both the pollutant from wastewater. The adsorption capacity was maximum when binary dye system was used in reaction with value of 80 mg/g and 53 mg/g for MB and AR183, respectively, while, in the case of single dye system, it was 59.7 mg/g and 45.2 mg/g, respectively, for the same. Removal of neutral red and methylene blue dyes from aqueous media using magnetized acid-treated MWCNTs was developed and used by Qu et al. [54]. Because of the very small size of the iron oxide particles, these developed adsorbents possess superparamagnetic property. The functionalized MWCNTs show 42.3 mg/g and 77.5 mg/g adsorption capacity for MB and neutral red, respectively, at pH 6.0. Ai and Jiang [55] have used self-assembled cylindrical graphene-CNT compounds as hybrid adsorbents for the removal of MB from aqueous solution. The developed adsorbent achieved maximum adsorption capacity of 81.97 mg/g. Adsorption of methyl orange onto MWCNTs was studied by Yao et al. [56] and the adsorption of methyl orange on MWCNTs was strongly depends on pH of the medium. The maximum adsorption capacity was reported to be 41.9 mg/g (at 40 mg/L dye concentration). Farghali et al. [57] synthesized composite of $CoFe_2O_4$-MWCNTs for the removal of MB from aqueous medium. In the study, authors confirm uniform distribution of $CoFe_2O_4$ on the MWCNTs through characterization processes. The maximum adsorption capacity was found to be 11.95 mg/g at 50 °C. MWCNTs functionalized by addition of calcium alginate was used by Sui et al. [58] in adsorption study for MB and methyl orange dyes from aqueous medium. The maximum adsorption capacity of developed adsorbents was found to be very high (606.1 mg/g) for MB as compared to methyl orange (12.5 mg/g). In 2019, Boukhalfa et al. used oxidized MWCNTs for the adsorptive removal of MB form wastewater. A very high adsorption capacity was calculated form the experimental data and its value was 905.5 mg/g. Hosseinzadeh et al. [59] have developed a very complex adsorbent after fabrication of MWCNTs with acrylic acid and N-isopropyl acrylamide compounds in the presence of Fe_2O_3 nanoparticles. The complex mixture was successfully used by the authors for the effective removal of rhodamine B, MB, and crystal violet dyes. The equilibrium adsorption data show maximum adsorption capacities of 287 mg/g, 302 mg/g, and 231 mg/g for crystal violet, MB, and rhodamine B, respectively.

The MWCNTs functionalized with carboxylic group (MWCNTs-COOH) and carboxyl group along with cysteamine (MWCNTs-COOH-Cysteamine) have been synthesized and applied for the remediation of amide black by Sadegh et al. [60]. The comparative study of two developed adsorbents showed maximum adsorption capacity of 90 mg/g and 131 mg/g by MWCNTs-COOH and MWCNTs-COOH-Cysteamine, respectively. MWCNTs and SWCNTs were used by Machado et al. [61], for the removal of reactive blue 4 from aqueous solution. At pH 2.0, the comparative study of the two adsorbents shows maximum adsorption capacity of 442 mg/g and 487.6 mg/g, respectively, for MWCNTs and SWCNTs.

Propylene imine dendrimer functionalized MWCNTs was used by Eskandarian et al. [62] for the adsorption of direct red 23 and direct blue 86 form wastewater. The experimental data show very efficient removal capacity of both the dyes by developed adsorbent and its value was 464.73 mg/g and 554.64 mg/g, respectively, for 50 mg/L concentration of both direct red 23 and direct blue 86. A very complex MWCNTs composite after modification with glycine-β-cyclodextrin was utilized for the removal of MB, acid blue 113, methyl orange, and disperse red 1 dye form aqueous environment [63]. The batch experiment shows influence of various parameters on dye removal efficiency of developed adsorbent. The calculated maximum adsorption capacity was 90.90 mg/g, 172.41 mg/g, 96.15 mg/g, and 500 mg/g for MB, acid blue 113, methyl orange, and disperse red 1, respectively. Abbasi [143] used glutaraldehyde cross-linked chitosan/ SiO_2/ MWCNT magnetic nanocomposites for the removal of direct blue 71 and reactive blue 19. The best pH for maximum adsorption of direct blue 71 was 6.8 and, for reactive blue 19, it was 2.0. At this pH value, maximum adsorption capacity was 69.93 mg/g and 106.38 mg/g, for direct blue 71 and reactive blue 19, respectively.

The adsorption capacities of various unmodified and modified CNTs as reported by various researchers are listed in Tables 3 and 4.

Table 3 Adsorption of dyes on to unmodified CNTs

Adsorbent	Adsorbate	Maximum adsorption capacity (mg/g)	References
MWCNTs	Acid red 183	45.2	[120]
MWCNTs	Methylene blue	59.49	[120]
MWCNTs	Reactive blue 4	69	[53]
MWCNTs	Blue 116	170	[121]
MWCNTs	Yellow 81	197	[121]
MWCNTs	Red 159	208	[121]
MWCNTs	Congo red	231.18	[122]
MWCNTs	Reactive red M-2BE	312.3	[123]
MWCNTs	Direct blue 53	334.8	[124]
MWCNTs	Reactive blue 4	442	[61]
MWCNTs	Methylene blue	6.96	[157]
MWCNTs	Methyl orange	5.46	[157]
MWCNTs	Reactive blue	69.0	[53]
MWCNTs	Acid red 183	45.0	[53]
MWCNTs	Acid red 18	166.67	[125]
MWCNTs	Alizarin red S	161.29	[126]
SWCNTs	Reactive blue 29	496.0	[127]
MWCNTs	Congo red	352.11	[122]
MWCNTs	Maxilon blue	260.7	[128]
MWCNTs	Reactive red M-2BE	335.7	[123]
MWCNTs	Acid blue 161	458.33	[129]
MWCNTs	Reactive blue 4	58.8	[53]
MWCNTs	Acid red 183	45.3	[53]
SWCNTs	Reactive red 120	426.49	[130]
MWCNTs	Malachite green	142.85	[125]

Table 4 Adsorption of dyes on to modified CNTs

Adsorbent	Adsorbate	Maximum adsorption capacity (mg/g)	References
Oxidize MWCNTs	Bromothymol blue	55	[131]
f-MWCNTs	Congo red	148	[132]
f-MWCNTs	Reactive green HE4BD	152	[132]
CNTs/activated carbon fiber	Basic violet 10	220	[133]
Magnetic MWCNTs	Methylene blue	48.06	[134]
MWCNTs/Fe_3C	Direct red 23	172	[135]
Calcium alginate/MWCNTs	Methylene blue	606.1	[58]
Calcium alginate/MWCNTs	Methyl orange	12.5	[58]
Chitosan/Fe_2O_3/MWCNTs	Methyl orange	66.9	[136]
G-CNTs	Methylene blue	81.79	[55]
MWCNTs/Fe_2O_3	Natural red	77.5	[54]
MWCNTs/Fe_2O_3	Methylene blue	42.3	[54]
Oxidized SWCNT	Basic red 46	49.45	[137]
HNO_3-oxidized MWCNTs	Bromothymol blue	55	[131]
Magnetite MWCNTs-Fe_3C	Direct red 23	85.5	[135]
SWCNT-COOH	Malachite green	22.33	[138]
SWCNT-NH_2	Malachite green	29.36	[138]
SWCNT-COOH	Methyl orange	25	[138]
SWCNT-NH_2	Methyl orange	27.15	[138]
Oxidized MWCNTs	Methyl orange	10	[139]
Functionalized CNTs loaded TiO_2	Methyl orange	42.85	[140]
MWCNT-SH	Methylene blue	166.67	[141]
$CoFe_2O_4$-MWCNTs	Methylene blue	14.28	[57]
HNO_3/MWCNTs	Jenus green	56	[142]
Magnetite/Chitosan/SiO_2/MWCNTs	direct blue 71	61.35	[143]
Glycine-β-cyclodextrin-MWCNTs	Methylene blue	90.90	[63]
Glycine-β-cyclodextrin-MWCNTs	Methylene orange	96.15	[63]
Glycine-β-cyclodextrin-MWCNTs	Acid blue 113	172.41	[63]
Magnetite MWCNTs	Crystal violet	227.7	[144]
MWCNTs/chitosan/ poly2-hydroxyethyl methacrylate	Methyl orange	400	[145]
Gelatin/MWCNTs/iron oxide	Direct red 80	380.7	[146]
Gelatin/MWCNTs/iron oxide	Methylene blue	465.5	[146]

(continued)

Table 4 (continued)

Adsorbent	Adsorbate	Maximum adsorption capacity (mg/g)	References
Gelatin/MWCNTs/iron oxide	Direct red 80	321.4	[146]
Gelatin/MWCNTs/iron oxide	Methylene blue	348.4	[146]

7 Conclusion

The toughest challenge of present era is to purify polluted water bodies and make them ready to be used for others. Many techniques have been used by the researchers for water purification and adsorption is one of the most powerful methods among others. Various adsorbents have been widely and effectively applied for the remediation purpose and, in recent years, carbon nanotubes have gained much attention from the scientific communities in adsorption studies. Literature studies revealed that CNTs have been under scientific investigation for more than 25 years back due to their inimitable belongings that preordain them for countless potential applications. The present review article explores the potential of CNTs in the remediation of heavy metal ions and dyes from the wastewater effluents. Both types of CNTs (SWCNTs and MWCNTs) in its original form and after fabrication and/or modification shows admirable adsorption capacities and effectually work in confiscating inorganic, organic and even biological contaminants from wastewater. However, the thorough literature study reflects small-scale application of CNTs in research work due to its high cost and chances to impose toxic effects on living organism. Thus, more research is still looked-for to produce cost-effective and efficient CNTs that have the least or no harmful effect in the surroundings for large-scale remediation applications.

Acknowledgements The authors are thankful to the Head, Department of Botany, Co-ordinator, Centre of Advanced Study, Botany, Institute of Science, Banaras Hindu University and Department of Chemical Engineering, Indian Institute of Technology (Banaras Hindu University), India, for providing all the necessary facilities to undertake the present work.

References

1. Fierascu RC, Brazdis RI, Fierascu I, Baroi AM, Fistos T, Avramescu SM (2020) Green technologies for wastewater depollution. Multidiscip Dig Publ Instit Proc 57(1):30
2. Tripathi M, Singh P, Singh R, Bala S, Pathak N, Singh S, Chauhan RS, Singh PK (2023) Microbial biosorbent for remediation of dyes and heavy metals pollution: a green strategy for sustainable environment. Front Microbiol 14:1168954
3. Srivastava S, Agrawal SB, Mondal MK (2015) A review on progress of heavy metal removal using adsorbents of microbial and plant origin. Environ Sci Pollut Res 22(20):15386–15415
4. Ab Latif Wani AA, Usmani JA (2015) Lead toxicity: a review. Interdiscip Toxicol 8(2):55

5. Ahmaruzzaman M (2011) Industrial wastes as low-cost potential adsorbents for the treatment of wastewater laden with heavy metals. Adv Coll Interface Sci 166(1–2):36–59
6. Sharma P, Kaur H, Sharma M, Sahore V (2011) A review on applicability of naturally available adsorbents for the removal of hazardous dyes from aqueous waste. Environ Monit Assess 183(1):151–195
7. Gong R, Zhang X, Liu H, Sun Y, Liu B (2007) Uptake of cationic dyes from aqueous solution by biosorption onto granular kohlrabi peel. Biores Technol 98(6):1319–1323
8. Safarikova M, Ptackova L, Kibrikova I, Safarik I (2005) Biosorption of water-soluble dyes on magnetically modified Saccharomyces cerevisiae subsp. uvarum cells. Chemosphere 59(6):831–835
9. Al-Tohamy R, Ali SS, Li F, Okasha KM, Mahmoud YAG, Elsamahy T, Jiao H, Fu Y, Sun J (2022) A critical review on the treatment of dye-containing wastewater: ecotoxicological and health concerns of textile dyes and possible remediation approaches for environmental safety. Ecotoxicol Environ Saf 231:113160
10. Hosny NM, Gomaa I, Elmahgary MG (2023) Adsorption of polluted dyes from water by transition metal oxides: a review. Appl Surf Sci Adv 15:100395
11. Behnajady MA, Vahid B, Modirshahla N, Shokri M (2006) Evaluation of electrical energy per order (EEO) with kinetic modeling on photooxidative degradation of C.I. acid orange 7 in a tubular continuous-flow photoreactor. Ind Eng Chem Res 45:553–557
12. Mokri HG, Modirshahla N, Behnajady MA, Vahid B (2015) Adsorption of CI Acid Red 97 dye from aqueous solution onto walnut shell: kinetics, thermodynamics parameters, isotherms. Int J Environ Sci Technol 12(4):1401–1408
13. Carolin CF, Kumar PS, Saravanan A, Joshiba GJ, Naushad M (2017) Efficient techniques for the removal of toxic heavy metals from aquatic environment: a review. J Environ Chem Eng 5(3):2782–2799
14. Lim AP, Aris AZ (2014) A review on economically adsorbents on heavy metals removal in water and wastewater. Rev Environ Sci Bio/Technol 13(2):163–181
15. Srivastava S, Agrawal SB, Mondal MK (2017) Synthesis, characterization and application of Lagerstroemia speciosa embedded magnetic nanoparticle for Cr (VI) adsorption from aqueous solution. J Environ Sci 55:283–293
16. Ouni L, Ramazani A, Fardood ST (2019) An overview of carbon nanotubes role in heavy metals removal from wastewater. Front Chem Sci Eng 1–22
17. Wen M, Li G, Liu H, Chen J, An T, Yamashita H (2019) Metal–organic framework-based nanomaterials for adsorption and photocatalytic degradation of gaseous pollutants: recent progress and challenges. Environ Sci Nano 6(4):1006–1025
18. Lima EC (2018) Removal of emerging contaminants from the environment by adsorption. Ecotoxicol Environ Saf 150:1–17
19. Hebbar RS, Isloor AM, Asiri AM (2017) Carbon nanotube-and graphene-based advanced membrane materials for desalination. Environ Chem Lett 15(4):643–671
20. Zare K, Gupta VK, Moradi O, Makhlouf ASH, Sillanpää M, Nadagouda MN, Sadegh H, Shahryari-Ghoshekandi R, Pal A, Wang ZJ, Tyagi I (2015) A comparative study on the basis of adsorption capacity between CNTs and activated carbon as adsorbents for removal of noxious synthetic dyes: a review. J Nanostruct Chem 5(2):227–236
21. Machado FM, Fagan SB, da Silva IZ, de Andrade MJ (2015) Carbon nanoadsorbents. In: Carbon nanomaterials as adsorbents for environmental and biological applications. Springer, Cham. Nano/Micro Engineered and Molecular Systems, January 20–23. Xiamen, China, pp 11–32
22. Terrones M, Botello-Méndez AR, Campos-Delgado J, López-Urías F, Vega-Cantú YI, Rodríguez-Macías FJ, Elías AL, Munoz-Sandoval E, Cano-Márquez AG, Charlier JC, Terrones H (2010) Graphene and graphite nanoribbons: Morphology, properties, synthesis, defects and applications. Nano Today 5(4):351–372
23. Aqel A, El-Nour KMMA, Ammar RAA, Al-Warthan A (2012) Carbon nanotubes, science and technology part (I) structure, synthesis and characterisation. Arab J Chem 5:1–23

24. Thostenson ET, Ren Z, Chou TW (2001) Advances in the science and technology of carbon nanotubes and their composites: a review. Compos SciTechnol 61(13):1899–1912
25. Iijima S, Ichihashi T (1993) Single-shell carbon nanotubes of 1-nm diameter. Nature 363:603–605
26. Jensen KA, Bøgelund J, Jackson P, Jacobsen NR, Birkedal R, Clausen PA, Vogel UB (2015) Carbon nanotubes—types, products, market, and provisional assessment of the associated risks to man and the environment. Environmental project No. 1805
27. Yang QH, Hou PX, Bai S, Wang MZ, Cheng HM (2001) Adsorption and capillarity of nitrogen in aggregated multi-walled carbon nanotubes. Chem Phys Lett 345(1–2):18–24
28. Rengaraj S, Yeon JW, Kim Y, Kim WH (2007) Application of Mg-mesoporous alumina prepared by using magnesium stearate as a template for the removal of nickel: kinetics, isotherm, and error analysis. Ind Eng Chem Res 46(9):2834–2842
29. Singhal S, Dixit S, Shukla AK (2019) Structural analysis of carbon nanospheres synthesized by CVD: an investigation of surface charges and its effect on the stability of carbon nanostructures. Appl Phys A 125(2):80
30. Maser WK, Benito AM, Muñoz E, de Val GM, Martínez MT, Larrea Á, de la Fuente GF (2001) Production of carbon nanotubes by CO2-laser evaporation of various carbonaceous feedstock materials. Nanotechnology 12(2):147
31. Bethune DS, Kiang CH, De Vries MS, Gorman G, Savoy R, Vazquez J, Beyers R (1993) Cobalt-catalysed growth of carbon nanotubes with single-atomic-layer walls. Nature 363(6430):605–607
32. Hutchison JL, Kiselev NA, Krinichnaya EP, Krestinin AV, Loutfy RO, Morawsky AP, Muradyan VE, Obraztsova ED, Sloan J, Terekhov SV, Zakharov DN (2001) Double-walled carbon nanotubes fabricated by a hydrogen arc discharge method. Carbon 39(5):761–770
33. Li H, Guan L, Shi Z, Gu Z (2004) Direct synthesis of high purity single-walled carbon nanotube fibers by arc discharge. J Phys Chem B 108(15):4573–4575
34. Guo T, Nikolaev P, Thess A, Colbert DT, Smalley RE (1995) Catalytic growth of single-walled manotubes by laser vaporization. Chem Phys Lett 243(1–2):49–54
35. Kingston CT, Simard B (2003) Fabrication of carbon nanotubes. Anal Lett 36(15):3119–3145
36. Jose-Yacaman M, Miki-Yoshida M, Rendon L, Santiesteban JG (1993) Catalytic growth of carbon microtubules with fullerene structure. Appl Phys Lett 62(6):657–659
37. Kosa SA, Al-Zhrani G, Salam MA (2012) Removal of heavy metals from aqueous solutions by multi-walled carbon nanotubes modified with 8-hydroxyquinoline. Chem Eng J 181:159–168
38. Lu C, Chiu H (2006) Adsorption of zinc (II) from water with purified carbon nanotubes. Chem Eng Sci 61(4):1138–1145
39. Sobhanardakani S, Zandipak R, Cheraghi M (2017) Adsorption of Cu2+ ions from aqueous solutions using oxidized multiwalled carbon nanotubes. Avicenna J Environ Health Eng 2(1):790–790
40. Al Hamouz OCS, Adelabu IO, Saleh TA (2017) Novel cross-linked melamine based polyamine/CNT composites for lead ions removal. J Environ Manag 192:163–170
41. Elmi F, Hosseini T, Taleshi MS, Taleshi F (2017) Kinetic and thermodynamic investigation into the lead adsorption process from wastewater through magnetic nanocomposite Fe_3O_4/CNT. Nanotechnol Environ Eng 2(1):1–13
42. Jiang L, Li S, Yu H, Zou Z, Hou X, Shen F, Li C, Yao X (2016) Amino and thiol modified magnetic multi-walled carbon nanotubes for the simultaneous removal of lead, zinc, and phenol from aqueous solutions. Appl Surf Sci 369:398–413
43. Jiang L, Yu H, Zhou X, Hou X, Zou Z, Li S, Li C, Yao X (2016) Preparation, characterization, and adsorption properties of magnetic multi-walled carbon nanotubes for simultaneous removal of lead (II) and zinc (II) from aqueous solutions. Desalin Water Treat 57(39):8446–18462
44. Zhao X, Jia Q, Song N, Zhou W, Li Y (2010) Adsorption of Pb (II) from an aqueous solution by titanium dioxide/carbon nanotube nanocomposites: kinetics, thermodynamics, and isotherms. J Chem Eng Data 55(10):4428–4433

45. Velickovic Z, Vukovic GD, Marinkovic AD, Moldovan MS, Peric-Grujic AA, Uskokovic PS, Ristic MD (2012) Adsorption of arsenate on iron (III) oxide coated ethylenediamine functionalized multiwall carbon nanotubes. Chem Eng J 181:174–181
46. Chen PH, Hsu CF, Tsai DDW, Lu YM, Huang WJ (2014) Adsorption of mercury from water by modified multi-walled carbon nanotubes: adsorption behaviour and interference resistance by coexisting anions. Environ Technol 35(15):1935–1944
47. Pillay K, Cukrowska EM, Coville NJ (2013) Improved uptake of mercury by sulphur-containing carbon nanotubes. Microchem J 108:124–130
48. Zhang X, Chen M, Yu Y, Yang T, Wang J (2011) Polyelectrolyte-modified multi-walled carbon nanotubes for the adsorption of chromium (vi). Anal Methods 3(2):457–462
49. Scindia YM, Pandey AK, Reddy AVR, Manohar SB (2002) Selective preconcentration and determination of chromium (VI) using a flat sheet polymer inclusion sorbent: potential application for Cr (VI) determination in real samples. Anal Chem 74(16):4204–4212
50. Ansari MO, Kumar R, Ansari SA, Ansari SP, Barakat MA, Alshahrie A, Cho MH (2017) Anion selective pTSA doped polyaniline@ graphene oxide-multiwalled carbon nanotube composite for Cr (VI) and Congo red adsorption. J Colloid Interface Sci 496:407–415
51. Luo C, Tian Z, Yang B, Zhang L, Yan S (2013) Manganese dioxide/iron oxide/acid oxidized multi-walled carbon nanotube magnetic nanocomposite for enhanced hexavalent chromium removal. Chem Eng J 234:256–265
52. Yao Y, Xu F, Chen M, Xu Z, Zhu Z. Proceedings of the 2010 5th IEEE international conference
53. Wang S, Ng CW, Wang W, Li Q, Li L (2012a) A comparative study on the adsorption of acid and reactive dyes on multiwall carbon nanotubes in single and binary dye systems. J Chem Eng Data 57(5):1563–1569
54. Qu S, Huang F, Yu S, Chen G, Kong J (2008) Magnetic removal of dyes from aqueous solution using multi-walled carbon nanotubes filled with Fe2O3 particles. J Hazard Mater 160(2–3):643–647
55. Ai L, Jiang J (2012) Removal of methylene blue from aqueous solution with self-assembled cylindrical graphene–carbon nanotube hybrid. Chem Eng J 192:156–163
56. Yao Y, Bing H, Feifei X, Xiaofeng C (2011) Equilibrium and kinetic studies of methyl orange adsorption on multiwalled carbon nanotubes. Chem Eng J 170:82–89
57. Farghali AA, Bahgat M, El Rouby WMA, Khedr MH (2012) Decoration of MWCNTs with CoFe2O4 nanoparticles for methylene blue dye adsorption. J Solution Chem 41:2209–2225
58. Sui K, Li Y, Liu R, Zhang Y, Zhao X, Liang H, Xia Y (2012) Biocomposite fiber of calcium alginate/multi-walled carbon nanotubes with enhanced adsorption properties for ionic dyes. Carbohyd Polym 90(1):399–406
59. Hosseinzadeh S, Hosseinzadeh H, Pashaei S, Khodaparast Z (2018) Synthesis of magnetic functionalized MWCNT nanocomposite through surface RAFT co-polymerization of acrylic acid and N-isopropyl acrylamide for removal of cationic dyes from aqueous solutions. Ecotoxicol Environ Saf 161:34–44
60. Sadegh H, Zare K, Maazinejad B, Shahryari-Ghoshekandi R, Tyagi I, Agarwal S, Gupta VK (2016) Synthesis of MWCNT-COOH-Cysteamine composite and its application for dye removal. J Mol Liq 215:221–228
61. Machado FM, Bergmann CP, Lima EC, Royer B, de Souza FE, Jauris IM, Calvete T, Fagan SB (2012) Adsorption of Reactive Blue 4 dye from water solutions by carbon nanotubes: experiment and theory. Phys Chem Chem Phys 14(31):11139–11153
62. Eskandarian L, Arami M, Pajootan E (2014) Evaluation of adsorption characteristics of multiwalled carbon nanotubes modified by a poly (propylene imine) dendrimer in single and multiple dye solutions: isotherms, kinetics, and thermodynamics. J Chem Eng Data 59(2):444–454
63. Mohammadi A, Veisi P (2018) High adsorption performance of β-cyclodextrin-functionalized multi-walled carbon nanotubes for the removal of organic dyes from water and industrial wastewater. J Environ Chem Eng 6(4):4634–4643
64. Ihsanullah, Al-Khaldi FA, Abu-Sharkh B, Abulkibash AM, Qureshi MI, Laoui T, Atieh MA (2016b) Effect of acid modification on adsorption of hexavalent chromium (Cr (VI)) from

aqueous solution by activated carbon and carbon nanotubes. Desalin Water Treat 57(16):7232–7244
65. Jung C, Heo J, Han J, Her N, Lee SJ, Oh J, Ryu J, Yoon Y (2013) Hexavalent chromium removal by various adsorbents: powdered activated carbon, chitosan, and single/multi-walled carbon nanotubes. Sep Purif Technol 106:63–71
66. Kumar ASK, Jiang SJ, Tseng WL (2015) Effective adsorption of chromium (VI)/Cr (III) from aqueous solution using ionic liquid functionalized multiwalled carbon nanotubes as a super sorbent. J Mate Chem A 3(13):7044–7057
67. Bandaru NM, Reta N, Dalal H, Ellis AV, Shapter J, Voelcker NH (2013) Enhanced adsorption of mercury ions on thiol derivatized single wall carbon nanotubes. J Hazard Mater 261:534–541
68. Gupta A, Vidyarthi SR, Sankararamakrishnan N (2014) Enhanced sorption of mercury from compact fluorescent bulbs and contaminated water streams using functionalized multiwalled carbon nanotubes. J Hazard Mater 274:132–144
69. Yaghmaeian K, Mashizi RK, Nasseri S, Mahvi AH, Alimohammadi M, Nazmara S (2015) Removal of inorganic mercury from aquatic environments by multi-walled carbon nanotubes. J Environ Health Sci Eng 13(1):1–9
70. Alimohammady M, Jahangiri M, Kiani F, Tahermansouri H (2017) A new modified MWCNTs with 3-aminopyrazole as a nanoadsorbent for Cd (II) removal from aqueous solutions. J Environ Chem Eng 5(4):3405–3417
71. Wu CH (2007) Studies of the equilibrium and thermodynamics of the adsorption of Cu2+ onto as-produced and modified carbon nanotubes. J Colloid Interface Sci 311(2):338–346
72. Lasheen MR, El-Sherif IY, Sabry DY, El-Wakeel ST, El-Shahat MF (2015) Removal of heavy metals from aqueous solution by multiwalled carbon nanotubes: equilibrium, isotherms, and kinetics. Desalin Water Treat 53(13):3521–3530
73. Stafiej A, Pyrzynska K (2007) Adsorption of heavy metal ions with carbon nanotubes. Sep Purif Technol 58(1):49–52
74. Abdel-Ghani NT, El-Chaghaby GA, Helal FS (2015) Individual and competitive adsorption of phenol and nickel onto multiwalled carbon nanotubes. J Adv Res 6(3):405–415
75. Goswami S, Ghosh UC (2005) Studies on adsorption behaviour of Cr (VI) onto synthetic hydrous stannic oxide. Water SA 31(4):597–602
76. Moosa AA, Ridha AM, Hussien NA (2016) Removal of zinc ions from aqueous solution by bioadsorbents and CNTs. Am J Mater Sci 6(4):105–114
77. Salam EA, Abou El-Nour KM, Awad AA, Orabi AS (2020) Carbon nanotubes modified with 5, 7-dinitro-8-quinolinol as potentially applicable tool for efficient removal of industrial wastewater pollutants. Arab J Chem 13(1):109–119
78. Velickovic ZS, Marinkovic AD, Bajic ZJ, Markovic JM, Peric-Grujic AA, Uskokovic PS, Ristic MD (2013) Oxidized and ethylenediamine-functionalized multi-walled carbon nanotubes for the separation of low concentration arsenate from water. Sep Sci Technol 48(13):2047–2058
79. Sankararamakrishnan N, Gupta A, Vidyarthi SR (2014) Enhanced arsenic removal at neutral pH using functionalized multiwalled carbon nanotubes. J Environ Chem Eng 2(2):802–810
80. Mageswari S, Kalaiselvan S, Shabudeen PS, Sivakumar N, Karthikeyan S (2014) Optimization of growth temperature of multi-walled carbon nanotubes fabricated by chemical vapour deposition and their application for arsenic removal. Mater Sci-Pol 32(4):709–718
81. Hu J, Chen C, Zhu X, Wang X (2009) Removal of chromium from aqueous solution by using oxidized multiwalled carbon nanotubes. J Hazard Mater 162(2–3):1542–1550
82. Di ZC, Ding J, Peng XJ, Li YH, Luan ZK, Liang J (2006) Chromium adsorption by aligned carbon nanotubes supported ceria nanoparticles. Chemosphere 62(5):861–865
83. Atieh MA (2011) Removal of chromium (VI) from polluted water using carbon nanotubes supported with activated carbon. Procedia Environ Sci 4:281–293
84. Huang ZN, Wang XL, Yang DS (2015) Adsorption of Cr (VI) in wastewater using magnetic multi-wall carbon nanotubes. Water Sci Eng 8(3):226–232

85. Lu W, Li J, Sheng Y, Zhang X, You J, Chen L (2017) One-pot synthesis of magnetic iron oxide nanoparticle-multiwalled carbon nanotube composites for enhanced removal of Cr (VI) from aqueous solution. J Colloid Interface Sci 505:1134–1146
86. Salam MA (2017) Preparation and characterization of chitin/magnetite/multiwalled carbon nanotubes magnetic nanocomposite for toxic hexavalent chromium removal from solution. J Mol Liq 233:197–202
87. Parlayici S, Eskizeybek V, Avcı A, Pehlivan E (2015) Removal of chromium (VI) using activated carbon-supported-functionalized carbon nanotubes. J Nanostruct Chem 5(3):255–263
88. Barakat MA, Al-Ansari AM, Kumar R (2016) Synthesis and characterization of Fe−Al binary oxyhydroxides/MWCNTs nanocomposite for the removal of Cr (VI) from aqueous solution. J Taiwan Inst Chem Eng 63:303–311
89. Deb AKS, Dwivedi V, Dasgupta K, Ali SM, Shenoy KT (2017) Novel amidoamine functionalized multi-walled carbon nanotubes for removal of mercury (II) ions from wastewater: combined experimental and density functional theoretical approach. Chem Eng J 313:899–911
90. Hadavifar M, Bahramifar N, Younesi H, Li Q (2014) Adsorption of mercury ions from synthetic and real wastewater aqueous solution by functionalized multi-walled carbon nanotube with both amino and thiolated groups. Chem Eng J 237:217–228
91. Zhang D, Yin Y, Liu J (2017) Removal of Hg2+ and methylmercury in waters by functionalized multi-walled carbon nanotubes: adsorption behavior and the impacts of some environmentally relevant factors. Chem Speciat Bioavailab 29(1):161–169
92. Moghaddam HK, Pakizeh M (2015) Experimental study on mercury ions removal from aqueous solution by MnO2/CNTs nanocomposite adsorbent. J Ind Eng Chem 21:221–229
93. Ramana DV, Yu JS, Seshaiah K (2013) Silver nanoparticles deposited multiwalled carbon nanotubes for removal of Cu (II) and Cd (II) from water: Surface, kinetic, equilibrium, and thermal adsorption properties. Chem Eng J 223:806–815
94. Liang J, Liu J, Yuan X, Dong H, Zeng G, Wu H, Wang H, Liu J, Hua S, Zhang S, Yu Z (2015) Facile synthesis of alumina-decorated multi-walled carbon nanotubes for simultaneous adsorption of cadmium ion and trichloroethylene. Chem Eng J 273:101–110
95. Kanthapazham R, Ayyavu C, Mahendiradas D (2016) Removal of Pb2+, Ni2+ and Cd2+ ions in aqueous media using functionalized MWCNT wrapped polypyrrole nanocomposite. Desalin Water Treat 57(36):16871–16885
96. Mubarak NM, Sahu JN, Abdullah EC, Jayakumar NS, Ganesan P (2015) Microwave assisted multiwall carbon nanotubes enhancing Cd (II) adsorption capacity in aqueous media. J Ind Eng Chem 24:24–33
97. Kabbashi N, Jamal Ibrahim D, Rosli NF (2011) Statistical analysis for removal of cad- mium from aqueous solution at high pH. Austr J Basic Appl Sci 5(6):440–446
98. Ge Y, Li Z, Xiao D, Xiong P, Ye N (2014) Sulfonated multi-walled carbon nanotubes for the removal of copper (II) from aqueous solutions. J Ind Eng Chem 20(4):1765–1771
99. Dichiara AB, Webber MR, Gorman WR, Rogers RE (2015) Removal of copper ions from aqueous solutions via adsorption on carbon nanocomposites. ACS Appl Mater Interfaces 7(28):15674–15680
100. Mubarak NM, Sahu JN, Abdullah EC, Jayakumar NS, Ganesan P (2015) Novel microwave-assisted multiwall carbon nanotubes enhancing Cu (II) adsorption capacity in water. J Taiwan Inst Chem Eng 53:140–152
101. Yang W, Ding P, Zhou L, Yu J, Chen X, Jiao F (2013) Preparation of diamine modified mesoporous silica on multi-walled carbon nanotubes for the adsorption of heavy metals in aqueous solution. Appl Surf Sci 282:38–45
102. Popuri SR, Frederick R, Chang CY, Fang SS, Wang CC, Lee LC (2014) Removal of copper (II) ions from aqueous solutions onto chitosan/carbon nanotubes composite sorbent. Desalin Water Treat 52(4–6):691–701
103. Zhao XH, Jiao FP, Yu JG, Xi Y, Jiang XY, Chen XQ (2015) Removal of Cu (II) from aqueous solutions by tartaric acid modified multi-walled carbon nanotubes. Colloids Surf, A 476:35–41

104. Tofighy MA, Mohammadi T (2016) Copper ions removal from aqueous solutions using acid-chitosan functionalized carbon nanotubes sheets. Desalin Water Treat 57(33):15384–15396
105. Saadat S, Karimi-Jashni A, Doroodmand MM (2014) Synthesis and characterization of novel single-walled carbon nanotubes-doped walnut shell composite and its adsorption performance for lead in aqueous solutions. J Environ Chem Eng 2(4):2059–2067
106. Ma X, Yang ST, Tang H, Liu Y, Wang H (2015) Competitive adsorption of heavy metal ions on carbon nanotubes and the desorption in simulated biofluids. J Colloid Interface Sci 448:347–355
107. Ouni L, Mirzaei M, Ashtari P, Ramazani A, Rahimi M, Bolourinovin F (2016) Isocyanate functionalized multiwalled carbon nanotubes for separation of lead from cyclotron production of thallium-201. J Radioanal Nucl Chem 310(2):633–643
108. Diva TN, Zare K, Taleshi F, Yousefi M (2017) Synthesis, characterization, and application of nickel oxide/CNT nanocomposites to remove Pb2+ from aqueous solution. J Nanostruct Chem 7(3):273–281
109. Farghali AA, Tawab HA, Moaty SA, Khaled R (2017) Functionalization of acidified multi-walled carbon nanotubes for removal of heavy metals in aqueous solutions. J Nanostruct Chem 7(2):101–111
110. AlOmar MK, Alsaadi MA, Hayyan M, Akib S, Ibrahim RK, Hashim MA (2016) Lead removal from water by choline chloride based deep eutectic solvents functionalized carbon nanotubes. J Mol Liq 222:883–894
111. Yang S, Li J, Shao D, Hu J, Wang X (2009) Adsorption of Ni (II) on oxidized multi-walled carbon nanotubes: effect of contact time, pH, foreign ions and PAA. J Hazard Mater 166(1):109–116
112. Lu C, Liu C, Su F (2009) Sorption kinetics, thermodynamics and competition of Ni2+ from aqueous solutions onto surface oxidized carbon nanotubes. Desalination 249(1):18–23
113. Kandah MI, Meunier JL (2007) Removal of nickel ions from water by multi-walled carbon nanotubes. J Hazard Mater 146(1–2):283–288
114. Chen C, Hu J, Shao D, Li J, Wang X (2009) Adsorption behavior of multiwall carbon nanotube/iron oxide magnetic composites for Ni (II) and Sr (II). J Hazard Mater 164(2–3):923–928
115. Lu C, Liu C (2006) Removal of nickel (II) from aqueous solution by carbon nanotubes. J Chem Technol Biotechnol: Int Res Process, Environ Clean Technol 81(12):1932–1940
116. Mubarak NM, Alicia RF, Abdullah EC, Sahu JN, Haslija AA, Tan J (2013) Statistical optimization and kinetic studies on removal of Zn2+ using functionalized carbon nanotubes and magnetic biochar. J Environ Chem Eng 1(3):486–495
117. Lu C, Chiu H (2008) Chemical modification of multiwalled carbon nanotubes for sorption of Zn2+ from aqueous solution. Chem Eng J 139(3):462–468
118. Ntim SA, Mitra S (2012) Adsorption of arsenic on multiwall carbon nanotube–zirconia nanohybrid for potential drinking water purification. J Colloid Interface Sci 375(1):154–159
119. Liné C, Larue C, Flahaut E (2017) Carbon nanotubes: Impacts and behaviour in the terrestrial ecosystem—A review. Carbon 123:767–785
120. Wang S, Ng CW, Wang W, Li Q, Hao Z (2012b) Synergistic and competitive adsorption of organic dyes on multiwalled carbon nanotubes. Chem Eng J 197:34–40
121. Vuono D, Catizzone E, Aloise A, Policicchio A, Agostino RG, Migliori M, Giordano G (2017) Study of adsorption behavior of multi-walled carbon nanotubes towards dyes applied in textile applications. Adv Sci Lett 23(6):5851–5854
122. Zare K, Sadegh H, Shahryari-Ghoshekandi R, Maazinejad B, Ali V, Tyagi I, Agarwal S, Gupta VK (2015) Enhanced removal of toxic Congo red dye using multi walled carbon nanotubes: kinetic, equilibrium studies and its comparison with other adsorbents. J Mol Liq 212:266–271
123. Machado FM, Bergmann CP, Fernandes TH, Lima EC, Royer B, Calvete T, Fagan SB (2011) Adsorption of Reactive Red M-2BE dye from water solutions by multi-walled carbon nanotubes and activated carbon. J Hazard Mater 192(3):1122–1131
124. Prola LD, Machado FM, Bergmann CP, de Souza FE, Gally CR, Lima EC, Adebayo MA, Dias SL, Calvete T (2013) Adsorption of Direct Blue 53 dye from aqueous solutions by multi-walled carbon nanotubes and activated carbon. J Environ Manag 130:166–175

125. Shirmardi M, Mesdaghinia A, Mahvi AH, Nasseri S, Nabizadeh R (2012) Kinetics and equilibrium studies on adsorption of acid red 18 (Azo-Dye) using multiwall carbon nanotubes (MWCNTs) from aqueous solution. E-J Chem 9(4):2371–2383
126. Ghaedi M, Hassanzadeh A, Kokhdan SN (2011) Multiwalled carbon nanotubes as adsorbents for the kinetic and equilibrium study of the removal of alizarin red S and morin. J Chem Eng Data 56(5):2511–2520
127. Jahangiri-Rad M, Nadafi K, Mesdaghinia A, Nabizadeh R, Younesian M, Rafiee M (2013) Sequential study on reactive blue 29 dye removal from aqueous solution by peroxy acid and single wall carbon nanotubes: experiment and theory. Iran J Environ Health Sci Eng 10(1):1–8
128. Alkaim AF, Sadik Z, Mahdi DK, Alshrefi SM, Al-Sammarraie AM, Alamgir FM, Singh PM, Aljeboree AM (2015) Preparation, structure and adsorption properties of synthesized multiwall carbon nanotubes for highly effective removal of maxilon blue dye. Korean J Chem Eng 32(12):2456–2462
129. Ruiz-García C, Pérez-Carvajal J, Berenguer-Murcia A, Darder M, Aranda P, Cazorla-Amorós D, Ruiz-Hitzky E (2013) Clay-supported graphene materials: application to hydrogen storage. Phys Chem Chem Phys 15(42):18635–18641
130. Walker DJ, Clemente R, Bernal MP (2004) Contrasting effects of manure and compost on soil pH, heavy metal availability and growth of Chenopodium album L. in a soil contaminated by pyritic mine waste. Chemosphere 57(3):215–224
131. Ghaedi M, Khajehsharifi H, Yadkuri AH, Roosta M, Asghari A (2012) Oxidized multiwalled carbon nanotubes as efficient adsorbent for bromothymol blue. Toxicol Environ Chem 94(5):873–883
132. Mishra AK, Arockiadoss T, Ramaprabhu S (2010) Study of removal of azo dye by functionalized multi walled carbon nanotubes. Chem Eng J 162(3):1026–1034
133. Wang JP, Yang HC, Hsieh CT (2010) Adsorption of phenol and basic dye on carbon nanotubes/carbon fabric composites from aqueous solution. Sep Sci Technol 46(2):340–348
134. Ai L, Zhang C, Liao F, Wang Y, Li M, Meng L, Jiang J (2011) Removal of methylene blue from aqueous solution with magnetite loaded multi-wall carbon nanotube: kinetic, isotherm and mechanism analysis. J Hazard Mater 198:282–290
135. Konicki W, Pełech I, Mijowska E, Jasińska I (2012) Adsorption of anionic dye Direct Red 23 onto magnetic multi-walled carbon nanotubes-Fe3C nanocomposite: kinetics, equilibrium and thermodynamics. Chem Eng J 210:87–95
136. Zhu HY, Jiang R, Xiao L, Zeng GM (2010) Preparation, characterization, adsorption kinetics and thermodynamics of novel magnetic chitosan enwrapping nanosized γ-Fe2O3 and multiwalled carbon nanotubes with enhanced adsorption properties for methyl orange. Biores Technol 101(14):5063–5069
137. Moradi O (2013) Adsorption behavior of basic red 46 by single-walled carbon nanotubes surfaces. Fullerenes, Nanotubes, Carbon Nanostruct 21(4):286–301
138. Derakhshan MS, Moradi O (2014) The study of thermodynamics and kinetics methyl orange and malachite green by SWCNTs, SWCNT-COOH and SWCNT-NH2 as adsorbents from aqueous solution. J Ind Eng Chem 20(5):3186–3194
139. Zhao D, Zhang W, Chen C, Wang X (2013) Adsorption of methyl orange dye onto multiwalled carbon nanotubes. Procedia Environ Sci 18:890–895
140. Ahmad A, Razali MH, Mamat M, Mehamod FSB, Amin KAM (2017) Adsorption of methyl orange by synthesized and functionalized-CNTs with 3-aminopropyltriethoxysilane loaded TiO2 nanocomposites. Chemosphere 168:474–482
141. Robati D, Mirza B, Ghazisaeidi R, Rajabi M, Moradi O, Tyagi I, Agarwal S, Gupta VK (2016a) Adsorption behavior of methylene blue dye on nanocomposite multi-walled carbon nanotube functionalized thiol (MWCNT-SH) as new adsorbent. J Mol Liq 216:830–835
142. Sobhanardakani S, Zandipak R, Sahraei R (2013) Removal of Janus Green dye from aqueous solutions using oxidized multi-walled carbon nanotubes. Toxicol Environ Chem 95(6):909–918
143. Abbasi M (2017) Synthesis and characterization of magnetic nanocomposite of chitosan/SiO2/carbon nanotubes and its application for dyes removal. J Clean Prod 145:105–113

144. Madrakian T, Afkhami A, Ahmadi M, Bagheri H (2011) Removal of some cationic dyes from aqueous solutions using magnetic-modified multi-walled carbon nanotubes. J Hazard Mater 196:109–114
145. Mahmoodian H, Moradi O, Shariatzadeha B, Salehf TA, Tyagi I, Maity A, Asif M, Gupta VK (2015) Enhanced removal of methyl orange from aqueous solutions by poly HEMA–chitosan-MWCNT nano-composite. J Mol Liq 202:189–198
146. Saber-Samandari S, Saber-Samandari S, Joneidi-Yekta H, Mohseni M (2017) Adsorption of anionic and cationic dyes from aqueous solution using gelatin-based magnetic nanocomposite beads comprising carboxylic acid functionalized carbon nanotube. Chem Eng J 308:1133–1144
147. Al-Khaldi FA, Abusharkh B, Khaled M, Atieh MA, Nasser MS, Saleh TA, Tyagi I, Gupta VK (2015b) Adsorptive removal of cadmium (II) ions from liquid phase using acid modified carbon-based adsorbents. J Mol Liq 204:255–263
148. Bhutani R, Tharani K, Sudha K, Tomar Y (2020) Design of a cogeneration plant for sugar industries using renewable energy resources. J Stat Manag Syst 23(1):181–190
149. Boukhalfa N, Boutahala M, Djebri N, Idris A (2019) Kinetics, thermodynamics, equilibrium isotherms, and reusability studies of cationic dye adsorption by magnetic alginate/oxidized multiwalled carbon nanotubes composites. Int J Biol Macromol 123:539–548
150. Chen B, Zhu Z, Ma J, Qiu Y, Chen J (2013a) Surfactant assisted Ce–Fe mixed oxide decorated multiwalled carbon nanotubes and their arsenic adsorption performance. J Mater Chem A 1(37):11355–11367
151. Chen R, Chai L, Li Q, Shi Y, Wang Y, Mohammad A (2013b) Preparation and characterization of magnetic Fe3O4/CNT nano particles by RPO method to enhance the efficient removal of Cr(VI). Environ Sci Pollut Res 20(10):7175–7185
152. Iijima S (1991) Helical microtubules of graphitic carbon Nature 354:56–58. Find this article online
153. Li YH, Wang S, Luan Z, Ding J, Xu C, Wu D (2003a) Adsorption of cadmium (II) from aqueous solution by surface oxidized carbon nanotubes. Carbon 41(5):1057–1062
154. Li YH, Ding J, Luan Z, Di Z, Zhu Y, Xu C, Wu D, Wei B (2003b) Competitive adsorption of Pb2+, Cu2+ and Cd2+ ions from aqueous solutions by multiwalled carbon nanotubes. Carbon 41(14):2787–2792
155. Mubarak NM, Sahu JN, Abdullah EC, Jayakumar NS (2016a) Rapid adsorption of toxic Pb (II) ions from aqueous solution using multiwall carbon nanotubes synthesized by microwave chemical vapor deposition technique. J Environ Sci 45:143–155
156. Mubarak NM, Sahu JN, Abdullah EC, Jayakumar NS, Ganesan P (2016b) Microwave-assisted synthesis of multi-walled carbon nanotubes for enhanced removal of Zn (II) from wastewater. Res Chem Intermed 42(4):3257–3281
157. Robati D, Bagheriyan S, Rajabi M, Moradi O, Peyghan AA (2016b) Effect of electrostatic interaction on the methylene blue and methyl orange adsorption by the pristine and functionalized carbon nanotubes. Physica E 83:1–6
158. Shadbad MJ, Mohebbi A, Soltani A (2011) Mercury (II) removal from aqueous solutions by adsorption on multi-walled carbon nanotubes. Korean J Chem Eng 28(4):1029–1034
159. Wang HJ, Zhou AL, Peng F, Yu H, Chen LF (2007a) Adsorption characteristic of acidified carbon nanotubes for heavy metal Pb (II) in aqueous solution. Mater Sci Eng, A 466(1–2):201–206
160. Wang H, Zhou A, Peng F, Yu H, Yang J (2007b) Mechanism study on adsorption of acidified multiwalled carbon nanotubes to Pb (II). J Colloid Interface Sci 316(2):277–283
161. Wang SG, Gong WX, Liu XW, Yao YW, Gao BY, Yue QY (2007c) Removal of lead (II) from aqueous solution by adsorption onto manganese oxide-coated carbon nanotubes. Sep Purif Technol 58(1):17–23

Removal of Dyestuffs from Dyeing Industry Effluents by Microbe-Mediated Oxide Nanoparticles (NPs) Via Adsorption for Sustainable Development

Jyotirmoy Sarma, Nirupamjit Sarmah, Neha Dhar, Roopa Kumari, and Sanchayita Rajkhowa

Abstract Dyes from dyeing industrial effluents cause serious threats to the environment. Hence, the removal of such pollutants should be of utmost interest. In this context 'adsorption process' plays a significant role. Although a plethora of research work on the mitigation of dyes from industrial wastes has been carried out across the world, the availability of literature on the use of oxide nano-materials synthesized from microorganisms for the removal of toxic dyes is lacking. Microbial synthesis of oxide nanoparticles (both magnetic and non-magnetic) is not only an environmentally benign and green process, but is also as significantly efficient as other conventional chemical methods. The unique properties exhibited by nanoparticles (NPs) are associated with their size and shape which can easily be fabricated as per the requirement making them specific for various applications. Through this chapter, an attempt has been made to provide collective information about the adsorption science involved during the uptake of toxic dyes by biosynthesized oxide nanoparticles.

Keywords Adsorption · Pollutant dyes · Microorganisms · Oxide nanoparticles (NPs)

J. Sarma
Department of Chemistry, Assam Don Bosco University, Tapesia Gardens, 782402 Guwahati, Assam, India

N. Sarmah
Department of Environment and Forests, Government of Assam, Assam, India

N. Dhar
Chemist/Microbiologist, M/S. Fresho Beverages, Assam, India

R. Kumari
Department of Applied Sciences and Humanities, Indira Gandhi Delhi Technical University for Women, Delhi, India

S. Rajkhowa (✉)
Department of Chemistry, Haflong Government College, Dima Hasao, Assam, India 788819
e-mail: sanchayita.rajkhowa@gmail.com

P. Singh (ed.), *Dye Pollution from Textile Industry*, SDGs and Textiles,
https://doi.org/10.1007/978-981-97-5341-3_8

1 Introduction

Dyes are one of the essential chemicals which play a major role in modern-day lifestyle. Although it has been in use since ancient times, the extensive application and the synthetic processes involved during its manufacturing in recent years have imposed a threat to ecological balance at a global level. By definition "*Dyes are substances that, when applied to a substrate provide colour by a process that alters, at least temporarily, any crystal structure of the coloured substances*" [5, 65]. The application of dyes is not only limited to textile industries, but also extended to food, plastics, cosmetics, pharmaceutical, paper and pulp, and photographic industries. Dyes have versatile characteristics in terms of their mode of action, e.g. adherence to the surface, formation of complexes/covalent bonds with metals and salts, and physisorption or mechanical retention for imparting colour to the substrate [5, 65]. Dyes may be natural or synthetic depending on their origin. Nevertheless, dyes are generally classified as acid, basic, direct, disperse, metal complex, mordant, reactive, sulphur dyes, and vat dyes on the basis of their chemical structure, application, and nature of functional groups. These functional groups known as chromophores are responsible for displaying a particular colour or shade by a dye. These groups are composed of a group of atoms like C, H, N, O, S, etc. and fall into categories such as azo, nitro, carbonyl, methane, arilmethane, anthraquinone, and so on. Other than chromophores, several other electron-donating or withdrawing groups may also be present in the chemical structure of dye molecules. Such groups are known as auxochromes which intensify the colour of chromophores, e.g. carboxyl, hydroxyl, amine, and sulphonate [3–77]. An annual production of around 800,000 tons of dyes worldwide with more than 10,000 varieties of dyes are used solely by textile industries, out of which 70% fall into the category of azo dyes [6–56]. Being of synthetic origin, azo dyes contain one or more azo groups, i.e. –N = N– attached to other aromatic groups like phenols or amines via double bonds in their structure [49, 87]. The azo group acts as a chromophore which is responsible for imparting colour to the compound. About 10–15% of these synthetic dyes are released into various water bodies as industrial effluents during dyeing practices [68]. These dyes remain in the environment for a longer period of time owing to their high stability under light, temperature, and inertness towards water, other chemicals like detergents, bleaching agents, etc. [21]. In addition, their complex structure makes it difficult to undergo biodegradation [27, 69]. Apart from azo dyes, most of them are readily soluble in water thus making the process of separation of dyes more cumbersome.

From earlier times, textile industries have made a huge contribution to the economic growth of a nation due to their significant share in creating job opportunities and employment and direct or indirect involvement with other small/medium enterprises. On the basis of the nature of fibre industries produce, there are two major fabric processes involved: dry and wet processes. In the former process, the wastes produced are generally solids while the latter generates liquid wastes. The wet process is found to be the major source of water pollutants like dyes, acids, NaOH, chrome, starch, and other toxic chemicals. Nevertheless, textile industries consume a relatively larger volume of potable water and various chemicals than other similar industries. The textile industry consumes an average of about 200 L of potable water per Kg of the product during manufacturing with an estimated wastewater generation of ~17–20% in the dyeing and finishing process [43, 94]. Textile wastewater (TWW) is characterized by a high content of organic compounds and salinity, low biodegradability, high fluctuation in pH and colour, as well as chemical oxygen demand (COD), and biochemical oxygen demand (BOD) imbalances. The coloured organic toxic compounds, surfactants, and salts are the major pollutants in such recalcitrant effluents [55]. Additionally, a trace amount of dyes, i.e. less than 1 mgL^{-1} (in the case of some dyes) if present in water can significantly affect the quality of water [6, 95]. Therefore, direct discharge of untreated, semi-treated, or poorly treated TWW into any water bodies greatly influences the natural underwater habitants. Several notable impacts are

- Depletion of dissolved oxygen (DO) due to overgrowth of aquatic plants due to consumption of organic compounds and nutrients.
- Eutrophication of aquatic system.
- Imbalance in the oxygen-depleting effects of waste contaminants.
- Decrease in water transparency and thus inhibition in the photosynthetic process.
- Extinction of several aquatic species due to high toxicity.
- Presence of elements copper, zinc, arsenic, and chromium in trace quantities.

There is a variation in the quantity of water consumed by various textile industries depending on the process employed for dyeing and the type of fabric used. Figure 1 illustrates the amount of water used for individual processes in such industries. The water discharged from these industries as effluents is substandard containing dyes, chemicals, suspended solids (SS), etc. Wastewater containing dyes poses multiple health disorders like nausea, vomiting, breathlessness, skin irritation and allergies, mental confusion, and can also cause cancer [84].

An urgent need for the development and implementation of an efficient, cost-effective, and environmentally benign TWW treatment process has been realized to safeguard the ecosystem from further deterioration. A number of technique like advanced oxidation processes, electrochemical, and valorization processes are amongst the most frequently employed sustainable techniques to minimize the toxicity of industrial effluents [35]. The major disadvantage associated with these processes is the inability to bear their high cost by all industries and the effect will reflect on the final price of the product. Hence, nanotechnology and nanoscience have opened doors for WWT in a cost-effective and sustainable way. WWT via nanotechnological pathways was found to be efficient as compared to other methods due to the fact that the nanoparticles (NP) are smaller in size with enhanced surface area and better physicochemical properties than macro-particles [10]. Although there's no universal process to treat all kinds of dye effluents, our aim is to explore the removal of dyes from industrial wastes by adsorption onto microbe-mediated oxide nanoparticles with minimum environmental pollution. However, complete removal of dyestuffs from industrial effluents with a single process is difficult. A number of physical–chemical and microbial processes have emerged over the years to achieve

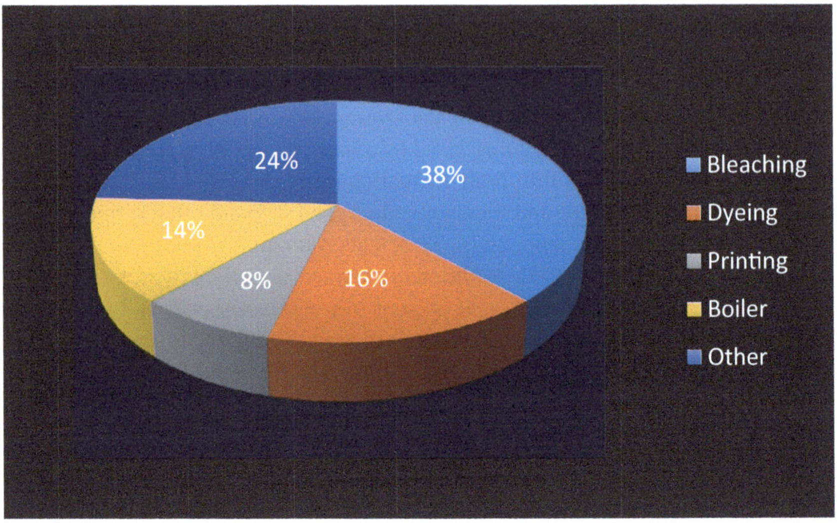

Fig. 1 Amount of water consumed in various processes involved in textile industry [64]

the highest satisfactory output with minimum toxic material present in the post-treated wastewater. Two or more processes may be carried out successively for this. Figure 2 depicts a number of such methods and processes that could be adopted for TWW treatment.

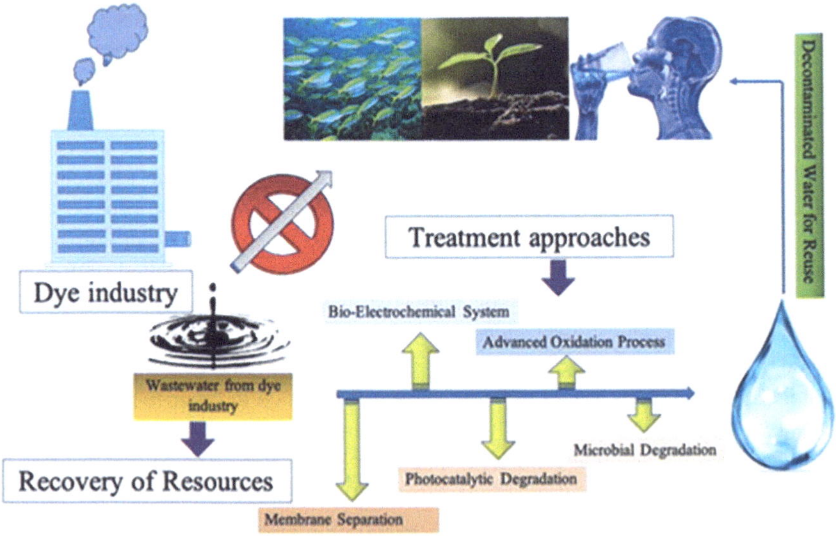

Fig. 2 Classification of wastewater treatment. *Source* Shindhal et al. [82] (Open access)

Nevertheless, there are several metal oxide nanoparticles (MONPs) extensively prepared and applied for WWT, e.g. Copper oxide (CuO), cerium oxide (CeO_2), iron oxide (Fe_2O_3), manganese oxide (MnO_2), titanium dioxide (TiO_2), vanadium oxide (V_2O_5), and zinc oxide (ZnO) [25–100]. The synthesis of such MONPs from microbes and extraction of other microorganisms is a green process for the remediation of wastewater. Having significantly unique yet better stability, and magnetic and electrical properties, MONPs showcase higher catalytic activity, reactivity, and adsorptivity in contrast to other conventional materials [14]. For instance, iron-NPs being environmentally benign are extensively applicable in WWT for their high reactivity with water and magnetic susceptibility [15]. Although there is limited information on microbial-mediated oxide NPs, this chapter will discuss their role in the remediation of dyeing industrial effluents and the removal of dyes from the same.

1.1 Microorganism-Assisted Nanotechnology

Biofabricated NPs along with the simultaneous use of microbes add sustainability and eco-friendliness to the field of nanotechnology. In contrast with the chemically synthesized NPs, biofabricated ones are produced from microorganisms (bacteria and viruses), fungi, or plant extracts. These sources can readily form metallic NPs by reducing salts of the metal complex. NPs thus formed are highly stable and easily solidify in aqueous media due to co-precipitation or the addition of bioactive substances on the surface of NPs. Figure 2 shows the general route to synthesize NPs by microbes that involves 3 primary steps: (i) metal arrest, (ii) reduction in the presence of an enzyme, and finally (iii) capping. Some biological molecules like amino acids, polysaccharides, and proteins actively participate in microbial extraction which eventually enhances NP stabilization. Figure 3 depicts a schematic representation of various NPs from a number of sources, their applications as well as characterization.

Aspergillus tubingensis (STSP 25), isolated from the rhizosphere of *Avicennia officinalis* in the Sundarbans, India, has successfully facilitated the synthesis of iron oxide nanoparticles through microbe-assisted processes [51]. Moreover, copper NPs synthesized from microbe 'Escherichia sp. SINT7'-mediated processes were found effective in the decomposition of hazardous dyes like reactive black-5, congo red, direct blue-1, and malachite green up to ~83, 97, 88, and 90%, respectively, with merely 25 mgL^{-1} concentration of NPs from industrial effluents [62]. Iron-sulphur NPs synthesized from *Pseudoalteromonas sp.* CF10-13 could readily reduce dye like naphthol green B by supplying electrons outside the cell [52]. Besides being a superior technology and environmentally benign, the biogenic NPs have several other advantages such as the production of catalytic enzymes by the microorganism which further helps in wastewater remediation. For example, microorganisms could be a source of catalytic enzymes in combination with NPs in waste effluent remediation. The interaction of inorganic entities with biological species is the basis of the

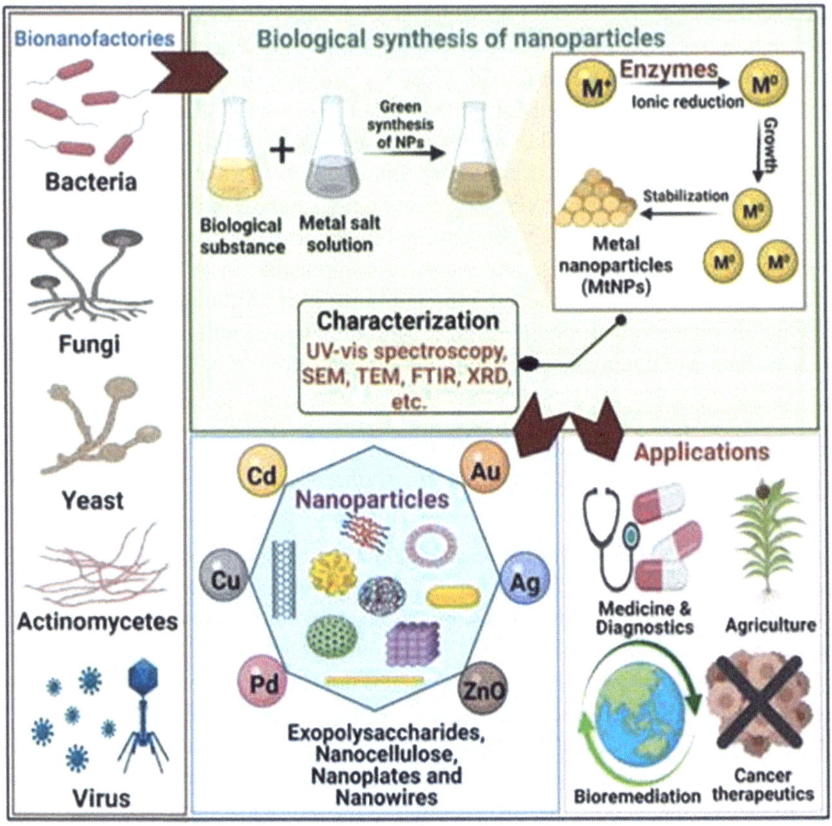

Fig. 3 Microbial synthesis of various nanoparticles with characterization and applications [47] (Open access)

survival of life on earth. The smaller the size of NPs is more suitable for their application in wastewater remediation. Their specific physical, chemical, and biological characteristics improve their usability in numerous fields.

1.1.1 Bacteria-Induced Production of NPs

Bacterial microorganisms which can reduce metals are used as green catalysts in bioremediation and synthesis. Genus *Shewanella* synthesizes a wide range of metal oxides via microbial respiration. Transfer of electrons occurs from the reduced organic substrates to oxidized inorganic materials by means of microbial anaerobic dissimilation process leading to crystallization during bioremediation technique. Studies have also revealed that the genus Shewanella acts as an electron donor during the oxidation of organic acids whereas electron acceptors in inorganic metal reduction

[36, 38]. During the bioreduction process, the genus Shewanella excretes nanowires and flavins extracellularly [11]. A few reports have been documented on the biosynthesis of spherical copper NPs using dead *Rhodotorula mucilaginosa* biomass by a biosorptive method that turns out to be an effective pollutant remediation technique. Nordmeier et al. [63] reported metallic molybdenum NP synthesis by *Clostridium pasteurianum* [63]. In addition, silver NPs having antibacterial properties were also synthesized using *Shewanella oneidensis* MR-1 [88].

S. Oneidensis MR-1 interestingly uses the toxic, water-soluble tellurite (Te) (that acts as an electron acceptor) for the synthesis of needle-shaped Te nanorod intracellularly and/or extracellularly [45, 46]. NPs containing gold (Au) and Te can also be biosynthesized from the crude extracts of 'Enterobacter cloacae MF01' microbe both under aerobic and anaerobic conditions [20]. A different work reports the involvement of gram-negative bacteria in the formation of biogenic selenium (Se) nanostructures under aerobic conditions [67]. Several reports as well advocate the significant role of volatile sulphur compounds (VSCs) in the biosynthetic formation of cadmium sulphide quantum dots (QDs) using *Pseudomonas fragi* GC01 [28]. *Escherichia* coli and *Enterobacter sp.* are also used in synthesizing ruthenium (Ru), ruthenium-palladium (Ru-Pd) NPs, and mercury (Hg) NPs, respectively [32, 86].

1.1.2 Fungus-Mediated Synthesis of NPs

NPs synthesised by fungus are preferred over those by microbes because of their higher resistance of fungal mycelial mesh against enhanced flow and agitation in the bioreactors. Chitin present in fungal cells is responsible for the synthesis of NPs by complexation with heavy metals. Fungi are found to be more productive in NP synthesis as they are involved in strong aggregate formation, and various metabolic reactions with higher efficiency [79]. *Penicillium chrysogenum* NG85 and *Fusarium chlamydosporum* NG30 could successfully synthesize silver (Ag) NPs using cell-free filtrate prepared from biomass. The enzymes and proteins present in that filtrate facilitated the formation of AgNPs by two steps: 1. Reduction of Ag^+ ions into Ag atoms by enzymes; 2. Ag atoms were stabilized by encapsulation with proteins [44]. The extract of '*Salacia chinensis*' leaves commonly called 'Phomopsis liquidambaris extracellular filtrate' provides the necessary biomolecules for reduction with encapsulation of AgNPs bestowed with antibacterial properties [81]. Biomass extract filtrates obtained from *Penicillium ochrochloron*, *Aspergillus niger*, and *Aspergillus tamarii* are proven to be green nanofactories for the production of fungus. Zinc (Zn) NPs are synthesized from endophytic fungus *Periconia sp.* extracted from *Balanites aegyptiaca* leaves [29]. Copper oxide NP could also be synthesized from copper sulphate solution using *Trichoderma asperellum* water extract [78].

2 Nature of Dyes from Dyeing Industry and Their Toxicity

Dyeing industry effluents contain a number of coloured toxic chemicals, suspended solid particles, dissolved solids, salts, and acids. Many of these dye molecules are non-biodegradable, interfering with sunlight penetration into aquatic environments that retards the growth and development of aquatic plants and animals. Many a time, dyes may also induce abnormal growth in such plants which can disturb the equilibrium of the aquatic ecosystem. Commonly used industrial dyes can be categorized into various classes as listed in Table 1.

Despite being commercially viable, synthetic dyes have adverse effects on all living species. Textile dyes primarily contain sulphur, acetic acid, enzymes, naphthol, nitrates, vat dyes, surfactants, and several auxiliary chemicals, and heavy metals such as arsenic, cadmium, copper, cobalt, lead, mercury, and nickel. Nevertheless some hazardous textile effluents including dye fixing agents (formaldehyde-based), chlorine-based stain removers, and hydrocarbons as softeners along with non-biodegradable toxic dyes are also noteworthy. In addition, these organic compounds are reactive in nature and form a number of carcinogenic products by reacting with disinfectant materials. The overall formation of colloidal suspension imparts turbidity and reduces sunlight penetration which ultimately interferes with the oxygen transport process at the air–water interface. This also results in water clogging in the fields and loss of soil productivity. Direct disposal of untreated water streams into

Table 1 Classification of industrial dyes and their properties [34]

Types of dyes	Chemical class	Properties
Acid dyes	Azo (including premetallized), anthraquinone, triphenylmethane, azine, xanthene, nitro, and nitroso	Water-soluble
Basic dyes	Diazahemicyanine, triarylmethane, cyanine, hemicyanine, thiazine, oxazine, and acridine	Water-soluble
Disperse dyes	Azo, anthraquinone, styryl, nitro, and benzodifuranone	Water-insoluble and nonionic dyes
Direct dyes	Polyazo compounds, along with some stilbenes, phthalocyanines, and oxazines	Water-soluble and anionic dyes
Reactive dyes	Chromophoric groups such as azo, anthraquinone, triarylmethane, phthalocyanine, formazan, and oxazine	Water-soluble
Solvent dyes	Predominantly azo and anthraquinone, but phthalocyanine and triarylmethane are also used	Solvent-soluble while water-insoluble and nonpolar or little polar
Sulphur dyes	thiazoles, thiazone, thianthrene, and phenothiazonethioanthrone	Water-soluble
Vat dyes	Anthraquinone (including polycyclic quinones) and indigoids	Water-insoluble

drains, rivers, and oceans will make the potable water unfit for consumption. It is thus necessary to remove these toxic species before releasing the effluents into the environment.

3 Microbe-Mediated Oxide Nanoparticles

3.1 Synthesis Involving Bacteria

In the natural environment, bacteria thrive abundantly, exhibiting adaptability to challenging conditions, rapid mutation, and cost-effective cultivation, making them ideal green precursors for the synthesis of metal oxides through microbe-mediated processes. The growth conditions as temperature, oxygenation, etc. can be controlled as per the nature of the bacteria. Controlled use of growth conditions is important in view of the size variations of nanoparticles that are required for various applications such as catalysis, antimicrobials, and also in optics [66]. Bacteria, widely prevalent in nature, exhibit diverse shapes such as bacilli (rod-shaped), cocci (spherical), and spirochaetes (spiral), with sizes spanning the microscale range. The utilization of bacteria in synthesizing nanoparticles, as an eco-friendly alternative to traditional chemical and physical methods, has witnessed significant growth, emerging as a prominent research domain in the field of green nanotechnology [40]. Over the past decade, bacteria have proven to be highly effective in various commercial biotechnology applications, including bioremediation and bioleaching [31]. In recent analyses, researchers have employed biomass or cellular extracts from diverse bacterial strains to fabricate a range of metal oxide nanoparticles, each exhibiting distinct morphologies. Certain microorganisms frequently employed for nanoparticle synthesis encompass magnetotactic bacteria and S-layer bacteria, distinguished by protein-coated cell walls [40]. The majority of metal particles are poisonous to these bacteria, so, as a protective measure, these organisms convert these metal particles to nanoparticles. Under significant conditions, bacteria release specific proteins that are subsequently employed to stabilize the nanoparticles. The ability of these bacterial strains to transform metal particles into nanoparticles has been attributed to intracellular proteins, enzymes, and various biochemical species [40]. The advantage of producing metal oxide nanoparticles through bacterial synthesis lies in its capacity to minimize the reliance on harmful and costly chemicals, generate particles devoid of additional chemical residues, and allow for the scalability of the method. Nevertheless, this approach comes with certain limitations, such as the time-consuming process of screening microbial preparations, the meticulous assessment of bacterial cultures to prevent contamination, challenges in controlling the morphology and crystallinity of particles, and the typically elevated cost associated with bacterial culture media [1].

3.1.1 Bacterial-Controlled Formation of Metal Oxide Nanoparticles for Sustainable Applications

Biomineralization, synonymous with bacterial-controlled agglomeration, is predominantly observed in microbes with iron-reducing capabilities, such as magnetotactic microorganisms. Magnetotactic microbes constitute a diverse group of prokaryotes exhibiting a range of cellular morphologies, encompassing vibrio, cocci, spirilla, and bacilli shapes [24]. These species produce phospholipid film-bound vesicles that are encircled by a direct cluster of magnetite nanocrystals (magnetosome). The intricate, multistep process involves the organization of phospholipid membrane-bound magnetosome vesicles, the transportation of iron into these vesicles, and culminates in the crystallization of iron oxide nanoparticles, constituting the fundamental steps in the formation of iron oxide nanoparticles, specifically magnetite. Facilitated by iron reductases and iron-chelating siderophores, the transport of iron into magnetosome vesicles is promoted. The reduced iron migrates from the cellular surface into the phospholipid layer. Finally, ferrous particles are oxidized, dried, and accelerated to produce magnetite [9]. The magnetotactic microbe-mediated synthesis results in iron oxide nanoparticles with elevated chemical purity, minimal size dispersion, and a distinct arrangement within their crystallographic morphology for sustainability [54].

3.1.2 Bacterial-Mediated Preparation of Metal Oxide Nanoparticles

In this type of synthesis, microorganisms alter their surroundings to establish favourable conditions for the extracellular synthesis of inorganic nanoparticles. To reduce the toxicity, they use a specific safeguard component against high metal particle concentration thus creating nanoparticles. The process involves altering the solvency of metal particles, modifying their redox state and extracellular complexation, and inducing the precipitation of metal particles. The zinc oxide nanoparticles created by *Rhodococcus sp.* were tolerably stable with hexagonal and spherical shapes, with an average range of 100–120 nm diameters [1]. In a separate investigation, TiO_2 nanoparticles were synthesized using a culture arrangement of Lactobacillus cells through heating at temperatures up to 60 °C for a duration of 10–20 min.

3.2 Fungal Synthesis

Studies have revealed that, beyond bacteria, fungi showcase impressive capabilities in synthesizing diverse metal oxide nanoparticles with a range of sizes and shapes, presenting an opportunity to develop novel, environmentally friendly biosynthesis techniques [42]. A fungus is a category of eukaryotes, encompassing yeasts, moulds, and other microorganisms. Fungi possess biomolecules like proteins and enzymes on their cell surfaces that facilitate the synthesis of metal oxide nanoparticles in comparison to other microorganisms [59]. Under gentle reaction conditions, fungi can serve as stabilizing agents, enabling the creation of diverse metal oxide nanoparticles from their respective metal ion precursors. When fungi synthesize metal oxide nanoparticles in a green way, the secreted enzymes from the fungal cell support the conversion of metal ions into their counter metal oxide nanoparticles. Likewise, metal ions trigger specific proteins in fungi that hydrolyze the metal ion complexes, leading to the formation of metal oxide nanoparticles [47]. Additionally, the fungi's proteins serve as coatings for metal oxide nanoparticles, making them more stable and environmentally friendly [70].

In general, the fungus produces metal oxide nanoparticles that are economically viable, environmentally friendly, and simple to process downstream. Fungal tolerance and metal bioaccumulation are superior to those of bacteria. The presence of mycelia in these organisms provides fungi with a substantial surface area, facilitating extensive interactions with metal salts [83]. The growth of fungal biomass is also faster than that of bacteria, which contributes to the faster formation of nanoparticles in fungi. Fungi also produce more proteins under similar synthesis conditions (Edebo et al. 2003). Additionally, the majority of the metal oxide nanoparticles produced by fungi-induced synthesis have a monodisperse morphology [97]. In a standard synthesis protocol, the cultivated fungal biomass is introduced to the metal ion precursors and agitated for the necessary duration and temperature to produce the corresponding metal oxide nanoparticles (MO-NPs) [70]. The zinc nitrate can be converted into ZnO nanoparticles effectively in the presence of *Aspergillus fumigatus* TFR-8. In addition, nanoparticles of silica (SiO_2), titania (TiO_2), and magnetite (Fe_2O_3) were synthesized successfully with the fungus Fusarium oxysporum (Sastry et al. 2006). Similarly, Fusarium Oxysporum was used to create metal oxide nanoparticles such as bismuth oxide (Bi_2O_3) [93] and barium titanate ($BaTiO_3$) (Sastry et al. 2006). Utilizing the Saccharomyces cerevisiae fungus to prepare TiO_2 nanoparticles at room temperature was another similar innovation.

4 Adsorption Process

The phenomenon of adsorption can be characterized as the process wherein contaminants present in a solution undergo a transfer onto the surface of a solid adsorbent. Adsorption is a widely practiced technique for wastewater remediation or treatment, to control environmental pollution, or clean potable water. To understand the adsorption processes, the essential prerequisite is to study the kinetics and isotherms. Fundamentally, adsorption is the process of accumulation of a substance at a surface or an interface. Accordingly, in the case of water treatment, adsorption occurs at an interface between solid adsorbent and contaminated water. The polluting contaminants of water can be removed via adsorption onto different adsorbents of natural or synthetic sources.

4.1 Adsorption Kinetics

The study of the rate of an adsorption process can be achieved via the so-called 'adsorption kinetics'. The information about the time required for an adsorption process helps in the study of adsorption isotherms for a particular system concerned. More specifically, the surface (adsorbent) saturation time in an adsorption process is found to be dependent upon several factors like (a) the nature of the adsorption couple, i.e. adsorbate/adsorbent, (b) experimental pH, and (c) experimental temperature. In the scope of industrial application, as an example during the removal of toxic dyes from wastewater, the solution carrying the adsorbate, i.e. the dye molecules are allowed to be in contact with the surface, i.e. the adsorbent for a sufficient period of time. During this time interval, the following kinetic steps are found to govern the overall adsorption process:

(a) Rate of transport of adsorbate from the bulk phase to an adsorbent surface
(b) Rate of diffusion across the film surrounding adsorbent particles
(c) Rate of diffusion of adsorbate in the pores of an adsorbent
(d) Rate of adsorption of adsorbate onto the adsorbent surface.

Amongst these, the slowest step is helpful for finding the overall rate of the reaction. So the rate determining step (RDS) of an adsorption process is commonly the slowest step. Kinetic equations are formulated in relation to the rate-determining step (RDS) and then tailored to align with the observed experimental data. Some commonly used kinetic models in the context of the kinetics of the adsorption process are Lagergren's first-order model, the pseudo-second-order model, and the Elovich model. However, Lagergren's first-order equation and the pseudo-second-order equation have found extensive application in the adsorption of a solute from a liquid solution (Table 2) [22, 74]. Moreover, the pseudo-second-order rate expression was also utilized in describing the chemisorption process involving the exchange of electrons between the adsorbent and adsorbate.

4.2 Adsorption Isotherms

'Adsorption isotherms' means the study of variation in the adsorption density of an adsorbate upon an adsorbent surface with respect to the change in its equilibrium concentration at a fixed temperature. This type of study helps in predicting the mechanisms of adsorption and the thermodynamic feasibility of the adsorption process. The information about the various adsorption models is essential for the proper application of the adsorption process in industry, especially for dye removal. Various isotherm models, including Freundlich, Langmuir, BET, Redlich-Peterson, Toth, Dubinin-Radushkevich (D-R), Sips, Frumkin, Harkins-Jura (H-J), Halsey, Henderson, and Temkin isotherms, have proven to be relevant and applicable throughout the evolution of adsorption science. Nevertheless, the prevalence of the Langmuir and Freundlich

Table 2 Common adsorption kinetics and isotherm models (self-made)

Kinetic models	Equation	Parameters
Lagergren first order	$\frac{dq_t}{dt} = k_1(q_e - q_t)$	q_e and q_t are the amount of an adsorbate adsorbed onto an adsorbent at equilibrium and at time, t, and k_1 is the first-order rate constant
Pseudo first order	$\ln(q_e - q_t) = \ln q_e - K_1 t$	do
Pseudo second order (linear)	$\frac{t}{q_t} = \frac{1}{k_2 q_e^2} + \frac{t}{q_e}$	q_e and q_t are the amount of an adsorbate adsorbed onto an adsorbent at equilibrium and at time, t, and k_2 is the second-order rate constant
Pseudo second order (non-linear)	$q_t = \frac{q_e^2 k_2 t}{q_e k_2 t + 1}$	do
Isotherm models		
Freundlich	$\log q_e = \log K_f + n \log C_e$	C_e and q_e are the equilibrium concentration and amount of adsorbate adsorbed respectively, K_f is the Freundlich co-efficient representing adsorption capacity and n is the adsorption intensity, $n < 1$ represents favourable adsorption
Langmuir	$\frac{1}{\Gamma} = \frac{1}{\Gamma_{max}} + \frac{1}{\Gamma_{max} C_e K_L}$	C_e is the residual concentration of adsorbate, K_L is the adsorption coefficient, $K_L = K_a/K_d$, and K_a and K_d are the rate constant for adsorption and desorption respectively. Γ and Γ_{max} are the concentrations of an adsorbate per unit area of an adsorbent at equilibrium and after saturation of the adsorbent surface respectively

isotherm models surpasses that of others, owing to their efficacy in characterizing the equilibrium region [13]. Their mathematical forms are shown in Table 2. In the context of adsorption upon nanoparticles, these isotherm models also play a vital role. The study of adsorption thermodynamics is needed to determine the feasibility of a reaction. The parameters like change in free energy, enthalpy, and entropy for checking the spontaneity of adsorption processes can be evaluated by carrying out the adsorption reactions at different temperatures and finally comparing the experimental data with theoretical equations.

4.3 *Adsorption Mechanism*

The adsorption mechanism can be understood by the nature of the interaction between the adsorbate and the adsorbent. Mainly two types of interactions exist, one is chemical and the other is physical. In addition, interactions by hydrogen bonding and ion exchange are also relevant. However, understanding the mechanism of adsorption is a complex task because of the associated challenges. A fundamental prerequisite for comprehending the adsorption mechanism is a thorough understanding of the structure of both the adsorbate and the adsorbent. In this context, the influence of surface charge is also noteworthy. Different spectroscopic as well as advanced microscopic techniques like IR/attenuated total reflection (ATR) spectroscopy, X-ray diffraction (XRD), scanning electron microscope (SEM), tunnelling electron microscope (TEM), and atomic force microscope (AFM) are used for characterizing the adsorbate/adsorbent complex. Amongst the various possible mechanisms of adsorption, chemisorption is mostly associated with the type of adsorbates bearing a variety of functional groups [18]. The type of surface complexation between the adsorption couple can be viewed pictorially by the development and application of several surface complexation models. In recent years several complexation models have been developed computationally.

It has been noted that the uptake of organic dyes by any surface is also influenced via a variety of noncovalent interactions, including hydrogen bonding, pi-pi, and ion–dipole interactions, predominantly relying on Van der Waals forces [17]. On the other hand, chemisorption involves the formation of a chemical bond between the dye molecules and the adsorbent by mutual sharing/exchange of electrons. The diffusion of dye molecules onto the adsorbent takes place through both film diffusion and particle diffusion models, integral components of the adsorption process.

Weak types of Van der Walls forces operate when an adsorbate like organic dye undergoes adsorption via physisorption mechanisms. Illustratively, the adsorption of the dye crystal violet (CV) occurs on the modified rice husk through the establishment of hydrogen bonds, where the rice husk surface hydroxyl groups interact with the nitrogen atoms of the dye molecule [17]. The surface hydrogen bonds were formed between the hydroxyl groups of the adsorbent surface and O/N-containing groups of the adsorbate. On the other hand, chemisorptions also occur by the ion exchange mechanism with the formation of chemical bonds between the adsorbate and the adsorbent. Interactions by ion exchange mechanism were found to exist between the adsorbate dyes like methylene blue (MB) and malachite green (MG) with the natural and oxalic acid-modified rice husk (RH) during adsorption [26]. The removal mechanism involves surface adsorption followed by pore diffusion. The two dyes 'direct red' and 'direct orange' were found to undergo diffusion onto the surface via 'chemisorption mechanism' [74].

4.4 Desorption Studies for Regenerating the Adsorbent

The recovery of the adsorbent material is an essential task during the adsorption experiment as it falls under the concept of recycling. Desorption of the adsorbate from the adsorbent surface can give back the adsorbent material or can be recovered. Desorption includes the breaking of the bonds between the adsorbate and the adsorbent as it is basically the reverse of adsorption. This may be achieved by thermal treatment (steam and hot water) as well as chemical processes. Recovery of adsorbent is a cost-minimizing task in wastewater treatment plants. Sometimes, the understanding of the desorption mechanism also in turn helps knowing the adsorption mechanism. The selectivity of desorbents along with the pH of the medium is essential for the chemical desorption process. In this context, several crucial organic desorbents, such as alcohols and carboxylic acids or a combination thereof, along with essential inorganic desorbents like hydrochloric acid (HCl), sodium hydroxide (NaOH), and sulphuric acid (H_2SO_4), are commonly employed [80]. The sustainability of an adsorption process is related to the proper regeneration of the used adsorbent. Agricultural waste like rice husk (RH) is an advantageous adsorbent in terms of its production, disposal, and most importantly its comparatively easy regeneration. The desorption mechanism depends upon the nature of the adsorbates. More specifically the interactions between the adsorbates like metals, dyes, and organic compounds with the respective adsorbent surface govern the desorption mechanisms. Ensuring the effectiveness of the regenerated adsorbent post-desorption is imperative for evaluating the success of the overall process. However, the adsorption and desorption processes are independent in the case of continuous adsorption systems [41]. The choice of desorbents is dependent upon several factors like the nature of adsorbates/adsorbents, eco-friendly reasons, and processing plant requirements. As an example, organic alcohol-based desorbents and surfactants are found effective for dye-loaded rice husks. In another case, interestingly deionized water along with $CaCl_2$ and NaOH was found as effective desorbents for organic pollutants like pesticides and phenolic compounds [33].

5 Removal of Pollutant Dyes by Adsorption onto Oxide Nanoparticles

Amongst an array of technologies developed and implemented for dye scavenging, adsorption stands out as the most versatile and superior method. Owing to its sludge-free clean operation, low initial cost, production of nontoxic by-products, ability for large-scale application, possibility of several reuse, and complete removal of dyes even from dilute solutions, the process has emerged as an appropriate technique for wastewater treatment [71, 73]. Nevertheless, the nanoparticles (NPs) also found application as adsorbents in the purification of wastewater from industry [53]. Moreover, the scope of nano-materials as adsorbents is under-explored, however, if explored these can open up new avenues for dye removal from water effluents prior to their discharge into water bodies due to their large surface area [73, 53]. The nano-adsorbents due to their reduced particle size and thereof higher surface area make them the most suitable adsorbents as compared to their conventional counterparts for the removal of pollutant dyes [96, 53]. Though nano-adsorbents can be produced via different physico-chemical methods, as an alternative pathway the biological route (involves formation of NPs from microorganisms, plants and animals) that occurs at ambient temperature, pressure, and neutral pH [96, 72] is mostly adopted because of its low cost, ease of rapid synthesis, controlled toxicity, controlled size characteristics, and environment-friendly approach [96]. The metal oxide-based nano-adsorbents are superior to their analogues like metal NPs and commonly carbon-based NPs, in terms of their minimal environmental impact, low solubility, and no secondary pollution [73, 53]. The biological synthesis of NPs is performed by using plant extracts (viz. different parts of the plant such as leaves, roots, stem, fruit, and flowers) and microbes including bacteria, fungi, and yeast [60]. In addition, the approach to microbial synthesis of oxide NPs is further supported by the fact that the method is not only an environmentally benign and green process, but also as significantly efficient as other conventional chemical methods. Certain factors like morphology, size, and aggregation make the metal oxide nanoparticles (MONPs) unique for having high dye removal capacity and heavy metal selectivity [60].

Biosynthesized metal oxide nanoparticles (MONP) can be reviewed from two different aspects: magnetic oxide NPs (Magnetic magnetite, Fe_3O_4, and magnetic maghemite, Fe_2O_3) and nonmagnetic oxide NPs (mainly the oxides of titanium, manganese, zinc, cerium, aluminium, magnesium, and zirconium) which have effectively been used for the removal of dyes from wastewater [50, 98]. The removal of dye effluents by magnetic nanoparticles as adsorbents is an emerging field of water and wastewater treatment, particularly for the removal of magnetic contaminants. For instance, the super magnetic Fe_3O_4 nanoparticles when coated with green tea polyphenols were found to exhibit higher dye removal efficiency than that of the unmodified one [85]. The magnetic iron oxide nanoparticles are also found to act as excellent adsorbent over a wide pH range for recovering *azo* dyes from mixtures. Nano-materials prepared from a Fe_3O_4 mesoporous carbon shell as well as NPs coated with sugarcane bagasse can be used to degrade methylene blue dye [71, 60]. Optimization of pH is important for achieving better efficiency of the adsorption process. The dye '*Rhodamine B*' can be adsorbed up to 95% by humic acid (HA) coated Fe_3O_4 nano-adsorbents at specific pH conditions [73]. Furthermore, the application of non-magnetic oxide nanoparticles has garnered significant interest in the realm of sustainable treatment, particularly for their efficacy as potent adsorbents in the removal of various hazardous dyes. This heightened interest stems from their notable surface reactivity and impressive adsorption capacity, making them valuable contributors to dye removal processes. For instance, MgO NPs are greatly effective in removing Direct red 23 and Reactive black 8 from synthetic coloured wastewater [30]. Nano-MgO encapsulated activated carbon NPs were shown to be an effective adsorbent for the removal of toxic dyes viz. Reactive black 5 (RB 5), Reactive orange 122 (RO 122), Malachite green, and Congo red [73]. Cr-doped ZnO NPs can remarkably remove MO (Methyl Orange) molecules from water [39]. ZnO-zeolite has a much higher adsorption capacity for eliminating MB dye [23]. The ZnO NPs display remarkable efficiency towards the removal of widely used azo dyes, methyl orange (MO), and amaranth (AM) from aqueous systems [99]. Furthermore, the investigation into the synergistic potential of ZnO nanoparticles functionalized with chitosan (CS) for the removal of dyes from synthetic and textile industrial effluents revealed promising outcomes. The results indicated that the composite ZnO-CS system, serving as an effective adsorbent, plays a pivotal role in eliminating dyes from industrial wastewater. Similarly, noteworthy contributions were observed with the utilization of functionalized nickel oxide (NiO) and copper oxide (CuO) nanoparticles, specifically in the removal of malachite green oxalate (MGO) and methyl orange (MO) [48, 47]. There are many other oxide nanoparticle-based adsorbents which engender a highly efficient dye removal capacity.

5.1 Effect of Different Parameters upon Adsorption

5.1.1 PH (Surface Charge)

The pH is the key parameter that controls the adsorption process of dyes onto adsorbents (viz. MNOPs) affecting the surface charge of adsorbents as well as the molecular stability of dye [16, 57]. The pH level of the aqueous solution exerts a profound influence on the surface sites of the adsorbent, significantly shaping the sorptive capacity for dye uptake [16]. Specifically, when examining the impact of pH on the elimination of toxic azo dyes such as Methyl Orange (MO) and Amaranth (AM), it was observed that ZnO-NPs exhibited optimal removal efficiency for each dye at a pH of 6 [99]. ZnO-zeolite has a much higher adsorption capacity for eliminating MB dye, and the maximum removal efficiency is found to be at pH 3 [23]. The effect of pH was investigated in a broad range of 2.6–14.0, for the adsorption of some anionic dyes onto adsorbents keeping all other variables constant. The findings indicated that the removal efficiency surpassed 85% at a pH of 2.6, progressively diminishing as the pH elevated to 13.2 for the dyes Nylosan Blue and Chromazurol S. On the other hand, in the case of basic red 2, the removal percentage was found to possess a direct relationship with solution pH. The results indicated that the adsorption of dyes is dependent upon the surface charge or pH of the medium [57]. At different surface charges of the adsorbents, the dyes interact differently. Furthermore, an investigation into the adsorption process of Maxilon Blue GRL dye revealed a discernible impact of each unique solution pH (ranging from 3 to 10) on the equilibrium uptake capacity. It was observed that the dye uptake diminished with an escalation in pH, with the highest adsorption capacity recorded at pH 3 [2]. The removal efficiency of Reactive Black 8 exhibited an upward trend with increasing pH levels, rising notably from 5 to 9. Conversely, for Direct Red 23, the removal efficiency demonstrated an increase as the pH decreased from 9 to 5 [30]. Hence the relation of dye adsorption with pH is dependent upon the nature of the dye as well as the adsorbent surfaces [48].

5.1.2 Initial Concentration of Dye

Adsorption also has a great dependence on the initial concentration of dye. The rate of pollutant dye degradation exhibited a gradual decline as the initial dye concentration rose from 200 to 600 ppm, followed by a steep decrease beyond that concentration level [58], the outcome of the initial concentration of trichloroethylene (TCE) vapour was observed to be proportional to adsorption capacity. As the concentration of inlet TCE increases, at increasing pressures, the adsorption capacity of activated carbon also increases steadily [61]. Furthermore, the examination of adsorption isotherms for methylene blue onto the surfaces of iron oxide NPs (IONPs) with and without citrate coating, as well as for methyl orange on the surface of unmodified IONPs, revealed a consistent trend. In all three scenarios, the quantity of adsorbed dye demonstrated an incremental rise with an increase in dye concentration until reaching a saturation point, indicative of the maximum adsorption capacity of IONPs [91]. Moreover, the percentage adsorption of methylene blue was found to decrease with an increase in initial dye concentration. But the actual amount of methylene blue dye adsorbed per unit mass of carbon increased with an increase in dye concentration [92]. On the other hand, at elevated concentrations, the dye 'BR2' exhibited a high percentage of dye removal (>93%), while dyes like 'NB' and 'ChS' demonstrated removal rates exceeding 50%. However, a very high percentage (> 95%) of dye removal was acquired even at low concentrations [57].

5.1.3 Temperature

Adjustment of temperature may be required in the adsorption process to predict the spontaneity of the adsorption process. Different thermodynamic parameters can be evaluated by carrying out the adsorption reaction at varying temperatures. Temperature serves as a crucial factor influencing the dual facets of adsorption—specifically, its impact on the equilibrium position, linked to the exothermic nature of the process, and the swelling capacity of the adsorbent [2]. The adsorption density of malachite green oxalate (MGO) and methyl orange (MO) dyes onto CuO and NiO surfaces is detrimentally influenced by temperature. As the temperature ascends from 303 to 323 K, the uptake capacity of both dyes experiences a decline [48]. The amount of adsorption of dye was found to be directly proportional to temperature in the case of ZnO-CS nanocomposites leading to 99% removal efficiency at optimum temperature constructed at 40 °C [58]. The equilibrium percentage removal of MO and AM onto ZnO-NPs varying the temperature from 35 to 55 °C in an increasing order leads to an increase in the adsorption percentage removal of ZnO-NPs [99]. The effect of the temperature on cationic dye adsorption was investigated at three different temperatures (30, 45, and 60 °C), and it was experienced that the increase in temperature negatively influenced the adsorption of dye [25].

6 Conclusion

The principal objective of industrialized wastewater treatment is to effectively mitigate the presence of toxic chemicals before releasing them into the environment, aligning with the principles of sustainable process development. Various treatment processes applied for the removal of such chemicals must not create secondary pollutants during the remediation process. Numerous methods developed for the removal of dyes produced in tons annually from the dyeing or textile industries are usually categorized under physical, chemical, and biological treatment techniques. The selection of a method is also influenced by the efficacy (i.e. complete degradation of dyes, minimum by-product formation, and cost of the method). In this context the application of adsorption techniques is immense. Importantly, the microbe-mediated adsorption processes are preferred over the physical–chemical techniques due to the negligible production of solid by-products, low cost, mineralization of toxic dyes, and self-recovery of microorganisms. The biological adsorption processes are greener and sustainable. To minimize the load on microbe-mediated remediation processes with satisfactory output, biological treatments are many times succeeded by suitable chemical treatments. There is an urgent need to integrate chemical pre- or post-treatments such as advanced oxidation processes (AOPs) with bioreactor models to achieve effective removal of the dyes from industry effluents. The microbe-mediated routes of dye removal using MONPs have been gaining momentum over the past several years. The synthesis of such MONPs from microbes and extraction of other microorganisms is a green process for the remediation of wastewater. Having significantly unique yet better stability, and magnetic and electrical properties, MONPs showcase higher catalytic activity, reactivity, and adsorptivity in contrast to other conventional materials. For instance, iron-NPs being environmentally benign are extensively applicable in WWT for their high reactivity with water and magnetic susceptibility. The elucidation of the adsorption mechanism and adsorption kinetics also enables us to fabricate the NPs with desired size, shape, monodispersity, and effectiveness for large-scale implementation with minimum environmental hazards.

References

1. Agarwal H, Venkat Kumar S, Rajeshkumar S (2017) A review on green synthesis of zinc oxide nanoparticles—an eco-friendly approach. Resour Technol 1–7
2. Aljebori AMK, d Alshirifi AN (2012) Effect of different parameters on the adsorption of textile dye maxilon blue GRL from aqueous solution by using white marble. Asian J Chem 24(12):5813–5816
3. Ananthashankar R (2012) Treatment of textile effluent containing reactive red 120 dye using advanced oxidation. MSc thesis. Dalhousie University, Halifax, Canada
4. Arun Prasad AS, Bhaskara Rao KV (2010) Physico chemical characterization of textile effluent and screening for dye decolorizing bactéria. Global J Biotech Biochem 5(2):80–86
5. Bafana A, Devi SS, Chakrabarti T (2011) Azo dyes: past, present and the future. Environ Rev 19:350–370
6. Banat IM, Nigam P, Singh D, Marchant R (1996) Microbial decolorization of textile-dyecontaining effluents: a review. Bioresour Technol 58(3):217–227
7. Bansal V, Poddar P, Ahmad A, Sastry M (2006) Room-temperature biosynthesis of ferroelectric barium titanate nanoparticles. J Am Chem Soc 128:11958–11963
8. Bansal V, Poddar P, Ahmad A, Sastry,M (2006) Room-temperature biosynthesis of ferroelectric barium titanate nanoparticles.J Am Chem Soc 128:11958–11963.

9. Banzylinski D, Schubbe S (2007) Controlled biomineralization by and applications of magnetotacticbacteria. Adv Appl Microbiol 62:21–62
10. Baruah A, Chaudhary V, Malik R, Tomer VK (2019) Nanotechnology based solutions for wastewater treatment. Nanotechnol Water Wastewater Treat 337–368
11. Beblawy S, Bursac T, Paquete C, Louro R, Clarke TA, Gescher J (2018) Extracellular reduction of solid electron acceptors by Shewanella oneidensis. Mol Microbiol 109:571–583
12. Bharde A, Rautaray A, Bansal D, Ahmad V, Sarkar A, Yusuf I, Sanyal S, Sastry M (2006) Extracellular biosynthesis of magnetite using fungi. Small 2(1):135–141
13. Bhattacharya AK et al (2006) Adsorption of Zn(II) from aqueous solution by using different adsorbents. Chem Eng J 123:43–49
14. Biju V (2014) Chemical modifications and bioconjugate reactions of nanomaterials for sensing imaging, drug delivery and therapy. Chem Soc Rev 43(3):744–764
15. Bolade OP, Williams AB, Benson NU (2020) Green synthesis of iron-based nanomaterials for environmental remediation: a review. Environ Nanotechnol Monit Manag 13:100279. https://doi.org/10.1016/j.enmm.2019.100279
16. Boumediene M, Benaïssa H, George B, Molina S, Merlin A (2018) Effects of pH and ionic strength on methylene blue removal from synthetic aqueous solutions by sorption onto orange peel and desorption study. J Mater Environ Sci 9(6):1700–1711
17. Chakraborty S et al (2011) Adsorption of crystal violet from aqueous solution onto NaOH-modified rice husk. Carbohydr Polym 86:1533–1541
18. Chowdhury S et al (2011) Adsorption thermodynamics, kinetics and isosteric heat of adsorption of malachite green onto chemically modified rice husk. Desalination 265:159–168
19. Christie R (2001) Colour chemistry. The Royal Society of Chemistry, Cambridge, United Kingdom
20. Contreras F, Vargas E, Jimenez K, Munoz-Villagran C, Figueroa M, Vasquez C et al (2018) Reduction of gold (III) and tellurium (IV) by Enterobacter cloacae MF01 results in nanostructure formation both in aerobic and anaerobic conditions. Front Microbiol 9:3118
21. Couto SR (2009) Dye removal by immobilised fungi. Biotechnol Adv 27(3):227–235
22. El-Said AG (2010) Biosorption of Pb(II) ions from aqueous solutions onto rice husk and its ash. J Am Sci 6:143–149
23. Elfeky AS, Youssef HF, Elzaref AS (2020) Adsorption of dye from wastewater onto ZnO nanoparticles-loaded zeolite: kinetic, thermodynamic and isotherm studies. J Phys Chem 234(2):255–278
24. Faivre D, Schüler D (2008) Magnetotactic bacteria and magnetosomes. Chem Rev 108:4875–4898
25. de Farias Silva CE, da Gama BMV, da Silva Gonçalves AH, Medeiros JA, de Souza Abud AK (2020) Basic-dye adsorption in albedo residue: effect of pH, contact time, temperature, dye concentration, biomass dosage, rotation and ionic strength. J King Saud Univ Eng Sci 32(6):351–359
26. Foo KY, Hameed BH (2011) Preparation of activated carbon from date stones by microwave induced chemical activation: application for methylene blue adsorption. Chem Eng J 170:338–341
27. Forgacs E, Cserháti T, Oros G (2004) Removal of synthetic dyes from wastewaters: a review. Environ Int 30(7):953–971
28. Gallardo-Benavente C, Carrión O, Todd JD, Pieretti JC, Seabra AB, Durán N et al (2019) Biosynthesis of CdS quantum dots mediated by volatile sulfur compounds released by Antarctic Pseudomonas fragi. Front Microbiol 10:1866
29. Ganesan V, Hariram M, Vivekanandhan S, Muthuramkumar S (2020) Periconium sp.(endophytic fungi) extract mediated sol-gel synthesis of ZnO nanoparticles for antimicrobial and antioxidant applications. Mater Sci Semicond Process 105:104739
30. Gayathri S, Manjunath NT, Shetty R (2019) Magnesium oxide nanoparticle as adsorbent for removal of synthetic dyes. Int J Eng Res Dev 7(3):319–323
31. Gericke M, Pinches A (2006) Microbial production of gold nanoparticles. Gold Bull 39:22–28

32. Gomez-Bolivar J, Mikheenko IP, Orozco RL, Sharma S, Banerjee D, Walker M et al (2019) Synthesis of Pd/Ru bimetallic nanoparticles by Escherichia coli and potential as a catalyst for upgrading 5-hydroxymethyl furfural into liquid fuel precursors. Front Microbiol 10:1276
33. Gupta VK et al (2000) Design parameters for the treatment of phenolic wastes by carbon columns (obtained from fertilizer waste material). Water Res 34:1543–1551
34. Gupta VK (2009) Application of low-cost adsorbents for dye removal–a review. J Environ Manag 90(8):2313–2342
35. Gupta GK, Shukla P (2020) Insights into the resources generation from pulp and paper industry wastes: challenges, perspectives and innovations. Bioresour Technol 297:122496. https://doi.org/10.1016/j.biortech.2019.122496
36. Harris HW, Sánchez-Andrea I, McLean JS, Salas EC, Tran W, El-Naggar MY et al (2018) Redox sensing within the genus Shewanella. Front Microbiol 8:2568
37. Hassaan MA (2016) Advanced oxidation processes of some organic pollutants in fresh and seawater. PhD thesis, Faculty of Science, Port Said University
38. Heidelberg JF, Paulsen IT, Nelson KE, Gaidos EJ, Nelson WC, Read TD et al (2002) Genome sequence of the dissimilatory metal ion–reducing bacterium Shewanella oneidensis. Nat Biotechnol 20(11):1118–1123
39. Homaeigohar S (2020) The nanosized dye adsorbents for water treatment. Nanomaterials 10(2):01–43
40. Iravani S (2014) Bacteria in nanoparticle synthesis: current status and future prospects. Int Sch Res Notices 1–18
41. Jain AK et al (2004) Removal of Chlorophenols using industrial wastes. Environ Sci Technol 38:119–125
42. Jeevanandam J, Chan Y, Danquah M (2016) Biosynthesis of metal and metal oxide nanoparticles. Chem Bio Eng Rev 3:55–67
43. Kant R (2011) Textile dyeing industry an environmental hazard. Nat Sci 4(1):22–26
44. Khalil NM, Abd El-Ghany MN, Rodriguez-Couto S (2019) Antifungal and anti-mycotoxin efficacy of biogenic silver nanoparticles produced by Fusarium chlamydosporum and Penicillium chrysogenum at non-cytotoxic doses. Chemosphere 218:477–486
45. Kim D-H, Kanaly RA, Hur H-G (2012) Biological accumulation of tellurium nanorod structures via reduction of tellurite by Shewanella oneidensis MR-1. Bioresour Technol 125:127–131
46. Kim D-H, Kim M-G, Jiang S, Lee J-H, Hur H-G (2013) Promoted reduction of tellurite and formation of extracellular tellurium nanorods by concerted reaction between iron and Shewanella oneidensis MR-1. Environ Sci Technol 47:8709–8715
47. Koul B, Poonia AK, Yadav D, Jin JO (2021) Microbe-mediated biosynthesis of nanoparticles: Applications and Future Prospects. Biomolecules 11(6):886
48. Kumar KY, Archana S, Raj TV, Prasana BP, Raghu MS, Muralidhara HB (2017) Superb adsorption capacity of hydrothermally synthesized copper oxide and nickel oxide nanoflakes towards anionic and cationic dyes. J Sci Adv Mater Dev 2(2):183–191
49. Leub H (2003) Industrial dyes, chemistry, properties and applications. Wiley VCH, Wenheim, Germany
50. Li X, Xu H, Chen ZS, Chen G (2011) Biosynthesis of nanoparticles by microorganisms and their applications. J Nanomat 1–16
51. Mahanty S, Chatterjee S, Ghosh S, Tudu P, Gaine T, Bakshi M et al (2020) Synergistic approach towards the sustainable management of heavy metals in wastewater using mycosynthesized iron oxide nanoparticles: biofabrication, adsorptive dynamics and chemometric modeling study. J Water Proces Eng 37:101426. https://doi.org/10.1016/j.jwpe.2020.101426
52. Mandeep PS (2020) Microbial nanotechnology for bioremediation of industrial wastewater. Front Microbiol 11:1–8
53. Mandeep, Shukla P (2020) Microbial nanotechnology for bioremediation of industrial wastewater. Front Microbiol 11:.590631
54. Mann S et al (1990) Biomineralization of ferrimagnetic greigite (Fe3O4) and iron pyrite (FeS2) in a magnetotactic bacterium. Nature 343:258–260

55. Mansour HB, Houas I, Montassar F, Ghedira K, Barillier D, Mosrati R et al (2012) Alteration of in vitro and acute in vivo toxicity of textile dyeing wastewater after chemical and biological remediation. Environ Sci Pollut Res 19(7):2634–2643
56. Manzoor J, Sharma M (2020) Impact of textile dyes on human health and environment. In: Manzoor J, Sharma M (eds) Impact of textile dyes on public health and the environment. IGI Global, pp 162–169
57. Muntean SG, Nistor MA, Muntean E, Todea A, Ianoş R, Păcurariu C (2018) Removal of colored organic pollutants from wastewaters by magnetite/carbon nanocomposites: single and binary systems. J Chem 1–16
58. Nakkeeran E, Varjani SJ, Dixit V, Kalaiselvi A (2018) Synthesis, characterization and application of zinc oxide nanocomposite for dye removal from textile industrial wastewater. Ind J Exp Biol 56:498–503
59. Narayanan K, Sakthivel N (2010) Biological synthesis of metal nanoparticles by microbes. Adv Colloid Interface Sci 1–2:1–13
60. Naseem T, Durrani T (2021) The role of some important metal oxide nanoparticles for wastewater and antibacterial applications: a review. J Environ Chem Ecotoxicol 3:59–75
61. Nikam S, Mandal D (2020) Experimental study of the effect of different parameters on the adsorption and desorption of trichloroethylene vapor on activated carbon particles. ACS Omega 5(43):28080–28087
62. Noman M, Shahid M, Ahmed T, Niazi MB, Hussain S, Song FE (2020) Use of biogenic copper nanoparticles synthesized from a native Escherichia sp. as photocatalysts for azo dye degradation and treatment of textile effluents. Environ Pollut 257:113514. https://doi.org/10.1016/j.envpol.2019.113514
63. Nordmeier A, Merwin A, Roeper DF, Chidambaram D (2018) Microbial synthesis of metallic molybdenum nanoparticles. Chemosphere 203:521–525
64. Ntuli F, Omoregbe DI, Kuipa PK, Muzenda E, Belaid M (2009) Characterization of effluent from textile wet finishing operations. In: Proceedings of the world congress on engineering and computer science (WCECS 2009). San Francisco, USA, WCECS 2009
65. Othmer K (2004) Encyclopedia of chemical technology, vol 7, 5th edn. Wiley-Interscience
66. Pantidos N, Horsfall LE (2014) Biological synthesis of metallic nanoparticles by bacteria, fungi and plants. J Nanomed Nanotechnol 5:1000233
67. Piacenza E, Presentato A, Ambrosi E, Speghini A, Turner RJ, Vallini G et al (2018) Physical-chemical properties of biogenic selenium nanostructures produced by Stenotrophomonas maltophilia SeITE02 and Ochrobactrum sp. MPV1. Front Microbiol 9:3178
68. Pourbabaee AA, Malekzadeh F, Sarbolouki MN, Najaf F (2006) Aerobic decolorization and detoxification of a disperse dye in textile effluent by a new isolate of Bacillus sp. Biotechnol Bioeng 93(4):631–635
69. Przystaś W, Zabłocka-Godlewska E, Grabińska-Sota E (2012) Biological removal of azo and triphenylmethane dyes and toxicity of process by-products. Water Air Soil Pollut 223(4):1581–1592
70. Raliya R, Tarafdar J (2013) ZnO nanoparticle biosynthesis and its effect on phosphorous-mobilizing enzyme secretion and gum contents in clusterbean (Cyamopsistetragonoloba L.). Agric Res 2:48–57
71. Razali SZ, Aziz MY, Edinur HA, Ishak AR (2020) Adsorption of methylene blue onto iron oxide magnetic nanoparticles coated with sugarcane bagasse. IOP Conf Ser Earth Environ Sci 596(1):1–8
72. Revati K, Pandey BD (2011) Microbial synthesis of iron-based nanomaterials—A review. Bull Mater Sci 34(2):191–198
73. Ruan W, Hu J, Qi J, Hou Y, Zhou C, Wei X (2019) Removal of dyes from wastewater by nanomaterials: a review. Adv Mater Lett 10(1):9–20
74. Safa Y, Bhatti HN (2011) Kinetic and thermodynamic modeling for the removal of Direct Red-31 and Direct Orange-26 dyes from aqueous solutions by rice husk. Desalination 272:313–319
75. Salman AD, Juzsakova T, Al-Mayyahi MA, Ibrahim RI, Abdullah TA, Khader EH (2020) Improvement of mechanical properties of oil well cement by incorporate nano-$CaCO_3$

prepared from eggshell waste. IOP Conf Ser. https://doi.org/10.1088/1757899X/765/1/012006
76. Sanoop PK, Anas S, Ananthakumar S, Gunasekar V, Saravanan R, Ponnusami V (2016) Synthesis of yttrium doped nanocrystalline ZnO and its photocatalytic activity in methylene blue degradation. Arab J Chem 9:S1618–S1626
77. Dos Santos AB, Cervantes FJ, van Lier JB (2007) Review paper on current technologies for decolourisation of textile wastewaters: Perspectives for anaerobic biotechnology. Biores Technol 98(12):2369–2385
78. Saravanakumar K, Shanmugam S, Varukattu NB, Mubarakali D, Kathiresan K, Wang M-H (2019) Biosynthesis and characterization of copper oxide nanoparticles from indigenous fungi and its effect of photothermolysis on human lung carcinoma. J Photochem Photobiol B Biol 190:103–109
79. Saravanan A, Kumar PS, Karishma S, Vo DV, Jeevanantham S, Yaashikaa PR et al (2021) A review on biosynthesis of metal nanoparticles and its environmental applications. Chemosphere 264:128580
80. Sathishkumar P et al (2012) Utilization of agro-industrial waste Jatropha curcaspods as an activated carbon for the adsorption of reactive dye remazol brilliant blue R (RBBR). J Clean Prod 22:67–75
81. Seetharaman PK, Chandrasekaran R, Gnanasekar S, Chandrakasan G, Gupta M, Manikandan DB et al (2018) Antimicrobial and larvicidal activity of eco-friendly silver nanoparticles synthesized from endophytic fungi Phomopsis liquidambaris. Biocatal Agric Biotechnol 16:22–30
82. Shindhal T, Rakholiya P, Varjani S, Pandey A, Ngo HH, Guo W et al (2021) A critical review on advances in the practices and perspectives for the treatment of dye industry wastewater. Bioengineered 12(1):70–87
83. Siddiqi KS, Husen A (2016) Fabrication of metal and metal oxide nanoparticles by algae and their toxic effects. Nanoscale Res Lett 11:1–15
84. Siddiqui SI, Rathi G, Chaudhry SA (2018) Acid washed black cumin seed powder preparation for adsorption of methylene blue dye from aqueous solution: thermodynamic, kinetic and isotherm studies. J Mol Liq 264:275–284
85. Singh KK, Senapati KK, Sarma KC (2017) Synthesis of superparamagnetic Fe_3O_4 nanoparticles coated with green tea polyphenols and their use for removal of dye pollutant from aqueous solution. J Environ Chem Eng 5(3):2214–2221
86. Sinha A, Khare SK (2011) Mercury bioaccumulation and simultaneous nanoparticle synthesis by Enterobacter sp. cells. Bioresour Technol 102:4281–4284
87. Stolz A (2001) Basic and applied aspects in the microbial degradation of azo dyes. Appl Microbiol Biotechnol 56(1–2):69–80
88. Suresh AK, Pelletier DA, Wang W, Moon JW, Gu B, Mortensen NP et al (2010) Silver nanocrystallites: biofabrication using Shewanella oneidensis, and an evaluation of their comparative toxicity on gram-negative and gram-positive bacteria. Environ Sci Technol 44:5210–5215
89. Suzuki Y, Kelly S, Kemner K, Banfield J (2002) Radionuclide contamination: nanometre-size products of uranium bioreduction. Nature 419:134–134
90. Taherzadeh MJ, Fox M, Hjorth H, Edebo L (2003) Production of mycelium biomass and ethanol from paper pulp sulfite liquor by Rhizopus oryzae. Bioresour Technol 88:167–177
91. Talbot D, Queiros Campos J, Checa-Fernandez BL, Marins JA, Lomenech C, Hurel C, Godeau GD, Raboisson-Michel M, Verger-Dubois G, Obeid L, Kuzhir P (2021) Adsorption of organic dyes on magnetic iron oxide nanoparticles. Part I: mechanisms and adsorption-induced nanoparticle agglomeration. ACS Omega 6(29):19086–19098
92. Teka T, Enyew S (2014) Study on effect of different parameters on adsorption efficiency of low cost activated orange peels for the removal of methylene blue dye. Int J Innov Sci Res 8(1):106–111
93. Uddin I, Adyanthaya S, Syed A, Selvaraj K, Ahmad A, Poddar P (2008) Structure and microbial synthesis of sub-10 nm Bi2O3 nanocrystals. J Nanosci Nanotechnol 8:3909–3913

94. Wang Z, Xue M, Huang K, Liu Z (2011) Textile dyeing wastewater treatment. Adv Treat Text Effl 5:91–116
95. Wijetunga S, Li XF, Jian C (2010) Effect of organic load on decolourization of textile wastewater containing acid dyes in upflow anaerobic sludge blanket reactor. J Hazard Mater 177(1–3):792–798
96. Yadav KK, Singh JK, Gupta N, Kumar VJJMES (2017) A review of nanobioremediation technologies for environmental cleanup: a novel biological approach. J Mater Environ Sci 8(2):740–757
97. Yadav KK, Singh JK, Gupta N, Kumar VJJMES (2017) A review of nanobioremediation technologies for environmental cleanup: a novel biological approach. J Mater Environ Sci 8(2):740–757
98. Yusof HM, Mohamad R, Zaidan UH (2019) Microbial synthesis of zinc oxide nanoparticles and their potential application as an antimicrobial agent and a feed supplement in animal industry: a review. J Anim Sci Biotechnol 10(1):1–22
99. Zafar MN, Dar Q, Nawaz F, Zafar MN, Iqbal M, Nazar MF (2019) Effective adsorptive removal of azo dyes over spherical ZnO nanoparticles. J Mater Res Technol 8(1):713–725
100. Zhao J, Zhao Z, Li N, Nan J, Yu R, Du J (2018) Visible-light-driven photocatalytic degradation of ciprofloxacin by a ternary $Mn_2O_3/Mn_3O_4/MnO_2$ valence state heterojunction. Chem Eng J 353:805–813

Sustainable Treatment Technologies

Hybrid Type Processes and Materials for Textile Effluent Treatment Towards Sustainable Development

Bhavana Balachandran and P. C. Sabumon

Abstract Textile effluent must be appropriately handled before being dumped into the sewer system or water bodies to meet the regulatory standards and to protect the environment. Due to the complexity of textile effluent, the treatment by conventional technologies is restricted. Therefore, hybrid treatment systems are needed to treat this effluent. When utilizing a hybrid technique, such as adsorption-advanced oxidation process (AOP), electrocoagulation-AOP, coagulation-biodegradation, AOP-membrane bioreactor, hybrid materials, etc., the treatment performance can be enhanced. One of the finest treatments for treating textile effluent is a hybrid AOP-biological process. This process gives better treatment along with less economic burden. In addition to hybrid processes, hybrid materials are gaining popularity and they also give the same benefits as a hybrid process. However, research must concentrate on the profitable manufacture of these hybrid materials. In this chapter, various hybrid systems and hybrid materials are examined, with an emphasis on their effectiveness, sustainability, and future potential in the treatment of textile effluent.

Keywords Textile effluent · Hybrid treatment systems · Hybrid materials · Adsorption-AOP · AOP-membrane bioreactor · Sustainable development

1 Introduction

Dyes are colored and complex chemicals present in the textile effluent. Dyes have two components; chromophores which impart color and auxochromes which intensify the color. Based on the wavelength absorbed by these two components, dyes will give different colors. Due to the complexity of the chemical structure, dyes are resistant to degradation under normal conditions. Around 60–70% of the dye produced in the world comes under the azo category. When these azo dyes decolorize in anaerobic conditions, they generate more toxic and cancer-causing by-products which are

B. Balachandran · P. C. Sabumon (✉)
School of Civil Engineering, Vellore Institute of Technology, Chennai Campus, Chennai 600127, India
e-mail: pcsabumon@vit.ac.in

P. Singh (ed.), *Dye Pollution from Textile Industry*, SDGs and Textiles,
https://doi.org/10.1007/978-981-97-5341-3_9

colorless compounds. Therefore, anaerobic process alone may not be suitable for azo dye degradation and hybrid types process combining with biological can be a successful alternative.

The treatments for textile effluent in the 1990s included were only equalization and primary sedimentation. However, more intensive treatment of textile effluent came to light later when environmental protection legislation in many countries got stricter. The textile effluent consists of toxic dyes and other chemicals. These chemicals are also toxic, mutagenic, and carcinogenic in nature even at low concentrations. Heavy metal contents in the textile effluent are an example of this category. Lead, cadmium, and chromium are the three principal heavy metals found in textile effluent [4].

Also, the textile effluent is characterized by substantial total organic content measured as chemical oxygen demand, suspended solids, dissolved salts, and color. The treatment of textile effluent began with the implementation of physical, chemical, and biological approaches. Due to the complexity of textile effluent, conventional methods fail in most of the cases [50]. Table 1 represents the advantages and disadvantages of single methods in treating textile effluent. Every conventional approach has its own merits and demerits. Therefore, combining the suitable conventional methods can eliminate the demerits and increase their treatment efficiency.

Studies have shown that combining many techniques can overcome the drawbacks of using just one technique and produce better outcomes. So people have developed hybrid processes. More than one treatment method is combined in this process to provide an improved single process. It is capable of overcoming every constraint imposed by the individual process. A hybrid process and a combined process are two distinct types. A hybrid process is not what we would refer to as a combined process. Multiple treatment processes are integrated into a combined procedure; these processes may operate in parallel or sequential order. Also, we can state that a hybrid approach lowers reactor cost and time by combining multiple treatment choices into a single reactor [50].

Hybrid materials are becoming more popular in textile effluent treatment procedures in addition to hybrid reactors or processes. Hybrid materials combine two or more different materials to solve the challenging difficulties of textile effluent treatment. Hybrid materials can be included in electrochemical processes, photocatalysis, membrane filtration, adsorption processes, etc. These hybrid materials will have a larger surface area than traditional adsorbents when utilized as an adsorbent. In photocatalysis, normally TiO_2 and ZnO are used as catalysts. However, scientists are working on novel hybrid materials to replace these conventional catalysts. If a hybrid material can function better than these TiO_2 and ZnO and reduce the cost of treatment that will be a better choice for textile effluent treatment. The other options are hybrid membranes. Normally, to create this hybrid membrane, graphene or carbon-based nanotubes are mixed with polymers. They are incredibly effective in removing harmful elements from textile effluent. Compared to conventional membrane approaches, they have enhanced performance because of their greater permeability and selectivity. In biological systems, these hybrid materials will help to immobilize the microorganisms, thereby enhancing the treatment. Another group is hybrid coagulants. It is possible to enhance the coagulation-flocculation process

Table 1 Benefits and drawbacks of individual treatment process

Process	Benefits	Drawbacks	References
Biological methods	Economical, eco-friendly, high adsorption on to microbes in the case of yeast and fungi. Anaerobic degradation in the case of Azo dyes	Less tolerant to high concentration of dyes and other toxic materials, formation of colorless toxic aromatic amines	[1, 24, 45]
Adsorption	Good color removal for all types of dyes	pH dependent, need regeneration, costly disposal of used adsorbent	[55]
Biosorption	Green technology, cost effective	pH and temperature dependent, specific to certain contaminant	[15]
Advanced oxidation processes	Completely mineralize dyes	If carboxylic acid is the by product, mineralisation needs more time and money	[37]
Membrane filtration processes	Ultra filtration-Advanced treatment technology Nano filtration-excellent advantages over ultrafiltration and reverse osmosis, low investment, low operation and maintenance cost	Trapped solute forms cake on the filter and continuous removal of these solute is necessary	[3]
Ion exchange	Regeneration of resin is easy	Regeneration is expensive	[3]
Chemical precipitation	Relatively simple technique and economical	Forms insoluble precipitate, large quantity of sludge	[3, 19]
Fenton reagent	Removes both soluble and insoluble dyes	Prefers pH below 3.5 and sludge generation	[19]
UV/H_2O_2	Less treatment time and high efficiency	Undesirable byproduct	[3]
Photocatalysis	No sludge production and also removes COD	pH dependent, depends on nature of oxidizing agent used	[54]
Electrochemical	Decolorize soluble and insoluble dyes	High cost and sludge production in the case of electrocoagulation	[35]
Sonolysis	No chemical requirement	Requirement of high DO content	[19]

and more effectively remove suspended particles and color from textile effluent by combining conventional coagulants like alum or ferric chloride with polymer-based coagulants. Similarly, many other hybrid materials are reported in the literature for textile effluent treatment.

Numerous researchers have concentrated on various physicochemical treatment combinations, which are frequently used by textile industries in straightforward, stand-alone ways. Combining the AOPs with traditional physicochemical methods has also emerged as a desirable option. A wide range of advanced oxidation and physicochemical processes are being researched together with biological systems, in addition to their different combinations. Studies on novel hybrid materials are also gaining interest nowadays. They can be incorporated with conventional methods of treatment and enhance the treatment capacity. Figure 1 shows such a combination. A review of the literature reveals that numerous studies have been done on the effectiveness of hybrid treatment systems for the treatment of textile effluent. Some of these hybrid systems are AOP-MBR/MR, MBR-ultrafiltration, ozonation-biological process, SBR-nanofiltration, UV radiation-membrane bioreactors, biological process-AOPs, etc.

Because hybrid treatment technologies are so versatile, textile industries can customize solutions to the unique properties of their effluent, giving wastewater management a framework that is adaptable and flexible. Concurrent usage of multiple treatment stages provides robustness against changes in effluent composition and flow rates, while also optimizing the effectiveness of pollutant removal. This flexibility is essential for sectors with changing effluent characteristics and dynamic production processes. Hybrid treatment technology adoption also helps achieve sustainability objectives by encouraging water reuse and reducing the negative environmental effects of industrial discharges. In line with a circular economy strategy,

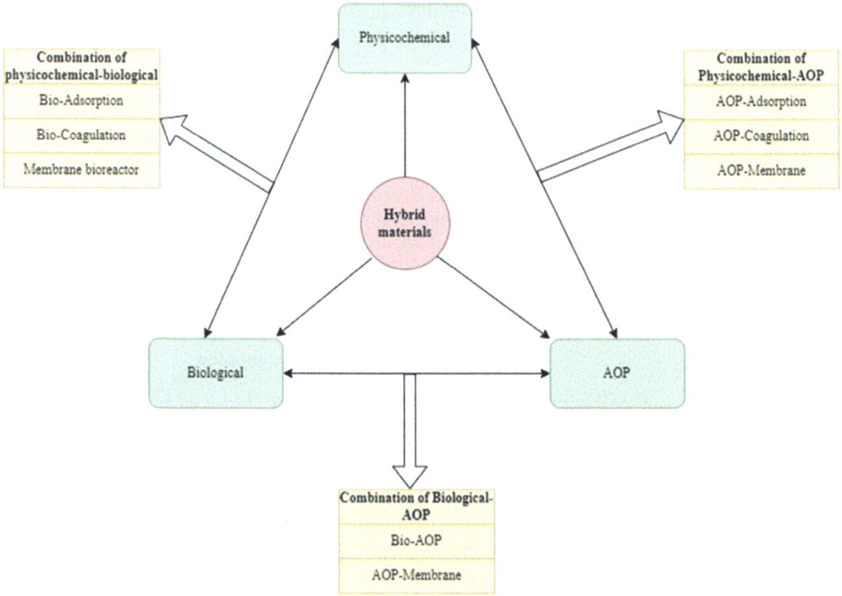

Fig. 1 Various combinations of hybrid processes using conventional methods

the focus is on resource recovery which means recovering valuable by-products or using treated water for non-potable uses which promotes water stewardship. When it comes to industrial effluent treatment, hybrid treatment technologies essentially provide a game-changing solution that strikes a balance between sustainability, adaptability, and efficiency. The incorporation of hybrid treatment technology emerges as a possible road toward achieving both operational excellence and environmental stewardship, as industries continue to traverse the intricate landscape of wastewater management. Figure 1 shows the various combinations of hybrid processes using conventional methods.

This book chapter deals with the following selected hybrid technologies used to treat textile effluent:

- Hybrid biological and AOP systems
- Hybrid AOP and MR/MBR systems
- Hybrid biological and physiochemical systems
- Hybrid biological systems
- Novel hybrid materials in textile effluent treatment.

Additionally, an attempt is made to recommend suitable hybrid process(es) for textile effluent treatment and to identify the future scope of such technologies.

2 Hybrid Treatment Technologies in Textile Effluent Treatment

Hybrid technologies are advanced wastewater treatment that can give good efficiency in textile effluent treatment processes. This book chapter reviews the selected commonly used hybrid technologies in the textile effluent treatment process and is described in the following sections.

2.1 Hybrid Biological-Advanced Oxidation Process

It is difficult to provide treated effluent that complies with requirements using biological processes alone. On the other hand, the cost of treatment goes up with an advanced oxidation process even though they could completely remove some of the synthetic dyes. AOP is a group of chemical processes used to treat water and wastewater with the goal of removing anthropogenic and biologically resistant contaminants by oxidation by interactions with hydroxyl radicals (OH). AOP can clean, decolorize, and deodorize the contaminants by converting them into biodegradable chemicals with little/no sludge production, which erases the need for another treatment stage. Primary oxidants, such as ozone, hydrogen peroxide, and oxygen, as well as energy sources, such as UV radiation, or catalysts, such as titanium dioxide, are used to create hydroxyl radicals. AOP occasionally generates carboxylic acid as a byproduct. Hydroxyl radicals cannot harm these carboxylic acids. So, in this instance, mineralizing the chemicals in the effluent demands a lot of time and money using AOP [37]. However, their combination or hybridization can overcome the limitations of both the technology and improve the efficiency. So, using biological methods as a pre-treatment and AOP as a polishing step gives the best results [2]. The O_3/UV/ H_2O_2/Fe^{2+} system removes more pollutants while using less energy than other photo and ozone-based AOPs, according to the findings of [5].

There are numerous reports of biological-AOP combinations in the literature. Pre-oxidation followed by biological processing, according to research, yields better outcomes [36]. COD and toxicity are successfully eliminated by this combination. The right AOP method and duration of AOP treatment must be chosen in order to achieve the best results [19]. The effectiveness of eliminating color from wastewater containing reactive red-198 was investigated using a combination UV/O_3/SBR reactor. The hybrid reactor had the highest color removal efficiency and TOC removal efficiency at pH 8 [33]. Table 2 provides a reported combination of biological-AOP or AOP-biological hybrid processes/reactors.

2.2 *Hybrid AOP-Membrane Reactors (AOP-MR/MBR)*

The treatment of industrial wastewater as well as resource recovery are two reasons why hybrid reactors have attracted so much attention. Other than these points, specific aims and objectives are established before selecting a hybrid reactor. The use of hybrid reactors that combine membrane and AOP reactors has many benefits. The benefits are less energy and fewer chemicals used, improved membrane flux, less fouling of membranes, and the hybrid process improves purifying effectiveness. The membrane filtration plus AOP-based hybrid reactors showed the best results in textile effluent treatment. These hybrid reactors can run at different temperatures with the aid of some special catalyst. Therefore, these hybrid reactors' key benefits include decreased space requirement, anti-fouling capability, self-cleaning capability, acceptable permeate fluxes, and the ability to perform both oxidation and filtration at the same time. These systems can benefit from having a single AOP as well as several AOPs [43].

A potential approach for treating textile wastewater is hybrid AOP-MR/MBR. If we can make this hybrid reactor design better, it will perform better than the available technology. Only a very small number of works have been done in actual textile effluent; currently, the majority of works in this hybrid reactor are based on synthetic effluent. The hybrid reactor should use less energy and should have lower operating costs over a longer period of time and more research to be done in this aspect. To extend the lifespan of the membrane, the cleaning and regeneration processes need to be improved. Furthermore, it is important to look at how MR (MBR)-AOPs' effluent affects the environment. Table 3 provides a reported combination of AOP-MR/MBR hybrid reactors.

Table 2 Combination of biological-AOP hybrid processes/reactors

Type of colored water	Biological process	AOPs	Performance/ removal efficiency	References
Real textile effluent	Activated sludge process	UV/H_2O_2 H_2O_2 250 mg/L, UV radiation 60 min	Color removal of 96% and removal of aromatic compounds of 86%	[34]
Cotton industry effluent	Biological aerated filter (2.5 h HRT)	Ozonation (0.25 mg O_3/mg COD) along with activated carbon (30 min HRT)	Average color removal efficiency of 90%	[59]
Dyeing and finishing industry effluent	Hydrolysis Acidification (3 h HRT), aeration tank (9 h HRT) and secondary sedimentation tank (3 h HRT)	Ozonation/zero valent iron experiment (30 min HRT)	Achieved COD < 80 mg/L	[32]
Textile effluent	Activated sludge reactor	UV/H_2O_2 and photo-Fenton process	Color removal of 86% efficiency in polishing steps	[48]
Dye industry effluent	UASB and ZVI	Electrochemical oxidation (12 h HRT)	COD reduction of ~ 80%	[58]
Procion Scarlet	Aerobic biodegradation	Electrochemical oxidation	COD removal of 80% and Color removal of 96%	[46]
RR-198	SBR	UV/O_3	Color Removal of above 90% in 3 h	[33]
Direct black 22	Bacterial consortium	Photocatalysis	Decolorization of 95.7%, TOC removal of 99.8%	[10]

2.3 *Biological and Physicochemical Hybrid Reactors*

By combining biological and physiochemical approaches, the hazardous pollutants found in textile wastewater may be successfully eliminated. This hybrid reactor can remove both organic and inorganic contaminants. Biological-coagulation, biological-adsorption, biodegradation-oxidation, comes under this category [19].

The choice between a coagulation-biological process and a biological-coagulation process relies on the kind and quantity of the coagulant, the amount of sludge, and the quantity of non-biodegradable and inhibiting compounds in the wastewater. For alkaline wastewater, coagulation could be useful before biological treatment [19].

Some researchers mixed activated carbon with the activated sludge due to the limitations of biological systems. As a result, the COD, TN, and TOC elimination

Table 3 Combination of AOP-MR/MBR hybrid reactors

AOP type	MR/ MBR	Type of wastewater	Performance/removal efficiency	References
Electrocatalytic	MR	Dye wastewater	COD reduction of 83.65% and color removal of 99.94%	[38]
Fenton oxidation	MBR	Dyeing plant effluent	TOC removal of 39.3%, color removal of 69.5%	[16]
Fenton oxidation	MR	Textile washing wastewater	COD reduction of 84.5% and TOC reduction of 70.5%	[7]
Photocatalytic	MR	RB 5 dye	Degraded 88.89% of dye by ZnO/PMR (photocatalytic membrane reactor)	[6]
Photocatalytic	MR	Congo Red	Rejection of 25% dye	[22]
Photocatalytic	MR	Acid Red B	Rejection of 65% dye	[25]
Photocatalytic	MR	Industrial textile effluent	Rejection of 78% pollutant	[14]

was enhanced. Adsorption isotherms can be used to explain the adsorbent's contribution in this case. Adsorbent will function as a modulator in this scenario, enhancing the system's capacity for treatment. The toxic chemicals will be swiftly absorbed by the adsorbent (modulator), which will also provide a rich environment for the metabolism of microbes on the liquid–solid surface. This will control the concentration of free harmful material. Microbial degradation, which outperforms adsorption on both activated carbon and biomass, is the primary stage in the dye removal for the biological process modified by activated carbon. This reactor can remove COD and color simultaneously from the effluent with a single step [19].

Biological and membrane filtration is another technique in textile effluent treatment. This will be mainly used when the reuse of water is needed. The main drawback of this reactor is the disposal of rejects from the membrane reactor. It is typically advised to do a site-specific techno-economic analysis to determine the ideal combination. Table 4 gives a description of these hybrid reactors for textile wastewater.

2.4 Hybrid Biological Processes

Modern textile effluent treatment uses a number of conventional biological methods and their combinations. Activated sludge systems may successfully remove acidic, dispersion, and vat dyes from low-concentration textile effluent. They are easily adsorbed onto the sludge due to their large molecular weight and lack of water solubility. However, azo dyes, which are naturally water soluble, make up the

Table 4 Combination of biological-physicochemical hybrid reactors

Biological process	Physicochemical process	Dye type	Performance/ removal efficiency	References
Membrane sequencing batch reactor	Adsorption using Granular Activated Carbon (GAC)	Acid Red 18	Dye removal of 77%	[51]
Bioaugmented membrane bioreactor	Adsorption using GAC	Structurally different azo dyes	Stable decolorization along with TOC and TN removal	[20]
Membrane bioreactor	Adsorption using GAC	Acid Orange II	Color removal >85%	[21]
Biodegradation	Coagulation/ Flocculation	Industrial textile wastewater	Decolorization of 82%, COD reduction of 76%	[8]
Biodegradation	Coagulation	Industrial textile effluent	Color removal of 98%	[9]
Biodegradation	Coagulation	Wastewater from dyeing and laundry industry	Color removal of 96%	[12]

majority of the colors manufactured worldwide. The activated sludge is incapable of adsorbing them. The reduction process under anaerobic conditions can destroy the azo bond in the dye. Following this anaerobic reduction, the colorless amines that were created can be processed by aerobic processes. Compared to their parent compounds, these colourless amines have a higher level of toxicity. The sequential anaerobic/aerobic method is a potentially effective way to lessen the harmful effects of textile wastewater.

The first large-scale anaerobic/aerobic facility in the Netherlands was built in 1999 to treat wastewater from textile factories. They initially built up an anaerobic facility to lower the COD of textile effluent. Since the effluent contains sulfuric chemicals, an aerobic system was added after the anaerobic system to lessen the formation of odorous sulphide. There are few toxicity data on anaerobic–aerobic systems that treat wastewater from textile production in the literature, notably data on full-scale installations. So, we can be sure that a full-scale anaerobic/aerobic process is a promising first step [17]. Table 5 provides a combination of reported hybrid biological reactors.

Table 5 Hybrid biological reactors

Effluent type	Reactor	Experimental setup	Performance/ removal efficiency	References
Real textile effluent	Hybrid anaerobic-aerobic reactor	Anaerobic SBR HRT 48 h, aerobic SBR HRT 6 h, pH 6.8–7.2, MLSS 6000–8000 mg/L	COD removal of 99.5%, color removal >40%	[47]
Desizing unit wastewater	Hybrid anaerobic baffled reactor	Mesophilic condition, HRT 168 h, flow rate 0.48 ± 0.01 L/h	Stable pH, alkalinity, degraded product was acetate and butyrate	[31]
Printing and dyeing wastewater	Hybrid anaerobic baffled reactor	HRT 12–13 h	COD removal of 47%, color removal of 56%	[23]
Dyeing wastewater	Pilot-scale anaerobic baffled reactor	HRT 12 h, OLR 1.5–2.0 kg COD/ (m^3·d)	COD reduction of 33.7%, sulfate removal of 22.5%	[39]
Textile effluent	Hybrid anaerobic baffled reactor	HRT 5 d	COD reduction of 42%	[42]
Dyeing unit wastewater	Hybrid column upflow anaerobic fixed bed reactor	HRT 96 h	COD reduction of 84.8%, color removal of 90%	[44]
Azo dye removal	Up-flow hybrid anaerobic digestion reactor with built in bio-electrochemical system	HRT 12 h	Color removal of 90.4%	[13]

3 Novel Hybrid Materials Used in Dye Effluent Treatment

Hybrid materials are an emerging class of innovative materials that are on the bleeding edge of technological development. By combining inorganic and organic components in a synergistic way at the molecular level, a hybrid material is created. By allowing the creation of a material that possesses the properties of both components, these concepts ultimately help conventional materials overcome their structural limitations. Wastewater treatment is the typical application for these materials. We will give a few of the significant hybrid materials reported in the literature in this part.

(i) Hybrid polymer coagulant-flocculants: In general, these hybrid materials are coagulants with physical or chemical incorporations of polymeric molecules. This hybrid material concentrates on the ability to enhance floc aggregation during operations. A hybrid substance performs better because of its interaction between the original coagulants and newly incorporated functional chemical groups or elements [50]. Inorganic-inorganic polymer-based hybrid materials showed best results in research [56].

(ii) Composite membranes: The creation of thin film composite membranes made use of the interfacial polymerization theory. The process of interfacial cross-linking between reactive monomers on a porous supporting membrane is known as interfacial polymerization. Reverse osmosis and nanofiltration membranes, two ultrathin selective layers with remarkable separation capabilities, were created using this technique for thin film production. The permeation and solute separation rates of these thin films can be adjusted to fit the function [28]. The removal of color, turbidity, hardness, and TOC from textile effluent by nanofiltration produced positive results. However, because it is run under pressure, a considerable amount of energy is used, and the membranes become fouled. These issues can be resolved by utilizing submerged nanofiltration technology [50]. The direct blue 51 containing wastewater was treated using a composite membrane made of chitosan, PEG, MWCNT, and iodine. The antibacterial activity, selectivity, and effectiveness of the composite membranes were superior. In this instance, the dye rejection was 83.7% [27].

(iii) Polymeric nanocomposite: When two or more types of nanoparticles are mixed, they form nanocomposites, which are potentially useful materials with unique physical properties. The creation of polymeric nanoparticles, which improve the usability and functioning of nanocomposites, involves mixing polymers with nanoparticles. The system is made more flexible by the polymers while colloidal stability or mechanical strength is provided by the nanoparticles. Adsorbents for wastewater treatment are typically made of polymeric nanocomposites. They are highly advised due to their varied pore size. As a result of their high specific surface area, there will be more adsorption sites. Additionally, the interparticle spacing is smaller in composites. The nanocomposite is widely favored because of these properties [40]. Wastewater from industrial dye production was treated using titanium dioxide-zeolite nanocomposite. At lower dye concentrations, it was discovered that the TiO_2-zeolite nanocomposite exhibits good results. So it will be helpful in the advanced industrial dye wastewater treatment stage for eliminating traces and untreated dye compounds [11].

(iv) Hybrid/tribid photocatalyst: Photocatalysis is thought to be a workable alternative for dye removal or degradation. In the photocatalysis procedure, TiO_2 is frequently utilized as a photocatalyst. Due to its broadband gap, TiO_2 is exclusively active in the UV light spectrum [41]. Many researches are happening to find out more photocatalytic catalysts that will be more efficient than the TiO_2 catalyst. Hybrid copper oxide nanoneedles were studied in the removal of coomassie brilliant blue, methylene blue, congo red, and amidoblack 10B. This hybrid material has environments that are both hydrophilic and hydrophobic, which is believed to improve the electrostatic interaction between the dye molecules in the adsorbent and favor the adsorption process [41].

Zhang et al. [57] studied a novel hybrid material $Na_2Mo_4O_{13}/\alpha\text{-}MoO_3$ for the degradation of dye. Color removal of 100% was achieved within 30 min of reaction time for 100 mg/L concentration of dye solution. In 2007, Liu and Sun announced the development of the first catalytic wet air oxidation catalyst (Fe_2O_3-CeO_2-TiO_2/-Al_2O_3) with high activity at ambient temperature and atmospheric pressure. In 2.5 h at 25 °C and 1 bar, 98.1% of the color and 96.1% of the total organic carbon (TOC) were removed from a 100 mL, 500 mg/L methyl orange dye solution [30]. Reactive Blue 19 (RB-19) and Reactive Green 19 (RG-19) dyes were taken up by an organic–inorganic hybrid clay material called N1-(3-trimethoxysilylpropyl) diethylenetriamine (TMSPDETA) that was used as an adsorbent in aqueous wastewaters. Electrostatic interaction is the primary mechanism by which reactive dyes interact with hybrid clay. The clay@TMSPDETA treats synthetic dye-textile effluent quite effectively. Using distilled water and wastewater from the plastics sector as the solvents, the removal rates of the simulated wastewater were up to 97.67% and 88.34%, respectively [52].

4 Cost Analysis and Recommendations

The sum of the expenses for capital, operation, and maintenance is the overall cost. Costs are typically very site-specific. While running costs include items like electricity costs, skilled workers' salaries, cost of regeneration and reject disposal, etc., capital costs include items like the land where it is located and the cost of implementation, etc. Therefore, these expenses will change depending on the process. Therefore, it is inappropriate to compare the costs of various reactors directly. Combined biological and AOP (Fe_3O/H_2O_2) cost around 0.57 US\$/m^3 [49]. While Hybrid photo-Fenton and membrane process cost around 0.736 US\$/m^3 [29] and Electrocoagulation, H_2O_2/Fe^{2+}/UV with adsorption 2.3 US\$/m^3 [18], Chemical and ion exchange process cost around 3.5 US\$/m^3 [53], integrated electro coagulation and Fenton process cost around 1.65 US\$/m^3 [26].

Studies have generally shown that biological processes are less expensive than AOP, membrane-related reactors. Although biological and physiochemical processes are less expensive, the results of their treatments might not be sufficient. Hybrid membrane bioreactors are more expensive, but they will provide effective treatment, water reuse ability, and less membrane fouling. Combinations of AOPs are difficult to produce acceptable outcomes at a reasonable price. Compared to AOP with other treatment methods including adsorption, membrane filtration, and other physiochemical approaches, AOP with biological will be less expensive. AOP as a prior treatment to biological processes is found to be more cost-effective and provides complete mineralization. It is interesting to note that integrated systems that facilitate the reuse of water and/or supplementary chemicals seem to be doable, as evidenced by the short payback period for capital costs. These systems may combine a wide variety of technologies, such as membrane, AOP, wet air oxidation, and biological processes. Separate treatment will be less expensive than treating the effluent from each unit of the textile sector together. Although hybrid materials may initially cost more, their long-term advantages, such as increased treatment effectiveness and environmental compliance, can make them a sensible choice for many industrial applications.

5 Future Scope

This book chapter reviewed a number of studies on integrated techniques for treating textile effluent. A variety of COD removal and color removal treatment efficiencies are demonstrated in the literature under study; these have been noted to be either fully or partially carried out in the existing literature. Future research must therefore concentrate on parameter optimisation and the application of several textile effluent treatment techniques. Future research should examine the viability of alternative and improved treatment methods in industrial settings, with a particular emphasis on upholding the integrity and effectiveness of treatment processes. Resource recovery must be considered in addition to wastewater treatment. Only a small number of investigations have been conducted on dye recovery. Numerous investigations are needed on this issue. Since dyes can have a wide range of chemical structures, we haven't been able to retrieve them all yet. Innovative technologies must be created, and industry and research collaboration must improve. Also, the range of hybrid materials and their applications in the treatment of textile wastewater must also be expanded by continued research and development in the field of hybrid materials. The future scope of hybrid technologies in textile effluent treatment is listed below:

(i) Improved toxic materials removal: Hybrid technology can combine physical chemical and biological processes to enhance the treatment capacity. Since the textile effluent consists of toxic dyes, heavy metals, waxes, etc. more comprehensive hybrid reactors are to be designed to remove these pollutants.
(ii) Resource recovery: Hybrid systems like membrane-based technology should be improved for resource recovery. Dyes, salts other chemicals, if possible biodegraded products that can be recovered and used again. That will support the circular economy. Most of the azo dyes degraded to form aniline-based compounds that can be used in the medical field. Also, utilization of used water can reduce the water demand of the textile industry.
(iii) Sustainability: Hybrid reactors should focus on less energy consumption and less chemical requirements. Designs that conserve energy and the use of renewable energy sources will be crucial.
(iv) Real-time monitoring and control: Real-time monitoring of the treatment process will be made possible via sensor technologies and automation. This makes it possible to precisely manage and optimize the treatment parameters, resulting in effective and reliable performance.
(v) Decentralized treatment: The on-site or regional treatment of wastewater by textile factories. As a result, less extensive infrastructure and no longer transportation routes are required.

(vi) Water reuse and zero discharge systems: Hybrid systems will let textile plants reuse water more frequently and make it easier to achieve zero liquid discharge (ZLD) targets, which will lessen their impact on the environment and conserve water.

(vii) Implementation of artificial intelligence: Helps forecast system performance, enhance hybrid treatment procedures, and quickly identify problems. This makes it possible for running the treatment systems economical and less maintained.

6 Conclusion

Hybrid treatment processes are the preferred choice to treat textile effluent due to its complex nature. Hybrid processes have the ability to totally mineralize harmful pollutants and can also be utilized to recover materials from textile effluent. The use of hybrid processes has shown improved efficiency in terms of color removal and COD reduction, as well as operational simplicity and time effectiveness when compared with individual treatment methods and combination methods The textile industry therefore requires customized treatments based on the effluent type rather than a broad strategy. But, it is critical to recognize that rigorous planning, consistent monitoring, and ongoing optimisation are necessary for hybrid treatment systems to be successful. Ensuring that the hybrid method selected satisfies industry-specific criteria and environmental compliance standards necessitates close collaboration among regulatory agencies, environmental engineers, and industry stakeholders. Even though there is a lot of study being done on hybrid processes, there is not enough research being done on the economic viability of these hybrid processes, which needs to be done in order to apply these improved hybrid processes in the industry. Based on this review, a biological system-AOP integration is the best in terms of treatment ability and economic perspective. It can repurpose effluent by removing color, organic pollutants, and harmful by-products with comparatively less amount of money. Along with hybrid processes, hybrid materials are also gaining attention in the effluent treatment field. The economic production of hybrid materials is yet to be studied.

Acknowledgements We gratefully acknowledge the Department of Science and Technology (DST), Government of India, for supporting this work through the research grant DST/TM/WIC/WTI/2K17/82(G4).

References

1. Abd El-Rahim WM, Moawad H, Azeiz AZA, Sadowsky MJ (2021) Biodegradation of azo dyes by bacterial or fungal consortium and identification of the biodegradation products. Egypt J Aquat Res 47(3):269–276. https://doi.org/10.1016/j.ejar.2021.06.002

2. Adane T, Adugna AT, Alemayehu E (2021) Textile industry effluent treatment techniques. J Chem 1–14. https://doi.org/10.1155/2021/5314404
3. Ahmad A, Mohd-Setapar SH, Chuong CS, Khatoon A, Wani WA, Kumar R, Rafatullah M (2015) Recent advances in new generation dye removal technologies: novel search for approaches to reprocess wastewater. RSC Adv 5(39):30801–30818. https://doi.org/10.1039/C4RA16959J
4. Al-Tohamy R, Ali SS, Li F, Okasha KM, Mahmoud YAG, Elsamahy T, ... Sun J (2022) A critical review on the treatment of dye-containing wastewater: ecotoxicological and health concerns of textile dyes and possible remediation approaches for environmental safety. Ecotoxicol Environ Saf 231:113160. https://doi.org/10.1016/j.ecoenv.2021.113160
5. Asaithambi P, Yesuf MB, Govindarajan R, Hariharan NM, Thangavelu P, Alemayehu E (2022) A review of hybrid process development based on electrochemical and advanced oxidation processes for the treatment of industrial wastewater. Int J Chem Eng. https://doi.org/10.1155/2022/1105376
6. Ashar A, Bhatti IA, Ashraf M, Tahir AA, Aziz H, Yousuf M, ... Bhutta ZA (2020) Fe3+@ ZnO/polyester based solar photocatalytic membrane reactor for abatement of RB5 dye. J Clean Prod 246:119010. https://doi.org/10.1016/j.jclepro.2019.119010
7. Aydiner C, Mert BK, Dogan EC, Yatmaz HC, Dagli S, Aksu S, ... Balci E (2019) Novel hybrid treatments of textile wastewater by membrane oxidation reactor: performance investigations, optimizations and efficiency comparisons. Sci Total Environ 683:411–426. https://doi.org/10.1016/j.scitotenv.2019.05.248
8. Ayed L, Ksibi IE, Charef A, Mzoughi RE (2021) Hybrid coagulation-flocculation and anaerobic-aerobic biological treatment for industrial textile wastewater: pilot case study. J Text Inst 112(2):200–206. https://doi.org/10.1080/00405000.2020.1731273
9. Bedekar PA, Bhalkar BN, Patil SM, Govindwar SP (2016) Moringa oleifera-mediated coagulation of textile wastewater and its biodegradation using novel consortium-BBA grown on agricultural waste substratum. Environ Sci Pollut Res 23:20963–20976. https://doi.org/10.1007/s11356-016-7279-8
10. Ceretta MB, Vieira Y, Wolski EA, Foletto EL, Silvestri S (2020) Biological degradation coupled to photocatalysis by ZnO/polypyrrole composite for the treatment of real textile wastewater. J Water Process Eng 35:101230. https://doi.org/10.1016/j.jwpe.2020.101230
11. Chong MN, Tneu ZY, Poh PE, Jin B, Aryal R (2015) Synthesis, characterisation and application of TiO_2–zeolite nanocomposites for the advanced treatment of industrial dye wastewater. J Taiwan Inst Chem Eng 50:288–296. https://doi.org/10.1016/j.jtice.2014.12.013
12. Costa AFS, Albuquerque CDC, Salgueiro AA, Sarubbo LA (2018) Color removal from industrial dyeing and laundry effluent by microbial consortium and coagulant agents. Process Saf Environ Prot 118:203–210. https://doi.org/10.1016/j.psep.2018.03.001
13. Cui MH, Cui D, Lee HS, Liang B, Wang AJ, Cheng HY (2016) Effect of electrode position on azo dye removal in an up-flow hybrid anaerobic digestion reactor with built-in bioelectrochemical system. Sci Rep 6(1):25223. https://doi.org/10.1038/srep25223
14. Desa AL, Hairom NHH, Ng LY, Ng CY, Ahmad MK, Mohammad AW (2019) Industrial textile wastewater treatment via membrane photocatalytic reactor (MPR) in the presence of ZnO-PEG nanoparticles and tight ultrafiltration. J Water Process Eng 31:100872. https://doi.org/10.1016/j.jwpe.2019.100872
15. Elgarahy AM, Elwakeel KZ, Mohammad SH, Elshoubaky GA (2021) A critical review of biosorption of dyes, heavy metals and metalloids from wastewater as an efficient and green process. Clean Eng Technol 4:100209. https://doi.org/10.1016/j.clet.2021.100209
16. Feng F, Xu Z, Li X, You W, Zhen Y (2010) Advanced treatment of dyeing wastewater towards reuse by the combined Fenton oxidation and membrane bioreactor process. J Environ Sci 22(11):1657–1665. https://doi.org/10.1016/S1001-0742(09)60303-X
17. Frijters CTMJ, Vos RH, Scheffer G, Mulder R (2006) Decolorizing and detoxifying textile wastewater, containing both soluble and insoluble dyes, in a full scale combined anaerobic/aerobic system. Water Res 40(6):1249–1257. https://doi.org/10.1016/j.watres.2006.01.013

18. GilPavas E, Dobrosz-Gómez I, Gómez-García MÁ (2019) Optimization and toxicity assessment of a combined electrocoagulation, H2O2/Fe2+/UV and activated carbon adsorption for textile wastewater treatment. Sci Total Environ 651:551–560. https://doi.org/10.1016/j.scitotenv.2018.09.125
19. Hai FI, Yamamoto K, Fukushi K (2007) Hybrid treatment systems for dye wastewater. Crit Rev Environ Sci Technol 37(4):315–377. https://doi.org/10.1080/10643380601174723
20. Hai FI, Yamamoto K, Nakajima F, Fukushi K (2011) Bioaugmented membrane bioreactor (MBR) with a GAC-packed zone for high rate textile wastewater treatment. Water Res 45(6):2199–2206. https://doi.org/10.1016/j.watres.2011.01.013
21. Hai FI, Yamamoto K, Nakajima F, Fukushi K (2012) Application of a GAC-coated hollow fiber module to couple enzymatic degradation of dye on membrane to whole cell biodegradation within a membrane bioreactor. J Membr Sci 389:67–75. https://doi.org/10.1016/j.memsci.2011.10.016
22. Hairom NHH, Mohammad AW, Kadhum AAH (2014) Effect of various zinc oxide nanoparticles in membrane photocatalytic reactor for Congo red dye treatment. Sep Purif Technol 137:74–81. https://doi.org/10.1016/j.seppur.2014.09.027
23. Huang R, Liu Z, Zhou X, Mo J, Liu X (2015) Research on treatment of printing and dyeing wastewater by hybrid anaerobic baffled reactor. Desalin Water Treat 54(3):590–597. https://doi.org/10.1080/19443994.2014.891463
24. Jafari N, Soudi MR, Kasra-Kermanshahi R (2014) Biodecolorization of textile azo dyes by isolated yeast from activated sludge: Issatchenkia orientalis JKS6. Annals Microbiol 64:475–482. https://doi.org/10.1007/s13213-013-0677-y
25. Jiang H, Zhang G, Huang T, Chen J, Wang Q, Meng Q (2010) Photocatalytic membrane reactor for degradation of acid red B wastewater. Chem Eng J 156(3):571–577. https://doi.org/10.1016/j.cej.2009.04.011
26. Kaur P, Sangal VK, Kushwaha JP (2019) Parametric study of electro-Fenton treatment for real textile wastewater, disposal study and its cost analysis. Int J Environ Sci Technol 16:801–810. https://doi.org/10.1007/s13762-018-1696-9
27. Khoerunnisa F, Rahmah W, Ooi BS, Dwihermiati E, Nashrah N, Fatimah S, ... Ng EP (2020) Chitosan/PEG/MWCNT/Iodine composite membrane with enhanced antibacterial properties for dye wastewater treatment. J Environ Chem Eng 8(2):103686. https://doi.org/10.1016/j.jece.2020.103686
28. Lau WJ, Ismail AF, Misdan N, Kassim MA (2012) A recent progress in thin film composite membrane: a review. Desalination 287:190–199. https://doi.org/10.1016/j.desal.2011.04.004
29. Lebron YAR, Moreira VR, Maia A, Couto CF, Moravia WG, Amaral MCS (2021) Integrated photo-Fenton and membrane-based techniques for textile effluent reclamation. Sep Purif Technol 272:118932. https://doi.org/10.1016/j.seppur.2021.118932
30. Liu Y, Sun D (2007) Development of Fe_2O_3-CeO_2-TiO_2/γ-Al_2O_3 as catalyst for catalytic wet air oxidation of methyl orange azo dye under room condition. Appl Catal B 72(3–4):205–211. https://doi.org/10.1016/j.apcatb.2006.10.015
31. Liu RR, Tian Q, Yang B, Chen JH (2010) Hybrid anaerobic baffled reactor for treatment of desizing wastewater. Int J Environ Sci Technol 7:111–118. https://doi.org/10.1007/BF03326122
32. Ma J, Chen Y, Nie J, Ma L, Huang Y, Li L, ... Guo Z (2018) Pilot-scale study on catalytic ozonation of bio-treated dyeing and finishing wastewater using recycled waste iron shavings as a catalyst. Sci Reports 8(1):7555. https://doi.org/10.1038/s41598-018-25761-6
33. Mahdizadeh H, Dadban Shahamat Y, Rodríguez-Couto S (2021) Discoloration and mineralization of a textile azo dye using a hybrid UV/O 3/SBR process. Appl Water Sci 11:1–9. https://doi.org/10.1007/s13201-021-01479-1
34. Nagel-Hassemer ME, Carvalho-Pinto CRS, Matias WG, Lapolli FR (2011) Removal of colored compounds from textile industry effluents by UV/H2O2 advanced oxidation and toxicity evaluation. Environ Technol 32(16):1867–1874. https://doi.org/10.1080/09593330.2011.566893

35. Paquini LD, Marconsini LT, Profeti LPR, Campos OS, Profeti D, Ribeiro J (2023) An overview of electrochemical advanced oxidation processes applied for the removal of azo-dyes. Braz J Chem Eng 1–31. https://doi.org/10.1007/s43153-023-00300-7
36. Paździor K, Wrębiak J, Klepacz-Smółka A, Gmurek M, Bilińska L, Kos L, ... Ledakowicz S (2017) Influence of ozonation and biodegradation on toxicity of industrial textile wastewater. J Environ Manag 195:166–173. https://doi.org/10.1016/j.jenvman.2016.06.055
37. Punzi M (2015) Treatment of textile wastewater by combining biological processes and advanced oxidation. Department of Biotechnology, Lund University, Lund
38. Qaseem S, Dlamini DS, Zikalala SA, Tesha JM, Husain MD, Wang C, ... Li J (2020) Electrocatalytic membrane anode for dye removal from wastewater. Colloids Surf A: Physicochem Eng Aspects 603:125270. https://doi.org/10.1016/j.colsurfa.2020.125270
39. Qi Z, Xiang G, Xiong D (2019) Performance evaluation of pilot-scale hybrid anaerobic baffled reactor (HABR) to process dyeing wastewater based on grey relational analysis. Appl Sci 9(10):1974. https://doi.org/10.3390/app9101974
40. Qu X, Alvarez PJ, Li Q (2013) Applications of nanotechnology in water and wastewater treatment. Water Res 47(12):3931–3946. https://doi.org/10.1016/j.watres.2012.09.058
41. Rajesh R, Iyer SS, Ezhilan J, Kumar SS, Venkatesan R (2016) Graphene oxide supported copper oxide nanoneedles: an efficient hybrid material for removal of toxic azo dyes. Spectrochim Acta Part A Mol Biomol Spectrosc 166:49–55. https://doi.org/10.1016/j.saa.2016.05.002
42. Rongrong L, Xujie L, Qing T, Bo Y, Jihua C (2011) The performance evaluation of hybrid anaerobic baffled reactor for treatment of PVA-containing desizing wastewater. Desalination 271(1–3):287–294. https://doi.org/10.1016/j.desal.2010.12.044
43. Rostam AB, Taghizadeh M (2020) Advanced oxidation processes integrated by membrane reactors and bioreactors for various wastewater treatments: a critical review. J Environ Chem Eng 8(6):104566. https://doi.org/10.1016/j.jece.2020.104566
44. Sandhya S, Swaminathan K (2006) Kinetic analysis of treatment of textile wastewater in hybrid column upflow anaerobic fixed bed reactor. Chem Eng J 122(1–2):87–92. https://doi.org/10.1016/j.cej.2006.04.006
45. Saroj S, Dubey S, Agarwal P, Prasad R, Singh RP (2015) Evaluation of the efficacy of a fungal consortium for degradation of azo dyes and simulated textile dye effluents. Sustain Water Resour Manag 1:233–243. https://doi.org/10.1007/s40899-015-0027-2
46. Senthilkumar S, Basha CA, Perumalsamy M, Prabhu HJ (2012) Electrochemical oxidation and aerobic biodegradation with isolated bacterial strains for dye wastewater: combined and integrated approach. Electrochim Acta 77:171–178. https://doi.org/10.1016/j.electacta.2012.05.084
47. Shoukat R, Khan SJ, Jamal Y (2019) Hybrid anaerobic-aerobic biological treatment for real textile wastewater. J Water Process Eng 29:100804. https://doi.org/10.1016/j.jwpe.2019.100804
48. Soares PA, Silva TF, Arcy AR, Souza SMGU, Boaventura RA, Vilar VJ (2016) Assessment of AOPs as a polishing step in the decolorisation of bio-treated textile wastewater: technical and economic considerations. J Photochem Photobiol, A 317:26–38. https://doi.org/10.1016/j.jphotochem.2015.10.017
49. Solmaz SKA, Birgül A, Üstün GE, Yonar T (2006) Colour and COD removal from textile effluent by coagulation and advanced oxidation processes. Color Technol 122(2):102–109. https://doi.org/10.1111/j.1478-4408.2006.00016.x
50. Su CXH, Low LW, Teng TT, Wong YS (2016) Combination and hybridisation of treatments in dye wastewater treatment: a review. J Environ Chem Eng 4(3):3618–3631. https://doi.org/10.1016/j.jece.2016.07.026
51. Taheri M, Fallah N, Nasernejad B (2023) Comparison of high-concentration azo dye removal by long HRT in MSBRs' bioaugmented with GAC and sponge media. Environ Sci Pollut Res 30(1):1201–1215. https://doi.org/10.1007/s11356-022-22055-3
52. Teixeira RA, Lima EC, Benetti AD, Naushad M, Thue PS, Mello BL, ... Seliem MK (2023) Employ a Clay@ TMSPDETA hybrid material as an adsorbent to remove textile dyes from wastewater effluents. Environ Sci Pollut Res 1–15. https://doi.org/10.1007/s11356-023-28568-9

53. Üstün GE, Solmaz SKA, Birgül A (2007) Regeneration of industrial district wastewater using a combination of Fenton process and ion exchange—A case study. Resour Conserv Recycl 52(2):425–440. https://doi.org/10.1016/j.resconrec.2007.05.006
54. Viswanathan B (2017) Photocatalytic degradation of dyes: an overview. Curr Catal 7:99–121. https://doi.org/10.2174/2211544707666171219161846
55. Yagub MT, Sen TK, Afroze S, Ang HM (2014) Dye and its removal from aqueous solution by adsorption: a review. Adv Coll Interface Sci 209:172–184. https://doi.org/10.1016/j.cis.2014.04.002
56. Zhang S, Du Z, Li G (2013) Metal-organic framework-199/graphite oxide hybrid composites coated solid-phase microextraction fibers coupled with gas chromatography for determination of organochlorine pesticides from complicated samples. Talanta 115:32–39. https://doi.org/10.1016/j.talanta.2013.04.029
57. Zhang Z, Yang R, Gao Y et al (2014) Novel $Na_2Mo_4O_{13}/\alpha$-MoO_3 hybrid material as highly efficient CWAO catalyst for dye degradation at ambient conditions. Sci Rep 4:6797. https://doi.org/10.1038/srep06797
58. Zhu X, Ni J, Wei J, Xing X, Li H (2011) Destination of organic pollutants during electrochemical oxidation of biologically-pretreated dye wastewater using boron-doped diamond anode. J Hazard Mater 189(1–2):127–133. https://doi.org/10.1016/j.jhazmat.2011.02.008
59. Zou XL (2015) Combination of ozonation, activated carbon, and biological aerated filter for advanced treatment of dyeing wastewater for reuse. Environ Sci Pollut Res 22:8174–8181. https://doi.org/10.1007/s11356-015-4423-9

Ozone as a Sustainable Option for the Treatment of Textile Dye Pollution

Abdullah Yasar, Iqra Wasif, Fatima Ahmad, Mahnoor Nadeem, Amtul Bari Tabinda, and Rameesha Tanveer

Abstract The textile industry is the second largest pollution causing industry. Many chemical and mechanical processes involved in the textile industry impose various health and environmental impacts. It is responsible for 20% of all freshwater pollution. The main problem with the textile industry is its untreated effluent discharge. The textile wet process produces approximately 80% discharge of industrial wastewater. Every kilogram of fabric that is generated utilizes up to 100–180 L of water. A textile mill unit generating 8000 kg of fabric per day consumes up to 1.6 million liters of water every day. The presence of different chemicals, dyes, and surfactants in textile wastewater imposes adverse health impacts. Various conventional treatment processes have been widely used by the industries but they have many shortcomings. The industries that treat dye-bath effluent usually use traditional treatment methods like trickling filters, activated sludge, flocculation, coagulation, and reverse osmosis. These processes do not degrade the dyes effectively and produce large volumes of sludge. Ozonation and advanced oxidation methods have recently gained immense popularity in the treatment of textile wastewater due to their high efficiency. AOPs include ozone (O_3), O_3/H_2O_2, UV, UV/H_2O_2, Fenton, and Photo-Fenton processes to treat textile effluent as they are effective for complete color and COD removal. Ozone-based treatment methods are a more sustainable option as ozone is environment friendly gas and provides us with better results compared to other methods. There is a dire need to make action plans or policies for the conservation and management of water and wastewater as the pollution levels are continuously rising.

A. Yasar (✉) · I. Wasif · F. Ahmad · M. Nadeem · A. B. Tabinda
Sustainable Development Study Centre (SDSC), Government College University, Lahore 54000, Pakistan
e-mail: yasar.abdullah@gcu.edu.pk

A. B. Tabinda
e-mail: amtulbaritabinda@gcu.edu.pk

R. Tanveer
School of Environment and Science, Australian Rivers Institute, Griffith University, Southport, QLD 4222, Australia

P. Singh (ed.), *Dye Pollution from Textile Industry*, SDGs and Textiles,
https://doi.org/10.1007/978-981-97-5341-3_10

Keywords Dye-bath effluent · Textile industry · Wastewater · Advance oxidation processes · Ozone · Sustainability

1 Introduction

Water is one of the natural resources on Earth, but a small portion is available for humans including its consumption, use, and its treatment [46]. Water is one of the most basic requirements of life that is being utilized to perform different activities of everyday life. Due to the rapid development of polluting industries such as textile, tanneries, pharmaceuticals, fertilizers, mining industries, etc., the whole world is facing water problems, and this has increased the living standards, growth of human population, and urbanization that ultimately leading to the production of recalcitrant and toxic pollutants in the water [2, 38]. These pollutants cause serious health effects. It is therefore required to treat industrial effluent on a priority basis. The textile industry is among the most polluting industries and is responsible for 20% of all fresh water pollution [20].

Farhana et al. [12] stated that the textile industry plays a pivotal role in the economy of developing countries like Bangladesh, India, Malaysia, Pakistan, and China. Wastewater discharge from the textile industry has high pH, temperature, toxic chemicals, COD, BOD, and color. The wastewater discharged from the textile industry has a strong smell, and color, and is alkaline in nature. Chemicals in wastewater usually contain pigments and dyes that are toxic in nature which can decrease the DO that damages the water quality and can threaten the aquatic life as well [3]. Solvent vapors, alkaline or acid mists, oil, cotton dust, dust from the wet process, dry process, and oxides of sulfur and nitrogen are all pollution sources that cause acute as well as chronic effects [27].

The production processes often use chemicals that may cause problems related to occupational health and safety. Toxic air emissions are also generated due to the presence of these chemicals in the final products or effluent. The main problem with the textile industry is its untreated effluent discharge. As it is water dependent industry, it generates 1000–3000 m^3 of wastewater against a product of 12–20 tons per day. When the wastewater enters the nearby freshwater channel, it reduces the light penetration resulting in a low DO level for aquatic life and thus also effects on the aesthetic beauty of the water channel [39, 46].

The conventional wastewater treatment processes are time-consuming, energy-intensive, consumption of capital, land, material, and manpower. There are many methods including membrane filtration, adsorption, and photocatalyst that are used to remove synthetic dyes, but these methods cannot stand alone due to the degradation of dyes [7]. For the sake of profit, industrialists show reluctance with the release of wastewater directly on the land that damages the agriculture and aquatic life because some treatment plants are not functional yet. It is essential to incorporate modern treatment technologies for the treatment of dye-bath effluent as untreated wastewater discharge in natural channels is a major industrial pollutant [9] (Fig. 1).

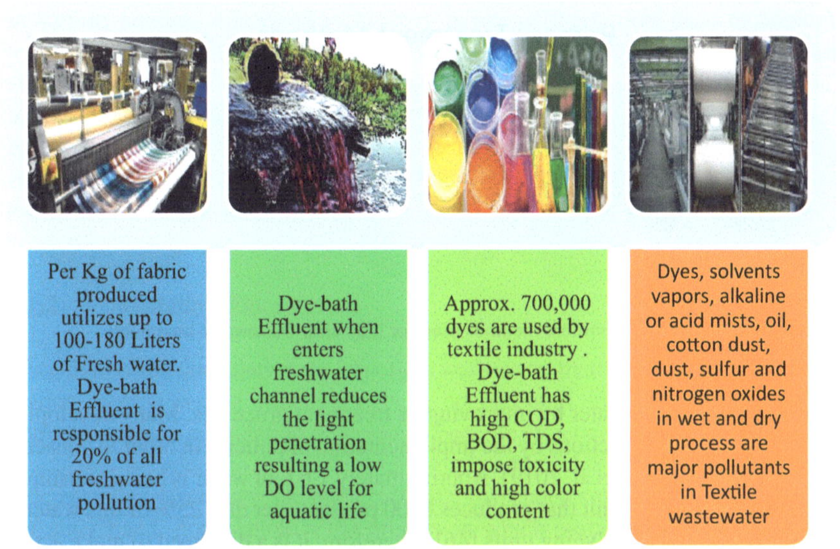

Fig. 1 Characteristics of textile industry wastewater

The conventional wastewater treatment technologies involve flow equalization, sedimentation, adsorption, membrane filtration, flocculation, and reverse osmosis are still used to treat wastewater effluent before its discharge [63]. Other more intensive methods include trickling filters and activated sludge to remove soluble organic matter. Moreover, sequential batch reactor, membrane bioreactor, biofilm reactor, and up-flow anaerobic sludge blanket reactor (UASB) are also secondary wastewater treatment methods. The chemical methods include coagulation and electrocoagulation. Lastly, the most advanced and latest treatment involves the advanced oxidation processes (AOPs) that remove inorganic compounds from wastewater but they have a high operating cost [27].

2 Treatment of Textile Wastewater

Textile wastewater is full of toxic substances and dyes which are harmful to the environment. When these dyes come in contact with a nearby water channel, the aesthetic beauty starts to decrease and, in this way, the aquatic plants cannot survive as these dyes impair photosynthesis [23]. Various conventional treatment processes have been widely used by the industries but they have many shortcomings. Figure 2 presents the primary treatment methods and advanced treatment technologies used for the treatment of dye-bath effluent.

The major issue in the textile industry is the treatment of wastewater prior to its discharge into the environment. The accessibility of water is low which is why it is

wastewater. The flocculant interacted with activated sludge. The maximum COD and total nitrogen (TN) removal was obtained at a 10 mg/L dose of aluminum. When the dose was increased to more than 10 mg/L, it decreased the microbial activity and hence for the removal of total phosphorus (TP), the effective dose was noted as 30 mg/L, respectively.

Commercial flocculants (PAC and PAM) can be used with paper sludge-based and lignin-based flocculants to treat reactive turquoise blue (RTB) and disperse dyes (DB) from wastewater. The jar test performed by [13] for the flocculation process used 1 L of dye wastewater for each flocculent. The physiochemical parameters were measured, and the result showed that the commercial coagulant/flocculant removed the color up to 50%. While paper sludge-based flocculant removed 90–92% of dyes from textile dye-bath effluent. In addition, lignin-based flocculant removed dyes up to 89%. It was shown that among all the flocculants, paper sludge-based flocculants were the most effective.

2.1.3 Biological Treatment

Shoukat et al. [42] treated textile wastewater by using biological treatment aerobic and anaerobic technology. The experiment emphasized system optimization in two phases. In the first phase, synthetic wastewater was optimized at the lab scale, and in the second phase, real textile wastewater was treated. For the combined process, the activated sludge was obtained from a membrane bioreactor (MBR), and it was adjusted with real textile effluent. The HRT was found to be 54 h for the removal of COD. The hybrid process proved effective as it showed removal efficiencies of 99% COD, 78% dyes, and 99% TKN. Moreover, the process was made more efficient when it was adapted to Sequential Batch Reactor (SBR) by the production of sustainable energy in the form of biogas.

The textile industry produces a variety of dyes in its effluent such as azo dyes, and anthraquinone. It is estimated that about 80% of textile wastewater is accounted for in the dyeing process. For color removal, biological wastewater treatment is insufficient, and degradation of dyes and physiochemical processes do not provide complete degradation of pollutants. Advanced treatment of ozonation is feasible for the removal of dyes and color from wastewater.

2.1.4 Membrane Bioreactor

The textile industry is one of the largest consumers of water and produces large amount of wastewater. For the treatment of textile waste effluent, membrane bioreactor (MBR) technology was employed. The study by [26] investigated microbial communities in order to decolorize the textile dye-bath effluent. The dye mass loading rate was set high for improvement of the performance of decolorization. In this way, dye concentration was set high and in the same way, biomass concentration was also increased. The result showed complete decolorization of the textile dye bath. Hence, it was observed that the removal efficiency of dye was increased with the increase of biomass loading rate made effective treatment of textile wastewater.

The membrane bioreactor treatment process is one of the cost effective and simple processes for the treatment of wastewater. But this process can cause fouling issues that could be resolved if MBR could be combined with any other treatment. Jegatheesan et al. [22] studied the combined process of MBR with ultrafiltration (UF), membrane distillation (MD), and nanofiltration (NF) for removing dye and recovering wastewater. It was indicated from the results that if MBR is combined with the UF process, only dyes will be removed from wastewater. If MBR combined with NF, dyes and some salt would be removed from textile effluent under optimal wastewater conditions. Thus, the combination of processes will depend upon the nature of the effluent and the chemistry of the dyes for the removal process.

2.1.5 Activated Sludge

Different types of chemicals and dyes are found in textile wastewater which makes it salty in nature. For the treatment of textile effluent, a sequencing batch reactor (SBR) was used by [33] in a time cycle of 24 h. The concentration of dye, TDS was 125 mg/L to 500 mg/L and 1000 mg/L to 10,000 mg/L in synthetic waste effluent. When the conc of dye was 500 mg/L, COD was removed up to 80%. When the TDS value was 5000 mg/L, low removal of COD was obtained as 59%. When the concentration of TDS was further increased, low removal of COD was achieved. Therefore, an inverse relation was observed between COD and TDS. When TDS was increased, only the removal of MLSS (mixed liquor suspended solids) and MLVS (mixed liquid volatile suspended solids) but COD removal did not decrease of microorganisms of activated sludge in the wastewater treatment system.

Industrial textile wastewater generates excess sludge that was recorded in a study of about 1.5 tons. From this perspective, a study was designed to consider the environmental issue of activated sludge and how to eliminate it. A pilot-scale study was carried out by [16] to optimize the efficiency of the biodegradation process by aerobically that would reduce the sludge content. Respirometric analysis was analyzed in which dye to biomass ratio was studied that indicated the ratio of 0.72 g of COD. As the ratio increased, the color and COD removal was also increased. The optimized environmental parameters were analyzed, and the degradation of contaminants was fully removed up to 95 after 5 days of retention time.

2.2 Advanced Treatment Technologies

AOPs have taken attention in industrial wastewater treatment particularly in textile wastewater because of the high risk of toxic pollutants to the ecological environment. AOPs are employed due to their efficiency in removing dyes and toxic substances from them. To determine the degradation of organic pollutants, the reacting species

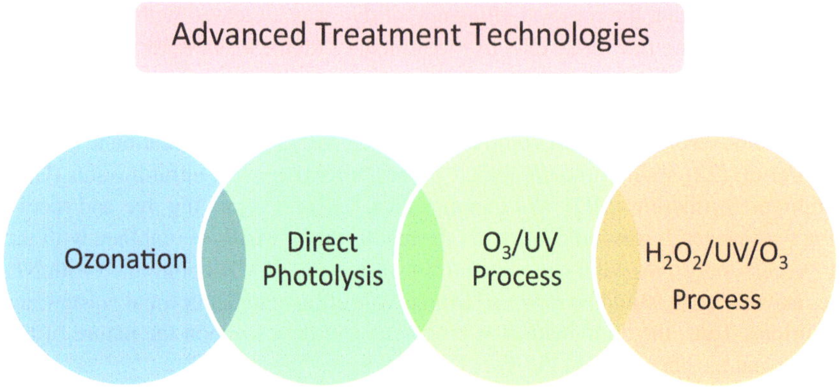

Fig. 3 Advanced treatment methods available for textile industry wastewater

such as hydroxyl radicals ($HO^{\bullet}$), radicals of sulfate (SO_4^{-2}), and superoxide ($O_2^{\bullet -}$) are formed by the various oxidants such as H_2O_2, ClO_3, persulfate, and oxygen. AOPs generate hydroxyl radicals which are strong oxidants. These radicals have the ability to degrade the toxic and persistent substances into their inert form that can remove from wastewater easily [53–56].

As textile wastewater is characterized by high COD, TDS, TSS, turbidity, color, and pH, it is important to handle such kind of wastewater by adopting methods that have the ability to degrade the dyes as well as lower the parameters in order to prevent the natural environment from degradation [58]. There are many methods that are associated with AOPs. These methods include ozonation, Fenton process, photo-Fenton process, UV/H_2O_2, UV along with catalyst (ZnO or TiO_2), electrochemical oxidation, and combination of ozone with other advanced processes. These advanced processes help in the degradation of contaminants from wastewater to make it reused as well (Fig. 3).

2.2.1 Ozonation

Ozone is being used as an oxidant for the removal of pollutants or micropollutants from drinking water as well as industrial wastewater. It is because of its high oxidation potential and its easy handling operation [50]. When ozone is decomposed, hydroxyl ions (OH) are generated which have the potential to remove pollutants from effluent. Ozone is a volatile gas and is composed of three atoms of oxygen that react with organic pollutants and then degrade them by decomposition. The following reactions (Eqs. 2.1–2.5) show the generation of hydroxyl radicals that speed up the reaction [52]:

$$O_3 + OH^- \rightarrow HO_2^- + O_2 \tag{2.1}$$

$$HO_2^- + O_3 \rightarrow O_2 + HO_2 \quad (2.2)$$

$$HO_2 \rightarrow O_2 + H^+ \quad (2.3)$$

$$O_2^- + O_3 \rightarrow O_3^- + O_2 \quad (2.4)$$

$$O_3^- + H_2O \rightarrow OH^\bullet + O_2 + OH^- \quad (2.5)$$

Ozone (O_3) has shown a high affinity toward compounds of olefins, phenols, amines, and sulfur groups. It has shown less affinity toward organic matter because of less OH. yield [59]. However it can be effective by adding any catalyst that could help in generating a high yield of OH. Oxide and ions of metals are a good option for catalytic ozonation. When Al- and Fe-based coagulants are added to the wastewater, the decomposition of ozone results in high production of $OH^\bullet$ which enhances the system efficiency through the hybrid process of coagulation and ozonation. These coagulants have the ability to degrade dissolved organic matter even having a high content of humic acid [64].

In ozonation-based treatment there is no need to add extra chemicals to wastewater or water. In this way, ozonation can be applied in textile wastewater without adjusting pH and chemical additives. Due to hydroxyl radicals, the toxic compounds can be destroyed. The process can be more effective by combining O_3 with UV, H_2O_2 with UV, H_2O_2 with Fe ions, and combining UV radiation with semiconductors and oxygen molecules. However, with the ability of the corona discharge generator inefficient production capacity is achieved, and the ozonation process becomes ineffective. By using other generators like an electrochemical ozone generator, the process could show better performance [21].

2.2.2 Direct Photolysis

To degrade the organic pollutants from textile effluent, direct UV with an oxidation process is a significant method. Basically, photolysis is a process in which organic pollutants are degraded by the direct use of UV radiation. UV lamps are used for this purpose that have low pressure and their wavelength is 254 nm to emit light. If the pressure is medium, a broad light spectrum is formed, H_2O_2 is used as a catalyst that makes the process more effective by the generation of $OH^\bullet$ radicals through UV photolysis [37]. The process is presented in the following reactions (Eqs. 2.6–2.9).

$$H_2O_2 + \lambda\,(254\ \text{nm}) \rightarrow 2OH \quad (2.6)$$

$$H_2O_2 + HO \rightarrow HO_2 + H_2O \quad (2.7)$$

$$H_2O_2 + HO_2 \rightarrow OH + H_2O + O_2 \quad (2.8)$$

$$HO_2 + HO_2 \rightarrow H_2O_2 + O_2 \quad (2.9)$$

Various catalysts such as TiO_2, $FeSO_4.7H_2O$, and H_2O_2 were used in the bioenergy production process and the process takes place under UV irradiance [1]. The UV rays generate $OH^•$ radical under 280 nm wavelength. When the UV radiation is reacted with H_2O_2, the $OH^•$ Radicals are formed. Moreover, when UV is integrated with an advanced process that has H_2O_2, the efficient results are figured out and it also helps in lowering the cost of the process [24]. For the removal of acid green dye from wastewater, a study was conducted in the presence of UV radiation. About 97% of the dye was removed from wastewater with the time duration of 3 to 4 h [32].

2.2.3 Ozone/UV Process

Hydrogen peroxide can be formed when ozone is photolyzed with UV radiation at the range of 200–280 nm. Hydroxyl radicles are generated by hydrogen peroxide under UV radiation. For UV light sources low pressure mercury vapor lamps are the most effective ones. The efficiency of the system is affected by pH, turbidity, temperature, pollutant type, and UV intensity [18].

$$O_3 + hv + H_2O \rightarrow H_2O_2 + O_2 \quad (2.10)$$

$$H_2O_2 + hv \rightarrow 2OH \quad (2.11)$$

$$2O_3 + H_2O_2 \rightarrow 2OH + 3\,O_2 \quad (2.12)$$

A study investigated the detoxification and decolorization of DB-86 blue dye solution in fresh and sea water by using O_3 and O_3/UV treatment processes. The result showed that O_3 was effective for complete color removal and about 62% COD removal as well. O_3/UV showed less efficiency because of the high temperature of the solution which affected the ozone half-life. For the detoxification, GC/MS was used that showed no toxic effect of compounds of phenol and amines because of dye degradation. The test performed on *Rotifer B. plicatilis* showed that the toxicity of zooplankton was reduced by the ozonation process [18].

Hassaan et al. [17] also investigated a study on seawater that had direct yellow 50 dye. The treatment of ozone itself and ozone combined with UV was performed. Dye concentration and time of reaction were also analyzed in a batch reactor. The individual treatment processes were analyzed first and then they were treated in combined form for further treatment. The result showed about 80–90% color removal by ozonation and partial removal of COD than the combined process. Thus, ozone treatment showed better results. Furthermore, GC/MS analysis of treated synthetic dye solution was performed that showed no toxic compounds.

2.2.4 $UV/O_3/H_2O_2$ Process

Among AOP's $O_3/UV/H_2O_2$ is considered an efficient option in which hydroxyl ions are generated with ozone that helps in the treatment and degradation of textile dye-bath effluent [40]. It has also studied that only ozonation could not provide effective removal of COD as the combined process could do so. For fiber dyeing wastewater, the integrated approach of $UV/O_3/H_2O_2$ was used by adding optimal doses of H_2O_2 and O_3. The result showed that the combined process removed the harmful pollutants from effluent and up to 99% of COD removal within 90 min. Moreover, before the treatment process, the inhibitory action was 47% which was reduced to 10% after the hybrid treatment process [60]. For the removal of azo dye Methyl Orange AOPs are effective. For this purpose, a study was designed in which ozone was facilitated with heterogeneous and homogeneous photo-catalytic processes for the degradation of dye and treatment of textile effluent. The combined system of $UV/H_2O_2/O_3$ was applied in a homogeneous medium while UV/TiO_2 was applied in a heterogeneous medium. It was shown that the hybrid process with ozone was much effective for dye degradation. Hence photo-catalytic process in a homogenous medium showed a better result in order to remove the contaminant from textile effluent as well as increasing the degradation of dye that was then removed from textile dye-bath effluent [44].

The photo-Fenton process is another advanced method that is similar to the Fenton process with the addition of UV radiation and is one of the most efficient processes for the treatment of textile wastewater. The efficiency of the process can be increased by adding UV radiation to the Fenton reagent which can be called a photo-Fenton oxidation process. When H_2O_2 is photolyzed with Fe^{+2} and Fe^{+3}, hydroxyl radicals are generated. As a result, the decomposition of organic pollutants increases which ultimately improves the removal efficiency of the treatment process. The complete process is as follows (Eqs. 2.13–2.15):

$$H_2O_2 + hv \rightarrow OH + OH \tag{2.13}$$

$$Fe^{+3} + H_2O_2 + hv \rightarrow Fe^{+2} + OH^- + OH \tag{2.14}$$

$$Fe\,(OH)^{+2} + hv \rightarrow Fe^{+2} + OH \tag{2.15}$$

In the photo-Fenton process, photolysis breaks the hydroxyl radical that increases the removal efficiency of the treatment process and produces less sludge in the effluent treatment process. When H_2O_2 reacts with metals of iron, the hydroxyl radical is formed. In the case of the electro-Fenton process, a reaction takes place on the cathode in the presence of H_2O_2 and iron catalyst in a solution to produce electrons and hydroxyl radicals [6]. Tekin et al. [49] performed the treatment of tartrazine

Fig. 4 The different types of Advanced Oxidation Processes (AOPs) discussed in this study

under the photo-Fenton process and UV irradiation. 0.25 g/L of bismuth oxyhalide was used under optimal conditions and it acted as a catalyst. The result showed a 91% degradation rate and 95% color removal efficiency (Fig. 4).

2.3 *Effect on Color and COD Removal*

Conventional biological treatment technologies are ineffective due to the presence of synthetic dyes and organic effluents. Therefore, ozone treatment is an effective option for it. For this azo reactive black-5 dye was used. The dye solution was prepared by using distilled water. Ozone gas was produced by using a corona discharge ozone generator along with dry air as feeding gas. Fine bubbles of ozone were produced in the bench-scale reactor. Different doses of ozone were taken by using the iodometry method. The different dye concentration was added. At the initial concentration of dye, the COD removal was up to 74% and the time was set at 120 min. As the concentration and time were increased, the removal was achieved at about 77.1%. In addition, the best color removal efficiency of 98% was achieved as the ozone dosage and incubation time were increased by about 300 min [45].

2.4 Effect on Turbidity

Turbidity is also an important physical parameter of wastewater that should be checked in order to remove UV light penetration into the water stream. A study was designed to remove turbidity by using chemical process of coagulation followed by an advanced oxidation process. 1000 mL wastewater was used and 5 different coagulants (SICOAG C-21, FLOCUSOL-PA/18, SIFLOC C 40 L PLUS, FLOCUSOL-CM/1 and SIFLOC C-30) were added with different dose concentrations. The removal efficiency was 99% for turbidity. Moreover, photo-Fenton process and UV/H_2O_2 were applied for the removal of total organic carbon (TOC) and the hybrid process removed 94% of TOC from textile dye-bath effluent [30].

2.5 Hybrid Treatment

Gil Pavas et al. [15] showed maximum removal efficiency of wastewater parameters like turbidity and COD up to 98% and 48%. The chemical process was further followed by a Fenton reagent ($FeSO_4$ and H_2O_2) that reduced the COD up to 74% with a time duration of 90 min. For further COD removal, the photo-Fenton process was used that reduce the COD by up to 87% and the BOD5/COD ratio was reached from 0.21 to 0.74 at the end of the treatment and degradation of harmful compounds within industrial effluent. For the treatment of this industry, a study was designed with chemical treatment assisted with nanofiltration. $FeSO_4$ was used as coagulant. Spiral wound polyamide nano filter was assisted with chemical coagulation. Response surface methodology was applied to check the operating parameters and complete mixing of coagulant. The result showed that the primary pretreatment process removed 40% of COD and turbidity at 8 pH. When the nanofiltration process was applied, the removal efficiency reached up to 98%. Hence the chemical coagulation and nanofiltration process is a promising technology for effective wastewater treatment [19].

When the biological process is combined with any other process, the efficiency of the process can be increased. For this purpose, a study was conducted on biological treatment that was integrated with the membrane process. The integrated process of membrane process was studied with a bioreactor as MBR. It was shown that MBR combined with NF, UF, and MD processes removed dyes, and color from wastewater and was successful in the reuse of treated water [22]. A combined treatment was also studied by [26], in which MBR was combined with NF after using anaerobic wastewater treatment. The result of the combined study showed that the color removal efficiency, turbidity, and COD were increased at 99%, 97%, and 96%, respectively.

2.6 Water Reuse

As the textile industry is among the major water-consuming industries, it is important to treat wastewater in order to reuse water within the premises of the industry. Many wastewater methods are applicable in industries to treat water that removes dyes and toxic substances from water, but the recovery of water is still an issue. Membrane technology is one of the promising processes that treat as well as reuse water. This technology has gained novelty for improving water quality. Textile dyes can also be removed by using this process. Reverse osmosis (RO), nanofiltration (NF), membrane filtration (MF), and ultrafiltration (UF) have been used as membrane technologies to treat wastewater [25].

Membrane filtration is used as a treatment of textile dye-bath effluent, and it can also be used as a pretreatment in order to recover membrane fouling issues. Ultrafiltration (UF) membrane has been used to treat chemicals and reactive dyes. UF is also being used for rinsing and washing purposes in the textile industry [5]. Moreover, nanofiltration (NF) and RO are used to treat salts, separation of organic compounds, and hydrolyze the reactive dyes in the effluent. The membrane process has lab scale, and pilot-scale applications to treat textile effluent. The pilot-scale application could be helpful to upgrade the process up to the full-scale level in order to increase the efficiency of the system [53].

Gadhi et al. [14] performed an experiment using the hybrid process of membrane filtration with pretreatment of coagulation, for the treatment of textile wastewater and sand filtration for the treatment of textile effluent and its reuse in the dyeing unit of the cotton industry. The pretreatment was carried out at a lab scale and was compared with ultrafiltration (UF). The UF hollow fiber was used (UF-HF). It showed the removal efficiencies of 99%, 93%, and 29% of color, turbidity, and total organic carbon (TOC). Further flat membrane sheets of UF of size 5, 10, 20, and 50 kDa were used. The 5 kDa membrane showed the result of removal efficiency of 54% COD, 84% turbidity, 94% TOC, 46% conductivity, and 82% color removal, respectively (Table 1).

The ozonation, photo-Fenton, and Fenton processes are effective for complete color and COD removal up to a greater efficiency level. However, combined processes are considered to be effective for the complete removal of contaminants from wastewater by complete color removal and make it efficient for reuse as well. Moreover, hybrid processes have been employed to overcome cost ineffectiveness, process limitations, and conventional treatment technologies [4]. Wastewater treatment technologies are being employed well but now researchers focus on the combined process for effective treatment of textile effluent to achieve maximum results.

Some of the processes include aerobic and anaerobic treatment technologies that require continuous treatment for the dye-bath effluent process. Due to the energy crisis, if the operation shuts down, the treatment processes will be required to work again from their initial stage. However, the AOPs do not face this issue as these are not continuous processes that require 24 h for processing, and these are time efficient as well. Due to space conflict, it is not feasible to install a conventional effluent treatment

Table 1 Impacts of different treatment methods on various parameters of dye-bath effluent

Treatment process	Physiochemical parameters(After treatment)	Optimized environmental conditions and oxidation time	Removal efficiencies (after treatment)	References
Application of combined coagulation and catalytic ozonation	Turbidity: 2.97 ntu	Temperature: 30 °C Oxidation time: 60 min PAC dose: 300 mg/L	Turbidity: 84%	[41]
	TSS: 157 mg/L		TSS: 86%	
	COD: 189 mg/L		COD: 89%	
	Color:0.0081 abs		Color: 98%	
Effect of pre-coagulation on ozonation	TSS: 54 mg/L	Initial pH: 8.5 Alum dose: 15 g Oxidation time: 60 min	TSS: 96%	[51]
	TDS: 625 mg/L		TDS: 92%	
	COD: 99 mg/L		COD: 91%	
	Color: 0.075 abs		Color: 93%	
Coagulation-flocculation and ozonation process	Color: 0.021 abs	Initial pH: 8.7 Oxidation time: 45 min ZrCl4 dose: 20 mg/L SnCl4: 15 mg/L	Color: 97%	[61]
	Turbidity: 2.3 ntu		Turbidity: 93%	
	Salinity: 0.16 ppt		Salinity: 97%	
	COD: 123 mg/L		COD: 88%	
Combined coagulation-ozonation and ultrafiltration	TSS: 18 mg/L	Initial pH: 9.2 Aluminum sulfate dose: 8–10 g Oxidation time: 60 min PAC dose: 200 mg/L	TSS: 96%	[36]
	TDS: 1590 mg/L COD: 132 mg/L		TDS: 78% COD: 93%	
	HM: 0.213 mg/L		HM: 98%	
Combined coagulation with ozonation	Color:0.083 abs COD: 120 mg/L	PAM dose: 500 mg/L Ph: 8.9 Temperature: 30 °C Oxidation time: 60 min	Color:87% COD: 80%	[35]
	Turbidity: 3.21 ntu		Turbidity: 85%	
Integrated coagulation-ozonation	COD: 94 mg/L	Initial pH: 8.2 PAC and PAM dose: 150 mg/L, 100 mg/L, respectively Oxidation time: 60 min	COD: 93%	[56]
	TN: 104 mg/L		TN: 92%	
	Turbidity: 3.2 ntu		Turbidity: 89%	
	Color: 0.001		Color: 99%	

(continued)

Table 1 (continued)

Treatment process	Physiochemical parameters(After treatment)	Optimized environmental conditions and oxidation time	Removal efficiencies (after treatment)	References
Integrated ozonation with coagulation	TOC: 42 mg/L	KMnO4 dose: 1.5 Mm Oxidation time: 45 min Temperature: 27.5 °C	TOC: 70%	[29]
	COD: 192 mg/L		COD: 80%	
	EC: 4.7 mS/cm		EC: 88%	
Pretreatment of combined coagulation and ozonation	Color: 0.0085 abs	Initial pH: 9 Polyferric sulfate: 1.5 g/L Ozone conc: 0.48 mg/L Oxidation time: 60 min	Color: 96.8%	[57]
	COD: 640 mg/L		COD: 83.4%	
	BOD5: 332 mg/L		BOD5: 77%	

plant (ETP). Moreover, sludge production during biological treatment also creates a problem. Thus, these are some gaps that require an integrated process of advanced treatment with conventional treatment.

3 Conclusion

Dye-bath effluent contains a variety of chemicals, dyes, heavy metals, and other contaminants that can severely affect aquatic habitats if released untreated. The vibrant colors and chemical compositions of dyes can upset aquatic life balance and cause water toxicity, harming biodiversity and endangering the health of neighboring ecosystems. Dye-bath effluent contains a variety of chemicals, dyes, heavy metals, and other contaminants that can severely affect aquatic habitats if released untreated. The vibrant colors and chemical compositions of dyes can upset aquatic life balance and cause water toxicity, harming biodiversity and endangering the health of neighboring ecosystems. Color, chemical oxygen demand (COD), and other contaminants in dye-bath effluent can be considerably reduced using effective treatment methods such as physicochemical treatments (such as coagulation, flocculation, and sedimentation) or sophisticated oxidation procedures.

Implementing such treatment techniques prior to discharge reduces environmental damage, safeguards water resources, and supports sustainable industrial practices in the textile industry. Color, chemical oxygen demand (COD), and other contaminants in dye-bath effluent can be considerably reduced using effective treatment methods such as physicochemical treatments like coagulation, flocculation, and sedimentation or sophisticated oxidation procedures. Implementing such treatment techniques prior to discharge reduces environmental damage, safeguards water resources, and supports sustainable industrial practices in the textile industry.

References

1. Abdurahman MH, Abdullah AZ (2020) Mechanism and reaction kinetic of hybrid ozonation-ultrasonication treatment for intensified degradation of emerging organic contaminants in water: a critical review. Chem Eng Process-Process Intensif 154:108047
2. Adeleye AS, Conway JR, Garner K, Huang Y, Su Y, Keller AA (2016) Engineered nanomaterials for water treatment and remediation: costs, benefits, and applicability. Chem Eng J 286:640–662. https://doi.org/10.1016/j.cej.2016.10.105
3. Ali A, Khan M, Ishaq A, Hussain A, Rehman SU, Khan IA, Shah SFA (2020) Pakistan textiles can bounce back vigorously. Int Rev Manag Mark 10(2):30. https://doi.org/10.32479/irmm.9377
4. Al-Qodah Z, Tawalbeh M, Al-Shannag M, Al-Anber Z, Bani-Melhem K (2020) Combined electrocoagulation processes as a novel approach for enhanced pollutants removal: a state-of-the-art review. Sci Technol Environ 744:140806
5. Barredo-Damas S, Alcaina-Miranda MI, Iborra-Clar MI, Mendoza-Roca JA (2016) Application of tubular ceramic ultrafiltration membranes for the treatment of integrated textile wastewaters. Chem Eng J 192:211–218
6. Bayrakdar M, Atalay S, Ersöz G (2021) Efficient treatment for textile wastewater through sequential photo Fenton-like oxidation and adsorption processes for reuse in irrigation. Ceram Int 47:9679–9690. https://doi.org/10.1016/j.ceramint.2020.12.107
7. Behera M, Nayak J, Banerjee S, Chakrabortty S, Tripathy SK (2021) A review on the treatment of textile industry waste effluents towards the development of efficient mitigation strategy: an integrated system design approach. J Environ Chem Eng 9(4):105277. https://doi.org/10.1016/j.jece.2021.105277
8. Bonilla-Petriciolet A, Mendoza-Castillo DI, Reynel-Ávila HE (eds) (2017) Adsorption processes for water treatment and purification, vol 256. Springer International Publishing, Cham
9. Çalık Ç, Çifçi Dİ (2022) Comparison of kinetics and costs of Fenton and photo-Fenton processes used for the treatment of a textile industry wastewater. J Environ Manage 304:114234
10. de Oliveira Neto GC, Tucci HNP, Correia JMF, da Silva PC, da Silva D, Amorim M (2021) Stakeholders' influences on the adoption of cleaner production practices: a survey of the textile industry. Sustain Prod Consum 26:126–145. https://doi.org/10.1016/j.jclepro.2019.04.334
11. Dotto J, Fagundes-Klen MR, Veit MT, Palacio SM, Bergamasco R (2019) Performance of different coagulants in the coagulation/flocculation process of textile wastewater. J Clean Prod 208:656–665. https://doi.org/10.1016/j.jclepro.2018.10.112
12. Farhana K, Kadirgama K, Mahamude ASF, Mica MT (2022) Energy consumption, environmental impact, and implementation of renewable energy resources in global textile industries: an overview towards circularity and sustainability. Mater Circ Econ 4(1):15
13. Feng Q, Gao B, Yue Q, Guo K (2021) Flocculation performance of papermaking sludge-based flocculants in different dye wastewater treatment: Comparison with commercial lignin and coagulants. Chemosphere 262:128416. https://doi.org/10.1016/j.chemosphere.2020.128416
14. Gadhi TA, Mahar RB, Bukhari SA (2019) Wastewater treatment and reuse to approach zero water discharge in Al-Rahim textile industries: substantial increase in wateruse efficiency in textile processing. U.S.-Pakistan Center for Advanced Studies in Water (USPCAS-W). MUET 56:213–234
15. GilPavas E, Dobrosz-Gómez I, Gómez-García MÁ (2017) Coagulation-flocculation sequential with Fenton or Photo-Fenton processes as an alternative for the industrial textile wastewater treatment. J Environ Manage 191:189–197. https://doi.org/10.1016/j.jenvman.2017.01.015

16. Haddad M, Abid S, Hamdi M, Bouallagui H (2018) Reduction of adsorbed dyes content in the discharged sludge coming from an industrial textile wastewater treatment plant using aerobic activated sludge process. J Environ Manage 223:936–946. https://doi.org/10.1016/j.jenvman.2018.07.009
17. Hassaan MA, El Nemr A, Madkour FF (2016) Application of Ozonation and UV assisted Ozonation for Decolorization of Direct Yellow 50 in Sea water. Pharm Chem J 3(2)
18. Hassaan MA, El Nemr A, Madkour FF (2017) Testing the advanced oxidation processes on the degradation of direct blue 86 dye in wastewater. Egypt J Aquat Res 43(1):11–19. https://doi.org/10.1016/j.ejar.2016.09.006
19. Hassanzadeh E, Farhadian M, Razmjou A, Askari N (2017) An efficient wastewater treatment approach for a real woolen textile industry using a chemical assisted NF membrane process. Environ Nanotechnol Monit Manag 8:92–96. https://doi.org/10.1016/j.enmm.2017.06.001
20. Hussain T, Wahab A (2018) A critical review of the current water conservation practices in textile wet processing. J Clean Prod 198:806–819. https://doi.org/10.1016/j.jclepro.2018.07.051
21. Hutagalung SS, Muchlis I, Khotimah K (2020) Textile wastewater treatment using advanced oxidation process (AOP). In: IOP conference series: materials science and engineering, vol 722, No. 1, p 012032. IOP Publishing. https://doi.org/10.1088/1757-899X/722/1/012032
22. Jegatheesan V, Pramanik BK, Chen J, Navaratna D, Chang CY, Shu L (2016) Treatmentof textilewastewaterwith membrane bioreactor:acritical review. Biores Technol 204:202–212. https://doi.org/10.1016/j.biortech.2016.01.006
23. Jones ER, Van Vliet MT, Qadir M, Bierkens MF (2021) Country-level and gridded estimates of wastewater production, collection, treatment and reuse. Earth Syst Sci Data 13(2):237–254
24. Jun BM, Elanchezhiyan SS, Yoon Y, Wang D, Kim S, Prabhu SM, Park CM (2020) Accelerated photocatalytic degradation of organic pollutants over carbonate-rich lanthanum-substituted zinc spinel ferrite assembled reduced graphene oxide by ultraviolet (UV)-activated persulfate. Chem Eng J 393:124733. https://doi.org/10.1016/j.cej.2020.124733
25. Keskin B, Ağtaş M, Ormancı-Acar T, Türken T, İmer DY, Ünal S, Koyuncu İ (2020) Halloysite nanotube blended nanocomposite ultrafiltration membranes for reactive dye removal. Water Sci Technol 83(2):271–283
26. Khouni I, Louhichi G, Ghrabi A (2020) Assessing the performances of an aerobic membrane bioreactor for textile wastewater treatment: influence of dye mass loading rate and biomass concentration. Process Saf Environ Prot 135:364–382. https://doi.org/10.1016/j.psep.2020.01.011
27. Kishor R, Purchase D, Ferreira L, Mulla S, Bilal M, Bharagava R (2020) Environmental and health hazards of textile industry wastewater pollutants and its treatment approaches 8762v
28. Lau YY, Wong YS, Teng TT, Morad N, Rafatullah M, Ong SA (2016) Coagulation-flocculation of azo dye Acid Orange 7 with green refined laterite soil. Chem Eng J 246:383–390. https://doi.org/10.1016/j.cej.2016.02.100
29. Liang J, Ning X, Sun J, Song J, Lu J, Cai H, Hong Y (2018) Toxicity evaluation of textile dyeing effluent and its possible relationship with chemical oxygen demand. Ecotoxicol Environ Saf 166:56–62. https://doi.org/10.1016/j.ecoenv.2018.08.106
30. López-López C, Martín-Pascual J, Leyva-Díaz JC, Martínez-Toledo MV, Muñío MM, Poyatos JM (2016) Combined treatment of textile wastewater by coagulation–flocculation and advanced oxidation processes. Desalin Water Treat 57(30):13987–13994
31. López-Vinent N, Cruz-Alcalde A, Ganiyu SO, Sable S, Messele SA, Lillico D, El-Din MG (2021) Coagulation-flocculation followed by catalytic ozonation processes for enhanced primary treatment during wet weather conditions. J Environ Manage 283:111975. https://doi.org/10.1016/j.jenvman.2021.111975
32. Ma D, Yi H, Lai C, Liu X, Huo X, An Z, Yang L (2021) Critical review of advanced oxidation processes in organic wastewater treatment. Chemosphere 275(1):130104. https://doi.org/10.1016/j.chemosphere.2021.130104
33. Mirbolooki H, Amirnezhad R, Pendashteh AR (2017) Treatment of high saline textile wastewater by activated sludge microorganisms. J Appl Res Technol 15(2):167–172. https://doi.org/10.1016/j.jart.2017.01.012

34. Moosavi S, Lai CW, Gan S, Zamiri G, Akbarzadeh Pivehzhani O, Johan MR (2020) Application of efficient magnetic particles and activated carbon for dye removal from wastewater. ACS Omega 5(33):20684–20697
35. Mukhlish MB, Khan MR, Islam MS, Nazir MI, Snigdha JS, Akter R, Ahmad H (2020) Decolorization of reactive dyes from aqueous solution using combined coagulation-flocculation and photochemical oxidation (UV/H_2O_2). Sustain Chem Eng 51–61
36. Naveed T, Ahmed N, Bhutto S, Tunyo N, Hashmi DR (2021) A sustainable solution to treat textile effluent by employing combined coagulation, oxidation and ultrafiltration techniques. J Appl Res Water Wastewater 8(1):66–70
37. Oliva G, Naddeo V, Zarra T, Belgiorno V (2018) Abatement of odour emissions by UV/ozone oxidation process. Glob Nest J 20(3):669–673. https://doi.org/10.30955/gnj.002798
38. Patel H, Yadav VK, Yadav KK, Choudhary N, Kalasariya H, Alam MM, Gacem A, Amanullah M, Ibrahium HA, Park JW, Park S, Jeon BH (2022) A recent and systemic approach towards microbial biodegradation of dyes from textile industries. Water 14(19):3163
39. Pathania D, Sharma A, Siddiqi ZM (2016) Removal of congo red dye from aqueous system using Phoenix dactylifera seeds. J Mol Liq 219:359–367. https://doi.org/10.1016/j.molliq.2016.03.020
40. Pourgholi M, Jahandizi RM, Miranzadeh M, Beigi OH, Dehghan S (2018) Removal of dye and COD from textile wastewater using AOP (UV/O3, UV/H2O2, O3/H2O2 and UV/H2O2/O3). J Environ Health Sustain Dev 3:621–629. https://doi.org/10.18502/jehsd.v3i4.223
41. Rizvi OS, Ikhlaq A, Ashar UU, Qazi UY, Akram A, Kalim I, Qi F (2022) Application of poly aluminum chloride and alum as catalyst in catalytic ozonation process after coagulation for the treatment of textile wastewater. J Environ Manage 323:115977
42. Shoukat R, Khan SJ, Jamal Y (2019) Hybrid anaerobic-aerobic biological treatment for real textile wastewater. J Water Process Eng 29:100804. https://doi.org/10.1016/j.jwpe.2019.100804
43. Sreeja PH, Sosamony KJ (2016) A comparative study of homogeneous and heterogeneous photo-Fenton process for textile wastewater treatment. Procedia Technol 24:217–223. https://doi.org/10.1016/j.protcy.2016.05.065
44. Srivastav M, Gupta M, Agrahari SK, Detwal P (2019) Removal of refractory organic compounds from wastewater by various advanced oxidation process-a review. Curr Environ Eng 6(1):8–16. https://doi.org/10.2174/2212717806666181212125216
45. Suryawan IWK, Helmy Q, Notodarmojo S (2018) Textile wastewater treatment: color and COD removal of reactive black-5 by ozonation. In: IOP conference series: earth and environmental science, vol 106, No 1, p 012102. IOP Publishing. https://doi.org/10.1088/1755-1315/106/1/012102
46. Tanveer R, Yasar A, Ikhlaq A, Nissar H, Nizami AS (2022) Comparison of ozonation, Fenton, and photo-Fenton processes for the treatment of textile dye-bath effluents integrated with electrocoagulation. J Water Process Eng 46:102547
47. Tanveer R, Yasar A, Ikhlaq A, Nissar H, Nizami AS (2022) Comparison of ozonation, Fenton, and photo-Fenton processes for the treatment of textile dye-bath effluents integrated with electrocoagulation. J Water Process Eng 46:102547. https://doi.org/10.1016/j.jwpe.2021.102547
48. Tanveer R, Yasar A, Nizami AS, Tabinda AB (2022) Integration of physical and advanced oxidation processes for treatment and reuse of textile dye-bath effluents with minimum area footprint. J Clean Prod 135366
49. Tekin G, Ersöz G, Atalay S (2018) Degradation of benzoic acid by advanced oxidation processes in the presence of Fe or Fe-TiO2 loaded activated carbon derived from walnut shells: a comparative study. J Environ Chem Eng 6:1745–1759. https://doi.org/10.1016/j.jece.2018.01.067
50. Tentscher PR, Bourgin M, von Gunten U (2018) Ozonation of para-substituted phenolic compounds yields p-benzoquinones, other cyclic α, β-unsaturated ketones, and substituted catechol. Environ Sci Technol 52(8):4763–4773

51. Wang H, Zhang S, He C, Yuan R, Wu X, Guo S, Van Hulle SW (2022) Effect of pre-coagulation on catalytic ozonation in the tertiary treatment of coking wastewater: Kinetic and ozone consumption analysis. J Water Process Eng 48:102856. https://doi.org/10.1016/j.jwpe.2022.102856
52. Wang J, Chen H (2020) Catalytic ozonation for water and wastewater treatment: recent advances and perspective. Sci Total Environ 704:135249. https://doi.org/10.1016/j.scitotenv.2019.135249
53. Wang J, Wang S (2018) Activation of persulfate (PS) and peroxymonosulfate (PMS) and application for the degradation of emerging contaminants. Chem Eng J 334:1502–1517. https://doi.org/10.1016/j.cej.2017.11.059
54. Wang J, Wang S (2020) Reactive species in advanced oxidation processes: Formation, identification and reaction mechanism. Chem Eng J 401:126158. https://doi.org/10.1016/j.cej.2020.126158
55. Wang J, Wang S (2021) Toxicity changes of wastewater during various advanced oxidation processes treatment: an overview. J Clean Prod 315:128202. https://doi.org/10.1016/j.jclepro.2021.128202
56. Wang J, Ji Y, Zhang F, Wang D, He X, Wang C (2019) Treatment of coking waste water using oxic-anoxic-oxic process followed by coagulation and ozonation. Carbon Resour Convers 2(2):151–156. https://doi.org/10.1016/j.crcon.2019.06.001
57. Wen Y, Zhang W, Zhu X, Zhang J, Wang L (2018b) Interfacial properties of high-order aggregation of organic dyes: a combination of static and dynamic properties. Energy 158:537–545. https://doi.org/10.1016/j.energy.2018.06.058
58. Yaseen DA, Scholz M (2019) Textile dye wastewater characteristics and constituents of synthetic effluents: a critical review. J Environ Sci Technol 16:1193–1226. https://doi.org/10.1007/s13762-018-2130-z
59. Yin H, Qiu P, Qian Y, Kong Z, Zheng X, Tang Z, Guo H (2019) Textile wastewater treatment for water reuse: a case study. Processes 7(1):34
60. Yonar TANER, Kurt A (2017) Treatability studies of hospital wastewaters with AOPs by Taguchi's experimental design. Glob Nest J 19:505–510. https://doi.org/10.30955/gnj.002302
61. Zakaria M, Shibahara K, Bhuiyan AH, Nakane K (2022) Preparation and characterization of polypropylene nanofibrous membrane for the filtration of textile wastewater. J Appl Polym Sci 139. https://doi.org/10.1002/app.52657
62. Zhang L, Zheng J, Guo J, Guan X, Zhu S, Jia Y, Zhang H (2019) Effects of Al^{3+}on pollutant removal and extracellular polymeric substances (EPS) under anaerobic, anoxic and oxic conditions. Front Environ Sci Eng 13(6):1–11. https://doi.org/10.1007/s11783-019-1169-5
63. Zhang Y, Shaad K, Vollmer D, Ma C (2021) Treatment of textile wastewater using advanced oxidation processes—a critical review. Water 13(24):3515
64. Zou R, Angelidaki I, Jin B, Zhang Y (2020) Feasibility and applicability of the scaling- up of bio-electro-Fenton system for textile wastewater treatment. Environ Int 134:105352. https://doi.org/10.1016/j.envint.2019.105352

Application of Natural Dyes in Textile Industry (A Sustainable Approach)

Sanchayita Rajkhowa, Jyotirmoy Sarma, Rakesh Dutta, Shahnaz Ahmed, and Ravindra Pratap Singh

Abstract Textile has played an important role in human civilization. From time immemorial, dyes have been an integral ingredient in styling any material especially fabrics by imparting various colours on to the material. Although dyes and textiles are two separate fields, many times they interfuse as a single domain due to the extensive application of dyes in textiles and fabrics. However, the wide use of synthetic dyes has started imposing an environmental threat over the years. Hence, there is an urge to switch over to natural alternatives for colouring fabrics and other materials in the textile industry. Natural dyes used in ancient times were extracted from plants and rocks. These were then soon replaced by readily available, cost-effective synthetic dyes due to their wide range of shades and textures. The growing environmental consciousness amongst consumers enforces the textile industries to shift towards natural dyes. In this chapter, a comprehensive description of the history of natural dyes, their extraction, characterisation and chemistry involved during the dyeing process is provided with a special emphasise on their use in textiles and their role in environmental sustainability.

Keywords Natural dyes · Dyeing process · Textiles · Fabrics · Environmental sustainability

S. Rajkhowa (✉)
Department of Chemistry, Haflong Government College, Dima Hasao 788819, Assam, India
e-mail: sanchayita.rajkhowa@gmail.com

J. Sarma
Department of Chemistry, Assam Don Bosco University, Guwahati 782402, Assam, India

R. Dutta
Department of Chemistry, Pandit Deendayal Upadhyaya Adarsha Mahavidyalaya, Behali, Biswanath 784184, Assam, India

S. Ahmed
Department of Chemistry, Barnagar College, Sorbhog 781317, Assam, India

R. P. Singh
Central Public Works Department, New Delhi, India

P. Singh (ed.), *Dye Pollution from Textile Industry*, SDGs and Textiles,
https://doi.org/10.1007/978-981-97-5341-3_11

1 Introduction

Natural dyes or colourants have paved their way from ancient times into the modern world through various forms and modifications. These are categorised under non-wood forest products and found to have extensive applications primarily in textiles, foods, paper, cosmetics, paints and ink, medicines, etc. By definition "*Dyes are substances that, when applied to a substrate provide colour by a process that alters, at least temporarily, any crystal structure of the coloured substances*" [1, 2]. Dyes have versatile characteristics in terms of their mode of action, e.g. adherence to the surface, formation of complexes/covalent bonds with metals and salts, physisorption or mechanical retention for imparting colour to the substrate [1, 2]. Dyes may be natural or synthetic depending on their origin. Nevertheless, dyes are generally classified as acid, basic, direct, disperse, metal complex, mordant, reactive, sulphur dyes and vat dyes on the basis of their chemical structure, application and nature of functional groups. These functional groups known as chromophores are responsible for displaying a particular colour or shade by a dye. These groups are composed of a group of atoms like C, H, N, O, S, etc. and fall into categories such as azo, nitro, carbonyl, methane, aryl methane, anthraquinone and so on. Other than chromophores, several other electron-donating or withdrawing groups may also be present in the chemical structure of dye molecules. Such groups are known as auxochromes which intensify the colour of chromophores, e.g. carboxyl, hydroxyl, amine and sulfonate [3–5]. An annual production of around 800,000 tonnes of dyes worldwide with more than 10,000 varieties of dyes are used solely by textile industries, out of which 70% falls into the category of azo dyes [6–8]. Being synthetic origin, azo dyes contain one or more azo groups, i.e. $-N=N-$ attached to other aromatic groups like phenols or amines via double bonds in their structure [9, 10]. The azo group acts as a chromophore which is responsible for imparting colour to the compound. About 10–15% of these synthetic dyes are released into various water bodies as industrial effluents during dyeing practices [11]. These dyes remain in the environment for a longer period of time owing to their high stability under light, temperature and inertness towards water [12]. In addition, their complex structure makes it difficult to undergo biodegradation [13, 14]. Apart from azo dyes, most of them are readily soluble in water thus making the process of separation of dyes more cumbersome.

The textile industry is one of the oldest industries having a large contribution to the economic growth of a nation due to its significant share in creating job opportunities and employment and direct or indirect involvement with other small/medium enterprises. On the basis of the nature of fibre industries produced, there are two major fabric processes involved: dry and wet processes. In the former process, the wastes produced are generally solids whilst the later generates liquid wastes. The wet process is found to be the major source of water pollutants like dyes, acids, NaOH, chrome, starch and other toxic chemicals. Nevertheless, textile industries consume a relatively larger volume of potable water and various chemicals than other similar industries. A textile industry consumes an average of about 200 L potable water per Kg of the product during manufacturing with an estimated waste water generation of ~17–20% in the dyeing and finishing process [15, 16]. It is therefore many textile industries are switching over to the use of natural dyes to make the entire process economically as well as environmentally sustainable.

1.1 History of Dyes

The evidence found in the prehistoric cave of Republic of Georgia around 36,000 BP confirms the use of dyed flax fibres in the ancient times [17, 18]. The use of different natural dyes such as yellow, black, white and reddish was traced in the cave painting around 15,000 BCE. The use of natural dye in the field of textiles was evident long back in the early age, i.e. the Neolithic Period or New Stone Age, which was around 10,200 BCE. The colour of natural dyes ranging from reds, browns and oranges were the first colour attached to different textile industries during the periods from the Neolithic to the Bronze Age across Egypt, Europe and Mesopotamia. It is followed by the colour blues and yellows. However, appearing of the colour green in textiles takes place in later periods. The oldest evidence of using dyeing in the textile industry was found in South Anatolia dating back to the Neolithic settlement periods. Remnant of red dyes was found at the site which is possibly from iron oxide from clay, i.e. Ochre [19]. In India and Phoenicia, archaeological evidence shows the use of dyeing has been carried out since 5000 years ago.

The dyeing fabrics found in Egyptian tombs have the evident in the use of dyeing since 4000 years ago. Multi-coloured dyes used in the purpose of fabrics were also been developed in the same periods [19]. Multi-coloured textiles were discovered with red-brown warp and yellow-ochre weft in the pyramids of Egypt in Sixth Dynasty around 2345–2180 BCE [20].

The application of natural dyes or dependence on these dyes continued to till the 1850s. After conducting a chemical analysis, it is possible to identify the dyes used in ancient textiles. But, dye such as indigo blue which can be detected from chemical analysis was difficult to determine the source from which plant parts was used [19]. The use of coloured dyes such as reds, blue and yellows produced from plant sources were commonly used items during the late Bronze Age and Iron Age. It was evidence from the colour of the available fragments of textile found in archaeological sites [19].

The origin or the source of the early natural dyes were from different types of animals, plants, vegetables, mineral sources, lichens and invertebrates such as insects that were found in nature without any processing. However, the plant kingdom is the primary source of these dyes. They were extracted mainly from the roots, berries, bark, leaves, fungi and wood [21]. Out of these, only a few of them are used for commercial purposes. Indigo, Logwood, Alizarin, yellow and Tyrian purple were a few notable natural dyes that have found their use since prehistoric times. The dyes such as indigo, saffron, madder and woad (Isatis tinctoria) which were mainly plant-based dyes were important commercial goods in the markets of Europe, Asia and Africa. During the same time in Asia, Africa and Europe, patterned fabrics were developed to control the colour absorption in cloth with the help of resist dyeing techniques. Dyes such as logwood were first introduced to Europe by the Spanish and it was carried by the colonists to America [22, 23].

The above natural dyes were mainly derived from parts of plants and invertebrates.

- Indigo: Indigo is known to be the oldest dye which had been extracted from the woad herbs leaves, isatistinctoria leaves and the Indigo plant.
- Logwood (Haematoxylumcampechianum): Haematoxylin is produced from the extract of logwood. After oxidation in the process of isolation, it converts to haematein. When it is combined with the Chromium metal, its colour gets transformed to grey and black from red. It was used in the process of silk and leather dyeing. It has been used today since ancient times.
- Alizarin: It is mainly extracted from the plant madder. It was a red coloured dye. Some other red shades were also extracted from insects such as kermes and cochineal.
- Yellow: It is derived from various parts of plants such as leaves (weld, quercetin) and the bark (oak). The compound present in green plants, carotenoids produces yellow and red dyes.
- Tyrian Purple: It is the dye which was produced from the glands of a particular species of snails. The production rate of these dyes was very low as well as the cost of production of these types of dyes was quite high. In an experiment in the year 1909, it was found that approximately 1.4 g of Tyrian Purple dye was produced from almost 12,000 snails. Due to this reasons, in the ancient times only the king, the emperors and the high officials had the power to use garments and other product which is dyed with this dyes [24].

One of the major challenges with the dyeing process is the persistence of colour onto the fibres where they are applied. The fibres are on constant exposure to light, wind and water. The amount of dyes retained by the fibres and the method of retention greatly depend on the chemical bond formed between the fibres and the dye molecules. From a dyeing perspective, dye materials are classified into three broad categories: (i) direct or substantive dyes, (ii) mordant dyes, and (iii) vat dyes.

1.2 The Natural Versus Synthetic Dyes

In the year 1856, English chemist William Henry Perkin unintentionally discovered the first synthetic 'mauve' dye whilst synthesising quinine from aniline [19]. It was a coal tar byproduct. The colour of that synthetic dye become the favourite of the royal family very quickly and it leads to the rising of a new industry, i.e. Dyestuff industry. The increase in the use of synthetic dye due to the good colouring properties and variable range of colour leads to the limitation of natural dye uses. The discovery of mauve dye leads to many more synthetic routes to the different aniline dyes such as safranine, fuchsine and induline [25–27]. The development of mauve has also encouraged the study of immunology and chemotherapy. Its invention directed the roadmap for future development & research of synthetic dyes. However, the use of mauve dye has been stopped after a short period.

In the year 1863, Paul Ehrich established the fact that a particular cell or a particular organism took up a selective dye only. He has also discovered that a particular amount of dose of dyes could be introduced into the human body which would kill the pathogenic microorganisms without affecting other cells in the target species. Ehrlich has used methylene blue dye to plasmodium which is responsible for malaria [19].

Synthetic dye has many more applications other than pigmentation. Some of its other application includes optical media (CD-R), camera sensors (colour filter array) and organic dye lesser [28].

The invention of the first synthetic dye in the middle of the nineteenth century contemporary to the Industrial Revolution leads to the socio-economic growth of the middle class population. The dyed garments become affordable to the common people only after the invention of synthetic dye. Earlier it was the symbol of royal families especially the Tyrian Purple dye. After its invention, scientists around the world during the mid to late nineteenth century started developing dyes for use in the fabric industry from different chemical compositions. In today's world, synthetic dye is used for the purpose of dyeing almost all types of garments with some exceptions to some communities in Asia, Africa and North America. The dyeing of fabrics leads to a new world of creativity and self-expression. Different coloured dyed fabrics brighten our world. In the recent times, we cannot imagine a world without dyed garments [29].

After the discoveries of synthetic dyes, the use of natural dyes decreased and at the same time, the use of synthetic dyes increased. The Industrial Revolution played a crucial role in the enhanced application of synthetic dyes which paved the way to the evolution of the textile industry. As a result, it increases the demand for dyes that are cost-effective, easily available and easy to apply on garments. However, for the production of natural dye, we needed a large area. In addition to its cumbersome extraction process, the solidity and firmness of the natural dye were less effective in comparison with the synthetic one. These economic limitations were the major factor that declined its use.

There are some other factors which led to the rise of synthetic dye use in garments along with some other purposes. The research on coal and tar enhances the use of synthetic dye. Earlier coal tar was not used in excess. In the late nineteenth century, coal targets the attention of researchers as a source of many new organic compounds. The textile manufacturers demanded new dyes as a result of the development of mauve dye. The researchers discovered that coal tar became the source of more than 50 chemical compounds that were used for the chemical dye industry during the start of the twentieth century. In today's world, these dyes are broadly used in various types of industries. Researchers making perpetual effort to develop the synthetic dyes as more sustainable, reduced its environmental impact and expand its use to other field [24].

However, the use of natural dyes has some advantages over synthetic ones. The major advantages of natural dyes are

- The natural dyes are eco-friendly, renewable and biodegradable in nature.
- The residue material remaining after the extraction of natural dyes can be used as fertiliser.
- These dyes could be act as a protection against harmful UV radiation [30].
- They also have different medicinal values such as antifungal, antibacterial, antiseptic, diuretic and laxative. Some of them are used in traditional medicinal system having curative such as turmeric, manjishtha root and Arjuna bark [31].
- Textile with cellulosic nature when treated with extract of plant to make it flame retardant [32].
- The natural dyes are skin-friendly as they absorb medicinal compounds throughout the skin.

However, there are some limitations of natural dyes in comparison with synthetic dyes. The major limitations are

- The dying process of natural dye takes longer time than the synthetic one.
- The cost of natural dye is higher than the synthetic dye.
- The shade range of natural dye is limited.
- The characterisation and identification of natural dyes are a major issue faced by the researchers during its study.

As a result of this limitation of natural dyes and the development of synthetic dyes after the Mouve dye has been discovered, the application of natural dyes becomes negligible. Only about 1% of the natural dyes have been used by some small industries at the cottage level. However, the synthetic dye has some disadvantage to the human health and to the environment. The drawbacks of synthetic dye led to the re-establishment the needs of natural dyes to human and human health [33]. The major drawbacks of synthetic dyes are their carcinogenic nature, non-biodegradable in nature and it is hazardous to the environment and human health.

In addition to the eco-friendly and bio-degradable nature, the application of natural dyes can also result in increased plantation of more dye-producing plants which eventually would lead to an escalation in carbon fixation from the atmosphere. As a result, natural dyes are not only eco-friendly but also provide a solution to the sustainability of the textile & garment industries [34].

2 Exploration of Plant-Based Dyes

Nature has an abundance of beautiful coloured objects such as plants, flowers, clay and rocks from where extraction of colours can be made by various means. Extraction of dyes and then application of those onto other surfaces requires an individual approach for each variety of dyes. It is essential to optimise the conditions prior to dyeing. These conditions may include the time of exposure to the dyeing stuff, concentration, temperature, type of mordant, dosage, preparation of dyestuff and also the nature of fibres onto which dye will be applied. Perhaps, this entire process

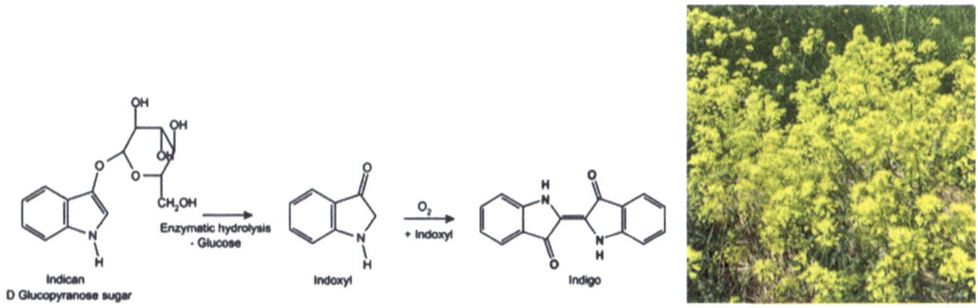

(a) (b)

Fig. 1 **a** Preparation of indigo from indicant, **b** Isatis (woad) plant

of natural dyeing would always be excitingly challenging with the knowledge of Alchemy.

Plants and vegetables are a rich source of colours although only a few numbers of these produce economically viable colourants. Almost every part of a plant, i.e. flowers, leaves, fruits, stems or even barks can serve as such source of excellent colours. However, not all colours extracted from plants or nature can be used to dye fabrics due to low adherence properties to the fabrics. For example, the only possible nature blue dye "indigo" could be obtained from *Indigofera* or *Isatis*. The simplest preparation method for indigo is given in Fig. 1. Direct extraction of indigo is not successful due to its presence as a precursor in the plants. Indigo is extracted from fermented leaves, evaporating the fluid and then oxidising the substrate to get indigo dye as a precipitate. The blue colour of indigo appears only after undergoing several reduction and oxidation processes. The woad plant (*Isatistinctoria*) is mostly grown in British Ishles and North Europe. The demand for indigo dye is estimated to be approximately 800 tonnes every year in the USA whose market value is around 20 million US dollars [35].

It is surprising to know that the global denim industry relies only on vat dyes, whether synthetic or natural indigo. The red colour used on fabrics is often extracted from a variety of plants like madder (*Rubiatinctorum*), Manjistha or Indian madder (*Rubiacordifolia*), Indian mulberry (*Morindacitrifolia*), Sappan wood (*Caesalpinasappan*) and Brazilwood (*Caesalpinaechinata*) are amongst the favourites. Madders are rich in alizarin, rubiadin, purpuroxanthin, purpurin and pseudopurpurin colourants. Each of these colourants slightly differs in their chemical structures although having anthraquinone as their basic skeleton as shown in Fig. 2.

Apart from plants, a variety of red dyes can also be extracted from insects. For example, purple-red carmine obtained from dried and then processed in water cochineal insects (*Dactylopius coccus*) in the form of carmine precipitate on addition of Al^{3+} ions.

Yellow colourants can be extracted from turmeric (*Curcuma longa*) rhizomes in the form of curcumin while saffron colour is obtained from dried stigmas of *Crocus sativus* flowers. The yellow colour matter flavonoid is produced from weld plants (*Reseda luteola*) whilst quercitrin (CI 75720/Natural Yellow) can be isolated from

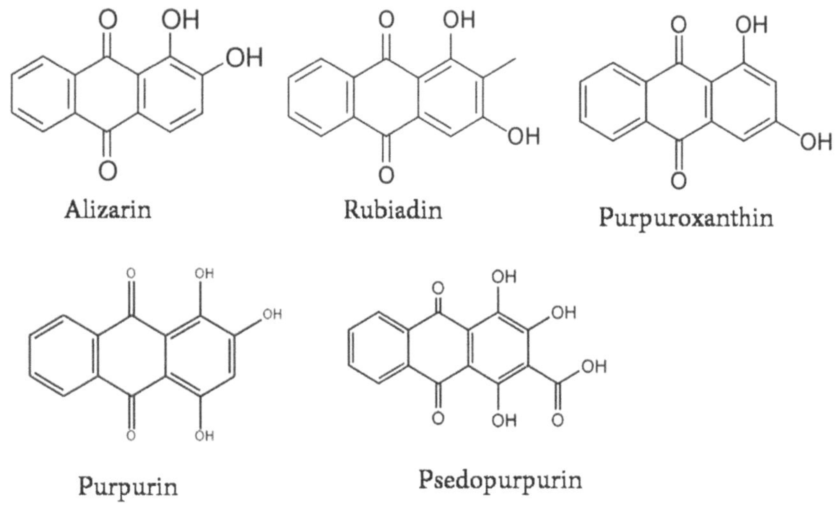

Fig. 2 Structure of colourants present in madder

black oak (*Quercus velutina*). Annatto, an orange-red condiment and food colour is extractable from *Bixaorellana* seeds [36].

The green dyes are often obtained from a mixture of indigo with other yellow pigments. The brown colour is extracted as quinine from lawsone (from henna, *Lawsoniainermis*) and juglone (from various walnuts). Black dye is achieved by treating the leaves and barks of shrubs or trees rich in tannins with iron salts. In combination with mordant salts, tannins may produce a variety of shades like brown, grey and reddish.

3 Extraction of Natural Dyes

In present days people are very much concerned about the use of biodegradable and eco-friendly materials. Researchers are inventing innovative technology to extract natural dyes and use them for various purposes. The main aim of the extraction of natural dyes is to reduce environmental pollution.

Natural dyes can be extracted from four sources- specialised plant and animal sources, dried bodies of some insects and minerals. In general, natural dyes are extracted from the leaves, stems, roots, flowers, and fruits of various plants. The colour shade provided by the plant parts will vary on the age of the plant, how the plant was grown and the condition of the soil where it was grown. Some of the sources of animal dyes are marine snails (Tyrian purple or crimson), cochineal insects (red), etc. Mineral dyes are extracted from coloured clays and earth oxides such as chromium and lead (chrome-red), chromium and oxygen (chrome-green), chromic acid and lead (chrome yellow), azurite and lapis lazuli (blue), malachite (green), cinnabar and lead oxide (red), ocher (red, brown, yellow), manganese (black), limestone or lime (white), iron and cyanide (Prussian blue) [37, 38].

3.1 Extraction Methods

Natural dye-containing substances contain very small amounts of colouring materials or dyes in combination with plenty of other components such as chlorophyll, protein, carbohydrates, fibres and tannins. Therefore, dye extraction is a crucial step in the preparation of purified natural dyes. Natural dyeing materials do not consist of any single chemical entity and apart from this the plant matrix consists of several non-colouring constituents; hence derivation of natural dyes from the mixture are complicated. Thus, before employing any standard procedure for the extraction of the colouring materials the nature and solubility properties of the dyes should be ascertained. The different methods employed in general for the extraction of natural dyes are [39–43]

- Aqueous extraction
- Alkali or acid extraction
- Solvent extraction
- Fermentation
- Enzymatic extraction
- Super critical fluid extraction
- Microwave and ultrasonic-assisted extraction.

3.2 Aqueous Extraction

Aqueous extraction is a traditional method of dye extraction from plants and other materials. In this method the material is crushed into powder and soaked in water for 12–15 hto slacken the cell structures. The material is then heated to boil so that dye molecules are extracted into the solvent used and finally, the dye solution is filtered to remove the non-dye scrapes. The process is repeated in sequence for several times to get the desired amount of dye. When large-scale extraction is necessary stainless steel vessels are used for soaking the material and soaking time is reduced by boiling the material for a longer period of time. Centrifuges are used to separate the remnants. This method is widely used in textile industries. The major drawbacks of this process include low dye yield; huge amount of water consumption, long time needed, elevated temperature requirements and only water-soluble dyes can be extracted. In this method, other water-soluble substances like sugars are also extracted which has to be removed if one wants the dye in powdered form or to be concentrated. Also, the yield of temperature-sensitive dyes is minimal near the boiling point.

3.2.1 Acid and Alkali Extraction Process

The basic principle of this process is that the extracted dye has to be soluble either in an acidic or basic medium to provide maximum yield. The dyes, which mainly occur as glycosides, are extracted in a dilute acidic or alkaline medium. Hydrolysis of glycosides is facilitated by acid or alkali addition which results in the extraction and yield of dyes effectively. Flavones are commonly extracted by acid hydrolysis which prevents their oxidative degradation whereas compounds with phenolic functional groups are preferred to be extracted using alkali medium where they are soluble. This extraction method is mostly useful for the extraction of radiant yellow-orange dye (annatto seeds), red dye (safflower petals) and lac dyes (lac insect secretions). The disadvantage of this method is that some of the pH-sensitive dyes may get destroyed under alkaline conditions. Most of the natural dyes are combination of different components and hence by adjusting the pH of the medium could produce a satisfactory yield of dye.

3.2.2 Solvent Extraction

Natural dyes can also be extracted using organic solvents like chloroform, methanol, ethanol, acetone, petroleum ether or by mixing two miscible solvents, for instance, water-alcohol, methanol-ethanol, and chloroform-CCl_4. Water-alcohol mixture is used for both water-soluble and water-insoluble dye materials. The yield of dye extracted is higher in the solvent extraction method as compared to that in the previous method discussed in Sect. 2.1.2. When alkali and acids are added to the ethanol solvent hydrolysis of glycosides is facilitated which releases dyes easily. The solvent extraction process is carried out at a lower temperature which lowers the chance of degradation. Solvents can be easily removed and reused and this makes the purification process easier. The disadvantages of this method are that some toxic residual solvents are present along with the dye after extraction and the extracted dyes are partially water-soluble hence dyeing process needs to be accomplished in an aqueous medium.

3.2.3 Fermentation Process

In this process of dye extraction, enzymes used are produced by the various microorganisms present in the environment. The most common type of extraction using the fermentation procedure is indigo extraction. In this extraction, fresh indigo leaves and branches are soaked in water at around 32 °C. Fermentation starts and the indigo-containing glucoside indican present in the indigo plants is broken down into glucose and indoxyl by the enzyme indimulsin present in the leaves. After 10–15 h of fermentation indoxyl containing yellow liquid is transferred to beating vats where indoxyl gets oxidised to blue-coloured indigotin dye. The dye is then collected, washed and pressed into cakes. This method is also used for the extraction of radiant yellow dyes obtained from annatto seeds. The disadvantages of this method are long extraction time, foul smell due to microbial action, need for immediate extraction of pigments after harvesting, etc.

3.2.4 Enzymatic Extraction

In the enzymatic extraction process, some commercially available enzymes such as amylase, cellulose and pectinase are used in milder conditions to loosen the binding materials present in plants such as cellulose, starches and pectin. This process is useful when roots, barks, etc. are the source materials.

3.2.5 Supercritical Fluid Extraction

Every gas acts as supercritical fluid above a certain value of temperature and pressure known as critical temperature and pressure, respectively, where the physical properties of the fluid are intermediate of a pure gas and pure liquid. These fluids possess much lower surface tension than liquids and are able to spread out more easily than a true liquid. Due to their low viscosity also they are excellent diffusive in nature and can interact with the substrate in a better way. Supercritical fluids can dissolve many substances in them as the solubility of any liquid increases in higher temperature and pressure in any solvent. For most of the supercritical fluid extraction method, carbon dioxide (CO_2) gas is used as it is cost-effective, easily available, non-toxic and leaves no residue. Critical pressure and temperature for CO_2 gas are 1070 psi and 31.4 °C, respectively. Using CO_2 as supercritical fluid extraction is carried out between pressure 1070–3500 psi and temperatures 32–49 °C. CO_2 acts as a non-polar solvent and sometimes a co-solvent is added to enhance the solubility of several ionic solutes. Obtaining light coloured dyes without the presence of any heavy metals and residual solvents is the great advantage of this method. Hence foods and pharmaceuticals are the popular industries to use this method extensively. Poor extraction of polar substances and the high cost of equipment used are disadvantages of this method.

3.2.6 Ultrasonic and Microwave Extraction

In this extraction method extraction efficiency is enhanced by using ultrasound and microwave which reduce solvent requirement, temperature and time of extraction. In this method, the plant material containing the natural dye is treated in aqueous or organic solvent medium in an ultrasonicator which produces cavities and small bubbles inside the liquid. These bubbles reach a certain size at a point and when these bubbles cannot retain their size these collapse-producing high pressure and temperature. In every second, such bubbles are formed in abundance and collapsed. Thereby, a very high pressure and temperature are produced which increases the process efficiency. This extraction can also be carried out even at low temperatures hence enabling the heat sensitive dye molecules to be easily extracted.

4 Characterisation of Natural Dyes

Natural dyes can be characterised by employing many methods. Broad ranges of modern instrumental techniques are available for analysis and characterisation of natural dyes. Some instrumental techniques include thin layer chromatography (TLC), high performance liquid chromatography (HPLC), UV–visible spectroscopy, infrared (IR) spectroscopy, scanning electron microscopy (SEM), X-ray fluorescence (XRF), X-ray powder diffraction (XRD), etc. [37, 42, 44–47].

Using the chromatographic method of analysis some physiochemical properties of natural dyes such as pH, solubility in polar and non-polar solvents, ash and organic contents can be analysed. ATLC technique proves to be useful in identifying the coloured components present in dyes. Dyes which are detected using the TLC technique are vegetable dyes and insect dyes. Red dyes such as purpurin carminic acid and alizarin are quantitatively analysed using a chromatographic technique.

UV–visible spectroscopy method is commonly used to gain information regarding the functional groups imparting colours to the natural dyes. The UV–visible spectral peaks of the dyes are obtained at their own predominating wavelength indicating their main colour. The UV–visible spectra of natural dyes especially give different peaks for the presence of mixed colourants. A reasonably concentrated dye solution is used in this technique.

X-ray fluorescence (XRF) is a quantitative and qualitative analytical technique for trace elements present in mineral dyes and geological materials. This method is very useful for analysing materials which are unstable in solution. The advantages of this method are accuracy, non-destructive, minute amount of sample required for analysis.

X-ray diffraction (XRD) method is an important technique for analysing and characterisation natural dyes. In this method, X-ray beams are directed towards finely powdered samples of the dyes. Depending on the lattice structure of the material, the X-ray beams are scattered in different directions and angles. From the diffraction patterns, the structures of the dye molecules are determined.

By the use of a scanning electron microscope (SEM) qualitative information about the texture, surface feature, shape and size of arrangement of constituting particles in the dye can be gathered.

Infrared (IR) spectroscopy is used to determine the chemical compounds of the dyes. IR spectroscopy can also be used to identify the structures of organic and inorganic compounds present in the dyes.

In addition to the above techniques, further analysis of colour components of natural dyes can be carried out using modern techniques like nuclear magnetic resonance (NMR), Fourier transform infrared spectroscopy (FTIR), atomic absorption spectroscopy (AAS), thermo gravimetric analysis (TGA), differential scanning calorimetry (DSC), etc.

5 Chemistry of Dyeing

The application of colour to the fabric, stock or yarn is called the process of dyeing. Dyes can be applied to fibres, vegetables or animals. In any class of dye application heat is supplied to the dye chamber and this heat is used in swelling of the fibre which makes the fibre more receptive to the dye molecule. This supplied heat energy is also used in diffusing the dye molecules from the solution phase to the fibre. Technically this process is known as exhaustion.

Another important term associated with the dyeing process is levelness which is the evenness of dye application in all forms of synthetic and natural fibres. Levelness can be obtained by following several controlled dyeing conditions, viz.,

i. By increasing agitation to enhance the collision between the dye solution and the substance to be dyed.
ii. By minimising the swelling of the substance being to be dyed.
iii. By increasing the rate of dyeing.
iv. By saving energy.

Levelling agents or retarding agents or retarders help in more uniform colour distribution in the fibres by slowing down the uptake of the fibres. Levelling agents possess surface active properties and these are related to synthetic detergents, soaps or wetting agents. Levelling agents may be of anionic or non-ionic character which increases the solubility of the dyes. Some of the commercially used levelling agents are Dispersol K paste, Dispersol VLH, Detergol D paste, Lyogen DK, Lyogen SMK, Lyogen DFT, Tenegal CVN, etc. [48–50].

- Colouration and absorption by the fabric are two factors determining the true dyeing process. A good dye must always give permanent colouration to the substance which must not fade away rapidly by washing or on exposure to light. The dye molecules stick on the surface of the substance by the process of absorption. Four kinds of forces help in the binding of the dyes to the substances, viz., (a) Ionic forces, (b) Covalent bonding, (c) Vander Wals' forces, and (d) Hydrogen bonding [42, 49, 51–53].

6 Methods of Dyeing

i. **Bale Dyeing**: This method of dyeing involves in introduction of the material into the dye chamber without being rubbed.

ii. **Beam Dyeing**: Here, the warp is coloured before weaving. The warp is kept in an array of perforated steel beams and the dyes are mechanically used to colour the warp through the perforations.
iii. **Batik Dyeing**: In this method of dyeing the material is covered with wax before dyeing and only the uncovered portions are dyed.
iv. **Yarn Dyeing**: This method incorporates the dyeing of the yarn before weaving by either a total or partial immersion process.
v. **Burl or speck dyeing**: In this method the coloured specks and scars are covered using different coloured links. This method is used to dye the woollens.
vi. **Piece Dyeing**: By this method the cut pieces of fabric are dyed.
vii. **Solution Dyeing**: In this method the dye is bonded in the liquor and coloured filaments are picked up from the solution. Both cellulose and non-cellulose-based fibres can be dyed using this process.
viii. **Cross Dyeing**: In this popular method of dyeing various colour effects are obtained for fibres with various affinities for the dyes. For example, nylon 6 and nylon 6,6 can obtain a dark blue and blue shade respectively from a single blue dyestuff.

7 Role of Natural Dyes in Textile Industry

Natural dyes have wide application in the textile industry. These dyes are used to colour natural fibres like cotton, wool, silk, linen and sometimes nylon and polyester synthetic fibres are also coloured using natural dyes. However, the most challenging issues of using natural dyes in textiles are reproducibility of shade, fadedness of colour under exposure to light and water, lack of standard procedure of application, etc. Researchers are trying to develop several dyeing methods for textiles using natural dyes, but still, now this information is inadequate in this regard. Therefore, further research is needed to develop a standard procedure for dye extraction and application in the textile industry [42, 54].

Here are a few examples of widely used natural dyes in the textile industry [55]

i. **Jack fruits**: The plant is cut into pieces, dried and crushed into a fine powder and the dye is obtained by subsequent boiling of the powder in water. Shades of yellow–brown are obtained. The colouring molecule present in this dye is morin. This dye is usually suitable for fabrics made of natural fibres like cotton and jute.
ii. **Indigo**: Seeds of this plant are used in the dyeing process. The indigo plants are kept in water for fermentation. After completion of hydrolysis of glucoside, the liquor is separated. On aeration of the liquor extract indoxyl is converted to indigotin. Cottons can be given a blue shade using indigo dye.
iii. **Tea waste**: Flavonols, flavonoids and phenolic acids are the main colour-imparting components of tea waste. Tea waste is a mordantable dye. The leftover waste of tea is combined with different mordants to produce dyes of yellowish brown to brown shade.

iv. **Madder or manjistha or Rubia**: From the root of this plant red coloured dye is extracted. Silk, cotton and wool fibre can be dyed using this dye.
v. **Hina**: Leaf of the hina tree is used to produce brown to mustard yellow coloured dye. The leaves are dried, crushed and boiled in water to obtain the desired dye. Wool, polyester and nylon can be dyed using this dye.
vi. **Safflower**: The petals of safflower are soaked and repeatedly boiled in an aqueous medium for about 2–3 h twice. The filtrate obtained from this process is vacuum dried to obtain dyes of cherry red to yellowish red shade.
vii. **Sappan wood**: Dyes from sappan wood can be extracted using the aqueous and alkali extraction method. This wood produces dye of bright red colour. When this dye is mixed with turmeric it produces a dye of orange shade and maroon shade dye when combined with catechu.
viii. **Saffron**: The stigma of the saffron flower is boiled in an aqueous medium to get the dye. Saffron provides shades of bright yellow dye. Silk, cotton and wool can be dyed using this dye.
ix. **Pomegranate rind**: the rind of pomegranate fruit produces a yellow colour dye. Pomegranate is a rich source of natural tannins. The colour-imparting component of pomegranate rind is flavogallol which is known as granatonine. Cotton, silk and wool can also be dyed using this natural dye.
x. **Mineral sources**: Some mineral sources like clays and ores are mixed with mineral salts to obtain natural dyes which can be used in textile industries.

8 Advantages of Natural Dyes

The toxic impacts associated with almost all synthetic dyes urge the application of natural dyes at the industrial level.

- Apart from imparting colours, natural dyes have medicinal values and are also more skin-friendly.
- Natural dyestuff obtained from a variety of natural sources may produce a wide range of shades. The colours can be customised by the use of different mordants or by changing the preparation technique which may not be possible with synthetic dyes.
- Natural dyes are plant or agro-based renewable as well as biodegradable materials.
- Wasteland can be economically useful for growing dye plants that can survive in such an environment.
- Waste produced during natural dye processing can be an ideal fertiliser for agricultural crops.
- Natural dye industry contributes towards the socio-economic upliftment by providing job opportunities on all grounds.
- These industries are proven to be a part of sustainable development both in rural and urban areas.
- Use of natural dyes can greatly reduce carbon footprint due to minimum consumption of fossil fuels.

9 Conclusion

The application of natural dyes has gained momentum on the grounds of sustainability and sustainable development in dyeing industries. The process of natural dyeing is an inclusive mechanism for safe ecological and human health, maintaining age-old traditions and socio-economic development. However, the availability of natural dyes on their extensive use may raise concerns about the future sustainability of this concept. Nevertheless, adequate cultivation of dye plants in a cost-effective route will eliminate the uncertainty associated with their future application. To achieve this, inexpensive production, affordable industrial application, less water/fossil fuel consumption, etc. are to be reviewed diligently. From the studies, it is found that only a few selected natural dyestuffs can match the synthetic dyes in terms of quality and stability. Indigo, tannins, alizarin dyes and flavonoids are such few natural dyes and pigments that could be in competition with their synthetic counterparts. At present, only a small fraction of world textile dyeing industries use natural dyes due to their limited economic viability. Application of coloured waste produced from food, timber and beverage industries is quite limited. Apart from these, the availability of suitable land for cultivation and production of dye plants at the expense of food crops is another constraint. Dye plants like indigo, woad and madder can easily cultivated on wastelands to maximise their production. In addition, the seasonal availability of natural dyes, tedious extraction and processing of dyes, transportation of large plant materials and production of waste in equivalent amounts further limit the application of natural dyes. Scientific development and implementation of novel methods of natural dye production would give a new dimension to their application with sustainability.

References

1. Bafana A, Devi SS, Chakrabarti T (2011) Azo dyes: past, present and the future. Environ Rev 19:350–370
2. Othmer K (2004) Encyclopedia of chemical technology, vol 7, 5th edn. Wiley-Interscience
3. Arun Prasad AS, Bhaskara Rao KV (2010) Physico chemical characterization of textile effluent and screening for dye decolorizing bactéria. Global J Biotech Biochem 5(2):80–86
4. Christie R (2001) Colour chemistry. The Royal Society of Chemistry, Cambridge, United Kingdom
5. Dos Santos AB, Cervantes FJ, van Lier JB (2007) Review paper on current technologies for decolourisation of textile wastewaters: perspectives for anaerobic biotechnology. Biores Technol 98(12):2369–2385
6. Ananthashankar R (2012) Treatment of textile effluent containing reactive red 120 dye using advanced oxidation. MSc thesis, Dalhousie University, Halifax, Canada
7. Hassaan MA (2016) Advanced oxidation processes of some organic pollutants in fresh and seawater. PhD thesis, Faculty of Science, Port Said University
8. Manzoor J, Sharma M (2020) Impact of textile dyes on human health and environment. In: Manzoor J, Sharma M (eds) Impact of textile dyes on public health and the environment. IGI Global, pp 162–169

9. Leub H (2003) Industrial dyes, chemistry, properties and applications. Wiley VCH, Wenheim, Germany
10. Stolz A (2001) Basic and applied aspects in the microbial degradation of azo dyes. Appl Microbiol Biotechnol 56(1–2):69–80
11. Pourbabaee AA, Malekzadeh F, Sarbolouki MN, Najaf F (2006) Aerobic decolorization and detoxification of a disperse dye in textile effluent by a new isolate of Bacillus sp. Biotechnol Bioeng 93(4):631–635
12. Couto SR (2009) Dye removal by immobilised fungi. Biotechnol Adv 27(3):227–235
13. Forgacs E, Cserháti T, Oros G (2004) Removal of synthetic dyes from wastewaters: a review. Environ Int 30(7):953–971
14. Przystaś W, Zabłocka-Godlewska E, Grabińska-Sota E (2012) Biological removal of azo and triphenylmethane dyes and toxicity of process by-products. Water Air Soil Pollut 223(4):1581–1592
15. Kant R (2011) Textile dyeing industry an environmental hazard. Nat Sci 4(1):22–26
16. Wang Z, Xue M, Huang K, Liu Z (2011) Textile dyeing wastewater treatment. Adv Treat Text Effl 5:91–116
17. Balter M (2009) Clothes make the (Hu) man. Science 1329–1338
18. Kvavadze E, Yosef OB, Cohen AB (2009) 30,000-year-old wild flax fibers. Science 325(5946):1359–1365
19. Adams DQ, Barber EJW (1996) Prehistoric textiles: the development of cloth in the neolithic and bronze ages, with special reference to the Aegean. Language 72(2):392–398
20. Rogers PW (2003) Dyes and dyeing. In: Jenkins, pp 25–29
21. Calderin J (2009) Form fit fashion. Rockport, pp 125–129
22. Asia A, World N (2015) Natural dye.Wikipedia 1–6
23. Dye (2021) Wikipedia. Wikipedia Foundation
24. KeyColour (2016) the evolution of textile dyes: history and development, KeyColour. https://www.keycolour.net/blog/the-evolution-of-textile-dyes-history-and-development/. Accessed 13 Oct 2021
25. Goethe JW von, Eastlake CL (1840a)Theory of color | Achtzehntes Jahrhundert Zweite Epoche. Taylor Institute. https://theoryofcolor.org/Achtzehntes+Jahrhundert+Zweite+Epoche?highlight=Jeremias#Jeremias_Friedrich_G_lich. Accessed 13 Oct 2021
26. Goethe JW von, Eastlake CL (1840b) Theory of color | Part V—Relation to Other Pursuits. Taylor Institute. https://theoryofcolor.org/tiki-index.php?page_ref_id=13#Relation_to_the_Technical_Operations_of_the_Dyer. Accessed 13 Oct 2021
27. Goodwin J (1982) A dyer's manual. Pelham Books Ltd., London, References-Scientific Research Publishing
28. Sisodiya K (2019) Microbial degradation of methylene blue dye bye bacterial strain. Int J Res Appl Sci Eng Technol 7(4):505–511
29. Patra R (2016) To dye for: a history of natural and synthetic dyes. Herit, Hist Sci Inf Fabr. https://blog.patra.com/2016/09/07/to-dye-for-a-history-of-natural-and-synthetic-dyes/. Accessed 13 Oct 2021
30. Chattopadhyay SN, Pan N, Roy A (2013) Development of natural dyed jute fabric with improved colour yield and UV protection characteristics. J Text Inst 104(8):808–818
31. Grifoni D, Bacci L, Lonardo SD (2014) UV protective properties of cotton and flax fabrics dyed with multifunctional plant extracts. Dye Pigment 105:89–96
32. Basak S, Samanta K, Arputhraj A (2012) Method of dyeing and protective finishing of cotton textiles using vegetable extract. Indian Patent
33. Khan MZH, Sarkar M, Imam FA (2014) Paper making from banana pseudo-stem: characterization and comparison. J Nat Fibers 11(3):199–211
34. Singh K, Kumar P, Singh NV (2020) Natural dyes : an emerging ecofriendly solution for textile industries. Pollut Res 39(January):87–94
35. Vankar PS (2000) Chemistry of natural dyes. Resonance 5(10):73–80
36. Hancock M, Boxworth AD (1997) Potential for colourants from plant sources in England & Wales. Arable Crops & Horticulture Division, Cambridge

37. Alemayehu T, Teklemarian Z (2014) Application of natural dyes on textile: a review. Int J Res Granthaalayah 2(2):61–68
38. Geetha B, Sumathy VJH (2013) Extraction of natural dyes from plants. Int J Chem Pharm Sci 1(8):502–509
39. Divya LRB, Ravi D (2013) Extraction of natural dyes from selected plant sources and its application in fabrics. Int J Text Fash Technol 3(2):53–60
40. Ezeokonkwo MA et al (2018) Preliminary characterization of some natural dyes. Afr J Pure Appl Chem 12(7):54–61
41. Mansour R (2018) Natural dyes and pigments: extraction and application. In: Handbook for renewable materials for coloration and finishing, pp 75–102
42. Samanta AK, Konar A (2011) Dyeing of textiles with natural dyes. In: Natural dyes, pp 3–56
43. Uddin MG (2015) Extraction of eco friendly natural dyes from mango leaves and their application on silk fabric. Text Cloth Sustain 1(7):1–8
44. Deveoglu O, Torgan E, Karadag R (2011) High-performance liquid chromatography of some natural dyes: analysis of plant extracts and dyed textiles. Color Technol 128:133–138
45. Klodzinska DJ et al (2021) Analysis of natural dyes from historical objects by high performance liquid chromatography and electromigration technique. Crit Rev Anal Chem 51(5):411–444
46. Septhum C, Rattanaphani V, Rattanaphani S (2007) Uv-vis spectroscopic study of natural dyes with alum as a mordant. Suranaree J Sci Technol 14(10):91–97
47. Torchani A, Gharbi R, Fathallah M (2014) Study of natural dyes for sensitized solar cells applications. Sens Transducers 27:185–189
48. Gulrajani ML, Lal S (1977) Evaluation of polymeric agents for dyeing of nylon. Ind J Text Res 3:8–11
49. Jesper et al (2014) Modelling, simulation and control of the dyeing process. Woodhead Publishing Series in Textiles, pp 206–221
50. Naebe M, Cookson PG, Rippon JA, Wang X (2010) Effect of leveling agent on the uptake of reactive dyes by untreated and plasma-treated wool. Text Res J 80(7):611–622
51. Reji RV, Kim TY, Son SW, Suryan A, Kim HD (2019) Computational investigation on the flow characteristics in beam dyeing process. J Text Inst 111(3):1–13
52. Uddin MK et al (2019) Effective cross-linking dyeing method for jute fabric with reactive dyes. Eur Sci J 15(27):125–135
53. Wahyuningsih S et al (2020) Chemistry of natural dyes of batik crafts colouring process. IOP Conf Ser.: Mater Sci Eng 858:012052
54. Jordeva S et al (2020) Dyeing of textiles with natural dyes. Tekst Ind 68(4):12–21
55. Gupta VK (2019) Fundamentals of natural dyes and its application on textile substrates. Chem Technol Nat Synth Dyes Pigments. https://doi.org/10.5772/intechopen.89964

Environmental Management System and Sustainable Textile Production

Manjeeta Kumari, Shiwani Berry, and Pinki Rani Agrawal

Abstract An environmental management system offers a structured approach to organizing and implementing environmental protection measures. The environmental management system is important because it monitors, evaluates, and enhances environmental performance and promotes effective adherence to environmental regulations. The textile industry is the oldest one; it began very early in the history of mankind. It first developed gradually, then quickly to meet other needs of mankind, including those for luxury, but this growth damaged the environment, necessitating remediation. For mill employees, customers, and the environment as a whole, the chemical contamination of the textile processes poses a health risk. Sustainable textile refers to a production process that is ecologically friendly, which means that all inputs, processes, materials, and outputs will be safe and healthy for people and the environment throughout the whole life cycle. A number of schemes have been put up in many nations to regulate the textile industry in order to improve the environment and prevent future ecosystem deterioration; however, developing nations must implement their designed policies from the beginning. Sustainability in the textile industry can be achieved by improving resource productivity, cost-efficiency, customer expectations, brand reputation, and environmental conditions and ensuring better health conditions for the wearers and the workers. According to the research, implementing an environmental management system can significantly diminish pollution in the textile industry.

Keywords Environmental management system · Sustainability · Resource efficiency · Textile · Impact assessment

M. Kumari · S. Berry (✉)
Depatment of Chemistry and Chemical Sciences, Central University of Himachal Pradesh, Shahpur, Kangra 176206, India
e-mail: shiwani.berry@hpcu.ac.in; shiwanipu@gmail.com

P. R. Agrawal
Deenbandhu Chhotu Ram University of Science and Technology, Murthal, Sonipat, Haryana, India

P. Singh (ed.), *Dye Pollution from Textile Industry*, SDGs and Textiles,
https://doi.org/10.1007/978-981-97-5341-3_12

1 Introduction

In order to effectively deal with ecological laws, an environment management system analyzes, assesses, and optimizes environmental performance. Businesses are being compelled to enhance their sustainability performance in a proactive manner as ecological issues become more critical [1]. Currently, organizations are widely aware that integrating management standards like EMS is a successful approach to achieve their social, environmental, and financial objectives [2]. EMS has developed into a vital tool for long-term organizational development in this evolving environment [3]. The textile industry, historically known for its significant environmental footprint, is undergoing a profound transformation driven by the urgent need for sustainability. Consumers are increasingly prioritizing eco-conscious and ethically produced textiles, leading to a shift in market dynamics. This article delves into the principles of sustainable textile production, focusing on environmentally friendly materials, cleaner production techniques, waste management, and embracing the circular economy concept. Furthermore, it explores the benefits of implementing Environmental Management Systems (EMS) in the textile industry, highlighting pollution reduction, improved compliance, and enhanced brand image. Federal laws are followed by organizations with the adoption of environmental management systems (EMS), which are defined in the ISO 14001 standard. Environmental management systems (EMS) help organizations in developing policies that correspond to the ISO 14001 standard and other legislative requirements. An EMS, or Environmental Management System, is a framework made up of a company's organized, recorded activities and regulations that directly affect and manage any current or potential environmental issues the business may have [4]. An EMS can be either certified or uncertified. Certified organization has met the standards for an EMS as specified by the system. One of the cornerstones of sustainable textile production is the choice of environmentally friendly materials. Traditional materials like cotton, polyester, and man-made cellulose have often posed ecological challenges. Cotton, for instance, demands extensive water and chemical usage in cultivation. To address this, the industry is increasingly turning to alternative natural fibers such as hemp, flax, linen, and nettle, which require fewer resources. Organic cotton, another sustainable option, reduces water pollution and offers enhancement in properties like hypoallergenic as well as soft textures. Polyester, often made from petroleum, poses a challenge due to its non-biodegradable nature. However, developments in bio-based polyester from renewable sources like sugar cane, beet, and plant oils are reducing the environmental impact associated with traditional polyester production. These bio-based alternatives offer lower water usage and the potential for reprocessing into virgin fibers. Man-Made Cellulose materials, despite being sourced unsustainably in the past, are evolving. Innovative materials like Lyocell, Bemberg, Cotton Liner, and Pinatex are derived from fast-growing plants with minimal irrigation and pesticide requirements. These sustainable alternatives offer biodegradability and minimize the ecological impact. Cleaner production techniques are pivotal in minimizing waste generation, optimizing energy usage, and reducing pollution during textile manufacturing. The textile industry is embracing cleaner production methods to enhance its environmental performance. These methods focus on minimizing the use of hazardous chemicals and optimizing manufacturing processes. This, in turn, reduces emissions and waste. Moreover, effective chemical management practices are being implemented to ensure responsible handling, disposal, and recycling of chemicals, aligning with sustainability goals. As stated by Chen and Burns [3], using dyes and other chemicals during the textile wet processing phase creates a negative influence on the environment. China is one of the influential participants in the universal value chain for clothing and textiles. Over 13.18% of Chinese manufacturing companies, according to the most recent census data, create goods related to textiles. The number of workers in this business is more than 12.5 million. In terms of textile and apparel exports, China topped the list in 2015 [5]. Effective waste management is crucial for sustainable textile production. Strategies such as recycling, upcycling, and waste reduction are essential in creating a closed-loop system where materials and products are reused, remanufactured, or repurposed. Embracing the circular economy concept not only minimizes waste but also maximizes resource efficiency. Circular design principles are becoming integral in textile product development. These principles promote eco-friendly materials and designs that facilitate recycling and reuse, extending product lifecycles and reducing environmental effects.

The application of EMS in the material industry yields numerous benefits. Firstly, EMS significantly reduces pollution and waste, promoting cleaner production techniques and responsible waste management practices. Secondly, EMS enhances compliance and risk management by ensuring that textile companies navigate complex environmental regulations and mitigate potential risks effectively. This proactive approach safeguards both the environment and the organization's financial stability. Thirdly, EMS builds a positive brand image and gains consumer trust in an era where consumers prioritize eco-conscious and socially responsible products.

Challenges and Progress

While progress is being made in the textile industry's sustainability journey, several challenges persist.

Waste Characterization and Segregation: Textile waste is diverse, including discarded materials from production processes (pre-consumer) and products that consumers no longer use (post-consumer). Properly characterizing and segregating these materials is challenging but essential for efficient recycling and upcycling.

Chemical Management: The textile industry employs various chemicals in processes like dyeing and finishing. Managing these chemicals' usage and disposal, while minimizing environmental harm, is complex.

Energy Efficiency: Textile manufacturing processes, especially dyeing and drying, consume substantial energy. Finding ways to reduce energy consumption and incorporate renewable energy sources can be a complex task.

Circular Design Principles: Integrating circular design principles into textile product development requires a shift in thinking. Designers and manufacturers must consider eco-friendly materials and designs that facilitate recycling and reuse.

Traceability and Transparency: Ensuring transparency and traceability of textile products' origins and impacts in a globalized market is challenging.

Innovative technologies and solutions are driving progress in overcoming these challenges:

- Advanced Waste Sorting Technologies: Integration of artificial intelligence (AI) and robotics in waste sorting facilities has improved the precision of waste characterization and segregation.
- Chemical Management Software: Innovative software solutions provide real-time monitoring of chemical usage, helping industries optimize chemical consumption, reduce waste, and minimize environmental impact.
- Renewable Energy Adoption: Many textile companies are investing in sustainable energy sources.
- Eco-Friendly Materials: There's a growing trend toward using sustainable and biodegradable materials in textiles.
- Blockchain for Traceability: Blockchain technology is being explored to create transparent supply chains in the textile industry.

Consumer preferences are evolving rapidly, with a growing emphasis on sustainability and ethical production practices. Sustainable fashion, characterized by durability, quality, and ethical sourcing, is replacing the fast fashion model. Transparency in the supply chain is becoming paramount as consumers seek information about product origins. Moreover, the circular economy model is gaining prominence, encouraging textile companies to explore recycling, upcycling, and repurposing textiles to reduce waste and enhance resource efficacy.

In summary, the textile industry is in the midst of a sustainability revolution driven by consumer demand, regulatory changes, and technological advancements. Embracing sustainable practices, including the use of eco-friendly materials, cleaner production techniques, effective waste management, and the adoption of EMS, is not only a necessity but also a strategic move to thrive in a rapidly changing market. As the industry continues to navigate these challenges and capitalize on market trends, it is poised for a more sustainable and environmentally responsible future, aligning with the global imperative to protect our planet for future generations.

1.1 Understanding of Environmental Management Systems

An environment management system is a system that integrates policy and procedure and manages the impact on the environment. It helps in providing an organized approach to develop and carry out environmental safeguards [6]. EMS comprises a systematic approach to managing and mitigating an organization's environmental impact. The key components include policy development, planning, implementation, monitoring, and continuous improvement.

1.2 Composition of an Environmental Management System

An organization needs to evaluate its environmental impacts, establish goals for minimizing their effects, and plan how to attain the targets to build an EMS. Organizational commitment is a particularly essential component of an EMS. To set up and run an effective EMS, dedication from the highest echelons of the business, as well as all workers, must be demonstrated [6] (Fig. 1 and Table 1).

The Positive Effects of EMS Strategies

- Product quality has been improved
- Enhanced staff dedication
- Improve community relations
- Positive group pressure relationships
- Greater publicity in the media
- Finance at a lower cost
- Low premiums for insurance
- Ensured current and future compliance
- Improved material efficiency.

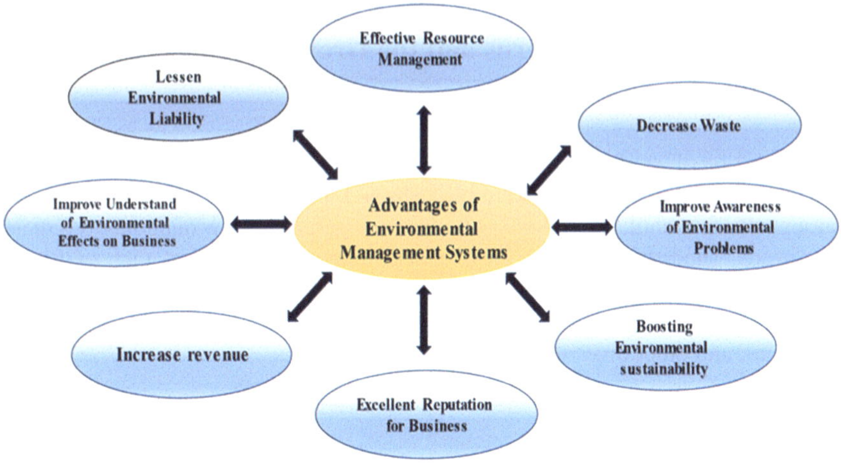

Fig. 1 Advantages of environment management system

Table 1 Elements of environment management system

Environmental policy	An environmental policy is a statement that states what an institution expects from EMS. It makes sure that all environmental actions are compatible with the aims of the firm
Identification of environmental impact	The real and prospective environmental consequences of a company's operations must be carefully identified and recorded. Performing an environmental audit can help with this
Goals and objectives	An audit of the environment acts as a base for determining an organization's environmental goals and objectives. Adopting more stringent long-term goals can help an organization by motivating it to better performance. To consistently progress goals should be examined on a regular basis
Environmental management program	The plan for environmental management outlines the methods and processes that an organization will use to attain its targets and objectives
Examinations and conformity evaluation	On a regular basis, audits should be done to make sure that the EMS is accomplishing its objectives and that operational processes are being improved to attain this goal. Regular environmental management is frequently necessary to make sure that regulatory and other requirements are met
Ongoing development	A critical component is continuously improved. When used for monitoring progress in achieving an organization's environmental priorities and objectives, an EMS is most successful. The procedures that are set in place for accomplishing these objectives should be studied frequently to find out whether or not they may be improved or whether improved processes can be implemented

2 Environmental Concerns and the Growth of the Textile Sector

The production of textiles is one of the ancient industries, spanning several 1000 years and beginning in a very basic and simple fashion. It is a critical firm activity since its major goal is to manufacture clothing for humans, offering a few of humanity's most basic needs. The textile business grows and develops slowly at first, then swiftly to meet a wide range of human requirements, including luxury applications. The textile industry began to adapt to new applications like firefighter cloth, space exploration, geotextiles, and other diversified purposes.

2.1 Environmental Issues Caused by Textile Manufacturing

Because a growing population necessitates economic growth, technological advancement has been accelerated to fulfill the demands of an expanding population [7]. After oil, the manufacturing of textiles and clothing business is the World's second most contaminated industry. It employs around 60 million people globally and generates over $450 billion in sales every year [8, 9]. Its products are diverse, but apparel accounts for 60% of the whole T&C market [10]. Textile product production, delivery, and usage necessitate and use vast amounts of non-renewable resources [10–14]. There is also a quick expansion of textile items, which has decreased their existence duration. For example, the number of times a garment has been worn before being discarded has decreased by 36% [10, 12, 15]. As a result of this tendency, "fast fashion" has emerged. Fast fashion refers to clothing that is easily available and reasonably priced. Outsourcing production to low and middle-income countries (LMIC) keeps prices down [15]. In 2015, the global textile consumption was around 95.6 million tons [16]. The usage of water, energy, and toxic substances, as well as carbon dioxide emissions and solid waste creation, are all environmental problems. The T&C business is the world's second-largest user of water [17]. The industry consumes around 93 billion metric tons of water every year. Its water usage accounts for around 4% of total yearly freshwater outflow [18]. In addition, it takes 55 gallons of water to manufacture 1 kg of cloth. Water is utilized for cleaning the materials, chlorine dioxide, and colorants, and cleaning the completed product. Most wastewater has not been re-processed before being discharged into the environment. As a result, they contaminate the subsurface water supply even more. The manufacturing process also adds to air pollution. The industry produces a significant amount of solid waste. Clothing sales are expected to treble from current levels by 2050 [17]. Developing marketplaces in Africa and Asia are driving the growing demand [13, 18–20]. These have major environmental consequences. For example, in 2015, 48 million tons (66%) of the 72,000,000 tons of polymer PET produced globally were amorphous PET utilized for fiber manufacture (polyester) [21]. Washing clothing annually releases 500 thousand metric plenty of plastic microfibers into the ocean, which is the equivalent of 50 billion plastic bottles. The majority of those threads include polyester, which is found in around 60% of clothing. Polyester manufacturing generates 2–3 times as much carbon as cotton production. Furthermore, polyester does not disintegrate in the water.

According to IUCN (International Union for Conservation of Nature) research, 35% of microplastics in the ocean are due to the washing of synthetic fabrics such as polyester [22]. Cotton cultivation consumes around 16% of pesticides and utilizes 2.5% of farmland globally. Ultimately Four percent of NPK fertilizer use. The runoffs pollute the water, resulting in eutrophication [21, 22]. Wastewater from textile industries is typically untreated and includes high levels of contaminants, which are released into bodies of water on a regular basis [16]. Wet procedures, such as dyeing, printing, and finishing, consume a lot of chemicals and water. According to estimates, the manufacture of 12–20 tons of textiles generates around 1000–3000 cubic meters of wastewater. The three principal sources of air pollution in the T&C sector are boilers that generate steam ovens, and storage tanks. Emissions are produced during the process of drying, which necessitates high temperatures. During the dyeing process, gases are also emitted [23]. The manufacturing sector accounts for around 10% of worldwide CO_2 emissions [9].

Textile manufacturing creates a variety of disposal streams, including water-based effluent, air pollutants, contamination of waterways, and toxic waste. The type of waste created is determined by the kind of manufacturing fabric facility, the methods and equipment employed, and the fibers and chemicals used [24]. Textile wet procedures employ large amounts of water and other chemicals sold as textile auxiliaries to improve the look, serviceability, and durability of the fabric. Chemical contamination of textile wet processes can endanger the health of mill workers, customers, and the environment. Many plans have been presented in various countries to manage textile wet operations in order to produce improvements in the environment and safeguard ecosystems from further deterioration; however, developing nations must implement their rules from the start. The safeguarding of the natural environment has become a major problem all over the world, and the worldwide community's attention has been drawn to environmental concerns and related issues that include workplace wellness and consumer product safety. Global warming exemplifies the gravity of environmental challenges. Textile wet processing generates a wide range of contaminants, both organic and inorganic, which may necessitate a variety of strategies to decrease or remove.

3 The Concept of Sustainable Textile Production

The 3R principles should be implemented through the whole cycle of resource generation, consumption, and return [25]. Therefore, the difficulties for the clothing and textile sectors will be to reduce energy and material intensity, reduce toxic substance dispersion, improve recycling, maximize the utilization of renewable resources, extend product durability, and increase service intensity. Sustainable textile production principles include resource efficiency, emissions reduction, ethical labor practices, and product longevity. It champions a circular economy to minimize waste and optimize resource use (Fig. 2).

By minimizing virgin material use, reusing textile and garment waste, and recycling them, efficient waste disposal will have a substantial impact on the 3R's.

Fig. 2. 3R processes

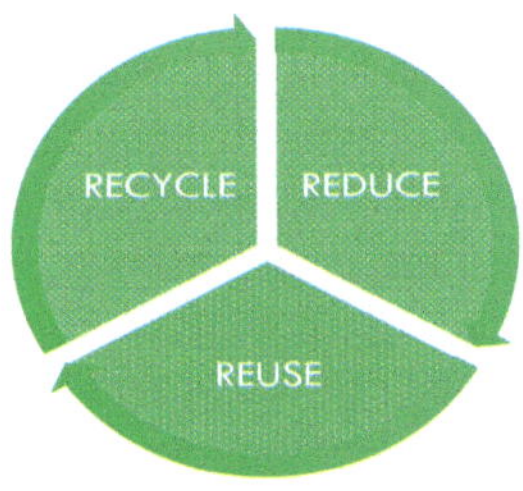

3.1 Importance of Sustainable Practices in the Textile Industry

Nowadays, one of the key goals in most sectors is to seek sustainability since sustainable implementation enhances trust among organizations and clients. Sustainability is described as "satisfying current needs without jeopardizing future generations' needs." Sustainable practices may take three forms: environmental, economic, and social [26]. The idea of "eco-friendly" or "eco-friendly production" is obtaining traction in a variety of industries. As an outcome, producers and merchants in the clothing and garment industries are progressively adopting environmentally friendly and ethical methods [27].

Many of the pesticides employed for conventional cotton growing contribute to soil degradation and acute toxicity. Organic cotton not only avoids soil and water pollution, but it also offers enhanced qualities such as hypoallergenic and a softer touch. As a result, organic cotton is an excellent choice for beginning materials used by the textile and garment industries [28].

In terms of eco-friendly procedures, novel materials and technologies (nanomaterials) are developing in sustainable apparel and textile production. Regarding green methods, some recently developed methods have advanced in traditional methods like turning (open-end rotor and air-jet spinning), creating, and that knitting (high-speed circular knitting and seamless knitting). These approaches benefit from utilizing less power, highly efficient, and creating fewer contaminants and noise. Generally, yarn and fabric manufacture is a mechanical process that consumes a large amount of energy and produces a lot of waste, dust, and sound. The total amount of energy used in the textile sector may be divided as follows: 23% for weaving, 34% for spinning, 38% for the use of chemicals, and 5% for other operations. The intriguing fact is that the energy spent when fabric reaches the user (care and upkeep phase) is about 4 times greater (75–80%) than the energy consumed during manufacture (20–25%) [29].

Other factors related to sustainability must be examined in addition to the complete manufacturing process. Several approaches to reducing energy and water can be used, including using greater amounts of water and energy-efficient machinery, training employees on energy-saving abilities, utilizing sensor systems for lighting, and improving collecting rainwater for non-drinking purposes, all of which contribute to more sustainable textile industry. Plastic, paper surplus textiles, as well as cardboard packaging might be recycled and repurposed as the residual trash created during production waste, textile waste is classified into three types based on its origin: post-consumer trash, pre-consumer garbage, and production waste. When clothing reaches the end of its usable life cycle, it produces post-consumer textile waste. With the rise of fast fashion came the concept of pre-consumer trash. This form of trash is produced at a higher rate than production and includes leftovers from shop sales, certain products originating from creative experiments made in business, and faulty returned goods. Approximately 95% of total textile waste may be recovered through recycling or upcycling [29]. In the words of Teli et al. [30], upcycling is the practice of transforming trash or obsolete things into fresh supplies or higher-quality products by including beneficial features and refurbishing. In the context of pre/post-consumer waste, it is one of the most cost-effective and stylish types of sustainability in the textile business [29]. Another method that must be applied is the cradle-to-cradle notion. In a circular economy, this approach employs recycling to conclude the cycle, allowing the manufacture of a future generation of goods from the originating pieces. The circular economy is a business concept whose goal is to maintain products/components/materials in the loop for as long as feasible [30]. In the past few years, different methodologies and indices have been created to analyze and examine the environmental effects of diverse systems. These instruments or statistics include a life cycle assessment (LCA), the assessment of environmental impacts (EIA), Strategy Environmental Assessments (SEA), cost–benefit evaluation (CBA), Materials Flow Analysis (MFA), and Environmental Footprint [31].

LCA is widely utilized within the textile and apparel industries. The life cycle evaluation is a system for estimating the entire environmental effect of a service, process, or good from cradle to grave (from basic material, manufacturing, and consumption to waste disposal) [32, 33]. The LCA method measures the effect that a good has on human well-being, the environment, and resources at each stage of production [33].

Sustainable practices are vital for the textile industry's long-term viability, mitigating environmental impact, enhancing reputation, and aligning with evolving consumer preferences.

4 Major CSR (Corporate Social Responsibility) Programs, Rules, and Standards, as Well as Their Success in Assuring Enterprise-Customer Communication

Governments and international organizations have introduced regulations addressing environmental issues in textiles, including restrictions on harmful chemicals and wastewater treatment standards. Numerous sustainable textile certifications and standards guide industry players toward environmentally responsible practices, emphasizing responsible sourcing, manufacturing, and labeling. The growing interest in the widely acknowledged impact of business operations on the general public and the environment as a whole has resulted in the rapid creation and dissemination of a variety of tools for measuring, assessing, enabling, and assuring effective communication of firms' CSR. International agencies, as well as members of the industrial community and business groups, have undertaken efforts with the goal of formalizing the basic norms of business behavior, with their reach expanding over time to encompass new sectors [34]. A wide range of criteria may be used to classify the instruments produced. Because the purpose of this article is to establish the impact of CSR-related activities on consumer choices as well as more broadly, their impact on product distinctiveness, classifying the instruments based on their primary receivers becomes especially beneficial. There are three primary categories of instruments targeted by the criterion:

- **Investors**—Investments with a social conscience
- **Corporations**—Management that is socially responsible
- **Clients—**Consumption that is socially responsible.

However, it should be emphasized that the borders dividing the categories are fairly fluid, as the majority of the initiatives related to CSR, legislation, and standards examined have a broad influence. Companies, for example, mostly employ reporting rules, and the reports assist both customers and investors. Aside from that, each of the primary categories has its subdivisions, such as general regulations, codes of conduct, recommendations for management and authorization systems, indexes for ranking firms based on their community involvement in order guidelines reports, social auditor guidelines, and environment-friendly and social labeling systems [34–38].

Management with Social Responsibility

Management with responsibility for society is made up of tools that define business practices along with leadership standards that allow businesses to include CSR in their plans and operations. This is the most diverse collection of conditions [39], which entails.

Corporate norms of behavior that a growing number of worldwide firms began internalizing in the mid-1980s. The fact that the codes are accepted freely is their greatest distinguishing feature. The majority of them concern fair corporate practices, environmental protection regulations, and labor circumstances. Some of them seek to prohibit cryptogenic activity or to establish guidelines for multinational firms' participation in community-benefiting projects. As a result, codes of conduct can be characterized as voluntarily established and formally unbinding norms governing safeguarding the environment, fair trade practices, customer and contractor collaboration, sustainable development, and labor conditions [40]. They can be categorized as follows as far as possible:

International terms created and implemented by international bodies, such as

OECD—Multinational Enterprise Guidelines,

ILO—Tripartite Declaration of Multinational Corporations and Social Policy Principles.

The European Commission's Green Paper on CSR. Promoting a European Corporate Social Responsibility Framework. The European Commission published a paper in 2001 outlining the laws and procedures for adopting corporate accountability at the European level. This is the first paper created at such a high level on social responsibility.

Codes developed by Non-Governmental Organizations to function as corporate code "models".

Global Compact of the United Nations—The UN Secretary-General, Kofi Annan, sponsored the formation of the Global Compact Corporate Responsibility Forum in 2000 to encourage commercial accountability among forum members. Above all, they are encouraged to follow nine fundamental principles that govern three primary areas: human rights, labor standards, and environmental standards.

Human Rights Principles for Businesses **from Amnesty International**

ICFTU (International Confederation of Free Trade Union)—The Basic Labor Standards Code

In June 2004, **ISO** agreed to begin producing social responsibility recommendations. ISO 26000, as opposed to the international standards ISO 9000 (quality administration) and the recently released ISO 14000 (environmental administration), will not serve as an umbrella norm for management systems that require certification.

"Joint" Codes—Generated collectively by diverse groups such as non-governmental organizations, trade unions, employer organizations, government officials, and so on.

Ethical Trading Initiative (ETI)

Is a coalition of businesses, non-governmental organizations, and labor unions that develops and promotes framework labor rules. The partnership ensures that employees manufacturing for the British market operate in circumstances that come together or surpass international standards.

Sustainability Management

Eco-management and audit scheme (EMAS) plans to encourage greater adherence from companies through ecological standards.

The International Organization for Standardization (ISO) 14000 is a set of environmental management standards and principles that enable enterprises to build organized control over the ecological effects of their operations, goods, and services. The International Organization for Standardization 300 series concerning emissions of compounds that pollute the air, water, or soil.

The International Organization for Standardization (ISO) 19011 standard gives standards for auditing environmental management systems (process rules, auditor credentials).

In the clothing and garment industries, standards of conduct are extremely widespread. Almost all major textile and garment companies use sophisticated CSR tools, namely the Standards of Best Practice, which are open to the public. However, in many situations, organizations' real behavior and the noble elements of their standards are diametrically opposed. International textile and garment companies want to position themselves as ethically responsible organizations, yet they are unwilling to make genuine faithfulness. It is a fascinating coincidence that laws concerned with labor difficulties are often found in the garment business, but codes dealing with environmental issues are mostly addressed by makers of industrial products (technical and protective fabrics, etc.) [41].

5 Estimation of Environmental Impact on Textile Production

The rise of quick fashion has altered society's perception of how we dress. Between 2000 and 2014, global garment manufacturing more than quadrupled, surpassing one hundred billion items per year in 2015 [42]. However, as output expanded, so did the environmental burden connected with the textile sector. This is nearly the second most polluted business, after oil and gas production. Aside from the large spending on water-related projects, there additionally exists the effect of dye or pigments, which frequently wind up poisoning water supplies. Textile treatments that have a negative influence on the environment have been proven to produce between 15 and 20% of water pollution [43]. Carbon and emissions of greenhouse gases, on the other hand, are continually emitted during the manufacturing, transportation, and usage processes [44]. Even after contacting the customer, the accompanying environmental effect is significant due to substantial recycling issues. When an item of garments is no longer usable, it gets discarded. A portion is put in boxes for reuse, while another portion will wind up in unsorted bins, with the two options being landfill or incineration [45]. Quick fashion has also boosted the compulsion to buy apparel. Consumers may buy new trends more easily because of easy access to internet businesses providing low-cost goods and the adoption of a fast fashion idea [45]. Furthermore, stores that have embraced the quick fashion concept sell fashion goods at more competitive costs but are engineered to last 10 times lower than textile products manufactured under the slow fashion philosophy [46]. According to Greenpeace, consumers buy 60% additional textile items than in 2000 and only keep them half as long [42], as a result of this there is an increase in garbage.

Eco-friendly programs prioritize the creation of new added-value commodities and the reduction, reuse, and recycling of industrial waste, grounded in a robust scientific methodology. To begin with, in terms of decrease, the total Cost of Ownership should be proactive, boosting product quality to encourage a longer lifespan. Greater pricing and margins are related to greater quality, which contributes to the viability of the industry in various situations with fewer outputs. Quality will also improve the possibilities of reuse, assisting in the reduction of waste. Finally, the Textiles and Clothing Institute should be able to utilize recycled textile materials more efficiently, not just by incorporating reused Fibers into production processes, but also by implementing innovative recycling procedures. These techniques might include recovering fresh fabrics from old textiles, such as the creation of new finishing goods. Aside from environmental benefits, these newly produced goods should improve the features and attributes of textiles, which are appreciated by the ultimate customer, in order to appropriately support the circular economy. It is feasible to expand new and improved goods while encouraging waste depletion and lowering the environmental effect of this business. This section focuses on evaluating how textile production affects the environment and how to measure and analyze these impacts. It involves two key aspects.

1. **Finding Environmental Hotspots in the Textile Supply Chain**

This involves identifying specific stages or processes within the textile supply chain that have the most significant environmental impacts. By pinpointing these "hotspots," companies can prioritize their efforts to reduce environmental harm. Techniques such as Life Cycle Assessment (LCA) are commonly used to provide a comprehensive view of where these hotspots exist.

2. **Analyzing Environmental Footprints and Impacts**

The purpose here is to quantify and assess the whole ecological impact of textile goods across their full existence. This involves evaluating elements like energy usage, water intake, emissions, along with trash formation. Companies may make educated decisions and apply sustainable practices to lessen their environmental effect by analyzing their environmental footprints. Life Cycle Analysis (LCA) techniques are frequently used for this purpose.

6 Developing Textiles Sustainable Production

6.1 Environment-Friendly Materials and Resource Efficiency

The relevance of using ecologically friendly materials such as cotton that's organic or recycled fibers. It also dives into the optimal utilization of resources during the manufacturing process. Sustainable textile manufacturing focuses on improving resource usage while decreasing waste output, hence lowering its ecological effect [42].

6.2 Cleaner Production Techniques

The adoption of cleaner production techniques. These techniques aim to reduce waste generation, optimize energy usage, and minimize pollution during textile manufacturing. By implementing cleaner production methods, the industry can enhance its environmental performance and reduce its ecological footprint.

6.3 Waste Management and Circular Economy

The importance of effective waste management and embracing the circular economy concept in textile production. It discusses strategies for recycling, upcycling, and waste reduction, creating a closed-loop system where materials and products are

Table 2 Ecological impacts and advantages of different materials

Material	Ecological impact	Sustainable adaptation	Advantages
Cotton	The large volume of water required for processing	Increased utilization of natural fibers including flax, linen, and nettle, which require low content of water, and fertilizer for development	It degrades naturally
	Fertilizers and insecticides are used extensively throughout cultivation		There is no need for UV light to cause discoloration
			The fabric is soft and breathable
	A large amount of cultivable land is needed		Non-allergenic, and simple to clean
Polyester	They are made of petroleum and do not biodegrade	Bio-based polyester (biosynthetic) from sustainable origin, such as starches and lipids derived from maize, bamboo, sugarcane, and beets	When compared to cotton, it requires less water
			Low-temperature washing is possible
	Polyester laundry produces microfibre plastics, which penetrate the food chain		It dries quickly and rarely has to be ironed
			Reprocessable into fresh (new) fibers
Man-made cellulose	Unsustainable cellulose is the source	Lyocell, Bemberg, Cotton Liner, and Pinatex are examples of innovative and sustainable materials that grow quickly and do not require watering or pesticides	Manufactured from renewable plants
	Chemicals utilized in the manufacturing process are not recovered		Biodegradable in a brief period of time

reused, remanufactured, or repurposed. This shift toward a circular economy is fundamental for achieving sustainability within the textile industry, as it minimizes waste and maximizes resource efficiency (Table 2).

7 Benefits of Environmental Management System in Textile Industry

EMS is a systematic approach that helps organizations manage and mitigate their environmental impact, and its implementation yields a wide range of benefits. These benefits can be categorized into three main areas.

7.1 Pollution Reduction and Waste Minimization

The textile industry is historically associated with significant pollution and waste generation. EMS implementation leads to substantial reductions in these negative environmental impacts. Firstly, EMS promotes the adoption of cleaner production techniques [20, 21]. These techniques focus on minimizing the use of hazardous chemicals and optimizing manufacturing processes to reduce emissions and waste. Through cleaner production, textile companies can significantly lower their environmental footprint. Secondly, EMS encourages responsible waste management practices. This involves the proper disposal and treatment of hazardous materials, as well as strategies for recycling and repurposing waste materials within the production process. By efficiently managing waste, the textile industry can minimize its contribution to landfills and reduce environmental harm.

7.2 Improved Compliance and Risk Management

Environmental regulations are becoming increasingly stringent, and non-compliance can result in legal consequences and damage to a company's reputation. EMS helps textile companies navigate this complex regulatory landscape. EMS ensures that textile companies comply with national and international environmental regulations [22, 37]. By systematically monitoring and managing environmental aspects, textile companies can identify areas where they may be at risk of non-compliance and take corrective actions promptly. Additionally, EMS enhances risk management by identifying potential environmental risks and liabilities [20]. This proactive approach allows companies to assess and mitigate risks before they escalate, safeguarding both the environment and the organization's financial stability.

7.3 Enhanced Brand Image and Consumer Trust

In an era of increasing environmental awareness, consumers are actively seeking products from environmentally responsible companies. EMS implementation helps textile companies build a positive brand image and gain consumer trust. Sustainable practices, guided by EMS, are attractive to environmentally conscious consumers [40]. When textile companies prioritize sustainability and communicate their efforts transparently, they differentiate themselves in the market and attract eco-conscious consumers. This can result in increased sales and brand loyalty. Moreover, EMS demonstrates a commitment to social responsibility [37]. By actively managing their environmental impact, textile companies convey their dedication to the well-being of the environment and communities. This resonates with consumers, who are more likely to support companies that align with their values. EMS not only reduces pollution and waste but also ensures compliance with regulations, minimizes risks, and enhances a company's reputation and consumer trust. These advantages make EMS a vital tool for textile companies aiming to thrive in a sustainable and environmentally responsible manner.

7.4 Challenges and Progress

The challenges facing the textile industry in achieving a sustainable circular economy through a smart waste management system are multifaceted. However, there is significant progress being made in waste sorting technology, chemical management, renewable energy adoption, eco-friendly materials, and traceability solutions. These advancements hold promise for a more sustainable and eco-conscious future in textile production. Continuous research and innovation will be essential to overcome these challenges effectively and transition to a truly sustainable textile industry [47] (Table 3).

8 Changing Consumer Preferences and Market Trends

The textile industry is witnessing a profound transformation as consumers increasingly prioritize sustainability and ethical production practices.

Technological and Financial Challenges

Implementing sustainable practices in the textile industry comes with a set of technological and financial challenges [48]. Adopting cleaner production techniques, sourcing environmentally friendly materials, and implementing circular economy strategies require significant investments in new technologies and processes. Companies often face initial capital expenses and may need to overhaul existing infrastructures, which can be financially demanding. However, the long-term benefits, including cost savings, regulatory compliance, and enhanced consumer appeal, often outweigh these initial challenges. Reducing energy consumption is essential for lowering greenhouse gas emissions and operational costs [49]. Many textile companies are investing in research and development to discover innovative, cost-effective solutions that promote sustainability [50].

Changing Consumer Preferences

Consumer preferences are undergoing a fundamental shift toward eco-conscious and socially responsible products [34]. Consumers nowadays are conscious of the environmental as well as social consequences of their purchases. So, they seek textiles that align with their values, such as products made from sustainable materials, produced using eco-friendly processes, and sourced ethically.

design principles into textile product development represents a shift in thinking, requiring a focus on eco-friendly materials and designs that facilitate recycling and reuse. This shift toward circularity is critical for minimizing waste, extending product lifecycles, and fostering a circular economy. Transparency and traceability present additional challenges, especially in a globalized market. Consumers increasingly demand information about a product's environmental and social footprint. Implementing systems like blockchain for traceability is one way to address this challenge and build consumer trust. Despite these challenges, there has been notable progress in various areas. Advanced waste sorting technologies, powered by artificial intelligence and robotics, have improved the precision of waste characterization and segregation. Chemical management software provides real-time monitoring of chemical usage, helping industries optimize consumption and minimize environmental impact. Renewable energy adoption is on the rise, with textile companies investing in sources such as solar and wind power. Sustainable materials, like as natural cotton and recycled polyester, are gaining popularity, lowering the industry's environmental imprint. Finally, the adoption of blockchain-based systems for traceability has raised transparency in the manufacturing process for textiles. Furthermore, the changing landscape of consumer preferences and market trends is shaping the industry's future. Consumers now prioritize eco-conscious and socially responsible products, which has given rise to sustainable fashion. This shift stresses longevity, quality, and ethical production, and it supports the use of naturally derived and recycled materials while decreasing waste in the manufacturing process. Transparency in the textile supply chain is a necessity, as consumers want to know the origins of their clothing. Lastly, the circular economy model is gaining prominence, prompting companies to explore innovative ways to recycle, upcycle, and repurpose textiles, thereby reducing waste and promoting resource efficiency. The textile industry is at a pivotal moment in its history, where sustainability, ethics, and environmental responsibility are no longer optional but integral to its future success. While challenges exist, progress in waste management technology, chemical management, renewable energy adoption, eco-friendly materials, and traceability solutions holds great promise for a more sustainable and eco-conscious future in textile production. Continuous research, innovation, and a commitment to sustainable practices will be essential for the industry to overcome these challenges effectively and transition to a genuinely sustainable and environmentally responsible future.

References

1. Epstein MJ (2018) Making sustainability work: best practices in managing and measuring corporate social, environmental and economic impacts. Routledge
2. Gianni M, Gotzamani K, Tsiotras G (2017) Multiple perspectives on integrated management systems and corporate sustainability performance. J Clean Prod 168:1297, e1311
3. Chen HL, Burns LD (2006) Environmental analysis of textile products. Cloth Text Res J 24:248–261
4. Benson CC (1996) The ISO 14000 international standards: moving beyond environmental compliance. NCJ Int L Com Reg 22:307
5. World Trade Statistical Review (2016) World Trade Organization 2016. https://www.wto.org/english/res_e/statis_e/wts2016_e/wts16_toc_e.htm. Accessed 4 Dec 2016
6. Khanna DR, Bhutiani R, Matta G (2009) Environmental management system. J Comp Toxicol Physiol 6(I):010–017. ISSN: 0972-8651-09-06-010 © Shivneri Publisher and Distributers
7. Chavan RB (2001) Indian textile industry—environmental issues. Department of Textile Technology, Indian Institute of Technology, Hauz Khas, New Delhi. Indian J Text Res 26:11–21
8. Malik A, Akhtar R, Grohmann E (2014) Environmental deterioration and human health: natural and anthropogenic determinants, pp 1–421
9. Snoek S (2017) Circular economy in the textile industry: transition theory in Dutch start-ups towards a circular economy. Published MSc thesis, Climate Studies, Environmental Policy Group
10. Enter, Expert Network on Textile Recycling, Strategic agenda on textile waste management and recycling, European Regional Development Fund. https://www.interregcentral.eu/Content.Node/Stategic-Agenda.pdf
11. Yasin S, Behary N, Perwuelz A et al (2017) Life cycle assessment of flame retardant cotton textiles with optimized end-of-life phase. J Clean Prod 172:1080–1088
12. Koszewska M (2018) Circular economy—challenges for the textile and clothing industry. AUTEX Res J 18:337–347
13. Patti A, Cucala G, Aciemo D (2020) Eco-sustainability of the textile production: waste recovery and current recycling in the composites world. Polymers 13:1–25
14. Koligkioni A, Parajuly K, Sørensen B et al (2018) Environmental assessment of end-of-life textiles in Denmark. Procedia CIRP 69:963–967
15. Bick R, Halsey E, Ekenga C (2018) The global injustice of fast fashion. Environ Health 17:1–4
16. Green of change, and, impact of the textile industry on the environment. https://www.greenofchange.com/textilepollution#:~:text=Due%20to%20the%20numbers%20that,pollution%20and%20solid%20waste%20-pollution
17. World Economic Forum (2020) Is fashion bad for the environment? https://www.weforum.org/agenda/2020/01/fashion-industry-carbon-unsustainable-environmentpollution/
18. Ellen MacArthur Foundation (2017) A new textiles economy: redesigning fashion's future. https://ellenmacarthurfoundation.org/a-new-textiles-economy
19. Lehner M, Mont O, Mariani G et al (2020) Circular economy in home textiles: motivations of IKEA consumers in Sweden. Sustain 12:1–23
20. Greenblue (2017) Chemical recycling: making fiber-to-fiber recycling a reality for polyester textiles. http://shift.tools/resources/1665
21. Kooistra KJ, Termorshuizen AJ, Pyburn R (2006) The sustainability of cotton: consequences for man and environment (No 223). Science Shop Wageningen UR
22. Margeta A, Šabalja Đ, Đorđević M (2021) The presence and danger of microplastics in the oceans. Pomorstvo 35(2):224–230
23. Ghaly A, Ananthashankar R, Alhattab M et al (2014) Production, characterization and treatment of effluent textiles: a critical review. J Chem Eng Process Dye: Tartrazine Ind Eng Chem Res 42:243–247
24. Best Management Practices for Pollution Prevention in the Textile Industry, Manual, Published by U.S. Environmental Protection Agency, Office of Research and Development, National Risk Management Research Laboratory, Center for Environmental Research Information, p 299 (1996)

25. Koszewska M (2018) Circular economy—challenges for the textile and clothing industry. AUTEX Res J 18(4). https://doi.org/10.1515/aut-2018-0023 © AUTEX
26. Purvis B, Mao Y, Robinson D (2019) Three pillars of sustainability: in search of conceptual origins. Sustain Sci 14(3):681–695. https://doi.org/10.1007/s11625-018-0627-5
27. Costa C, Azoia NG, Silva C, Marques EF (2020) Textile industry in a changing world: challenges of sustainable development. U Porto J Eng 6(2):86–97. https://doi.org/10.24840/2183-6493_006.002_0008. ISSN: 2183-6493
28. Avadí A, Marcin M, Biard Y, Renou A, Gourlot J-P (2020) Life cycle assessment of organic and conventional non-Bt cotton products from Mali. Int J Life Cycle Assess 25(4):678–697. https://doi.org/10.1007/s11367-020-01731-x
29. Nayak RK, Padhye R (2015) The care of apparel products. In: Sinclair R (ed) Textiles and fashion: materials, design and technology. Woodhead Publishing Series in Textiles, pp 799–822. https://doi.org/10.1016/B978-1-84569-931-4.00031-3
30. Teli MD, Valia SP, Maurya S, Shitole P (2014) Sustainability based upcycling and value addition of textile apparels. In: International conference on multidisciplinary innovation for sustainability and growth, vol 1, pp 91–97. https://globalilluminators.org/conferences/misg-2014-kuala-lumpur-malaysia/misg-full-paper-proceeding-2014/
31. In recent years, several tools and indices have been developed to assess and compare various systems' impact on the environment. These tools or indices include Life Cycle Assessment (LCA), Environmental Impact Assessment (EIA), Strategic Environmental Assessment (SEA), Cost-Benefit Analysis (CBA), Material Flow Analysis (MFA), and Ecological Footprint (Ness et al. 2007; Resta and Dotti 2015)
32. Kumar PS, Suganya S (2019) Systems and models for circular economy. In: Muthu SS (ed) Circular economy in textiles and apparel: processing, manufacturing, and design. The Textile Institute Book Series, pp 169–181. https://doi.org/10.1016/b978-0-08-102630-4.00008-x
33. ISO (International Organization for Standardization) (2006) Environmental management—life cycle assessment—principles and framework. ISO 14040:2006. https://www.iso.org/standard/37456.html
34. Walor C (2008) Can consumers buy responsibly? Analysis and solutions for market failures. J Consum Policy 31:315–326
35. Mapping instruments for Corporate Social Responsibility, European Commission, Directorate-General for Employment and Social Affairs (2003, April)
36. Montero MJ, Araque RA, Rey JM (2009) Occupational health and safety in the framework of corporate social responsibility. Saf Sci 47:1440–1445
37. ABC of the main instruments of Corporate Social Responsibility, European Commission, Directorate-General for Employment and Social Affairs (2004)
38. Contributing to sustainable development: the role of fair trade and nongovernmental trade-related sustainability assurance schemes. Commission of the European Communities Brussels, 5.5.2009 COM (2009)
39. Koszewska M (2010) CSR standards as a significant factor differentiating textile and clothing good. Department of Material and Commodity Sciences and Textile Metrology Analytical Team for the Innovative Products Market Technical University of Łódź. Fibres Text East Eur 18(6(83)):14–19
40. Pandit-Kochan R (2008) Consumer approach to the quality and safety of textile products. Part I. Quality of textile products from the point of view of consumers. Fibres Text East Eur 16(4(69)); Surdykowska B (2008) Kodeksy dobrych praktyk. Kontrola Państwowa, nr 4 (321)
41. Jean-Paul S. Business ethics in the textile, clothing and footwear (TCF) industries: codes of conduct, working papers. International Labour Office Geneva. http://www.ilo.org/public/english/dialogue/sector/papers/bzethics/index.html
42. Cobbing M, Vicaire Y (2016) Timeout for fast fashion. Greenpeace. https://storage.googleapis.com/planet4-international-stateless/2018/01/6c356f9a-fact-sheet-timeout-for-fast-fashion.pdf
43. Kant R (2012) Textile dyeing industry an environmental hazard. Nat Sci 4(1):22–26. https://doi.org/10.4236/ns.2012.41004

44. Loetscher S, Starmanns M, Petrie L (2017) Changing fashion: the clothing and textile industry at the brink of radical transformation. In: Environmental rating and innovation report 2017. WWF Switzerland. https://www.wwf.ch/sites/default/files/doc-2017-09/2017-09-WWF-Report-Changing_fashion_2017_EN.pdf
45. Morgan LR, Birtwistle G (2009) An investigation of young fashion consumers' disposal habits. Int J Consum Stud 33(2):190–198. https://doi.org/10.1111/j.1470-6431.2009.00756.x
46. Coste-Maniere I, Croizet K, Sette E, Fanien A, Guezguez H, Lafforgue H (2019) Circular economy: a necessary (r)evolution. In: Muthu SS (ed) Circular economy in textiles and apparel: processing, manufacturing, and design. The Textile Institute Book Series, pp 123–148. https://doi.org/10.1016/B978-0-08-102630-4.00006-6
47. Chowdhury NR, Paul SK, Sarker T, Shi Y (2023) Implementing smart waste management system for a sustainable circular economy in the textile industry. Int J Prod Econ 262:108876. https://doi.org/10.1016/j.ijpe.2023.108876
48. Pandit P, Nadathur GT, Jose S (2019) Upcycled and low-cost sustainable business for value-added textiles and fashion. In: Muthu SS (ed) Circular economy in textiles and apparel: processing, manufacturing, and design. The Textile Institute Book Series, pp 95–122. https://doi.org/10.1016/b978-0-08-102630-4.00005-4
49. Islam M, Mostafa M (2018) Textile dyeing effluents and environmental concerns: a review. J Environ Sci Nat Resour 11:131–144
50. Siddique K, Rizwan M, Shahid MJ, Ali S, Ahmad R, Rizvi H. Textile wastewater treatment options: a critical review. https://doi.org/10.1007/978-3-319-55423-5_6 © Springer International Publishing AG

Sustainable Textile Production in the Developing Countries: Major Challenges or Concerns and Their Possible Solutions

Priya, Shiwani Berry, and Pinki Rani Agrawal

Abstract Sustainable textile production has become a major concern in developing countries due to its huge contribution to environmental pollution. The textile industry continuously produces materials like cloths, bags, geotextiles, nets, rugs, etc. and the use of such material results presence of pollutants in the environment as heavy metals, soda, acids, dyes, aromatic hydrocarbons and other toxic substances, which result in water pollution, soil pollution, greenhouse gas emission, etc. To minimize such hazardous effects, presently this industry mainly focuses on the development of sustainable products. Some important components which are mainly focused on achieving sustainability are the use of suitable raw material (ecodesign), eco-friendly methods of production and packaging, life cycle assessment (LCA) of outcome and possible recyclability of used material/products, corporate policy and environmental management system (EMS), innovation in the business model. In developing countries new business model introduced in the textile industry, 3R's model: reduce, reuse, repair, which aims towards sustainability, includes less waste generation, environmentally friendly, enzymatic textile processing and storage and supply chain management. This chapter focused on the key factors involved in the production of sustainable textile products in developing countries.

Keywords Sustainable · Textile industry · Eco-friendly · Packaging · Developing countries

Priya
Government Degree College, Nagrota Surian, Kangra 176027, India

Priya · S. Berry (✉)
Department of Chemistry and Chemical Sciences, Central University of Himachal Pradesh, Shahpur, Kangra 176206, India
e-mail: shiwani.berry@hpcu.ac.in; shiwanipu@gmail.com

P. R. Agrawal
Deenbandhu Chhotu Ram University of Science and Technology, Murthal, Sonipat, Haryana, India

P. Singh (ed.), *Dye Pollution from Textile Industry*, SDGs and Textiles,
https://doi.org/10.1007/978-981-97-5341-3_13

1 Introduction

Textile production increased tremendously in past years to manage the production with an increase in the population globally, due to the new innovations this approach is beneficial also in developing countries due to low cost. Production of textiles in developing countries comes from Western countries which results in pollution but now its production approaches low cost and clean textile production. After a long journey textile production presently reaches developing countries such as Myanmar, Kenya, Ethiopia, Nicaragua and Madagascar. Raw materials for the textile industry such as paints, coatings, synthetic fibres, adhesives, synthetic drugs, perfumes and benzene (in dyes, pigments) as finishing agents.

Globally the textile industry provides employment to Tens of millions of people, after water and food this sector supplies the second most essential everyday necessities to the people's [1, 2]. The textile business has an extensive supply chain that starts with the production of fibre and ends with the making of clothing. According to the new global sustainability strategy for 2030, which contains 17 Sustainable Development Goals (SDGs). The SDGs clearly state the new goals of social, environmental and economic development, including promoting economic growth, ending poverty and safeguarding the environment. Lately, many technologies involved in sustainable textile production some of them are plasma technology, enzyme processing, ozone bleaching, supercritical carbon dioxide dyeing, ultrasonic treatments, nanotechnology, foam technology, electrochemical dyeing, laser technology, digital printing and even some advanced materials which are reusable (biodegradable material such as wool, cotton, silk and flax, recyclable nylon and polyester). Socially, the textile industry faces many problems that usually arise in the supply chain, where problems like child labour, low-wage labour and other immoral practices are common. About 80% of the entire finished textile industry production is transferred from developing to industrialized nations. Low labour cost, as well as less strict labour relations and environmental laws compared to established economies attracts manufacturers in the developing countries [1]. The textile industry uses a lot of energy and gives rise to a lot of pollutants, which raises a number of environmental problems. Many chemicals and fuels are used in this industry along with water as necessary resources, around the world about three trillion gallons of fresh water are annually used to manufacture fabrics [2]. Due to these various problems, many researchers have undertaken a range of studies in an effort to make the textile industry more ecologically friendly for example the handling of related pollutants [3, 4], the production process [5], obstacles faced [6], corporate social responsibility [7], product design [8] and the circular economy [1], are few examples of such difficulties. The textile industry is widely recognized for its numerous beneficial effects on the economy, but it is also well-known for its significant environmental and social problems with regard to long-term sustainability [9]. A variety of products is produced in this industry such as automotive textiles, industrial textiles, agrotextiles, geotextiles, protective textiles, sport textiles and medical textiles [9, 10]. Even though the textile industry is widely known for its ongoing need for innovation, there aren't many studies that look at the sustainability issue in industrial textiles through the lens of sustainability innovation. Innovation in the textile industry has long been a part of the industry and is important to other industries as well as the textile business [11].

The textile treatments and dyeing processes are responsible for approximately one-third of the chemicals released into the atmosphere. Despite the pandemic's disruption, the demand for textile products is expected to climb in the coming years, especially given the growth of e-commerce and digitization as well as consumer demand for a wide range of products from this sector. Globally in 2021, the textile industry's worth will be more than USD 1.04 trillion [1].

Developing nations such as India should not become complacent about luring and taking in Foreign Direct investment (FDI), particularly in the textile industry where there is a clear and pressing demand for state-of-the-art machinery. The primary factor influencing Bangladesh's export supply appears to be its gross capital formation. Because gross capital formation plays a major role in Bangladesh's export supply augmentation, increasing investment in the exportable sector could have a major impact on this sector [12].

2 Present Market of Textile Industry in Developing Countries

Barriers such as Tariff and non-tariff which affect the textile industry tariff barriers are visible and straight barriers and non-tariff barriers are non-visible and more elusive. The economy of developing countries in international trade is affected by the non-tariff barrier effect on the export market. The Indian clothing and textile industry is affected by more than 60% non-tariff barriers in the USA, Canada and EU-25 [12]. Due to Europe's high demand for fashion, many developing nations today have booming garment, textile and footwear sectors. For example, the economy of Cambodia expanded in past years due to an increase in exports. The textile industry employs about one million people which is more than a tenth of the working force and generates around 11% of the GDP of the nation. Other countries such as Myanmar and Bangladesh supply goods to the EU and their economy is rising fast [13]. With a market valuation of more than $1.5 trillion and over 75 million employees, the textile sector is an essential component of our economy. However, the supply chain does not follow the stringent guidelines established by the UN in 2016 [1].

Conversion of complex yarn and fibres into useful fabrics via using various processes leads to its use in various industries such as automobile industry textiles, medical textiles, industrial textiles, geotextiles, protective textiles, sport textiles and agro textiles. Although the pandemic has impacted the textile industry negatively but still it will expand very fast due to e-commerce and digitalization which help to meet the demand of customers within less amount of time which is the key role in textile commercialization [13].

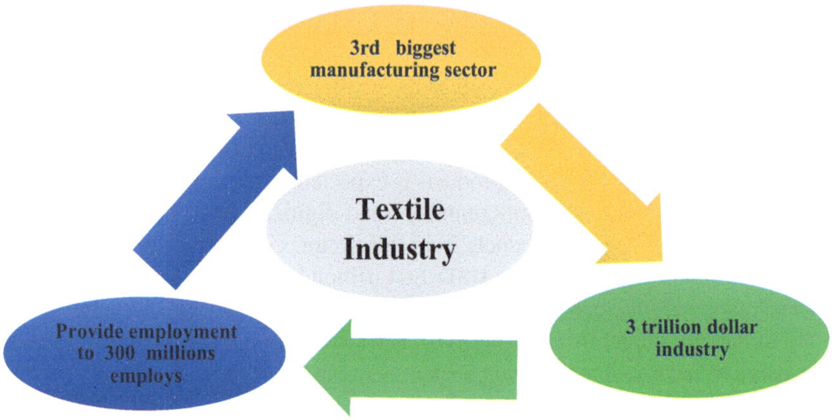

The textile industry emits about 3–10% of global carbon emissions yearly [14]. Production of textiles to provide the most required products within less time and with less harmful (degradable) waste are the basic need presently in this industry to meet sustainable production. World Index of the Most Sustainable Companies, which includes producers pre-selected according to the outcomes of four screening techniques that evaluated twelve critical performance indicators of environmental accountability (energy and resources consumption efficiency) and community involvement (employee turnover and diversity in leadership), announced the release of the Global 100 by Canadian communication and research firm Institutional Knights [15]. Currently, the score of sustainability in the fashion industry is 32. According to a UN study, a continuous increase in the production use of up to 26% of the global carbon will result in a 2 °C rise in the temperature by 2050 [14]. Currently, Asia is the centre of the textile industry globally, many Asian countries such as India, Bangladesh and China are the biggest textile producers but face many problems regarding the business [1]. Although all these concerns did not limit its production, according to a report the estimated projected global textile market growth is US $1.87 billion till 2029 with 12.90% annual growth [16]. According to UN estimates, the world's population will reach 8.5 billion by 2030, resulting in an increase in the need for the fashion industry from 62 million tonnes to 102 million tonnes. Europe is the fashion industry hub of twenty fashion stores, followed by the US with eight and Asia with one each. This illustrates the scope and complexity of the roles played by nations, individuals and businesses, in the global textile and garment sector. Companies such as GAP, H&M, SHEIN and Zara play a major role in the clothing industry. But follow an unethical path to increase the production and supply of the textile material i.e. clothing [17].

3 Challenges Faced by Textile Industry in the Developing Countries

The observed shift of chemical production to the developing nations from the developed countries, due to less enforcement and more flexible rules and regulations, was primarily observed during the Anthropocene period [18]. But these changes lead to some serious problems by affecting the sustainability, as mentioned below.

3.1 Lack of Scientific Knowledge

Lack of scientific knowledge which works for sustainable and green chemistry results in serious concern because the textile industry is involved in water consumption, energy consumption and gas emissions. Research is not supported properly for the green and sustainable practices is also an issue in this area. Scientific researchers who follow the sustainable approach do not communicate or interact directly with the government and industries leading to non-favourable results [18]. Numerous environmental issues are brought up by the textile industry's high energy consumption and pollution output [2]. According to a report the emission of greenhouse gases from this industry may increase >50% by 2030. Out of the total production only <1% of the textile waste is reused and the major part about >100 billion is still discarded per year [16]. In the absence of strong and strict environmental and social regulations in the textile industry, the production and manufacturing process of the textile industry consumes high energy, water and chemicals which cause adverse effects on the environment, by contributing significantly to greenhouse gas (GHG) emissions. Such GHG emissions are needed to reduce by extensive transportation but it also contributes significantly to carbon emissions. Uses of synthetic chemicals (with stain- and water-repellent qualities) such as perfluoroalkyl and polyfluoroalkyl substances (PFAS), result in immediate concern due to their long-term presence in the environment, which contaminate the groundwater and lead to negative health effects [19].

The textile sector uses a lot of chemicals, energy and water as essential resources; globally, the production of textiles uses over three trillion gallons of freshwater yearly [2]. According to a report about 92 billion tonnes of textile products (garments) are discarded in landfills per year. Out of the total >50 million metric tonnes of the fibres involved in cloth production are plastic based and contribute about 63% of the total amount. Although the clothes are also made from natural fibres (cellulose based: cotton, jute and linen) and semisynthetic fibres (viscose and rayon) which are mostly derived from natural gas, oil or coal and contribute about 33% of the clothes production [16]. Textile workers have frequently highlighted high material consumption, waste generation and wastewater issues amongst other environmental issues [9]. Both natural and synthetic textiles dyed by using chemicals which are discharged into the water, result in contamination of water by increasing the concentration of heavy metals and numerous other chemicals which can cause hazardous effects on the health of living beings and pollute the environment [20]. Although new trends like eco-friendly brands enter the textile industry. High clothing companies such as H&M and Zara are one of them. The eco-friendly tag is given to the products with 50% of sustainability, but lack of knowledge leads to unsustainable textile production resulting in hazardous effects on the environment [17].

The COVID-19 pandemic's adverse effect is also seen in the textile industry, about 60% of the textile consumption increased which leads towards harmful effects on the environment. Many developing countries face the effect of post-COVID adverse effects on the textile industry such as Bangladesh, where the readymade garment (RMG) sector supply chain disruption results in a rise in production costs [21]. Annually 1.7 billion tonnes of greenhouse gases are emitted and about 9% of the microplastic is discharged into the oceans [16], which is the result of the lack of scientific knowledge and has to be provided to industry to minimize its harmful effects.

3.2 *Free Trade Agreements, Multi-Fibre Agreements and Quota System Phase-Out*

Free trade agreements, multi-fibre agreements and quota system phase-out in the textile industry are mainly studied in developing nations such as Turkey, India, Bangladesh and Sri Lanka. According to a study on the textile industry in Bangladesh, large textile companies stand to gain the most from the MFA phase-out since they have the capacity, whereas small and medium-sized businesses are more at risk because they are having a harder time surviving the fierce competition. The percentage growth of exports in India during the pre-Multi Fibre Agreement (MFA) era increased at a declining rate year over year. However, in the year 2005–06, when the MFA was eliminated a sharp rise in export performance was observed. In Turkey's textile industry, free trade agreements (FTAs) affect intra-industry trade, as it has good performance in the textile industry. From the late 90s, the garment industry has grown to be the main export sector in Sri Lanka. According to a study, eliminating quotas will result in the closure of almost 50% of the world's textile manufacturers since they will no longer have that financial security. Still, it's anticipated that a few medium- and large-scale factories will make it by seizing chances presented by a free market [12]. Studies on the role of global buyers have also been conducted; these emphasize that local businesses operating in global value chains have limited upgrading capacity due to the power imbalance between them and their lead corporations. Upgrading is hindered by the limited information flow that local enterprises receive from their foreign clients. Certain studies assess how local firms emphasize and improve their agency in developing their own capabilities, learning and advancing technology. Local companies take an active role in knowledge absorption and upgrading by developing human resources, building connections with suppliers and component makers, and gathering market intelligence. The four case studies demonstrate how important a role foreign investment may have in upgrading. Although it has been difficult to raise the value chain, foreign direct investment (FDI) has been essential in starting the garment sector in Lesotho, Bangladesh, Cambodia and Madagascar. The primary reason Asian multinational businesses have invested in disembodied production units is usually to gain preferential market access and this expenditure has not resulted in any backward integration or upgrades [22].

3.3 *Labour System in the Developing Countries*

Labour system in developing countries faces many problems both socially and economically. The Bangladesh factory tragedy discloses crucial social issues that remain unnoticed in the textile sector and promotes public awareness of topics including worker rights, compensation, health and safety, and other problems [23]. On a social scale, the textile industry deals with problems that occur usually in the supply chain, where concerns like child slavery, low-income labour and other immoral practices are common [24], child and woman labour are affected mainly in this industry. However, this problem is seen in the manufacturing nations frequently due to a lack of laws pertaining to child labour, minimum wage requirements or occupational health and safety [19]. Manufacturers find developing countries to be attractive locations for production due to their low labour costs, as well as less strict labour relations and environmental laws compared to established economies. Many markets prohibit the use of child labour, especially those in Europe and North America. Since labour costs are cheaper in developing nations than in developed ones, and because labour relations and environmental regulations are laxer there, manufacturers find these regions to be appealing places for manufacturing. Many markets, particularly those in North America and Europe, forbid the use of child labour [25]. This industry also engages in other unethical practices, such as acquiring raw materials without considering environmental responsibility and social [24]. Thus, developed countries like the European Union, Japan and the United States became the consumers of 58.1% of global clothing, whereas developing countries like China, Vietnam and Bangladesh became the exporters of 43.8% of global clothing. However the waste generated from the textile industry results as a burden to the developed as well as developing countries. In developing countries production, processing, solid waste and wastewater result in a negative impact on the environment more than the developed countries. However, the garbage produced during production and transportation puts a strain on developed nations. The impact of the global apparel sector on climate change rose by 35% in 2005–2016 and is expected to rise by 49% between 2016 and 2030 if business continues as usual. Suppliers pushed for low labour costs and cheap manufacturing conditions in developing nations as a result of retailers' unreasonable demand for inexpensive production with shorter lead times, which resulted in catastrophic events like in Bangladesh the collapse of the Rana Plaza and similar incidents in India and Pakistan [13, 14, 26]. Countries such as India, Bangladesh and Ethiopia deployed labour from other places due to less mandatory labour laws. Cheap labour with overworking result increase in production and due to low wages companies or factories increase their production. According to the National Labour Committee Report 2001, young women between the ages of 16–25 made up 85% of the garment workforce and were overworked but paid minimum wages of USD 64/ month [17].

Previously, the 3R (reduce, reuse, recycle) framework represented the circular economy in various industries after that 6R (reuse, recycle, refuse, repair, rethink, recover) and then 7R (rethink, reduce, repair, reuse, refurbish, recycle, recover) came to existence but in the textile industry, the circular economy follows 9R framework as R0–R9 (reuse, reduce, refuse, rethink, repair, refurbish, remanufacture, repurchase, recycle, recover) [6].

3.4 Involvement of Media in the Textile Industry

The media has covered scandals involving big businesses that contract out their manufacturing to underdeveloped nations, violating human rights and causing harm to the environment and society at large. Companies' corporate reputations have suffered as a result of suppliers, unethical behaviour poisoning the entire supply chain. It is a mystery to certain corporations where their clothing is manufactured. Since most fashion firms do not own their manufacturing facilities, it is challenging to keep an eye on or regulate working conditions along the entire supply chain. One supplier receives the orders and then contracts out the work to other factories. Being transparent is a significant problem since supply chain employees become invisible. Should the manufacturer desire to tackle sustainability concerns within the supply chain [24].

Digitalization increases options but some policies such as no return policies and verities in style and sizes of the textile products result in easy access to the product consumption but at the end, we are left with only waste. The waste is the result of the cheapest material consisting of fabrics such as polyurethane, rayon and synthetic fabrics that cause environmental hazards such as deforestation and different kinds of pollution [27].

3.5 Corrupted or Unethical Behaviour in the Textile Industry

There has been a general tendency amongst economists to view corruption in the public sector as being more problematic than that in the private sector, as though one were an exception. Because of their self-legitimizing morality, carelessness and reliance on the ethical authority of the private sector, TNCs take advantage of this to avoid public scrutiny [12]. Aiming to exploit developing nations' underdeveloped labour forces and material resources, laws that support commercial development are designed to obscure the true requirements of society. The fact that TNCs deploy CSR procedures only as a business tactic is well known [28]. Since these firms have an influence on the environment, they fail to adopt effective environmental protection measures that are commensurate with those effects. In accordance with the United Nations Economic Commission for Europe [7]. Some factors such as costly production, shortage of manpower are responsible for increasing corruption in both local and international textile industries [29]. Some investigations show that the most commonly corrupted industries are petroleum, food, alcohol, clothing, leather material and transportation material (equipment and construction). Developing countries such as India, Bangladesh and Vietnam supply basic clothing materials that are cheap and readymade, resulting in the mostly corrupted clothing industry. Such corruption leads to major life risks and the most common example of it is the fire in Rana Plaza, Bangladesh, where many lives are taken in a garment factory [30]. Corruption is a major concern presently in developing countries, the measure of corruption in particular countries is found by the corruption perception index (CPI) annually published by transparency international [31]. According to the Transparency International Corruption, 2018 index, Bangladesh is at 149 rank out of 180 in corruption. All the local factories are controlled by their owner's leads to bankruptcy due to a bad working environment (overworking, wages cutting, etc.) and bad leadership [17]. For example, the 2020 Corruption Perceptions Index (CPI), released by Transparency International in 2021, indicates that, in contrast to rich economies, corruption is far more common in the least developed countries (LDCs) [32]. Simultaneously, there has been a prevalent tendency amongst economists to view corruption in the public sector as more relevant than that in the private sector, as though one might be considered a unique situation. TNCs take advantage of this situation to avoid being scrutinized by the public since they have a self-legitimizing morality, they are not liable, and they use the private sector's ethical authority. The goal of legislation that supports commercial development is to exploit developing nations' material resources and labour power while overshadowing the actual requirements of society. It is often known that TNCs view CSR procedures as merely another business tactic [7].

4 Possible Solutions to Control the Negative Part of the Textile Industry

4.1 Sustainable Textile Production

The textile industry is already dealing with the need for sustainability to limit environmental degradation and human or natural resource exploitation in order to meet consumer demands. For fabric fibres production 99% of the chemicals are used which are recycled in the closed loop production cycle, such fibres are Lyocell, manufactured from the bamboo cellulose [20]. Studies show that about 2700 L water is needed for the manufacturing of only a single cotton shirt and its degradation time is

200 years. So its use, reuse and recycle loop is needed to build sustainable production technology. Developed countries such as Canada, Norway, the UK, Germany, the USA, Singapore, Brazil and Australia already approach sustainable production [33]. Such sustainable production policies should be considered by developing countries so that they can minimize the impact of the waste generated from the textile industry and provide the material with good sustainability and environment-friendly material.

Some fashion brands are looking into ways to make their fabrics easily biodegradable after they are disposed of by using natural textile waste resources like wood, fruit, cactus and others. There are more ways to lessen clothing's carbon footprint [33]. By maximizing resource utilization, the circular textile value chain enables a self-sufficient and environmentally responsible manufacturing strategy. By utilizing the disposable nature of textiles, it enhances up-cycling and recycling by changing design, collection and reprocessing [6]. Many well-known companies approach the production of various products from recycled fibres for example: Econyl® Regeneration System for recovering Nylon 6 using both pre-consumer and post-consumer waste was introduced by Aquafil (Trento, Italy) and through a fibre-to-fibre closed loop technique based on chemical recycling, Teijin (Tokyo, Japan) has created Ecocircle®, which is beneficial in the markets for men's and women's sportswear. Post-consumer cotton clothing is recovered by Evrnu (Seattle, WA, USA) to create high-standing bio-based fibres. Bottles, tyres, plastic, fishing nets and other waste materials were transformed into coats, shoes and bags by EcoAlf (Barcelona, Spain). Textile trash is acquired by Martex Fiber (Spartanburg, SC, USA), which produces recycled cotton textiles (ECO2Cotton®) from post-industrial waste derived from cut-and-sew operations in the apparel manufacturing industry. The "Garment Collecting Initiative" was introduced by H&M (Stockholm, Sweden) in 2013 when it introduced denim styles composed of recycled cotton fibre from the clothing that was gathered. Later, in 2015, the company Aquafil (Trento, Italy), in partnership with Speedo (Nottingham, UK), promoted the recovery of fabric waste from cut-and-sew manufacturers that are transformed into the synthetic fibre Econyl®. The North Face (San Francisco, CA, USA) used three eco-friendly elements for the 2015 Denali jacket launch: yarns reused from leftover fabric and recycled plastic bottles. According to a study laminated composites created from cotton fibrous waste by introducing an unsaturated polyester resin, such composites were then compared to glass fibre-based composites in terms of mechanical resilience. The recycled reinforcements that were used came from two distinct sources: comber noil and knitting waste from spinning [4].

4.2 *Waste Management*

The total waste production increased tremendously in the past few decades. Studies show that from 1970 to 2018 municipal solid waste increased by 232% and textile trash climbed by 868%. The increase in textile waste by 55% [19] increased the concern and led the industry to find the solution to the problem. The concept of waste reduction introduced the processing of waste to reuse and recycle in the textile industry, which helps to lower the environmental load caused by the high consumption of water, energy and material during production and transportation. For example, during the cutting process, 15% of the fabric is wasted, and the waste generated by the textile industry ends up in landfills, and the amount increases with increased fashion demand. Every year about 100 billion garments are manufactured and disposed of as waste. Even daily clothes washing results in the release of fibres in the water bodies. Recycling or up-cycling of fashion waste offers various environmental benefits such as reducing greenhouse gases, polyester fibres and dyes introduction in the environment. Most of the waste, about 15% is reused, 95% is recycled and other amounts left are discarded [33]. Waste from aqueous systems (from leakages, faulty valves), cutting room fibres, bundling room Fabric pieces, waste from the production floor and waste chemicals (toxic chemicals) are the main wastes generated from the textile industry which are needed to be managed urgently, which is possible by following the 3R's: Reduce, Reuse and Recycle [34]. The possible ways to reduce the pollution created by the waste generated from the textile industry by using low polluting technologies, recycling waste before eliminating it in the environment and treatment of the effluents discharged from factories, etc. [35]. Briefly described methods to approach textile recycling are: (i) repurposing the products into industrial recycling scraps in their original form, (ii) Shredding and cutting of the waste textile combine with the new material via different methods (re-extruding, melting and re-blending), (iii) hydrolysis and pyrolysis involve in the conversion of waste material monomers, fuel and other chemicals, (iv) burning of waste material to recover heat and energy [4].

4.3 Consumer Behaviour

Post-consumer fashion wastes are the main concern in the present time globally about 39 million wastes are generated per year, and out of that about 57% go to landfill. Most of that is degraded and some of that undergo recycling or upgrading and some of it is destroyed. By following some important rules the waste generation by fashion industry can be reduced, such as (i) enhancing decision-making and accurate assessment with suppliers to reduce errors that encourage waste throughout creation, (ii) encouraging buyers to reuse textile products (clothes and other materials) through up-cycling initiatives, design trades, article assortment emphasis and material-reusing bearings, (iii) taking a "support-to-support" or whole lifecycle approach to item planning, creation and promotion, as well as identifying areas where waste may be reduced and resources can be recycled [36]. The various surveys show that the factors such as income, gender, marital status, living arrangement and type of housing, influence the recycling of textile products such as shoes and clothes [37]. Consumer behaviour is the main factor which affects the clothing industry the most because it is based on the demand of the consumers. Most consumers want low-priced or cheap products, and even ignoring the quality of the product is not considered. But presently it is found that consumer behaviour is more environmentally friendly so they approach sustainable product buying. From the time when sustainability entered in the textile industry, it started attracting the attention of consumers, they were interested in buying organic clothes or even clothes from recycled waste in the clothing industry, such interest led to great change in this industry [2]. Many international brands such as People Tree, Stella McCartney, Patagonia, Thought, Rent the Runway, Indigenous and H&M, have actually shifted their attention to approaching fashion which takes ethics, customers and the environment into consideration. Clothing designed and sold by Rustic Hue and similar but small companies such as Doodlage, Bodice, Raw Mango, 11.11 and [Ka] [Sha], approach ethical and fair methods by providing the workers with safe working experience, collaborations with craftspeople and handloom weavers to produce handcrafted goods, to obtaining organic or up-cycling post-production and post-consumer wastes and even using recycled raw materials [38].

4.4 Ethical Approaches

Unethical approaches already cause great damage in the apparel industry. Even the Construction and demolition waste management is not able to follow ethical behaviour in the textile industry due to the rigid legal framework, a lack of political will, a negative attitude and a lack of policies and even create barriers to zero waste generation in the construction industry [36]. According to Arangdad et al., the design industry is worth $2.4 trillion dollars which extends continuously and is expected to expand in the future also [37]. Trends in clothing are changing due to the development of cooperative actions and research, including effort occasions, strengthening synergies and knowledge sharing which help in understanding and developing the need for ethical fashion. In order to produce clothing that is usually used only once for a short period of time, the existing design framework uses large volumes of finite resources, such as oil, which are then typically lost to inland landfills or burned, so the present demand is to minimize the use of synthetic and chemical colouring substances. As we see that ethical practices in the textile industry are very much needed and many famous fashion brands focus on ethical practices in the industry such as Adidas have developed collections by using plastic bags from ocean waste. Similarly, Zara also concentrated on high-street fashion and plans to transform its shop into a sustainable one by 2025 [36]. The ethical fashion brand Conserve India, New Delhi, buys all types of plastic waste from waste pickers (about 80%) and has converted 12,000 tonnes of waste into wallets and belts that are globally sold in fair trade stores. Waste materials used include fire hoses, rice bags, seat belts, cement bags, tyre tubes and other recyclable packaging materials. Aeon Row uses recycled yarn to make their range, which has fantastic essentials and an even better concept. They use repurposed textiles and don't need chemicals, water, land or cotton dyes [38].

4.5 *Media Influence on Textiles*

Along with the negative effects of social media on the textile industry, good impacts are also seen. The adaptation of 3DVD technologies (involving complete digitalization) results in modification in fashion industry supply chains. Some of the positive impacts are reduction in costs, creative designing and fast production, such productions involved in designing products are as per users' priorities by increasing the perfection, and fast production of the customized and complicated material [39]. An example of fast production is a clothing empire named Shein, with 23 million followers on Instagram have 2nd ranking in the shopping Apple store app, further, it is seen that many scandals of design stealing have been seen [27]. After the conduct of the investigation, it was found that the scholarly literature and data interpretation gathered elaborated on the effects of social media on the fashion industry. A few guidelines for action include maintaining a regular online presence to build a brand's trust, reaching out to the fashion industry, a professional approach for potential partnerships and effectively communicating with customers by keeping up with the newest online trends and coming up with creative marketing campaigns. Media influence the fashion industry by maximizing its reach to the major part of the population by blogging, promotions, fashion influencers, etc. [40], which also leads to fast and mindless purchasing results as waste. Negative approaches in marketing by promoting sustainability via using some phrases and words such as 'made with love', 'fair' and 'green' to increase sales. The durability of such products is less and even non-repairable resulting in an unsustainable approach which leads to harmful effects [19, 41].

All the aspects of the textile industry are studied in this chapter and it is found that our day-to-day life is influenced or impacted deeply by both the positive and negative nature of this industry. The concern to approach sustainable textile production is now starting to increase and maybe we can see the completely sustainable textile industry in the next few years.

5 Conclusion and Future Scope

The textile industry in developing countries such as India, Sri Lanka, Bangladesh and Pakistan continuously focuses on increasing the industries to strengthen their economy. For this aim, they approach sustainability but also deal with the various hurdles. This chapter represents the present global market of the textile industry along with facing problems like lack of interest on the scientific scale, labour system in developing countries, corrupted or unethical behaviour in the textile industry, involvement of media in the textile industry and free trade agreements, multi-fibre agreements and quota system phase-out. Determining the social hotspots associated with the textile sector; determining the existing impacts on society managed by the textile companies belonging to different life cycle stages. In future, the studies concentrate on the unstudied elements responsible for the adverse effect on the textile industry.

For this type of research, variables including geographic, political and legal environments will also be considered. People utilize social media, which is a collection of online platforms and communication channels, to share media, profiles, observations, assumptions, perspicacity, anxiety and information. It promotes interaction and communication between different international communities. Corruption in the textile industry has a negative impact on the industry as well as on the life of the workers. Corruption in the textile industry occurs due to the overexploitation of the workers, by overworking regularly and low wages or income provided by the corrupted companies or factories. Such problems need to be solved urgently so that the manual power of the textile industry is maintained and the production increased.

In reality, various studies indicate that "health and safety" is one of the textile industry's social hubs. However, controversy occurs when businesses pay less attention to other worker-relevant concerns during the most crucial stages, such as Fabric production, collective bargaining or salary assessment. In a field where working in minimum wage nations with a high probability of uneven labour standards limits workers' capacity to assert their rights and diminishes their ability to empower, collective bargaining mishandling can have repercussions that extend beyond workers' employment need to be managed properly.

This study finds that we need to focus on the sustainable and green synthesis of textile products so that the industries, government and other possible areas can accept the new changes and can produce the products as much as required. And in the future, it is suggested that new innovations can lead us to produce as much as production to reach the highest value and help in the increasing economies of developing countries.

References

1. Harsanto B, Primiana I, Sarasi V, Satyakti Y (2023) Sustainability innovation in the textile industry: a systematic review. Sustainability 15(2). https://doi.org/10.3390/su15021549
2. Desore A, Narula SA (2018) An overview on corporate response towards sustainability issues in textile industry. Environ Dev Sustain 20(4):1439–1459. https://doi.org/10.1007/s10668-017-9949-1
3. Pattnaik P, Dangayach GS, Bhardwaj AK (2018) A review on the sustainability of textile industries wastewater with and without treatment methodologies. Rev Environ Health 33(2):163–203. https://doi.org/10.1515/reveh-2018-0013
4. Patti A, Cicala G, Acierno D (2021) Eco-sustainability of the textile production: waste recovery and current recycling in the composites world. Polymers (Basel) 13(1):1–22. https://doi.org/10.3390/polym13010134
5. Rahman M, Billah MM, Hack-Polay D, Alam A (2020) The use of biotechnologies in textile processing and environmental sustainability: an emerging market context. Technol Forecast Soc Chang 159:120204. https://doi.org/10.1016/j.techfore.2020.120204
6. Kazancoglu I, Kazancoglu Y, Yarimoglu E, Kahraman A (2020) A conceptual framework for barriers of circular supply chains for sustainability in the textile industry. Sustain Dev 28(5):1477–1492. https://doi.org/10.1002/sd.2100
7. Luque A, Herrero-García N (2019) How corporate social (ir)responsibility in the textile sector is defined, and its impact on ethical sustainability: an analysis of 133 concepts. Corp Soc Responsib Environ Manag 26(6):1285–1306. https://doi.org/10.1002/csr.1747

8. Salo HH, Suikkanen J, Nissinen A (2020) Eco-innovation motivations and ecodesign tool implementation in companies in the Nordic textile and information technology sectors. Bus Strateg Environ 29(6):2654–2667. https://doi.org/10.1002/bse.2527
9. Roy M, Sen P, Pal P (2020) An integrated green management model to improve environmental performance of textile industry towards sustainability. J Clean Prod 271:122656. https://doi.org/10.1016/j.jclepro.2020.122656
10. Sun Z, Sun L, Strang K (2018) Big data analytics services for enhancing business intelligence. J Comput Inf Syst 58(2):162–169. https://doi.org/10.1080/08874417.2016.1220239
11. Scott R (2018) Innovation by design. IHS Jane's Def Wkly 55(44):28–32. https://doi.org/10.4324/9780429402326-12
12. Yoganandan G, Jaganathan AT, Saravanan R, SenthilKumar V (2013) Factors affecting the export performance of textile industry in developing countries—a review of literature. 2(4):2319–2828
13. Martin R (2020) Textile workers in developing countries and the European fashion industry towards sustainability? 1–8
14. Patwary S (2020) Clothing and textile sustainability: current state of environmental challenges and the ways forward. Text Leather Rev 3(3):158–173. https://doi.org/10.31881/TLR.2020.16
15. Islam MT, Jahan R, Jahan M et al (2022) Sustainable textile industry: an overview. Non-Metallic Mater Sci 4(2). https://doi.org/10.30564/nmms.v4i2.4707
16. Campina (2022) Strengthening sustainability fundamental. Published online
17. Guido A (2023) Bangladesh and the fast fashion industry: injustices in the east & profits in the west. 5(1)
18. Barra R, González P (2018) Sustainable chemistry challenges from a developing country perspective: education, plastic pollution, and beyond. Curr Opin Green Sustain Chem 9:40–44. https://doi.org/10.1016/j.cogsc.2017.12.001
19. Schumacher KA, Forster AL (2022) Textiles in a circular economy: an assessment of the current landscape, challenges, and opportunities in the United States. Front Sustain 3. https://doi.org/10.3389/frsus.2022.1038323
20. Bick R, Halsey E, Ekenga CC (2018) The global environmental injustice of fast fashion. Environ Heal A Glob Access Sci Source 17(1):1–4. https://doi.org/10.1186/s12940-018-0433-7
21. Biswas MC, Samit Chakraborty CS (2020) Impact of COVID-19 on the textile, apparel and fashion manufacturing industry supply chain: case study on a ready-made garment manufacturing industry. J Supply Chain Manag 3(2):181–199
22. Calabrese L, Balchin N (2022) Foreign investment and upgrading in the garment sector in Africa and Asia. Glob Policy 13(S1):34–44. https://doi.org/10.1111/1758-5899.13059
23. Tyagia GK (2023) Recent sustainable development and innovations in textile industry. Trends Text Eng Fash Technol 8(2):941–943. https://doi.org/10.31031/tteft.2023.08.000684
24. Guarnieri P, Trojan F (2018) Decision making on supplier selection based on social, ethical, and environmental criteria: a study in the textile industry. Resour Conserv Recycl 2019(141):347–361. https://doi.org/10.1016/j.resconrec.2018.10.023
25. Loo BPY (2002) The textile and clothing industries under the fifth Kondratieff wave: some insights from the case of Hong Kong. World Dev 30(5):847–872. https://doi.org/10.1016/S0305-750X(02)00005-0
26. Chen X, Memon HA, Wang Y, Marriam I, Tebyetekerwa M (2021) Circular economy and sustainability of the clothing and textile industry. Mater Circ Econ 3(1):1–9. https://doi.org/10.1007/s42824-021-00026-2
27. Payne K (2022) The slow burn of fast fashion: unsustainable practices within the industry's supply chain. University of South Florida Katelyn Payne Spring. https://www.usf.edu/business/documents/undergraduate/honors/thesis-payne-katelyn.pdf
28. García-Rodríguez FJ, García-Rodríguez JL, Castilla-Gutiérrez C, Major SA (2013) Corporate social responsibility of oil companies in developing countries: from altruism to business strategy. Corp Soc Responsib Environ Manag 20(6):371–384. https://doi.org/10.1002/csr.1320
29. Dalal H (2019) Challenges: a study of textile industry in India. 9(5):423–429. https://pramanaresearch.org/

30. Xiao Y, Lenzen M, Benoît-Norris C, Norris GA, Murray J, Malik A (2018) The corruption footprints of nations. J Ind Ecol 22(1):68–78. https://doi.org/10.1111/jiec.12537
31. Rapp WV (2020) The darkside of pursuing growth in the product cycle—corruption, graft, and bubbles. (376). https://doi.org/10.7916/d8-1d2e-wr91
32. Guo S, Choi TM (2023) Risk management for second-hand clothing imports in least developed countries: legislations and perception of public-sector corruption. Risk Anal 1–18. https://doi.org/10.1111/risa.14134
33. Marinova V, Radev R (2023) Sustainability of the European textile industry. 20:1–13
34. Patel M, Sahu A, Rajak R (2022) Solid waste management in textile industry. In: Handbook of solid waste management: sustainability through circular economy, pp 1225–1256. https://doi.org/10.1007/978-981-16-4230-2_57
35. Ashfaq A, Khatoon A (2014) Waste management of textiles: a solution to the environmental pollution. Int J Curr Microbiol Appl Sci 3(7):780–787
36. Hussain CM, Singh S, Goswami L, Emerging trends to approaching zero waste. Environ Soc Perspect 1–16
37. Rezaei Arangdad S, Thoney-Barletta K, Joines J, Rothenberg L (2019) Influence of demographics and motivational factors on US consumer clothing and shoes disposal behavior. Res J Text Appar 23(3):170–188. https://doi.org/10.1108/RJTA-08-2018-0051
38. Khandual A, Pradhan S (2019) Fashion brands and consumers. Springer Singapore. https://doi.org/10.1007/978-981-13-1268-7
39. Casciani D, Chkanikova O, Pal R (2022) Exploring the nature of digital transformation in the fashion industry: opportunities for supply chains, business models, and sustainability-oriented innovations. Sustain Sci Pract Policy 18(1):773–795. https://doi.org/10.1080/15487733.2022.2125640
40. Ahmad N, Salman A, Ashiq R (2015) The impact of social media on the fashion industry: empirical investigation from Karachiites. J Resour Dev Manag 7:1–8. https://doi.org/10.4018/978-1-4666-4197-6.ch019
41. Ray S, Nayak L (2023) Marketing sustainable fashion: trends and future directions. Sustainability 15(7). https://doi.org/10.3390/su15076202

Water and Energy Efficiency in Textile Production

Md. Iftakharul Muhib and Md. Mostafizur Rahman

Abstract The textile industry is characterized by its complex nature as a manufacturing sector that exhibits fragmentation and heterogeneity, primarily due to the prevalence of small and medium enterprises. Energy consumption is a significant determinant of costs within the textile industry. In periods characterized by significant fluctuations in energy prices, the enhancement of energy efficiency emerges as a paramount priority for textile manufacturing facilities. Again, certain industries are significant contributors to the release of a wide range of hazardous organic and inorganic substances, including toxic chemicals. Numerous global studies have identified a range of sustainable technologies that have been successfully implemented in the textiles, both in specific sectors and across different areas. However, there is a lack of comprehensive and easily accessible information pertaining to this emerging concern. Therefore, the purpose of this chapter is to serve as a comprehensive reference on these emerging issues for engineers, researchers, investors, textile manufacturers, policymakers, and other individuals with a vested interest.

Keywords Textile industry · Water · Energy · Efficiency · Technologies

1 Introduction

Large volumes of power, fuel, and water are used by the textile sector, which results in the emission of greenhouse gas (GHGs) and toxic effluent. Regarding energy consumption, how the textile sector is structured in a given nation will determine

---

Md. I. Muhib · Md. M. Rahman (✉)
Department of Environmental Sciences, Jahangirnagar University, Dhaka 1342, Bangladesh
e-mail: rahmanmm@juniv.edu

Md. I. Muhib
General Education Department, City University, Birulia, Savar, Dhaka 1216, Bangladesh

Md. M. Rahman
Laboratory of Environmental Health and Ecotoxicology, Department of Environmental Sciences, Jahangirnagar University, Dhaka 1342, Bangladesh

P. Singh (ed.), *Dye Pollution from Textile Industry*, SDGs and Textiles,
https://doi.org/10.1007/978-981-97-5341-3_14

what proportion of the nation's overall final energy use is made up of fuel and electricity. For instance, fuels are the main energy source for wet processing of textiles while electricity is the primary energy source for spinning yarn. In addition to utilizing a lot of energy, the wet processing of materials during textile manufacture utilizes a lot of water and generates a lot of contaminated effluent. The industry's plan to make its production methods more ecologically friendly will also include water conservation and pollution prevention, especially in regions of the world where water is in short supply. The population of the world was 6.9 billion in 2010, and by 2050, it is projected to increase to 9.5 billion [1]. The majority of this expansion will occur in developing and impoverished nations. Residents of these nations will have more purchasing power as their economies develop, which will lead to an increase in per-capita consumption of products, especially textiles.

In brief, population and economic expansion in the future will spur substantial increases in textile production and consumption, which will in turn result in notable increases in the sector's overall usage of energy, water, carbon dioxide (CO_2), and other detrimental emissions. According to several studies [2–5], the global textile sector has the ability to save energy by implementing energy-efficient technology and practices. However, given the anticipated continued rise in absolute textile production, future reductions in absolute energy use and CO_2 emissions (by 2030 or 2050, for example) will necessitate innovation beyond currently available technology. The same is true for lowering emissions of other pollutants during production and conserving water. New methods and materials, as well as technologies that can cost-effectively cut down on the industry's consumption of energy and emissions, will probably be developed. For the industry's medium- and long-term climate change mitigation and environmentally friendly production plans, market deployment of these innovative technologies will be essential.

Other studies (such as [6]) have identified cross-cutting efficiency technologies that, although not specifically for the textile industry, are applicable to this industry. These studies have identified energy-efficiency technologies for the textile industry that are already commercially available. However, information on alternative and cutting-edge energy-efficiency, low-carbon, water conservation, and pollution-reduction solutions for the sector that have not yet reached full commercialization is few and difficult to find.

In order to help engineers, academics, investors, textile firms, politicians, and other interested parties, this book chapter compiles the knowledge that is currently accessible on developing technologies for increasing the energy and water efficiency in the textile sector. Although local factors such as economic viability, raw material availability, the type of energy utilized and its cost, and regulatory framework influence the adoption of particular technologies in each given region of the world, this research is meant to be useful to a broad audience.

2 Water Consumption in Textile Industries

One of the top 10 water-using industries is the textile industry, which is a significant water consumer. In various textile processing steps, a lot of water is consumed. The specific operations carried out at the plant, the machinery employed, and water management procedures all have a significant impact on the amount of water used in the textile sector [7]. A process house uses nearly three times as much water as all other units combined. When it comes to processing different types of natural textiles, wool and other felted fabrics use more water than other types of textiles, such as woven, knitted, stock, and carpet. When it comes to cotton textiles, 95% of the water is utilized to produce the raw materials, and the rest of 5% is consumed to process the cotton fabric. To process 1 kg of cotton cloth, it is predicted that 100–150 L of water will be required. Traditionally, a pair of cotton jeans requires more water to create (a bigger water footprint). Wool and silk, which are protein fibers, need more water overall because of scouring and coloring. According to recent reports, cellulosic fabric bleaching uses more water than other methods. In terms of the precise dyeing procedure, various kinds of colors require varying amounts of water. Additionally, the amount of water utilized varies depending on the dyeing machine and procedures used. In place of dyeing, coloration via printing technology, which uses significantly less water, has also been taken into consideration. In this regard, the development of digital printing, a virtually water-free method, is noteworthy [7].

2.1 Water Requirements in Fabrics, Dyeing, and Processing

The actual amount of water needed for wet processing textiles varies on the kind of processed materials, dyeing types, and the finishing materials being used. According to Uqaili and Harijan, processing 1 kg of textile material requires 50–100 L of water on average (global average). Water and the appropriate chemicals are required at practically every stage of the wet processing of textiles.

The textile industry uses a lot of water in its many operations, and after the process, almost as much effluent is expelled as was used in the processing medium. When compared to chemical textile processing (scouring, bleaching, dyeing, printing, and finishing), which requires water, mechanical processing (spinning and weaving) uses much less water. Aqueous media is used to apply almost all of the dyes and compounds to textile surfaces. A fabric unit that produces 20,000 lb per day uses 36,000 L of effluent. Water is primarily utilized in textile wet processing to chemically process fabrics and as a washing and rinsing medium. According to Shaikh and Ayaz, Rathore, Alexandra, boiler operating, providing reduced temperature water for drying steam, and cleaning are other uses for water in the textile sector. Phthalates, PBDEs, lead, organochlorines, etc. that have been associated with a number of health problems are frequently found in the effluents emitted by the textile industry.

2.2 Water Usage in Natural Fiber Processing

The production of natural fibers like cellulose (cotton), lignocellulose (jute, linen, hemp, banana), and proteinous materials (wool, silk) need a significant amount of water from the textile industry. In the cotton textile business, desizing, scoring, bleaching, dying, printing, and finishing are chemical processes that require a significant amount of water to properly size the warp yarn. Moreover, water is necessary for washing and rinsing treated materials. The water footprint of the aforementioned processes varies greatly, according to the particular mill operations, the processing equipment utilized, and the dominant water use management philosophy.

About 250–350 kg of water are used per kilogram of fabric when processing cotton, and 200–300 kg of water are used per kilogram of fabric when processing wool. 1. Preparing cotton fibers requires a significant amount of water. In a process house, the amount of water used is typically three times greater than the total amount of inputs used. It has been noted that, in reality, processing textiles requires a significant amount of water. However, it differs from mill to mill based on a number of variables, including the kind of fabric, the kind of machinery, the procedure, and the dyestuff utilized. The washing machine uses a significant amount of water at the conclusion of each procedure. Much water is also needed for the processing of natural fibers (cotton, jute, sisal, etc.). Both the use of newer types of gear, including continuous ranges of dyeing and the pretreatment process, and the use of old and new technologies result in a considerable difference in water consumption in the processing phases.

The material-to-liquor (water) ratio changes depending on the kind of dyeing machine being used for wet processing of natural fibers. An energy, chemical, and water-efficient dyeing machine might save a significant amount of these resources. The data have demonstrated that continuous bleaching and dyeing machines use less water than batch processing techniques. The numerous textile firms employ state-of-the-art continuous textile processing machines, which consume extremely little water. The installation of these water-efficient machines can provide challenges at times because of their high initial cost and limited operating lengths in a single processing shade. Additionally, they are not appropriate for materials that require tension-free handling, such knit fabrics and crepe fabrics [8].

2.3 Water Usage During the Synthetic Fiber Production Process

The estimated water usage and total amount needed for the manufacturing of different textiles derived from synthetic fibers are shown in Table 4. Compared to the natural textile processing sector, the synthetic textile industry uses less water. Wet processing of polyester requires just 100–200 kg/kg of water. There is a great deal of variance in the amount of water used, primarily due to the types of machinery, fiber, and old and modern technologies used [9, 10].

3 Ways to Conserve Water Resources in Textile Industry

The literature contains multifaceted activities related to water conservation in the textile industry that can be broadly categorized into four areas (Fig. 1): I. Reusing and treating textile wastewater to conserve water, II. Water conservation via textile processing machine advancements, III. Conservation of water via advancements in manufacturing techniques, IV. Conservation of water via advancements in chemicals.

3.1 Reusing and Treating Textile Wastewater to Conserve Water

3.1.1 Managing the Complexity of Textile Effluent Issues

Reusing and recovering wastewater is a viable strategy for the textile industry to become more ecologically conscious and sustainable. This complicated issue has many components, one of which is the enormous amount of water consumed in the textile sector. The water environment is contaminated by effluent discharge, which

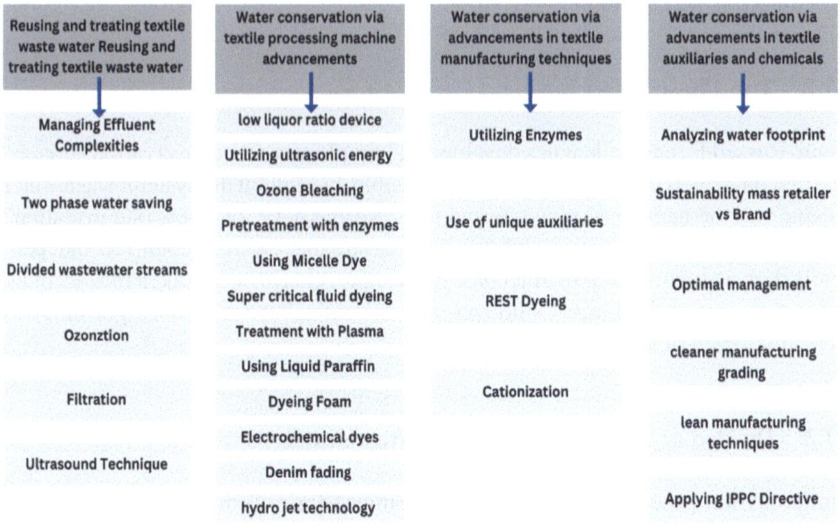

Fig. 1 Water conservation techniques in textile industries

has significantly increased the shortage of clean water. There are several obstacles to overcome in the recycling of textile wastewater, including elevated TDS levels, intense color, sharp pH fluctuations, elevated COD, and high temperatures [11]. Numerous academics have expressed interest in the potential recycling of textile wastewater. However, recovering textile effluent is a difficult task due to the vast range of colors, pH, heavy metal temperatures, and dissolved chemical diversity that exist in it. However, there is a chance to reuse textile wastewater up to 90% if it is handled adequately [12].

Wastewater effluent from a dyeing and finishing mill's secondary wastewater treatment plant can be treated using chemical ion exchange, coagulation, as well as electrochemical techniques to prepare it for potential reuse in textile dyeing or finishing. Methods of chemical and electrochemical coagulation successfully reduce the effluent's color, COD, and turbidity. Further lowering the COD, Fe content, total hardness, and conductivity can be achieved with success by using the ion exchange approach. Even at lower concentrations of 200 mg/L, hydrogen peroxide can increase the electrochemical treatment's efficiency by 100% [13]. Despite being highly effective at reusing textile effluent, a combination of the three above-discussed approaches has limited commercial feasibility at bulk scale because of the processes' complexity and high cost. The efficacy of effluent situated in the China Province of Zhejiang has been documented by Xu et al. 600 m^3 of textile wastewater might be processed daily by this wastewater treatment facility. Wastewater from this plant has undergone three rounds of treatment in order to be fit for reuse in textile finishing and dyeing:

(a) A biological procedure that involves aerobic and anaerobic treatment in two stages
(b) The process of biological aerated filters (BAF)
(c) Filtration by membrane [14].

3.1.2 Two-Phase Water Saving

Dunn, Russell F., and colleagues developed a two-phase water conservation strategy. The first phase was to use as little water as possible throughout the system; wastewater reusing was the next phase also referred to as "spring field" or "slow rate irrigation" [15]. "Source water streams" and "sink water streams" were recognized independently. For evaluation, four distinct process design scenarios were taken into account. Using the commercial Lingo Optimizer, a mathematical optimization algorithm was created for each of the four cases.

3.1.3 Divided or Separated Streams of Wastewater

Numerous wet processes used in the textile industry result in an intricate blend of chemicals and colors in the wastewater. Separating the wastewater streams makes it considerably simpler and more economical to clean and reuse them [16]. When compared to mixed flow treatment, the so-called "splitflow treatment" offers several benefits, including higher recycling rates, higher water recovery, and better water quality for a lower initial expenditure [17].

3.1.4 Ozonation in the Treatment of Wastewater

In order to eliminate color and reduce COD (chemical oxygen demand) to a level that allows water to be reused for dyeing in lighter hues, Gianluca Ciardelli and Nicola Ranieri employed ozonation. It has been documented that ozonation and electro-flocculation can be used in tandem to recycle textile wastewater for dyeing. According to Ciardelli and Ranieri [18], the color removal and COD reduction via ozonation are likewise expected to have reasonable operating costs.

It has been stated that in just 15 min, a bubbling reactor equipped with an ozone system may achieve 90% decolorization of textile wastewater. Compared to ozone combined with photocatalysis or photoelectrocatalysis, direct ozone produced the same degree of decolorization while using less energy. According to Cardoso et al. [19], the ozonation method is faster, less expensive, easier to use, more efficient, and capable of treating large amounts of textile effluent.

3.1.5 Filtration in the Treatment of Wastewater

To make wastewater reusable, the textile sector also uses a variety of filtration procedures. These consist of microfiltration, nano-filtration, ultrafiltration, and reverse osmosis. Because microfiltration's wider pores allow it to remove bigger contaminants from wastewater before reverse osmosis and more sensitive nano-filtration process, it is thought to be the most economical option for pretreatment of wastewater [20]. Reverse osmosis and ultrafiltration units combined with membrane separation techniques can cut the amount of freshwater used in a small-scale textile mill by more than 55% [21]. However, there are a number of significant obstacles preventing the textile sector from widely implementing membrane filtering technology, including its high cost, the need for regular membrane replacements, and the need for varied pretreatments depending on the influent water [22]. Compared to reverse osmosis membrane filtration, nanofiltration membranes exhibit superior COD removal efficiency, which may be explained by their sieving removal mechanism [23].

Reactive dyeing wastewater can be treated by crossflow nano-filtration to recover the electrolyte solution and remove color using a thin-film composite polysulfone membrane. The nano-filtration mechanism was investigated in relation to dye concentration, feed pressure, concentration of electrolyte, and crossflow velocity. According to Tang and Chen [24], the results demonstrated a decolorized permeate with a higher water flux recovery of 99% that could be used again in textile processing.

3.1.6 Using Ultrasound to Treat Wastewater

Tezcanli Guyer and Ince studied the degradation pattern of azo dyes in effluents. The probe dye that was employed was C.I. Acid Orange 7. To achieve the highest rate of absorption decay, various operation parameters were optimized, including the kind, injection technique, UV intensity, ozone flow rate, and ultrasonic power density of the bubble gas. It was discovered that textile azo dye solutions may be decolored with ultrasound; however, total dye breakdown or total dye mineralization would require the use of ultrasound in conjunction with ozone and/or UV irradiation [25]

While there have been some encouraging examples of the recycling of textile wastewater, there have also been some concerns expressed about potential contamination, insufficient level of treatment, and quality problems with the reuse of reclaimed water [26]. Owing to the intricacy of the wastewater originating from the processing sector, it is frequently deemed insufficient to render water reusable with a single water treatment procedure. Water treatment becomes more expensive and complex when a variety of various treatments are used [27].

3.2 Water Conservation Through Textile Processing Machine Advancements

3.2.1 Devices with Low Liquor Ratios

The ratio of textile material to liquor has the greatest effect on the amount of water used in dyeing. Reducing the liquor ratio to 1:8 (from 1:10) saves 15% on processing expenses and 20% on water usage. Low liquor ratio dyeing machines not only use less water when dying textile materials but also less of these ingredients [28]. Reducing the liquor ratio significantly enhances the cellulosic fibers' ability to absorb color, particularly when direct dyes are used [29].

One popular method for getting rid of unfixed dyes and dyeing aids after dying is overflow rinsing. This uses a significant amount of water. An effective and environmentally beneficial substitute for overflow rinsing is a pressure-based process [30].

3.2.2 Utilizing Ultrasonic Energy for Coloring

The previous evaluation makes it evident that a large number of scholars think that using energy in the form of radio waves, ultrasound, infrared radiation, and microwaves can lessen the harm that the textile processing industry causes to the environment [31].

Utilizing ultrasound technology seems to be a viable substitute for reducing the amount of water, energy, chemicals, and time used in various wet processing procedures for textiles. In textile wet processing, ultrasonic technology can be applied in two ways: (i) direct ultrasonic probes mixing (Horn) and (ii) indirect bath by ultrasonic process. For sono-dyeing and sono-finishing textiles, horn-type applicators are typically utilized [32].

3.3 Water Conservation Via Advancements in Textile Manufacturing Techniques

3.3.1 Ozone Bleaching

For correct dying, the textile substrates must be pretreated. In most cases, cleaning the fabric before dyeing it is sufficient to dye it in darker tones. However, cloth bleaching is typically necessary for hues that are lighter and brighter. Compared to the conventional method, ozone bleaching of textiles has been claimed to require extremely little water, as opposed to the use of free radical hydroxyl species. On the other hand, ozone generation setup is expensive and necessitates processing in enclosed, leak-proof chambers.

Additionally, a unique method of ozone-bleaching pre-wetted cotton cloth and then ultrasonically assisting the rinse has been suggested [33]. After applying ozone for 15 min in a closed bath and adjusting fabric pick-up to 50% using pH 7 water, the fabric was rinsed for 5 min at 40 °C without the use of chemicals. Comparable desizing efficiency, hydrophilicity, and degree of whiteness were obtained with the ozone-ultrasound combination method as compared to traditional chemical, water, and energy-intensive preparation procedures [33].

3.3.2 Pretreatment with Enzymes

Polonca Preša and Petra Forte Tavčer investigated the potential for conserving water and energy during the processing of cotton prior to dyeing. An environmentally beneficial substitute for traditional alkaline scouring is bio-scouring. Water usage for H_2O_2 and alkaline scouring has been compared to water used for bio-scouring and per acetic acid bleaching thereafter. It has been observed that compared to the alkaline process, the bio-scouring technique required 33.3% less water. The fabric needs to be neutralized following alkaline scouring and hydrogen peroxide bleaching. The value of pH in this method is somewhat close to the pH level [34].

3.3.3 Spun Dyeing of Yarns

A greener option to traditional PLA dyeing, which uses a lot of energy and water, is the polymerization of PLA (polylactic acid) combined with simultaneous pigmentation in one phase. This technique uses a catalyst that contains a chromophore to speed up the polymerization process and give the PLA polymer color [35]. The environmental impact of spinning dyed modal fibers with carbon black and standard jet dyeing modal fabric with black reactive dyes was compared by Terinte et al. According to their findings, compared to traditionally dyed fabric, spun-dyed modal fabric consumes 50% less water, 50% less energy, as well as a 60% reduced carbon footprint during manufacture [36].

3.3.4 Using Micelle Dye

According to Xie et al. [37], micelle dyeing has been proposed as a greener substitute for traditional dyeing methods. At a 1:5 liquor ratio, reactive dyeing was done using a dye micelle solution. The dye used for the control sample had a standard liquor ratio of 1:15. The characteristics of the cloth dyed with micelles at a low liquor ratio were contrasted with samples that had been dyed traditionally. It was discovered that the fixation, reactivity, dye exhaustion, and fastness characteristics, such as wet crocking, washing, and sweat fastness, were equivalent to those of fabric colored traditionally. There have been reports of 60% dyeing water savings, 50% steam savings, and 60% electrolyte saving during dyeing.

3.3.5 Supercritical Fluid Dyeing

A notable advancement in textile processing, the first dyeing of PET in supercritical fluid aims to completely substitute water as a dyeing medium [38].

The preferred fluid is supercritical carbon dioxide (SC-CO2) because of its unique properties beyond its critical point. In addition to having similar density and solubility to liquid solvents, SC-CO2 also offers the benefit of very quick diffusion and viscosity that is more akin to a gas. In addition to saving water, SC-CO2 dyeing boosts output because of shorter dyeing cycles, uses less energy as dyeing is done at a lower temperature, uses fewer chemicals and auxiliaries, and significantly lowers air emissions.

In contrast, the dye solubility on SC-CO2 falls with a controlled density drop in density-controllable disperse dyes [39]. The scarcity of specific dyes for waterless natural fiber dyeing is a significant barrier to the widespread use of SC-CO2 dyeing in industry. Since their debut in 1956, reactive dyes have emerged as the standard option for dying cotton [40]. Reactive dyes are currently the most common type of dye for cotton. Zhang et al. used the dichlorotriazine reactive group to create three unique Azo dispersion reactive dyes. Wool was waterless dyed in SC-CO2 medium with these dispersed reactive dyes for 90 min at 120 °C and 20.0 MPa of pressure. By diazotizing the alkalescent aryl amine of 2-chloro-4-nitroaniline in the presence of a safer and milder acid, hydrochloric acid, rather than concentrated sulfuric acid, in an environmentally friendly medium of polyethylene glycol 200 mixed with water, they were able to create a safer and more environmentally friendly method of synthesizing these special disperse reactive dyes. Zhang et al. reported medium-level washing fastness with good leveling capabilities and good rubbing fastness (4–5). It appears that the only viable method for dying cotton with SC-CO2 is to create synthetic dispersion reactive dyes especially for cotton and utilize the right swelling agents for the material. Due to the complex chemicals involved, some suggested solutions may drastically reduce the environmental benefit and greatly increase the complexity of the dyeing system [41].

3.3.6 Treatment with Plasma

Researchers are interested in the ecologically friendly procedure of plasma treatment of textiles because of its enormous potential for energy and water savings. Irving Langmuir made the groundbreaking discovery of partly ionized gas, known as plasma, the fourth state of matter, in 1929. Depending on the temperature, current, gaseous matter, pressure, and kind of textile fiber utilized, plasma changes the fabric surface more quickly [42].

Yan and Zheng revealed that the ideal conditions for SC-CO2 dyeing of silk cloth modified with plasma were found to be 90 °C, 40 MPa of pressure, and 60 min of dying time. They found that altering the silk fabric's surface before dying enhanced its rubbing and washing fastness [43]. In the presence of He/N_2 gaseous mixtures, Teli et al. applied plasma treatment to silk fabric at a frequency of 21–23 kHz and a discharge voltage of 5 kV. They applied a continuous nitrogen flow rate of 50 ml per minute to silk cloth and left it exposed for 1 to 10 min. Even at 40 °C, silk cloth treated with plasma showed quicker depletion and deeper hues of acid dye [44].

3.3.7 Using Liquid Paraffin for Dye

Out of 110 organic solvents, Suxin Xu et al. chose liquid paraffin as a waterless dyeing medium by applying a solvent assessment process that took each solvent's EHS properties and dye uptake into account. Cloth made of polyethylene terephthalate (PET) was colored with a liquid paraffin solution.

Since the pH of the dye bath and the usage of auxiliaries had no bearing on the outcome of liquid paraffin dyeing, they left these parameters alone. A temperature of 130 °C was employed for dyeing. Liquid paraffin offers superior EHS characteristics and a stronger dye absorption when compared to chlorinated solvents. Furthermore, it does not need costly high-pressure machinery, which is necessary for SC-CO2 dyeing. They stated that the surface oligomer content had dropped to 0.02% and that the color constancy remained outstanding even after the solvent had been reused seven times. It was discovered that the mechanical, color yield, and fastness characteristics of the paraffin-dyed fabric were similar to those of the aqueous-dyed fabric. There have also been reports of successful PLA dyeing in liquid paraffin media, with fastness characteristics similar to those of PLA materials dyed aqueously.

3.3.8 Denim Fading in the Ozone

Among younger shoppers, faded jeans are a hugely popular fashion trend. Beyond denim, knitted and other woven clothing is now in demand due to the bohemian and fading aesthetic. Traditional washing-based fade solutions call for a large amount of water and chemicals.

A chemical- and water-free substitute for cotton clothing's faded fashion effects is ozone fading. It has been observed that plasma-induced ozone treatment is more successful than basic ozone treatment [45].

3.3.9 The Use of Hydro-Jet Technology to Fade Denim

One timeless design trend is the appearance of the faded jeans. However, the typical method for achieving this fading is to use water-intensive techniques like $KMnO_4$ spray and/or stone wash. With less water used, hydro-jet treatment has recently been created to produce texture, surface finishing, and fading that are comparable. This treatment involves using specific nozzles to expose one or both of the garment's surfaces to a high-pressure water jet. The type of dye used and the amount of fluid impact energy given to the fabric determine how much of the dye fades. Chemicals are not employed in the process; instead, recycled water is used to create fading effects. Because of this, hydro-jet fading is a very green substitute for traditional denim washing techniques [46].

3.4 Water Conservation Through Advances in Chemicals and Auxiliary Used in Textile Processing

3.4.1 Utilizing Enzymes

Some of the prospective benefits of enzymatic processes include water and energy conservation as well as the avoidance of the use of dangerous or harmful compounds [47]. In order to prepare cotton for dyeing, scouring is a step in the pretreatment process that removes various natural impurities. Because the scouring process is typically carried out at high temperatures in highly alkaline conditions, it requires

Table 1 Enzyme use in the production of environmentally friendly textiles [50]

Processing of textiles	Objective	The Enzyme
Desizing	Elimination of sizing agents (polyvinyl alcohol, starch)	Pectinase, amylase, polygalacturonase
Scouring	Elimination of lignin and pectin	Protease, pectinase, lipase, cellulase
Bleaching	Growing amounts of whiteness	Laccase, glucose oxidase
Polishing	Smoothing out the fuzz and making it smoother	Cellulase

a significant amount of water and energy. The pectatelyase enzyme uses significantly less energy and water than the alkaline scouring method, making it a more environmentally friendly option [48].

Hydrolase and oxidoreductase are the two primary categories of enzymes utilized in cotton finishing and pretreatment. Amylase, cellulose, protease, cutinase, pectinase, and lipase are among the hydrolase enzyme group's applications. These include wool finishing, denim finishing, softening, felting, and biopolishing and bio-scouring of textiles [49]. Table 1 gives a summary of the most often utilized enzymes in wet processing of textiles [50].

3.4.2 The Creation of Unique Auxiliaries Aimed at Reducing the Production of Dye Agglomerates to Attain Level Dyeing at Low Liquor Ratios

Regarding the claim that low liquor ratio dyeing conserves water, there are no two views [47]. However, a risk connected to low liquor ratio dyeing is the propensity for dye molecules to aggregate in the dye bath, leading to uneven dying. Hassan and Bhagavandas investigated how three distinct dyes and various auxiliaries affected the way wool dyed at a liquid ratio of 1:10. According to Hassan and Bhagvandas [51], they discovered that "Teric G12A6" from Huntsman was ideally suited for preventing dye aggregation during low liquor ratio dying of wool in all three colors.

3.4.3 REST Dyeing

REST (Rapid Enzymatic Single-Bath Treatment) is the acronym given to the combined pretreatment and dyeing technique that Öner and Sahinbaskan devised. For the dyeing trials, they employed 100% cotton fabric that was starch-sized. No freshwater was added to the bath during the REST process, and the same bath was used for all procedures including enzymatic desizing, scouring, hydrogen peroxide bleaching, enzymatic anti-peroxide treatment, and dyeing. In comparison to the traditional dyeing procedure, they reported generating consistent hues with 66% water savings and substantial energy savings. Control of pH was found to be a crucial parameter because of the employment of enzymes, and when compared to fabric dyed conventionally, some tone-wise changes in the final hue were also noticed [52].

3.4.4 Cationization

It is said that cationization increases the amount of dye used, yields greater color values, and minimizes the number of rinse cycles required once cotton is dyed. In low-salt or salt-free cotton reactive dyeing, pre-cationization of cotton is also said to be beneficial [53]. According to reports, cationization can also aid in obtaining ultra-deep coloring in black [54]. It is possible to minimize the quantity of wash-off baths and the liquor ratio used in each wash-off bath when cotton is colored without the use of an inorganic electrolyte [55].

Without the use of salt or alkali, it has been discovered that the employment of unique reactive dyes with cationic soluble groups and nicotinic acid quaternary triazine reactive groups provides a high level of dye exhaustion and fixing on cotton fiber [56]. It has also been discovered that cationic starches can be used to improve dye fixing and fatigue [50].

It has also been noted that pre-cationizing the cloth to be bleached, without adding alkali to the bleaching solution, results in a whiteness index that is comparable to that of the traditional bleaching method. Additionally, the water-intensive neutralization and removal of residual alkali is done away with.

In addition to acting as a catalyst for the bleaching process, the cationic group on the cationized cotton fabric also offers the site required for the activation of the hydrogen peroxide bleaching bath in place of NaOH [57].

3.5 Sophisticated Water Analysis and Water-Saving Devices

3.5.1 Saving Water Through an Examination of the Water Footprint

The water footprint (WF) of various fabrics manufactured in Spain was examined by Chico et al. [58]. From the phases of wood and cotton production to the industrial operations, the estimated gray, blue, and green water footprints of the processes in the textile value chain were made. They released the findings of an evaluation of the water footprint of five different textile kinds that are frequently utilized in the production of jeans. This study examined five equivalent spinning, dying, and weaving techniques for two distinct fibers—cotton and Lyocell fiber. They discovered that the step that uses the most water is the cotton fiber production stage. They discovered that Lyocell fiber, which is based on cellulose, uses a lot less water than cotton fibers. In comparison to cotton, which requires 263, 2767, and 203 m^3/t of water, lyocell fiber requires, on average, 1384, 34.5, and 35.3 m^3/t [58].

3.5.2 Sustainability of the Supply Chain for Mass Retailer VS Brand

MacCarthy and Jayarathne [59] examined water-saving and sustainability tactics used in the supply chains of a supermarket and a major brand retailer. They observed that the brand retailer's manufacturers were more driven and engaged in several small- and large-scale water-related projects. These initiatives included desalination plants and water conservation. The brand merchant employed a sophisticated two-pronged campaign that urged end users to wash their clothes in an environmentally friendly manner while also assisting manufacturers in posting water-saving guidelines. A strong waste management system (GWM) had been established by the manufacturer within the supply chain of a large supermarket retailer. It was discovered that this GWM technology helped to reduce water waste during the production cycle.

It was discovered that the producers of goods for supermarket merchants complied with the bare minimum of national laws pertaining to worker welfare, safety, and the environment. On the other hand, McCarthy and Jayarathne found little evidence of innovative approaches to obtaining accreditations, conserving water, or protecting the environment.

3.5.3 Applying Optimal Management Techniques

Best management practices (BMPs) for water recycling and conservation in the textile sector were proposed by Chougule and Sonaje [60]. They found significant problems with the water use practices of the textile wet processing industry, such as the use of water in amounts that aren't even necessary for the process, the propensity to use too much water when rinsing after dyeing, and the use of drinking-grade water for all processes—even though not all of them call for it. According to Chougule and Sonaje [60], they recommended putting BMPs into practice by conducting routine water audits, establishing plant-wide water conservation goals, designating a special team for water conservation, obtaining water conservation data from vendors, research organizations, and national and international organizations, and informing human resources of the importance of water conservation.

3.5.4 Introduction of Cleaner Manufacturing Grading Scheme

In 2012, Tong et al. suggested creating a cleaner production indicator system and evaluating it through the application of the Analytical Hierarchy Process (AHP). The Delphi Method was used to determine the final judgment matrix, and Yaahp software was utilized to give each indicator a weight factor. This led to the development of an extensive production evaluation indicator system for waste reduction and water conservation for the textile printing and dyeing sectors. They determined the following four crucial markers for reducing waste and conserving water [61]. The indicators of water resources, main raw materials and additive consumption, pollution generation, and comprehensive utilization are the four types of indicators.

3.5.5 Using Lean Manufacturing Techniques

The US EPA created a toolkit to assist businesses in reducing water and energy use and increasing machine efficiency. Root cause analysis, 5S, Kaizen events, value stream mapping (VSM), 5Why, fishbone diagrams, and visual management are a few examples of lean manufacturing tools that are useful for locating issue sources and enhancing outcomes when it comes to lowering water use and environmental risk [62].

3.5.6 Application of the IPPC Directive by the European Union

The installation of water flow meters at water pipelines, semi-counter current rinsing in dyeing and finishing, reusing treated dyeing and finishing wash waters, and reusing compressor-cooling water in production were among the BREF (Best Available Techniques; BAT) recommendations put into practice by Kocabas et al. [63]. They succeeded in cutting their overall water use by 29.5% and their energy consumption by 9% [63]. Another study found that applying BAT solutions, such as optimizing the steam boiler, reusing steam condensate, implementing good management practices, modifying the machinery, and reusing wastewater, reduced water consumption in the textile mill by 35–65% [64]. The mill's overall water consumption may be lowered by 35–65% by implementing additional BAT options, such as steam boiler optimization, steam condensate reuse, and strong management techniques, in addition to wastewater reuse and mechanical changes.

An environmental performance evaluation (EPE) of a textile-woven fabric production mill in Bursa, Turkey, was carried out by Alkaya and Demirer [65]. Following EPE, the five sustainable production options listed below were determined and put into practice: Reusing sneezing and stenter cooling water, upgrading the water softening system, replacing overflow washing with drop-fill washing, and updating different water transmission system valves and fittings According to reports, water usage dropped by 40.2% and wastewater amount dropped by 43.4% when the aforementioned options were put into practice [65]. Table 2 lists each of these water-saving techniques along with the benefits and drawbacks that go with them.

4 Energy Consumption in Textile Industries

In general, the textile sector is not thought to be an energy-intensive one. However, a lot of plants are used in the textile business, and these plants use a lot of energy in total. The proportion of a nation's manufacturing energy that is utilized by the textile industry is determined by the manufacturing sector's organizational structure. For instance, in China, the textile sector uses almost 4% of the total energy used in manufacturing [69], compared to less than 2% in the United States [70]. Each country has a different proportion of the textile industry's overall product cost that

Table 2 Strategies for preserving water in textile wet processing: benefits and drawbacks

Techniques for conserving water	Benefits	Drawbacks	References
To treat wastewater effluent, a mix of ion exchange, chemical coagulation, and electrochemical techniques has been proposed	Water is made reusable	Extremely intricate and capital-intensive procedure	[13]
Divided or separated streams of water	Facilitated treatment of wastewater and potential for water recycling	Certain equipment become restricted to particular processes, separate infrastructure is needed, and the dye house may lose process flexibility	[17]
Wastewater ozonation	Water from wastewater could be recycled	Process that requires a lot of capital and is unable to manage complex wastewater	[19]
Filtration at the micro and nanoscale	Effective cleansing of water to enable its reuse	Changing filtration membranes frequently necessitates a significant financial outlay	[23]
Ultrasound application in wastewater treatment	Can be applied to textile azo dye solutions to decolorize them	It is impossible to get total dye breakdown or full dye mineralization	[27]
Low liquor ratio apparatuses	Significant chemical and water savings	High capitalization	[29]
Ultrasonic energy application in wet processing	Increased dye penetration and improved washing efficiency	There is no standardized bulk production machinery available	[31]
Ozone depletion	Enormous water savings	For the degree of whiteness to be improved and yield results that are comparable to those of traditional bleaching	[33]
Pretreatment with enzymes	Both economical and environmentally beneficial	When enzyme pretreatment is used instead of traditional pretreatment, the degree of absorbency is lower	[34]
Spun dyeing of yarns	Conserving water and reducing carbon emissions	Greater MOQ (minimum order quantity), fewer shades available, and longer lead times	[35]

(continued)

Table 2 (continued)

Techniques for conserving water	Benefits	Drawbacks	References
Supercritical fluid dyeing	Eco-friendly dyeing that uses less water and produces no wastewater	Significantly capital-intensive, machines still have unsolved engineering issues	[38]
Micelle dyeing	60% water savings, 50% steam savings, and 60% electrolyte savings	Enhancing the stability and dependability of micelle formation is necessary	[37]
Treatment by plasma	Water-efficient and environmentally friendly	Specialized equipment is needed	[42]
Dyeing by foam	Energy and water savings of over five times when compared to traditional dyeing methods	Specialized equipment is needed, and problems with foam stability must be resolved	[66]
Dyeing with electrochemical process	The electrode constantly regenerates the reducing agent, enabling complete recycling of the dye bath water	Process could not be carried out without specialized tools	[67]
Denim fading in the ozone	Conserving water and energy	If workers come into direct contact with ozone, they could be at risk. It is necessary to have specific sealed equipment in a well-ventilated location	[45]
The use of hydro-jet technology to fade denim	Conserving water	It is challenging to regulate the fading level	[46]
The creation of unique auxiliaries to reduce the production of dye agglomerates in order to attain level dyeing at low liquor ratios	Gradient coloring	The price of the particular chemical raises the total cost of processing	[51]
Rest dying: This method involved using the same bath for all steps, including enzyme desizing, scouring, hydrogen peroxide bleaching, enzymatic antiperoxide treatment, and dying, without adding any fresh water	66% less water and significant energy savings when compared to a traditional dying method	Controlling pH is an important factor, and the final shade can have some tone-wise variations	[52]

(continued)

Table 2 (continued)

Techniques for conserving water	Benefits	Drawbacks	References
Cotton's cationization	Lower usage of water and salt	Overuse of alkalis and differences in shade	[55]
Life cycle analysis and water footprint analysis	Increased supply chain visibility	Overuse of alkalis and differences in shade	[68]
Putting Lean principles into practice	Reduces waste and boosts water efficiency	Overuse of alkalis and differences in shade	[62]
Using the best available technology, or BAT	Preservation of water	High capitalization	[65]
Introduction of a more hygienic manufacturing grading scheme	Increased supply chain visibility	High capitalization	[61]

goes toward energy. The general cost factor proportions for 20 Tex[2] combed cotton yarn across multiple countries are displayed in Table 3. Costs associated with energy are frequently in the third or fourth place overall for products.

The textile business consumes a lot of fuel and power. The composition of the textile industry in a given nation determines the proportion of electricity and fuels used in the sector's overall final energy consumption. For instance, gasoline is the primary energy source in wet processing, but electricity is used mostly in spun yarn spinning. According to U.S. Manufacturing Census Data from 2002, fuel energy accounted for 61% and electricity for 39% of the total energy consumed in the country's textile industry. Among the 16 major industrial sectors examined in the United States, the textile industry is classified as the fifth-highest consumer of steam. According to the same study, onsite losses in boilers, motor systems, distribution, and other areas account for about 36% of the energy input used in the textile sector [72].

Table 3 Part of the production costs for 20 tex combed cotton yarn in 2003 across a number of nations [71]

Factors affecting costs	China (%)	Brazil (%)	Italy (%)	India (%)	Turkey (%)	Korea (%)	USA (%)
Raw material	61	50	40	51	49	53	44
Waste	11	7	6	7	8	8	6
Labor	2	2	24	2	4	8	19
Energy	8	5	10	12	9	6	6
Auxiliary material	4	4	3	5	4	4	4
Capital	14	32	17	23	26	21	21
Total	100	100	100	100	100	100	100

Table 4 Creative initiatives for various procedures [73]

Processes	Initiatives
Spinning	Utilizing a capacitor-based automatic power factor correction technology
	Swapping out ineffective motors in open ends, ring frames, and other spinning equipment
	For the speed frame, photocell
	Using ring frames with synthetic flat belts
Weaving	V-belt drives are swapped out with flat belts
	Replacement of outdated machines with new ones
	Replacing the humidification plant's fans
	Using an air pressure controller
Processing	Low liquid ratio jet dyeing machine to replace traditional quick jet dyeing machine
	RF dryer combined with steam dryer for yarn dyeing
	Utilizing leftover water for cloth dyeing and recovering it

5 Opportunities for Improvement

High capacity utilization, equipment fine-tuning, technological upgrades, and other re-engineering activities are prominent strategies for energy efficiency and conservation in the textile industry. In addition, the Indian textile industry heavily employs a few key subgroup-specific energy-saving strategies, which are included in Table 4.

The energy expenses of Indian textile mills have drastically increased because of higher spindle speeds and higher electrical utility bills. 70–80% of the energy is used by ring frames. Significant savings were obtained by using lightweight drum pulleys and a ring frame with a 30 mm tin roller pulley shaft. Increased productivity and spindle speeds are possible with an inverter drive without using more energy [73].

6 Opportunities for Improving Energy Efficiency in the Textile Sector

This review of potential energy-efficiency improvements in the textile sector covers both total replacement of the existing machinery with cutting-edge new technology and prospects for retrofit and process optimization. Retrofit efforts are given extra consideration, nevertheless, because cutting-edge new technologies have large upfront capital expenditures. In many situations, the energy savings from replacing old equipment with new equipment may not be worth the expense. Nonetheless, the new technologies are more economically justified when all the advantages—such as reduced waste, less water waste, fewer reworking, improved product quality, etc.—are considered after the installation of the new technologies [74].

7 Technologies for Energy Conservation and Management

Energy management affects many different departments in an organization, therefore it's important to raise awareness, increase understanding, and get everyone involved in the manufacturing process to participate and cooperate.

7.1 Increasing the Effectiveness of Electricity Consumption

Lighting: Owing to the way it works, a large portion of power is used for lighting. It's critical to reevaluate how well the light source is being used and implement energy-saving techniques.

Electric motor: A large number of comparatively tiny electric motors are used in the textile business. Many modern machines use multiple motors, each of which is directly coupled to a machine part to drive it independently from others. In contrast to conventional machines, which were driven by a single motor that transmitted the mechanical power it generated to various parts of the machine collectively.

Electric heating: In an effort to cut costs, other techniques (such as steam, gas heating, or direct indirect fired heating) have generally supplanted electric heating in the textile sector [75].

7.2 Enhancements in Fuel Efficiency

Fuel Selection: The process of choosing fuel involves looking for qualities such as high calorific value, easy burning, and appropriate flue gas characteristics in order to minimize air pollution.

Boiler Selection: Water-tube boilers have replaced Lancastrian- or Scotch-type tubular or smoke-tube boilers in the textile sector. Boiler efficiency has increased as a result [76].

7.3 Enhancement of Steam Usage Efficiency

Piping: Although the textile industry uses a relatively small amount of steam, the areas where it is needed are widely dispersed, which results in significant steam losses from pressure drops and heat radiation from steam transportation pipes. High-pressure, small-diameter piping is preferred over low-pressure, large-diameter piping for the long-distance transmission of steam.

Steam Accumulators: By converting excess steam into hot water, a steam accumulator can be inserted halfway through the heat-transporting pipe, between the boiler and the heat-consuming load, to store excess steam during light loads [77].

Using Heat Exchangers: Heating and cooling gases and liquids as heat media is often necessary in all textile industry production processes. This is accomplished through heat exchange between various fluids, and heat exchangers are employed to perform indirect heating and cooling in order to prevent contamination or chemical reactions from occurring as a result of their direct contact [76].

7.4 *Non-traditional Energy Sources*

Various alternative renewable energy sources include solar, wind, geothermal, biomass, and tidal energy. Controlling the technology is simple and easy, and it requires almost minimal maintenance. Air pollution won't have any negative effects. It denotes the absence of effluent and the waste of nothing.

7.5 *Energy Audit*

To lower the quantity of energy input into the system without adversely affecting the output, an energy audit involves surveying, inspecting, and analyzing energy flows in an enterprise with the goal of energy conservation [78].

7.5.1 Domains of the Energy Audit

Examine the observable aberrant symptoms of the inefficient energy transfer in the system by studying the energy audit of the operating equipment.

Lighting: sparks, loose joints, contactor switching, etc.

Heating: The result of misalignment, metal-on-metal contact, friction, and inadequate lubrication in motors and other load apparatus coupled to the motor [79].

8 Concluding Remarks

The textile industry will need to adapt to emerging water and energy efficiency technologies as a result of population and economic growth, which is predicted to drive a sharp rise in textile consumption over the next several decades, as well as an increase in the industry's overall energy consumption and emissions of greenhouse gases and other pollutants. This chapter provides an outline of the procedures used in the textile industry and gathers information on the savings of energy and water, as well as other advantages for the environment, expenses, the state of commercialization, and references. Information about emerging or advanced low-carbon and energy-efficiency technologies that are not yet commercialized or are in the very early stages of adoption is scarce and/or scattered, despite studies from around the world identifying a variety of sector-specific and cross-cutting water and energy-efficiency technologies that have already been commercialized for the textile industry. The purpose of this chapter is to provide engineers, researchers, investors, textile producers, policymakers, and other interested parties with information on these new technologies.

Credit Authorship Contribution Statement Data acquiring, writing original draft: MIM; writing, reviewing, editing, supervising: MMR.

Data Availability The datasets will be available from the corresponding author on reasonable request.

Declaration of Competing Interests The authors declare that they have no known competing interests.

References

1. UN/DESA (United Nations, Department of Economic and Social Affairs) (2012) World population prospects: the 2012 revision. http://esa.un.org/unpd/wpp/unpp/panel_population.htm
2. Carbon Trust (2005) Low cost heat recovery at W & J Knox Ltd, Ayrshire. Good Practice. www.carbontrust.co.uk
3. CIPEC (Canadian Industry Program for Energy Conservation) (2007) Benchmarking and best practices in Canadian wet-processing. http://oee.nrcan.gc.ca/industrial/technicalinfo/benchmarking/ctwp/index.cfm
4. ECCJ (Energy Conservation Center, Japan) (2007) Overview of energy saving technologies in textile industry. http://www.aseanenergy.org/download/projects/promeec/2007-2008/industry/eccj/ECCJ_SW03%20Overview%20of%20energy%20saving%20technology_TH.pdf
5. Hasanbeigi A (2010) Energy-efficiency improvement opportunities for the textile industry. Berkeley CA, Lawrence Berkeley National Laboratory Report LBNL-3970E. http://china.lbl.gov/publications/energy-efficiency-improvement-opportunities-textile-industry. https://www.osti.gov/biblio/991751
6. U.S. DOE/AMO (2012) U.S. Department of Energy, Advanced Manufacturing Office (AMO). Industries and Technologies. http://www1.eere.energy.gov/industry/industries_technologies/index.html

7. Samanta KK, Basak S, Chattopadhyay SK, Gayatri TN (2015a) Water-free plasma processing and finishing of apparel textiles. In: Handbook of sustainable apparel production. CRC Press, Boca Raton, Fl, pp 3–37
8. Saxena S, Raja ASM, Arputharaj A (2017) Challenges in sustainable wet processing of textiles. Textiles and clothing sustainability. Springer, Singapore, pp 43–79
9. Mansour HB, Houas I, Montassar F, Ghedira K, Barillier D, Mosrati R, ChekirGhedira L (2012) Alteration of in vitro and acute in vivo toxicity of textile dyeing wastewater after chemical and biological remediation. Environ Sci Pollut Res 19:2634–2643
10. World Bank (2010) A detailed analysis on industrial pollution in Bangladesh. Workshop Discussion Paper Transfer, EPA-625/3-74-004, World Bank Dhaka Office, Dhaka, Bangladesh
11. Banat IM, Nigam P, Singh D, Marchant R (1996) Microbial decolorization of textile-dyecontaining effluents: a review. Bioresour Technol 58:217–227. https://doi.org/10.1016/S0960-8524(96)00113-7
12. Wasatex EU (2016) Water saving processes for textile production. Eur Union Proj 30–32
13. Lin SH, Chen ML (1997) Treatment of textile wastewater by chemical methods for reuse. Water Res 31:868–876. https://doi.org/10.1016/S0043-1354(96)00318-1
14. Lu X, Liu L, Liu R, Chen J (2010) Textile wastewater reuse as an alternative water source for dyeing and finishing processes: a case study. Desalination 258:229–232. https://doi.org/10.1016/j.desal.2010.04.002
15. Dunn RF, Wenzel H, Overcash MR (2001) Process integration design methods for water conservation and wastewater reduction in industry. Clean Prod Process 3:319–329. https://doi.org/10.1007/s10098-001-0122-y
16. Kurt E, Koseoglu-Imer DY, Dizge N, Chellam S, Koyuncu I (2012) Pilot-scale evaluation of nanofiltration and reverse osmosis for process reuse of segregated textile dyewash wastewater. Desalination 302:24–32. https://doi.org/10.1016/j.desal.2012.05.019
17. Rott U, Minke R (1999) Overview of wastewater treatment and recycling in the textile processing industry. Water Sci Technol 40:137–144. https://doi.org/10.1016/S0273-1223(99)00381-9
18. Ciardelli G, Ranieri N (2001) The treatment and reuse of wastewater in the textile industry by means of ozonation and electroflocculation. Water Res 35:567–572. https://doi.org/10.1016/S0043-1354(00)00286-4
19. Cardoso JC, Bessegato GG, Boldrin Zanoni MV (2016) Efficiency comparison of ozonation, photolysis, photocatalysis and photoelectrocatalysis methods in real textile wastewater decolorization. Water Res 98:39–46. https://doi.org/10.1016/j.watres.2016.04.004
20. Pang YL, Abdullah AZ (2013) Current status of textile industry wastewater management and research progress in malaysia: a review. Clean - Soil, Air, Water 41:751–764. https://doi.org/10.1002/clen.201000318
21. Nandy T, Dhodapkar RS, Pophali GR, Kaul SN, Devotta S (2005) Application of chemical, biological and membrane separation processes in textile industry with recourse to zero effluent discharge—a case study. Environ Technol 26:1055–1063. https://doi.org/10.1080/09593332608618491
22. Verma AK, Dash RR, Bhunia P (2012) A review on chemical coagulation/flocculation technologies for removal of colour from textile wastewaters. J Environ Manag 93:154–168. https://doi.org/10.1016/j.jenvman.2011.09.012
23. Liu M, Lü Z, Chen Z, Yu S, Gao C (2011) Comparison of reverse osmosis and nanofiltration membranes in the treatment of biologically treated textile effluent for water reuse. Desalination 281:372–378. https://doi.org/10.1016/j.desal.2011.08.023
24. Tang C, Chen V (2002) Nanofiltration of textile wastewater for water reuse. Desalination 143:11–20. https://doi.org/10.1016/S0011-9164(02)00216-3
25. Tezcanli-Güyer G, Ince NH (2004) Individual and combined effects of ultrasound, ozone and UV irradiation: a case study with textile dyes. Ultrasonics 42:603–609. https://doi.org/10.1016/j.ultras.2004.01.096
26. Schäfer AI, Beder S (2006) Relevance of the precautionary principle in water recycling. Desalination 187:241–252. https://doi.org/10.1016/j.desal.2005.04.083

27. Hayat H, Mahmood Q, Pervez A, Bhatti ZA, Baig SA (2015) Comparative decolorization of dyes in textile wastewater using biological and chemical treatment. Sep Purif Technol 154:149–153. https://doi.org/10.1016/j.seppur.2015.09.025
28. Khatri A, Peerzada MH, Mohsin M, White M (2015) A review on developments in dyeing cotton fabrics with reactive dyes for reducing effluent pollution. J Clean Prod 87:50–57. https://doi.org/10.1016/j.jclepro.2014.09.017
29. Burkinshaw SM, Salihu G (2017a) The role of auxiliaries in the immersion dyeing of textile fibres: Part 4 theoretical model to describe the role of liquor ratio in dyeing cellulosic fibres with direct dyes in the absence and presence of inorganic electrolyte. Dye Pigment. https://doi.org/10.1016/j.dyepig.2017.08.052
30. Petek J, Glavic P (1996) An integral approach to waste minimization in process industries. Conserv Recycl 17:169–188. https://doi.org/10.1016/0921-3449(96)01151-2
31. Gulzar T, Adeel S, Hanif I, Rehman F, Hanif R, Zuber M, Akhtar N (2015) Eco-friendly dyeing of gamma ray induced cotton using natural quercetin extracted from Acacia Bark (A. nilotica). J Nat Fibers 12:494–504. https://doi.org/10.1080/15440478.2014.964445
32. Harifi T, Montazer M (2015) A review on textile sonoprocessing: a special focus on sonosynthesis of nanomaterials on textile substrates. Ultrason Sonochem 23:1–10. https://doi.org/10.1016/j.ultsonch.2014.08.022
33. Benli H, Bahtiyari MI (2015) Combination of ozone and ultrasound in pretreatment of cotton fabrics prior to natural dyeing. J Clean Prod 89:116–124. https://doi.org/10.1016/j.jclepro.2014.11.007
34. Presa P, Forte Tavcer P (2009) Low water and energy saving process for cotton pretreatment. Text Res J 79:76–88. https://doi.org/10.1177/0040517508092019
35. Hussain T, Tausif M, Ashraf M (2015) A review of progress in the dyeing of eco-friendly aliphatic polyester-based polylactic acid fabrics. J Clean Prod 108:476–483. https://doi.org/10.1016/j.jclepro.2015.05.126
36. Terinte N, Manda BMK, Taylor J, Schuster KC, Patel MK (2014) Environmental assessment of coloured fabrics and opportunities for value creation: spin-dyeing versus conventional dyeing of modal fabrics. J Clean Prod 72:127–138. https://doi.org/10.1016/j.jclepro.2014.02.002
37. Xie K, Cheng F, Zhao W, Xu L (2011) Micelle dyeing with low liquor ratio for reactive dyes using dialkyl maleic acid ester surfactants. J Clean Prod 19:332–336. https://doi.org/10.1016/j.jclepro.2010.10.006
38. Long JJ, Xu HM, Cui CL, Wei XC, Chen F, Cheng AK (2014) A novel plant for fabric rope dyeing in supercritical carbon dioxide and its cleaner production. J Clean Prod 65:574–582. https://doi.org/10.1016/j.jclepro.2013.08.008
39. Montero GA, Smith CB, Hendrix WA, Butcher DL (2000) Supercritical fluid technology in textile processing: an overview. Ind Eng Chem Res 39:4806–4812. https://doi.org/10.1021/ie0002475
40. Haque ANMA, Hannan M, Masud Rana M (2015) Compatibility analysis of reactive dyes by exhaustion-fixation and adsorption isotherm on knitted cotton fabric. Fash Text 2. https://doi.org/10.1186/s40691-015-0026-9
41. Banchero M (2013) Supercritical fluid dyeing of synthetic and natural textiles—a review. Color Technol 129:2–17. https://doi.org/10.1111/cote.12005
42. Kumar PS, Gunasundari E (2018) Detox fashion. Springer Nature Singapore Pte Ltd., Detox Fashion, Textile Science and Clothing Technology. https://doi.org/10.1007/978-981-10-4783-1
43. Yan J, Zheng LJ (2011) Study on dyeing of the plasma modification silk fabric in supercritical carbon dioxide. Adv Mater Res 175–176:661–666. https://doi.org/10.4028/www.scientific.net/AMR.175-176.661
44. Teli MD, Samanta KK, Pandit P, Basak S, Chattopadhyay SK (2015) Low-temperature dyeing of silk fabric using atmospheric pressure helium/nitrogen plasma. Fibers Polym 16:2375–2383. https://doi.org/10.1007/s12221-015-5166-4
45. Kan CW, Cheung HF, Chan Q (2016) A study of plasma-induced ozone treatment on the colour fading of dyed cotton. J Clean Prod 112:3514–3524. https://doi.org/10.1016/j.jclepro.2015.10.100

46. Khalil E (2015) Sustainable and ecological finishing technology for denim jeans. AASCIT Commun 2. https://doi.org/10.6084/M9.FIGSHARE.1480400
47. The European Commission (2003) Integrated pollution prevention and control. Reference document on best available techniques for the textiles industry 626
48. Hasanbeigi A, Price L (2015) A technical review of emerging technologies for energy and water efficiency and pollution reduction in the textile industry. J Clean Prod 95:30–44. https://doi.org/10.1016/j.jclepro.2015.02.079
49. Singh R, Kumar M, Mittal A, Mehta PK (2016) Microbial enzymes: industrial progress in 21st century. 3 Biotech 6. https://doi.org/10.1007/s13205-016-0485-8
50. Varadarajan G, Venkatachalam P (2016) Sustainable textile dyeing processes. Environ Chem Lett 14:113–122. https://doi.org/10.1007/s10311-015-0533-3
51. Hassan MM, Bhagvandas M (2017) Sustainable low liquor ratio dyeing of wool with acid dyes: effect of auxiliaries on agglomeration of dye molecules in a dyebath and dyeing uniformity. J Clean Prod 152:464–473. https://doi.org/10.1016/j.jclepro.2017.03.139
52. Öner E, Sahinbaskan BY (2011) A new process of combined pretreatment and dyeing: REST. J Clean Prod 19:1668–1675. https://doi.org/10.1016/j.jclepro.2011.05.008
53. Ali S, Mughal MA, Shoukat U, Baloch MA, Kim SH (2015) Cationic starch (Q-TAC) pretreatment of cotton fabric: Influence ondyeing with reactive dye. Carbohydr Polym 117:271–278. https://doi.org/10.1016/j.carbpol.2014.09.064
54. Fu S, Farrell M, Hauser P, Ankeny M (2016) Influence of liquor ratio and amount of dyestuff in producing ultradeep black dyeing using mercerised and cationised cotton. Color Technol 132:232–237. https://doi.org/10.1111/cote.12216
55. Burkinshaw SM, Salihu G (2017b) The role of auxiliaries in the immersion dyeing of textile fibres: Part 11 residual inorganic electrolyte levels present during the wash-off of commercial grade reactive dyes. Dye Pigment 0–1.https://doi.org/10.1016/j.dyepig.2017.12.006
56. Zhang Y, Zhang W (2015) Clean dyeing of cotton fiber using a novel nicotinic acid quaternary triazine cationic reactive dye: salt-free, alkali-free, and non-toxic by-product. Clean Technol Environ Policy 17:563–569. https://doi.org/10.1007/s10098-014-0821-9
57. Hashem M, El-Bisi M, Sharaf S, Refaie R (2010) Pre-cationization of cotton fabrics: an effective alternative tool for activation of hydrogen peroxide bleaching process. Carbohydr Polym 79:533–540. https://doi.org/10.1016/j.carbpol.2009.08.038
58. Chico D, Aldaya MM, Garrido A (2013) A water footprint assessment of a pair of jeans: the influence of agricultural policies on the sustainability of consumer products. J Clean Prod 57:238–248. https://doi.org/10.1016/j.jclepro.2013.06.001
59. MacCarthy BL, Jayarathne PGSA (2012) Sustainable collaborative supply networks in the international clothing industry: a comparative analysis of two retailers. Prod Plan Control 23:252–268. https://doi.org/10.1080/09537287.2011.627655
60. Chougule MB, Sonaje NP (2012) Novel techniques of water recycling in textile wet processing through best management practices (Bmp's). Int J Appl Sci Adv Technol 1:29–33
61. Chen L, Wang L, Wu X, Ding X (2017) A process-level water conservation and pollution control performance evaluation tool of cleaner production technology in textile industry. J Clean Prod 143:1137–1143. https://doi.org/10.1016/j.jclepro.2016.12.006
62. Maia LC, Alves AC, Leão CP (2013) Sustainable work environment with lean production in textile and clothing industry. Int J Ind Eng Manag 4:183–190
63. Kocabas AM, Yukseler H, Dilek FB, Yetis U (2009) Adoption of European Union's IPPC Directive to a textile mill: analysis of water and energy consumption. J Environ Manag 91:102–113. https://doi.org/10.1016/j.jenvman.2009.07.012
64. Ozturk E, Karaboyaci M, Yetis U, Yigit NO, Kitis M (2015) Evaluation of integrated pollution prevention control in a textile fiber production and dyeing mill. J Clean Prod 88:116–124. https://doi.org/10.1016/j.jclepro.2014.04.064
65. Alkaya E, Demirer GN (2014) Sustainable textile production: a case study from a woven fabric manufacturing mill in Turkey. J Clean Prod 65:595–603. https://doi.org/10.1016/j.jclepro.2013.07.008

66. Sarwar N, Mohsin M, Bhatti AA, Ahmmad SW, Husaain A (2017) Development of water and energy efficient environment friendly easy care finishing by foam coating on stretch denim fabric. J Clean Prod 154:159–166. https://doi.org/10.1016/j.jclepro.2017.03.171
67. Nieminen E, Linke M, Tobler M, Beke BV (2007) EU COST Action 628: life cycle assessment (LCA) of textile products, eco-efficiency and definition of best available technology (BAT) of textile processing. J Clean Prod 15:1259–1270. https://doi.org/10.1016/j.jclepro.2006.07.011
68. Sandin G, Peters GM, Svanström M (2013) Moving down the cause-effect chain of water and land use impacts: an LCA case study of textile fibres. Resour Conserv Recycl 73:104–113. https://doi.org/10.1016/j.resconrec.2013.01.020
69. Marbek resource consultants (2001) Identification and evaluation of best available technologies economically achievable (BATEA) for textile mill effluents. http://www.p2pays.org/ref/41/40651.pdf
70. United States Department of Energy (U.S. DOE) (2006) Industrial Assessment Center (IAC), Industrial Assessment Center (IAC) Database. Department of Energy, Washington, DC, USA. http://iac.rutgers.edu/database/index.php
71. Koc E, Kaplan E (2007) An investigation on energy consumption in yarn production with special reference to ring spinning. Fibres Text East Europe 15, 63, 4. https://www.infona.pl/resource/bwmeta1.element.baztech-1706ad9a-9539-469c-8b61-05f0054de3cb
72. United States Department of Energy (U.S. DOE) (2005) Improving ventilation system energy efficiency in a textile plant. http://www1.eere.energy.gov/industry/bestpractices/casestudyventilationtextile.html
73. Khude P (2017) A review on energy management in textile industry. Innov Ener Res 6, 169, 2. https://www.researchgate.net/profile/Prakash-Khude/publication/321133082_A_Review_on_Energy_Management_in_Textile_Industry/links/5a0ed4910f7e9bdf9111afa8/A-Review-on-Energy-Management-in-Textile-Industry.pdf
74. Sharma S (2012) Energy management in textile industry. Int J Power Syst Oper Energy Manag 2(1):2. ISSN (PRINT). https://www.idc-online.com/technical_references/pdfs/electrical_engineering/ENERGY%20MANAGEMENT%20IN.pdf
75. Dhayaneswaran Y, Ashokkumar L (2013) A study on energy conservation in textile industry. J Inst Eng (India): Series B 94:53–60. https://link.springer.com/article/10.1007/s40031-013-0040-5
76. Bhaskar MS, Verma P, Kumar A, Delhi N (2013) Indian textile industries towards energy efficiency movement. Int J Environ Sci Dev Monit (IJESDM) 4:36–39. https://citeseerx.ist.psu.edu/document?repid=rep1&type=pdf&doi=bb9a85fe52a38d027ecfa4dc3265390d4500f591
77. Worrell E, Bode JW, de Beer J (1997) Energy Efficient Technologies in Industry (ATLAS project for the European Commission). Utrecht University, Utrecht, the Netherlands. https://www.osti.gov/etdeweb/biblio/600451
78. Abdel-Dayem AM, Mohamad MA (2001) Potential of solar energy utilization in the textile industry—a case study. Renew Energy 23(3–4):685–694. https://www.sciencedirect.com/science/article/abs/pii/S0960148100001543
79. Goyal S (2011) Energy conservation in textile industry. Lambert Academic Publishing

Circular Economy and Future Prospective

Closing the Loop: Advancing Circular Economy Practices in the Global Textile Industry

Fizzah Ayub, Syeda Laiba Haider Naqvi, Syeda Hamayal Zahra Naqvi, Abdullah Yasar, and Rameesha Tanveer

Abstract This chapter discusses the revolutionary idea of a circular economy within the textile industry and highlights how it could entirely change the manner in which textiles are made, used, and disposed of. The second-most polluting industry in the world is textile industry, which is recognized for leaving a significant environmental footprint. The textile industry consumes 35% of all pesticides. A feasible way of dealing with these environmental issues while also generating economic benefits is to move towards a circular economy. Moreover, the chapter describes how a circular strategy might take the place of the traditional linear model, which is characterized by resource depletion and waste generation. The environmental effects of the industry are considerably reduced in a circular economy by the utilization of recycled or sustainably sourced materials, energy-efficient procedures, and eco-friendly colors. Furthermore, the circular economy encourages local manufacturing and distribution, which reduces distant travel's energy consumption and emissions. The requirement for frequent substitutions is reduced, resources are conserved, and waste is reduced by enhancing the lifespan of textiles through resilient design and maintenance procedures. Stronger governance, collaboration, and changes in consumer behavior are essential in accelerating this transition. Governments must establish laws and policies that encourage sustainability. While funding and cooperation are essential for innovation and systemic change. Developing the full potential of a circular economy in the textile industry will depend on changing customer behaviors through education and easily accessible alternatives that are environmentally friendly.

Keywords Circular economy · Textile industry · Sustainability · Closed loop · Environmental impact

---

F. Ayub (✉) · S. L. H. Naqvi · S. H. Z. Naqvi · A. Yasar
Sustainable Development Study Centre, Faculty of Physical Sciences, Government College University, Lahore, Pakistan
e-mail: fizzah.ayoub@gmail.com

A. Yasar
e-mail: yasar.abdullah@gcu.edu.pk

R. Tanveer
Australian Rivers Institute, School of Environment and Science, Griffith University, 10 Southport, West Leederville, Queensland 4222, Australia

P. Singh (ed.), *Dye Pollution from Textile Industry*, SDGs and Textiles,
https://doi.org/10.1007/978-981-97-5341-3_15

1 Introduction

The economy of a nation depends on the production and consumption of products and services as well as financial activity. Money, production, and consumption habits have traditionally influenced international relations, but creating sustainable global economic systems is a difficult task. Biological and technological phases are the focus of the circular economy [85]. A key component of sustainable development is the idea of a circular economy. Its goal is to maximize products' lifespans as much as feasible in order to minimize the need for new energy and material inputs and the consequent increase in a product's individual carbon footprint [63]. Instead of discarding the end-of-life product, a solution can be to reuse it or recycle it, updating or reusing it as needed to create a brand-new product that is identical, comparable, or completely distinctive. This new product can be recycled or reprocessed multiple times after it has been used, making the process circular [91]. The so-called linear economy, which traditionally uses goods only once and discards them when they are no longer useful, is opposed by the circular economy. Since practically all materials exist in cycles naturally and there are no dead ends, the circular economy concept is not actually new but rather a return to the natural order. Each process's final result serves as the raw material for the next step [103].

The worldwide textile industry has increased its output and consumption throughout time, but this growth additionally caused a number of problems because of linearity of the economy and production. There are numerous possibilities as well as difficulties presented by current efforts to decrease waste and achieve zero waste, which are affecting not just manufacturing and production standards and procedures but also promotional and marketing activities. Leading international companies are declaring their commitments to the circular economy and their goals will be achieved soon. The textile industry has started on a new journey to investigate how to use the concepts of circular economy at all the stages of product development [86]. Although the textile industry can be referred to as the driving force for industrialization, developing nations are currently responsible for a significant portion of the world economy. The industry contributed about a total of 877 billion dollars to global trade in 2019 and employed more than 70 million people within its supply chain thus increasing export revenues, creation of jobs, and the improvement of living standards. For instance, they have textile industries which are crucial for countries like Bangladesh, India, and even Vietnam which generates remarkable 18% of total export revenue or in Bangladesh it contributes an astonishing 80% towards their economies [96]. In spite of positive socioeconomic significance of this industry, the environment tolerates heavy costs for it. However, its negative environmental impact is also disturbing. It is the second most polluting industry after oil, with 10% of global greenhouse gas (GHG) emissions. Textile factories have been releasing wastewater containing dye effluents into rivers thereby causing adverse conditions in aquatic ecosystems [76]. With almost 20% of global industrial water pollution results from its unwashed dye about 200,000 tons. This has led to the contamination of the groundwater, which in turn results in a number of health problems including those related to consumption by both human beings and animals. Moreover, high use of water from cotton field irrigation coupled with manufacturing processes leads to serious consequences. For example, the drying up of Northern Aral Sea in Uzbekistan and displacement of people is a major challenge that has been associated with high use of water [57].

Apart from water-related issues, the textile industry has many other environmental matters to face such as degradation of soils, deforestation, microfiber pollution in seawater, use of chemical additives, and the disposal of millions of dollars ($3 billion) worth of used fashion logistics, estimating mistakes, unsustainable consumption creates a sustainability void [31]. The importance of eco-friendly solutions in the textile industry is highlighted by modern research with regard to environmental damage linked to social sustainment. However, there is what can be described as a value-chain driven by costs and efficiency that drives unchangeable ecological and societal destruction. It is ironical that though modernization of technologies is largely regarded as a way to cut back production costs and be able to produce more efficiently, it ultimately just generates additional requests for resources, increasing the rate extraction [51]. However, only technological solutions, ethical fashion consumption and sustainable raw materials' production are not sufficient efforts to tackle these environmental problems. For the industry to fit into the global warming constraint set at 1.5 °C, a paradigm shift from linear towards circular is predictable. Textile industry should adopt circular economy philosophy in order to move away from linear model that is characterized by waste thus promoting sustainability and environmental conservation [43].

The importance of shifting to the circular economy by emphasizing sustainability in the textile industry cannot be overlooked. It is inappropriate to ignore the fact that single-use products created to correspond with the linear economy have resulted in significant pollutants, releases of greenhouse gases, and mineral exhaustion. The textile industry is one of the most significant industries involved in the economic development of countries with limited resources, which have made an impact on the global economic stage through the manufacturing of goods and services for the unemployed people [110]. A strategy in reducing and mitigation of the environmental effects associated with textiles is the concept of the circular economy, stressing on the constant reusing, reprocessing, and recycling. International firms make commitments about their support of circular economy principles while efforts to restrict waste and introduce a zero-waste paradigm change in production standards. The transformative journey involves putting into place circularity at each step of the product development process enhancing sustainability and conservation of the environment. The growth of the economy has led to an increased focus on the environmental impact of the industry as a necessary step toward sustainability in harmony with the earth [54].

2 Circular Economy Fundamentals

2.1 Comparison of Linear and Circular Economy

The predominant model for economic development has been the linear economy approach, which has its origin in the industrial revolution of the eighteenth century. It is centered on mass production and consumption and is distinguished by a "use-dispose" philosophy. Although this paradigm was successful in promoting economic growth, its negative effects on the environment and society have made it unsustainable in the twenty-first century [76] (Fig. 1).

It causes pollution, the depletion of resources, and destruction to important natural resources like soil, water, and air. Furthermore, it contributes to socioeconomic problems like poverty, overcrowding, and health difficulties. The linear economy promotes a culture of disposability since it is basically extractive and wasteful. It often neglects the long-term effects of resource extraction due to its emphasis on short-term gains [97].

The circular economy approach, on the other hand, presents a revolutionary alternative. It is based on four key naturalistic principles: resilience, using waste as nutrients, utilizing renewable energy, and a systemic approach to thinking [62]. The natural cycle of waste becoming a resource for other processes is represented by waste as nutrients. This refers to recycling and modifying products in the textile industry to increase usability and reduce waste. Similar to biodiversity in nature, resilience encourages a broad pool of resources to provide adaptability to changing conditions [54] (Fig. 2).

A cooperative system that maintains resource flows must incorporate renewable energy sources. The systemic thinking focuses on how parts are interconnected, providing opportunities for positive effects on the environment when the circular economy is implemented [32].

The demands of the environment and society are satisfied by this holistic approach, which considers both living things and financial gains. Mass production and marketing that prioritize speed and scale define the linear economy. It has a significant negative impact on the environment since it speeds up production and wear-and-tear cycles. On the other hand, the circular economy is regenerative and

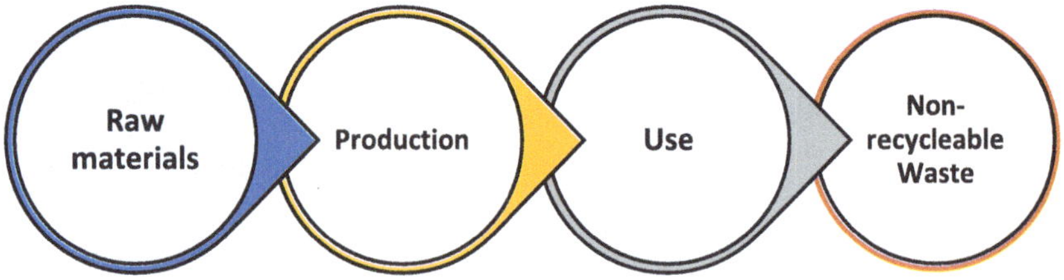

Fig. 1 Linear economy model (self-made)

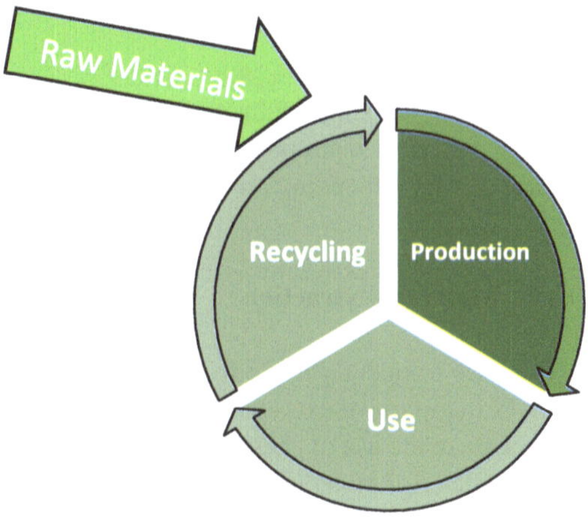

Fig. 2 Circular economy model (self-made)

restorative and follows natural cycles [15]. It attempts to ensure a renewable flow of resources and goods while protecting natural capital, maximizing resources, lowering risks, and reducing costs. Responsible resource usage and appropriate working conditions are necessary for the preservation of natural capital [9]. Optimization tries to prevent excessive utilization of resources that many sectors experience. Risk minimization considers the important function that resource regeneration plays in maintaining ecosystem equilibrium Furthermore, a renewable flow of resources and goods emphasizes the requirement for a more considerate and sustainable approach by considering time as a key aspect in production and consumption [47] (Fig. 3).

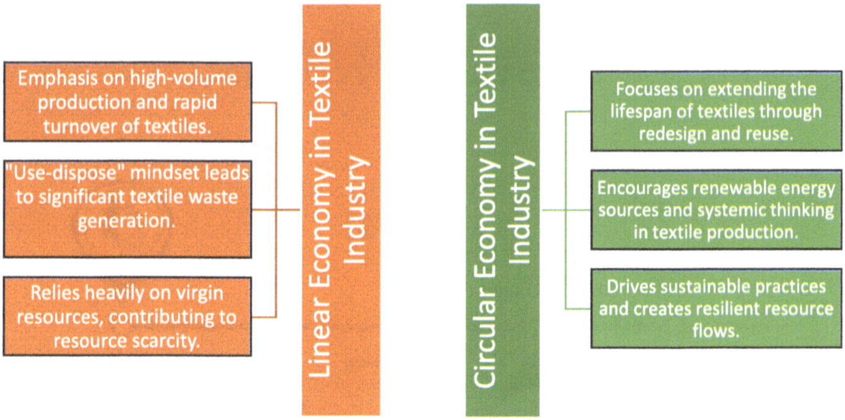

Fig. 3 Shifting paradigms linear and circular economy in textiles (self-made)

2.2 *Life Cycle of Textiles*

The lifecycle of textiles includes a number of stages, including the extraction of raw materials, product manufacturing, distribution, usage, and disposal. Environmental influences vary depending on the stage (Fig. 4).

2.2.1 Material Extraction

Textiles are frequently made from virgin raw materials in a linear economy, which involves the massive extraction of natural resources like cotton, wool, or synthetic fibers. Due to the use of chemicals and pesticides, this method causes habitat disruption, soil deterioration, and water contamination. The use of recycled or sustainably obtained materials is given importance in the circular economy, which lowers the demand for virgin resources. This lessens the environmental impact of resource extraction and assists in the preservation of natural environments [58].

2.2.2 Product Manufacturing

In a linear economy, mass production techniques frequently require the use of toxic chemicals for textile dyeing and finishing as well as highly energy-intensive operations. This increases releases of greenhouse gases, contamination of the water, and air pollution. The environmental impact of textile production can be considerably reduced by using energy-efficient processes, eco-friendly dyes, and cleaner production methods in a circular economy [28].

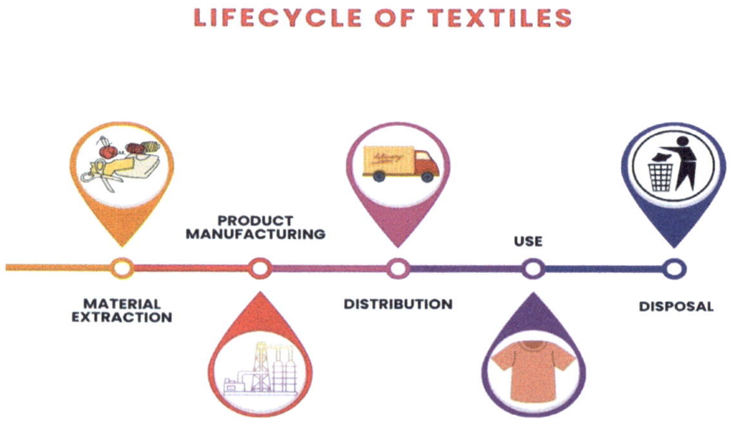

Fig. 4 Life cycle of textiles (self-made)

2.2.3 Distribution

In a linear economy, textiles are often manufactured centrally and distributed internationally, resulting in high energy and emissions related to transportation. Although in a circular economy, local manufacturing and distribution infrastructures can reduce energy use and emissions associated with transportation, encouraging more environmental friendly purchasing habits [43].

2.2.4 Use

In a linear model, products are frequently made for brief periods of time, which promotes quick disposal and more frequent replacements. This starts a loop of constant resource use. However, in the circular economy model, accelerating the lifespan of textiles through resilient design, replace, and maintenance methods minimizes the requirement for numerous replacements, preserving resources and lowering waste [77].

2.2.5 Disposal

In a linear economy, textiles that have reached the end of their useful lives are regularly discarded, polluting landfills and perhaps releasing harmful chemicals into the environment. Air pollution is also a result of incinerating waste. Implementing efficient recycling and upcycling procedures while utilizing a circular economy enables textiles to be reused or transformed into new goods, eliminating waste and eradicating the need for new raw materials [98].

2.2.6 Reuse and Recycling of Textiles

Reusing and recycling textiles can generally benefit the environment by reducing the amount of virgin textile fiber produced and by removing certain processes that appear later in a textile product's life cycle. Moreover, the recycling and reuse of textiles are more advantageous to the environment than landfilling and incineration [119]. On the other hand, reuse is often considered to be more beneficial than recycling, especially when the process of reusing phase lasts longer considerably. The term "textile reuse" describes a range of methods used to extend a textile product's useful life from one owner to the next. Renting, dealing, swapping, borrowing, and acquiring are common ways to achieve this. Garage sales, secondhand stores, online marketplaces, and charitable donations all make this possible [8]. Textile recycling is the process of reclaiming pre- and post-consumer textile waste to make new textile or non-textile products. Textile recycling is typically categorized as mechanical or chemical recycling. By using mechanical recycling, waste is decreased and made useful for landscaping, building, agriculture, and attractiveness. Chemical recycling involves the dispersion or degradation of polymers [112]. Chemical recycling can provide fibers that are on scale with virgin materials. Organized textile waste could be chemically treated to provide materials such as protein-based fibers for wood panel adhesives and cellulosic fibers for the production of bioethanol [60].

The textile recycling pathway can be divided into many categories based on the kinds of techniques employed or the extent of disintegration of the recovered materials. Fabric cannot be recycled unless it is recovered and applied to new products. However, fiber recycling involves separating a fabric while keeping the original fibers. Reusing polymers or oligomers involves breaking down fibers without destroying the polymers or oligomers. Furthermore, monomer recycling calls for breaking down polymers or oligomers while preserving the monomers [50]. Moreover, there are three types of textile recycling: upcycling, closed-loop, and open-loop. When a product made from recycled materials is more valuable or high-quality than its original design, it's referred to as upcycling, downcycling is the opposite of this. Reusing recycled materials to create a product that is almost identical to the original is known as closed-loop recycling. However, open-loop recycling makes use of materials from one product and utilizes them into another [88].

The closed-loop recycling method recovers the raw materials used to make polymer products and reprocesses them to create new products of comparable quality to those made from virgin materials. Technologies for recycling and reusing textiles are extensively accessible and have been improved through time to promote mixed fabrics [20]. Reuse and recycling are essential components of the textile industry's shift to a circular economy for reducing environmental impact. The textile industry adopts a more environmental friendly strategy by increasing the lifespan of textile products through activities [24]. Recycling provides a technique to reprocess textile waste for use in new goods while lowering the dependency on virgin resources, whether through mechanical or chemical processes. This change to closed-loop recycling is consistent with the ideas of the circular economy, where materials are constantly regenerated and reused, resulting in a textile industry that is more efficient with resources and environmentally responsible [111].

3 Circular Economy in Textile Industry

The need for a shift to a circular economy is clearly demonstrated by the textile industry. Its current linear model results in significant waste, depletion of resources, and environmental deterioration because it is one of the most polluting industries in the world. Every phase of the lifetime of a textile product produces significant waste, from pre-consumer textile scraps through post-consumer clothing that is discarded as a result of shifting fashion trends [92]. The industry can change its methods by adopting the circular economy. In addition to recycling, this also incorporates upcycling, improving final products, and lowering environmental impact. With an

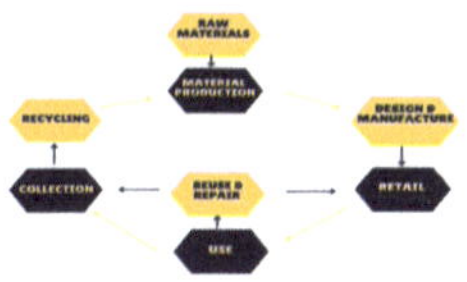

Fig. 5 Circular economy in the textile industry (self-made)

emphasis on transparency, resource efficiency, and responsible consumption, companies like Nudie Jeans, Patagonia, and Pratibha Syntex are already setting the standard for sustainable business practices [91].

A paradigm change in thinking about economics is represented by the switch from a linear to a circular economy. Although historically successful, the linear model now creates serious problems for sustainability and the well-being of society [6]. The circular economy offers a more comprehensive and regenerative approach because it is based on naturalistic principles [100]. Industries like textiles may lessen their environmental effect and promote a more substantial and inclusive economic system by giving priority to resource efficiency, less waste, and sustainable practices. The circular economy is not only a theoretical framework but also a real-world and revolutionary path towards a more sustainable future, as demonstrated by modern businesses and projects [94] (Fig. 5).

4 Global Perspectives on the Textile Industry

The economic circumstances of several countries are significantly affected by the textile industry. We will examine the historical growth and current situation of the textile industries in Georgia, India, Pakistan, China, England, and America, identifying significant advancements and difficulties.

4.1 Georgia

Georgia is a country in the former Soviet Republic that is situated where Asia and Europe converge. The state's population was significantly impacted by the textile industry as Georgia developed to become the South's top manufacturer of textiles. In the beginning, silk was produced, but once the cotton gin was created, cotton manufacturing took over as the leading industry [67]. Small mills started to appear, and the number increased from 19 in 1840 to 38 in 1850. Major increases occurred throughout the 1880s and 1890s as a result of the development of steam power, and in 1899 the Georgia Institute of Technology established the Textile Department to offer undergraduates a proper education. For vehicle business, carpet, bedspread, and tire fabric manufacturing began in the 1900s [13]. American military uniforms were in high demand during World War I, and several mills received government contracts. The Civil War, the boll weevil, and racial segregation all had an impact on the development of the textile sector. Nylon was used in World War II, and specialized fabrics were created. The North American Free Trade Agreement (NAFTA), which ended barriers and enabled the entrance of low-cost Asian textile businesses, changed the sector by means of modernization and technology in 1994 [23].

4.2 India

For both domestic use and export, textiles have historically been a major commodity in Indian Ocean civilizations. There were approximately 132 weaving centers in South India between 1000 and 1500 AD, according to inscriptional evidence [92]. The Industrial Period in the eighteenth century witnessed the development of some 70 mechanical technological developments. The old industries declined as a result of the opening of Suez Canal in 1869, the growth of shipping, and the construction of railroads, which created international competition [83]. Due to the East India Company's power in the seventeenth century, Manchester textiles were given up to be sold in India, while Indian commodities were marketed in British marketplaces. Cotton prices increased significantly as a result of the US Civil War, which was unfavorable for the cotton textile business [26]. To counter the Indian-owned textile factories, the British established mills. According to a statistic from 1936, India's two million handlooms produced 1400 million yards of khadi, or 26% of the entire amount used there [107]. In the 1970s, 1980s, and mid-1990s, cotton spinning had a rapid expansion, however, the industry later suffered from import taxes imposed by the government [25].

Due to an increase in exports to China, there was a boom in the years after World War I. 60% of the nation's employed people were working in Bombay, which at this time was the heart of the cotton industry [84]. Despite handlooms having issues, 818 million yards of cloth were produced during World War II as opposed to 177 million yards in 1938–1939 [3]. Following independence, the sector saw significant changes, including technological advancements in machinery, fibers, and the introduction of synthetics, which experienced a tenfold increase from 1.5% to 15% between 1965 and 1975 [82]. Power looms were introduced to the weaving business during the twentieth century. Decentralized small-scale power looms were created as a result of several changes to structures and legislation, and they turned out to be more efficient financially [81].

4.3 Pakistan

Humans require clothes and textiles for their survival; the global demand for textiles is estimated to be worth 18 trillion US dollars and is predicted to rise by 6.5%. Many other nations throughout the world mostly import textile products from Asia

[74]. Pakistan is now the world's main producer and exporter of textile products, particularly cotton textiles. However, since there is intense competition from other countries, Pakistani textile products are no longer as cheap as they once were [48]. The goal of the 5-year textile policy, which the Ministry of Textile Industry developed in 2004, was to raise the overall target for exports of textile goods from US$13.8 billion to US$25 billion at the end of the policy period. The policy's most significant development was the provision of funds, specifically the Textile Investment Support Fund and the Technology Upgradation Fund. The primary goal of granting this money was to address infrastructural and technical gaps by opening up new investment possibilities, but unfavorably this strategy has not yet been implemented [56].

The textile industry struggled to implement strategies and policies because of ineffective financial management. The sector today faces enormous liabilities under plans for the operational process, but it also faces difficulties with execution and the timely satisfaction of its fundamental demands [1]. In contrast to the allowed financing plan of Rs. 123 billion for 2009–12, a total of Rs. 24.75 billion was granted. However, the planning commission has been holding up projects worth more than Rs. 6 billion for the past 2 years on the grounds that there is a severe problem with Pakistan's textile industry which is caused by a lack of essential utilities like gas, which is crucial for the sector [61].

Due to the lack of gas and electricity in the textile industry, which led to the closure of 10% of small- and medium-sized textile manufacturing facilities, thousands of textile workers who were employed in the industry or were paid on a daily basis went on the streets of Faisalabad in March 2012 [52]. They burned tires and shouted anti-government slogans. Due to all of these factors, the textile sector is at least unable to meet home demand and cannot quickly consume its domestic cotton to meet export orders [66]. Due to old-fashioned textile machinery and equipment, Pakistan's textile sector has huge production difficulties. The sector requires new investments to handle this challenge [108]. Since a long time ago, this industry has not seen any fresh significant investments. Nobody wants to invest in Pakistan because of the country's present situation. Given this, it appears impossible for Pakistani textiles to compete in global markets [10].

4.4 China

In order to support its weaving industry in Northern China, China obtained aid from British India in the late 1800s by importing cotton yarn. Operating in China offered several benefits, including access to a substantial market, an abundance of raw materials, economical transportation, cost-effective labor, and longer working hours [117]. Chinese businesses owned locally coexisted with foreign businesses, primarily British and Japanese. These businesses profited from lessened political limitations and tariff allowances following the Sino-Japanese War [113]. As a result of the industrial pause brought on by the start of World War I, China and Japan both developed

Table 1 Global perspective on circular economy initiatives and challenges in textile industry

Country	Circular economy initiatives	Challenges	References
India	Historical emphasis on local trade; modern efforts in sustainable production and recycling	Competition with fast fashion; environmental pollution	[84]
Pakistan	Limited progress due to financial constraints and utility shortages	Shortage of utilities, policy implementation issues	[1]
China	Strong focus on industry upgradation and green practices; Shift towards value-added products	Transition from export-focused to domestic market; environmental concerns	[42]
England	Historical contributions to textile industry advancements; recent efforts towards sustainability	Industry consolidation and modernization; global competition	[19]
America	Historical significance in textile production; recent shifts towards sustainable practices	Offshoring of production; rebuilding of domestic industry	[100]

The US exported $26.3 billion worth of fiber, textiles, and clothing in 2016, and spent $ 2 billion on capital projects to produce such goods in 2015. It was also projected that 565,000 people worked in the country's textile supply chain in 2016 [104]. The expansion of the textile sector from previous periods to the present era demonstrates how the industry has advanced through a number of phases in terms of production, export, expertise, technology, economics, and upliftment [55]. Numerous causes have contributed to the sector's expansion, but in every case examined, this business has supported the lives of numerous families and civilizations [100]. Numerous data and analyses have been conducted to identify the factors contributing to growth, including the availability of inexpensive labor, a large market, technical advancements, human capital, and favorable conditions for the growth of raw materials [17] (Table 1).

5 Environmental Impacts of the Textile Industry

In terms of environmental contamination, the textile industry uses 35% of all pesticides, making it the second-most polluted industry globally. Concerns have been raised about the high amount of chemical pollution and the toxicity of utilizing synthetic dyes to color fabrics. An industry that utilizes a lot of water is the textile industry, which relates to water consumption [95]. As an example, 1 pound of cotton requires 20,000 gallons of water to manufacture. Unworn clothing makes up a significant portion of waste generation roughly 40% of overall output. 90% of customers cannot afford to purchase them, which results in the manufacture of unworn clothing, and 87% of that clothing is burnt or dumped in landfills [55].

The failure to recycle and reuse discarded textile goods, of which only 3% are done so, costs the textile sector $500 million annually. In relation to recycling, the linear manufacturing procedure generated large amounts of waste and a greater labor demand [21]. In America, people choose to wear clothing from other countries with recognized brands because it gives them a status signal that sets them apart from other people. Therefore, if no government has plans to address this issue, the quantity of used textiles will increase in the upcoming years [29]. The textile industry employs 75 million people, the majority of whom are young women. In addition, the textile industry employs 11 out of every 100 children worldwide [98] (Fig. 6).

American producers have a lot of potential for high-fiber raw materials like organic cotton and wool. This is mainly because this area of the world is attractive for promoting sustainable manufacturing because of its large landmass, enough farming space, and rich cultural and textile traditions [5]. Due to the effects of climate change on cotton and wool production, prioritizing local economic growth and encouraging inhabitants to adopt a resilient mindset will be significant strategies in the future [55].

The linear economy in which the textile industry operates has a substantial negative impact on the environment. This conventional paradigm emphasizes a "take-make-dispose" strategy, which causes significant environmental deterioration [63]. For example, the industry relies extensively on chemical processes, and the use of

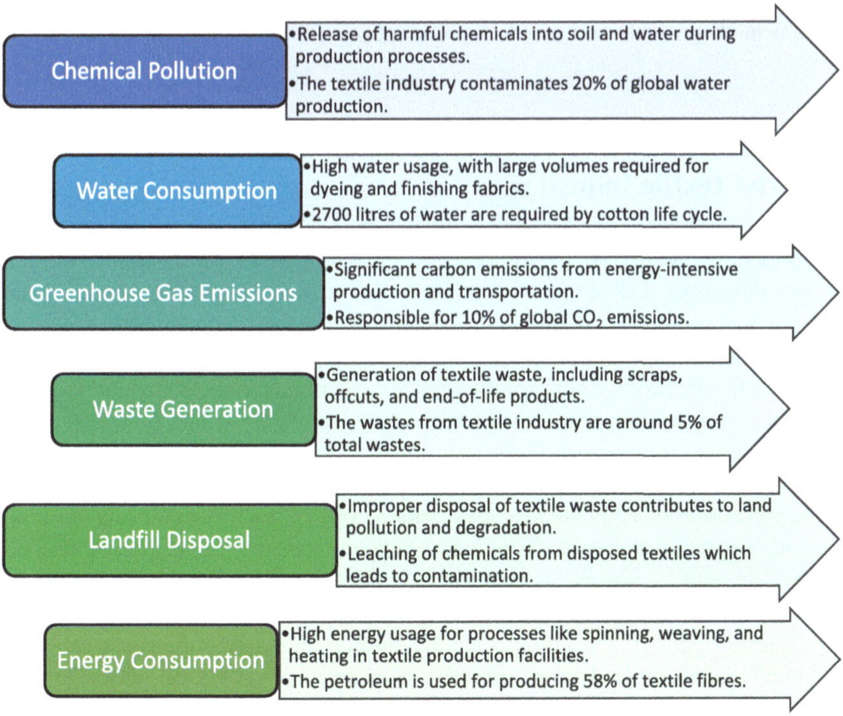

Fig. 6 Environmental impacts of the textile industry (self-made)

synthetic colors raises concerns over toxicity and chemical pollution. The industry is one of the biggest consumers of water, thus the use of water is another significant concern. Furthermore, there is an increasing amount of textile waste being produced, much of it is disposed of in landfills or burned because of financial limitations [97]. The circular economy structure, in contrast, presents a positive difference. By supporting practices like recycling, reusing, and regenerating, it promotes sustainability and resource efficiency. In the textile industry, this corresponds to using eco-friendly materials, implementing closed-loop production techniques, and increasing the lifespan of clothing [109]. The industry may reduce many of the environmental problems caused by the linear model by implementing circularity. As an example, recycling textiles may dramatically lower waste volumes and minimize the load on landfills. Moreover, making the switch to a circular economy has significant economic advantages [54].

The transition of the textile industry from a linear to a circular economy has enormous potential for sustainable growth in addition to being essential for environmental preservation [14]. The textile industry may enhance its capacity for producing premium raw materials like organic cotton and wool by encouraging local economic growth and resilience in the context of climate change consequences. This modification not only complies with global sustainability objectives but also assures that the textile industry will continue to prosper in a way that is beneficial to the environment. A more sustainable, commercially successful, and ecologically sensitive future is made possible by the industry's adoption of circularity [93] (Table 2).

6 Implementation of Circular Economy Principles in the Textile Industry

The circular economy specifically is a new industrial paradigm seeking to replace the prevalent, open-ended linear economy model built on "take-make-dispose" and emerged as an important issue in conversations among practitioners about alternative and sustainable economic production models [36]. A closed production system replaces a linear resource consumption model, where materials are held in loops in order to accomplish longer use and reuse, could significantly alter how businesses use resources to promote sustainable development. This system replaces the "end-of-life" idea with restoration. As a result, implementing circular economy is not simply an environmental strategy but also an economic one [22]. Opportunities to develop more sustainable and circular fashion are abundant considering the modern fashion industry's massive scale and its extremely significant environmental effect [30].

Literature is increasingly focusing on how the textile industry innovates or adapts its business structures to accommodate the circular approach [27]. The creation of sustainable business models is currently focusing on a variety of values through implementing the so-called triple bottom line perspective—economic, environmental, and social. Because business models have both positive and negative effects

Table 2 Overview of key differences of various aspects between linear and circular economy

Aspect	Linear economy	Circular economy
Strategy	Take-make-dispose	Recycle-reuse-regenerate
Environmental impact	The use of chemicals, water consumption, and textile waste all have significant adverse effects on the environment	A focus on environmental practices, resource conservation, and sustainability
Resource usage	A significant reliance on chemical dyes, synthetic substances, and water consumption results in pollution and depletion of natural resources	Closed-loop manufacturing, eco-friendly materials, and extending product life are prioritized
Economic benefits	Traditional structure, but with significant long-term environmental costs with possible financial implications	Reduces waste, encourages environmentally friendly behaviors, and ensures the long-term stability of the industry, all of which have economic benefits
Sustainability goals	The traditional method restricts sustainability efforts and makes environmental issues severe	Corresponds with global environmental goals and encourages resilient, sustainable growth
Future industry outlook	Long-term viability could be restricted by sustainability issues	Puts the industry on the path to a future that is more environmentally conscious, economically successful, and sustainable

on the environment in which they exist [2]. Furthermore, when examining the processes of circular developments, numerous factors are revealed by the literature across the textile value chain, from take-back and reusing to product design, that play a critical role in advancing or inhibiting a company's goals to create a circular product [34]. Circular economy-based sustainable business models may be challenging to put into practice in the textile industry due to issues with scalability, compatibility with value propositions of fashion customers, shifting consumer demands, a lack of appropriate technological understanding, and challenges with altering the supply chain [75].

The sustainable development objectives of economic performance, social inclusion, and resilience to environmental damage led to the establishment of the concept of the circular economy, which has been described as a beneficial system in which resource input and waste, emissions, and energy leakage can be reduced through slowing, separating, and limiting material and energy loops. Considering their similarities, the concepts of sustainability and the circular economy are not the same. The circular economy places an emphasis on environmental advantages and incentives for the economic competitors that adopt the system, while sustainability suggests a comprehensive framework that can be adjusted to numerous circumstances and targets [27]. The textile industry has seen a significant transition during the past 30 years. The circular economy has taken center stage in interactions about how to make businesses, especially the textile industry, more sustainable over the past

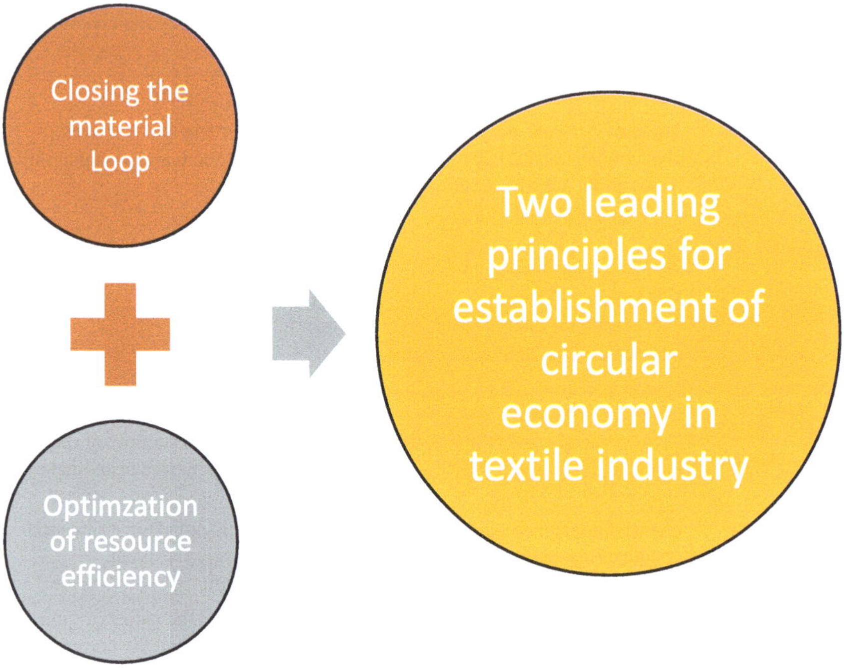

Fig. 7 Circular economy principles (self-made)

few years [110]. The circular economy is a concept of sustainability that attempts to reduce waste and increase resource efficiency. Closing the material loop and optimizing resource efficiency are the two leading principles in establishing a circular economy in the textile industry [18] (Fig. 7).

Considering the full lifespan of textiles, from manufacture to disposal, will be needed by the first principle. This includes utilizing eco-friendly materials, upcycling, and remanufacturing processes to increase the longevity of products, as well as designing items with durability and recyclability in mind [16]. Furthermore, it demonstrates how crucial it is to set up effective collecting and recycling systems in order to recover and reuse utilized textiles. The industry minimizes its dependency on virgin resources and minimizes the load on landfills by reducing waste and keeping products in use [22].

The second principle focuses on increasing resource efficiency throughout the process of production. Adopting advanced manufacturing processes that concentrate on minimizing resource inputs—such as water, energy, and chemicals—while aiming to remove hazardous compounds is necessary to achieve this [117]. It also includes getting alternatives that are sustainable and incorporating renewable energy into manufacturing operations. Adopting closed-loop manufacturing techniques, in which outputs are recycled back into the production cycle, minimizes the impact on the environment [64]. By following these principles, the textile sector not only reduces its environmental impact but also takes advantage of economic possibilities through cost reductions, improved resilience to changes in resource prices, and the development of new, sustainable business models [96].

7 Possible Drivers of Circular Economy in Textile Industry

Three essential wants drive the significant actions essential to advance sustainability and circularity in textile industry. These are

- Requirement of stronger governance to determine transformation
- Requirement for collaboration and funding for the implementation of solutions
- Need to transform the consumption habits.

7.1 Stronger Governance and Strategies

Governments are a key force behind change, but change to a more environmentally conscious and circular textile industry is being inhibited by gaps in policy and legislation. While it has been noted that some nations have implemented supportive programs and that participants in the textile industry are becoming more and more involved in establishing more sustainable and circular business models, more work needs to be done to develop convincing policy frameworks that promote sustainability and circularity in the textile industry [44]. It is necessary for there to be laws and policy frameworks that make it easy for enterprises to switch to new business models. Standards for production and eco-design help level the playing field [38]. Governments have to implement stricter regulations against hazardous materials and unfair work practices [99]. The entire industry, in particular the most problematic areas such as chemicals, resource efficiency, farming practices, and the utilization of non-renewable resources in the manufacture of synthetic fibers—as well as social and environmental issues, must be taken into consideration when formulating such supportive laws [106].

The role of governments is to facilitate a peaceful and equitable transition and incorporating the right groups, such as members of the impacted communities and workforces and their representatives, in the process. Particularly, there lack sufficient implementation mechanisms for encouraging action [46]. This comprises both disincentives, such as taxes on virgin materials and unsustainable behaviors, as well as benefits, including lower taxes on second-hand raw materials, expenditures on research and development, and sustainable public procurement [78]. The latter, in particular, shows potential for boosting request for sustainable textiles since it may allow local and national government agencies to use their public purchasing of textile products to investigate and encourage new business strategies like selling services rather than items [90]. If improved governance is to be achieved, it is also necessary to address the inability of countries to implement laws and the absence of international cooperation between states [89].

7.2 Collaboration and Funding

A sustainable textile industry depends on advanced solutions and innovative business representations, it is becoming increasingly apparent. Leveraging current approaches and best practices will not be sufficient. However, there are still certain gaps in technology, particularly when it comes to the systemic adjustments required to go beyond minor incremental developments [49]. There are also familiarity and experience gaps with regard to new business models, particularly in terms of how to switch from conventional to new circular and resource-efficient company models and how to give the training, support, and resources necessary for new business models to succeed [12] Involving public–private partnerships, new or strengthened events for collaboration, and other appropriate stakeholders will be necessary. All parties must work together to bring about the amount and intensity of change that the textile industry needs, especially the smaller producers that have not done much to advance sustainability [68]. It takes tremendous cooperation across the textile sector to build an effective structure of support, with significant financial commitment and coaching. The idea that circularity is a value chain-wide endeavor and that it is necessary to be incorporated at design stage will be established as a result of our collaboration [53].

Furthermore, there is not enough money available to create and expand the new business models. Moreover, there is an inadequate amount of capital for the industry's yarn, fabric, and textile manufacturing phases, particularly when these are small businesses operating in poor nations [7]. Thus, one essential activity is for partnerships to leverage financial institution investment, particularly in those regions of the world where capital is challenging to leverage. Spaces and systems that enable the intense level of collaboration required are needed [65]. Governments require these opportunities for collaboration in the situation of the requirement for greater governance and policies. The United Nations is well-positioned to offer this support by building on the strong foundation of already-existing networks and forums. The organization has a significant track record in programs that incorporate textiles as well as in the necessary related disciplines [4].

7.3 Change in Consumption Habits

Without significant advocacy, governments and brands/retailers are unlikely to take significant action. In order to achieve this, consumer awareness gaps must be filled, and consumers must be made aware of and inclined toward purchasing sustainable fabrics for usage in clothing and home items [80]. Particularly, there is a need to tackle rapid fashion and overconsumption, as well as to promote behaviors that lessen the effects of garment care on the environment and increase the lifespan of clothing. Furthermore, it necessitates consumer buy-in for new circular business models, such as clothing subscription-rental models [1]. On the other hand, re-use, repair/repurposing, and recycling models call for consumers to bring back their clothing to stores or collection depots and take part in sharing platforms. Consumer education and motivation to participate in the solution are essential [41]. It also indicates that there are enough forward-thinking companies and merchants offering customers sustainable options, allowing them to use their purchasing power [33]. There is a need for fresh, creative campaigns that broaden the impact of already-existing ones, like utilizing social media influencers and UN ambassadors to alter beliefs of what constitutes "fashionable" living. Other methods of motivating consumers should be used, such as discounts or refunds to encourage sustainable purchases, in addition to education and awareness-raising. Most significantly, circumstances must be created so that customers can select sustainable solutions with ease [45].

Moreover, customers require information in order to make moral and sustainable decisions. To help consumers make more informed choices, consumer information tools such as product labeling or reliable company-level analyses are crucial [15]. But for product labels to be less confusing, there needs to be better coordination, as well as steps to make them more applicable to a wider range of items and raise their dependability and relevance. In order to affect consumer behavior, governments, brands/retailers, and civil society organizations must all collaborate [4] (Table 3).

8 Conclusion

Change is not merely a possibility but also a requirement in the textile industry. The transition from a linear economy to a circular one involves more than just a change in strategy. But it also indicates the rise of a more environmentally responsible and sustainable textile industry. Every country deal with its own distinct textile industry background. However, they are all connected by the general requirement for sustainable practices. There are opportunities throughout the textile life cycle. It serves as a call to action for the use of eco-friendly materials, effective production methods, and responsible end-of-life process. The textile industry may considerably minimize its environmental impact by adopting circular economy concepts. The need for change is made more critical by environmental effects. Approaches in the circular economy offer a road map for reducing these negative consequences. The textile industry may reduce its damaging effects on the environment by recycling, reusing, and regenerating textiles. A collective effort is needed to put these concepts into practice. The shift of the textile industry to a circular economy is an investment to a more sustainable future, not just a trend. All parties involved in the journey such as governments, corporations, and customers must work together to make it successful. The textile industry may not only tackle its current difficulties by adopting circularity but also establish the way for a more profitable and more sustainable global economy. A future that values labor and the environment must be integrated immediately.

Table 3 Circular economy drivers in textile industry

Initiatives and actions	Challenges and considerations
1. *Stronger governance and strategies*	
Implementation of supportive programs	Gaps in policy and legislation
Development of convincing policy frameworks	Stricter regulations needed against hazardous materials
Standards for production and eco-design	Addressing social and environmental concerns
Stricter regulations against hazardous materials	Need for international cooperation and implementation mechanisms
Incentives and disincentives for sustainable practices	Balancing disincentives and benefits
2. *Collaboration and funding*	
Involvement of public–private partnerships	Gaps in technology and systemic adjustments for circularity
New or strengthened events for collaboration	Knowledge and experience gaps in transitioning to circular models
Leveraging financial institution investment	Insufficient capital for small businesses in certain regions
Creation of spaces and systems for collaboration	Need for intense cooperation across the textile sector
3. *Change in consumption habits*	
Filling consumer awareness gaps	Tackling rapid fashion and overconsumption
Promoting sustainable fabric usage	Encouraging behaviors that increase clothing lifespan
Offering sustainable options to customers	Providing consumer information tools for informed choices
Coordination for clearer product labels	Collaborative efforts required to affect consumer behavior

The textile industry's transformation to a circular economy is more than a trend. It's an essential investment in building a sustainable future. The effects of traditional linear practices on the environment highlight the need for change. Adopting circular economy principles, such reusing, recycling, and regenerating textiles, offers a practical road map to lessen these adverse effects. This shift represents a comprehensive strategy that corresponds with the increasing worldwide awareness of environmental responsibility rather than just a technical change. The worldwide dedication to sustainable methods unites the distinct histories of the textile industries in each nation. The textile life cycle presents numerous potentials for the implementation of sustainable materials, effective production techniques, and responsible end-of-life procedures. The successful implementation of these ideas involves cooperation between businesses, consumers, and governments. In the textile business, integrating the circular economy is a journey that calls for a convergence of activities and interests. It is critical to understand that this change is a means of achieving both a more profitable and sustainable global economy as well as a solution for current problems. In addition to reducing environmental effect, the circular economy emphasizes the immediate integration of a future that puts greater emphasis on moral and ecological behaviors by valuing both workers and the environment. Since the textile industry is leading the way in this change, it creates the way for more significant socio-economic change and establishes circularity as a pillar of a more ethical and sustainable global future.

References

1. Abbas S, Halog A (2021) Analysis of Pakistani textile industry: recommendations towards circular and sustainable production. Circular Econ: Assess Case Stud 77–111
2. Abdelmeguid A, Afy-Shararah M, Salonitis K (2022) Investigating the challenges of applying the principles of the circular economy in the fashion industry: a systematic review. Sustain Prod Consump 32:505–518
3. Aggarwal A (2021) Circular economy for textiles as engrained in the traditional Indian life. Environ Sci: Indian J 17:180
4. de Aguiar Hugo A, de Nadae J, da Silva Lima R (2021) Can fashion be circular? A literature review on circular economy barriers, drivers, and practices in the fashion industry's productive chain. Sustainability 13(21):12246
5. Aguiar GJA, Almeida LR, Fernandes BS, Gavazza S, Silva GL, Machado Santos S (2023) Use of life cycle assessment as a tool to evaluate the environmental impacts of textile effluents: a systematic review. Environ Sci Pollut Res 1–16
6. Albu AV, Sopota D, Caciora T, Herman GV, Berdenov Z, Ilies A, Ilies DC, Kecse G, Sturzu B, Ghergheles CG (2021) Digitalization of garment in the context of circular economy world 30:31
7. Aloini D, Dulmin R, Mininno V, Stefanini A, Zerbino P (2020) Driving the transition to a circular economic model: a systematic review on drivers and critical success factors in circular economy. Sustainability 12(24):10672
8. Amaral MCD, Zonatti WF, Silva KLD, Karam Junior D, Amato Neto J, Baruque-Ramos J (2018) Industrial textile recycling and reuse in Brazil: case study and considerations concerning the circular economy. Gestão & Produção 25:431–443
9. Angelova RA (2020) The circular economy: a new paradigm for the textile and clothing industries. In: E3S web of conferences, vol 207, p 03008. EDP Sciences
10. Arslan, A, Qayyum A, Ayubi S, Khan SA, Asad ullah M (2022) The impact of energy crisis and political instability on outsourcing: an analysis of the textile industry of Pakistan. J Asian Finance, Econ Bus 9(3):235–243
11. Azamov S (2022) The emergence of English and Uzbek terms used in the field of textiles and light industry and their characteristics. J Pharmaceut Negative Results 3598–3602
12. Barros MV, Salvador R, do Prado GF, de Francisco AC, Piekarski CM (2021) Circular economy as a driver to sustainable businesses. Cleaner Environ Syst 2:100006
13. Bedianashvili G, Jamieson A, Longford C, Martkoplishvili I, Paul J, Sagona C (2022) Evidence for textile production in Rabati, Georgia, during the Bedeni phase of the Early Kurgan period. J Archaeol Sci Rep 43:103467
14. Bocken N, Miller K, Evans S (2016) Assessing the environmental impact of new Circular business models. In: Proceedings of the "New Business Models"—Exploring a changing view on organizing value creation, Toulouse, France, 1, pp 16–17

63. Koszewska M (2019) Circular economy in textiles and fashion—the role of a consumer. In: Circular economy in textiles and apparel. Woodhead Publishing, pp 183–206
64. Kumar L, Kamil I, Ahmad M, Naqvi SA, Deitch MJ, Amjad AQ, Kumar A, Basheer S, Arshad M, Sassanelli C (2022) In-house resource efficiency improvements supplementing the end of pipe treatments in textile SMEs under a circular economy fashion. Front Environ Sci 10:1002319
65. Kurkela E (2020) Drivers and challenges of circular business models: comparative case study in textile industry
66. Kusi-Sarpong S, Gupta H, Khan SA, Chiappetta Jabbour CJ, Rehman ST, Kusi-Sarpong H (2023) Sustainable supplier selection based on industry 4.0 initiatives within the context of circular economy implementation in supply chain operations. Prod Plann Control 34(10):999–1019
67. Kvavadze E, Martkoplishvili I, Chichinadze M (2020) Ancient human activities and environment: palynological date from the early bronze age of Georgia, Caucasus. Shota Rustaveli National Science Foundation, Cezanne, Tbilisi
68. Lehmann C, Cruz-Jesus F, Oliveira T, Damásio B (2022) Leveraging the circular economy: investment and innovation as drivers. J Clean Prod 360:132146
69. Li Y, Lu L, Tan Y, Wang L, Shen M (2017) Decoupling water consumption and environmental impact on textile industry by using water footprint method: a case study in China. Water 9(2):124
70. Li SN, Li Y, Cao ML, Padhye R (2016) Current status of the UK textile industry and its sustainable development. In: Textile bioengineering and informatics symposium proceedings, vol 1, pp 756–764
71. Lin B, Chen Y, Zhang G (2018) Impact of technological progress on China's textile industry and future energy saving potential forecast. Energy 161:859–869
72. Liu C (2020) The effects of world war I on the Chinese textile industry: was the world's trouble china's opportunity? J Econ Hist 80(1):246–285
73. Luvsandondov O, Siqintana BG, Khurelchuluun B (2023) Evaluation of policy implementation to support innovation in industrial sector: a case study of the textile industry in inner Mongolia, China. Int J Sci Bus 29(1):34–55
74. Memon JA, Aziz A, Qayyum M (2020) The rise and fall of Pakistan's textile industry: an analytical view. Eur J Bus Manag 12(12):136–142
75. Mishra S, Jain S, Malhotra G (2021) The anatomy of circular economy transition in the fashion industry. Soc Responsib J 17(4):524–542
76. Muthu SS (2019) Circular economy in textiles and apparel
77. Muñoz-Torres MJ, Fernández-Izquierdo MÁ, Ferrero-Ferrero I, Escrig-Olmedo E, Rivera-Lirio JM (2022) Social life cycle analysis of textile industry impacts for greater social sustainability of global supply chains. Systems 11(1):8
78. Neves SA, Marques AC (2022) Drivers and barriers in the transition from a linear economy to a circular economy. J Clean Prod 341:130865
79. Ormerod A (2021) The decline of the UK textile industry: the terminal years 1945–2003. In: The cotton and textiles industry: managing decline. Routledge, pp 54–97
80. Ostermann CM, da Silva Nascimento L, Steinbruch FK, Callegaro-de-Menezes D (2021) Drivers to implement the circular economy in born-sustainable business models: a case study in the fashion industry. Revista de Gestão 28(3):223–240
81. Patel BS, Nagariya R, Singh RK, Sambasivan M, Yadav DK, Vlachos IP (2022) Development of the house of collaborative partnership to overcome supply chain disruptions: evidence from the textile industry in India. In: Production planning control, pp 1–24
82. Pattnaik P, Dangayach GS, Bhardwaj AK (2018) A review on the sustainability of textile industries wastewater with and without treatment methodologies. Rev Environ Health 33(2):163–203
83. Prakash Y, Charwak B, Kumar PV (2020) Textile industry in new India: challenges and opportunities. Int J Indian Cult Bus Manag 21(4):435–458

84. Radhakrishnan S (2021) Circular economy in textiles and fashion. In: Circular economy: assessment and case studies, pp 163–202
85. Radhakrishnan S (2022) Sustainable development goal: sustainable management and use of natural resources in textile and apparel industry. In: Sustainable approaches in textiles and fashion: circular economy and microplastic pollution. Springer Singapore, Singapore, pp 165–205
86. Radukić S, Petrović-Ranđelović M, Kostić Z (2023) Implementation of the circular economy principles in the textile industry. Econ Sustain Dev 7(1):1–14
87. Rahman SS, Siddiqua S, Cherian C (2022) Sustainable applications of textile waste fiber in the construction and geotechnical industries: a retrospect. Cleaner Eng Technol 6:100420
88. Rani S, Jamal Z (2018) Recycling of textiles waste for environmental protection. Int J Home Sci 4(1):164–168
89. Ranta V, Aarikka-Stenroos L, Ritala P, Mäkinen SJ (2018) Exploring institutional drivers and barriers of the circular economy: a cross-regional comparison of China, the US, and Europe. Resour Conserv Recycl 135:70–82
90. Raspini JP, Bonfante MC, Cúnico FR, Alarcon OE, Campos LM (2022) Drivers and barriers to a circular economy adoption: a sector perspective on rare earth magnets. J Mater Cycles Waste Manag 24(5):1747–1759
91. Reike D, Hekkert MP, Negro SO (2023) Understanding circular economy transitions: the case of circular textiles. Bus Strateg Environ 32(3):1032–1058
92. Repp L, Hekkert M, Kirchherr J (2021) Circular economy-induced global employment shifts in apparel value chains: job reduction in apparel production activities, job growth in reuse and recycling activities. Resour Conserv Recycl 171:105621
93. Ribul M, Lanot A, Pisapia CT, Purnell P, McQueen-Mason SJ, Baurley S (2021) Mechanical, chemical, biological: moving towards closed-loop bio-based recycling in a circular economy of sustainable textiles. J Clean Prod 326:129325
94. Rossi E, Bertassini AC, dos Santos Ferreira C, do Amaral WAN, Ometto AR (2020) Circular economy indicators for organizations considering sustainability and business models: plastic, textile and electro-electronic cases. J Clean Prod 247:119137
95. Roy Choudhury AK (2014) Environmental impacts of the textile industry and its assessment through life cycle assessment. In: Roadmap to sustainable textiles and clothing: environmental and social aspects of textiles and clothing supply chain, pp 1–39
96. Saha K, Dey PK, Papagiannaki E (2022) Implementing circular economy in the textile and clothing industry. In: Supply chain sustainability in small and medium sized enterprises. Routledge, pp 239–276
97. Salmi A, Kaipia R (2022) Implementing circular business models in the textile and clothing industry. J Clean Prod 378:134492
98. Sandin G, Peters GM (2018) Environmental impact of textile reuse and recycling—A review. J Clean Prod 184:353–365
99. Sandvik IM, Stubbs W (2019) Circular fashion supply chain through textile-to-textile recycling. J Fashion Mark Manag: Int J 23(3):366–381
100. Schumacher K, Forster AL (2022) Facilitating a circular economy for textiles workshop report. NIST Spec Publ 1500:207
101. Shen B, Ding X, Chen L, Chan HL (2017) Low carbon supply chain with energy consumption constraints: case studies from China's textile industry and simple analytical model. Supply Chain Manag Int J 22(3):258–269
102. Shen L, Sun C, Ali M (2021) Path of smart servitization and transformation in the textile industry: a case study of various regions in China. Sustainability 13(21):11680
103. Silobrit I, Jureviciene D (2023) Assessing circular textile industry development. Econ Cult 20(1):55–67
104. Suarez-Visbal L, Stuckrath C, Carreón JR (2023) Circular economy in the textile and apparel sector: an overview of global trends, challenges and opportunities. In: Accelerating sustainability in fashion, clothing and textiles, pp 257–267

105. Sugden K (2016) The location of the textile industry in England and Wales, 1813–1820. Text Hist 47(2):208–226
106. Tan J, Tan FJ, Ramakrishna S (2022) Transitioning to a circular economy: a systematic review of its drivers and barriers. Sustainability 14(3):1757
107. Thinakaran S, Chandravelu P, Ponnambalam SG, Sankaranarayanan B, Karuppiah K (2022) Analyzing the challenges to circular economy in Indian fashion industry. IEEE Access 11:711–727
108. Uddin F (2021) Environmental hazard in textile dyeing wastewater from local textile industry. Cellulose 28(17):10715–10739
109. Virgens ND, Silva S, Laranjeira E (2023) Applications of the circular economy to the second-hand textile and clothing market: the case of Humana in Portugal. Int J Fashion Des Technol Educ 16(2):214–223
110. Warwas I, Podgórniak-Krzykacz A, Przywojska J, Kozar Ł (2021) Going green and socially responsible–textile industry in transition to sustainability and a circular economy. Fibres Text Eastern Europe 3(147):8–18
111. Weber S, Weber O, Habib K, Dias GM (2023) Textile waste in Ontario, Canada: opportunities for reuse and recycling. Resour Conserv Recycl 190:106835
112. Xie X, Hong Y, Zeng X, Dai X, Wagner M (2021) A systematic literature review for the recycling and reuse of wasted clothing. Sustainability 13(24):13732
113. Xu CK, Cheng H, Liao ZJ (2018) Towards sustainable growth in the textile industry: a case study of environmental policy in China. Polish J Environ Stud 27(5)
114. Xu J (2017) The role of China in the UK relative imports from three selected trading regions: the case of textile raw material industry. Int J Environ Res Public Health 14(12):1481
115. Ye M, Shen JH, Golson E, Lee CC, Li Y (2022) The impact of Sino–US trade friction on the performance of China's textile and apparel industry. Int Finance 25(2):151–166
116. Zhao H, Lin B (2019) Resources allocation and more efficient use of energy in China's textile industry. Energy 185:111–120
117. Zhao H, Lin B (2020) Impact of foreign trade on energy efficiency in China's textile industry. J Clean Prod 245:118878
118. Zhu X, Zhang B, Yuan H (2022) Digital economy, industrial structure upgrading and green total factor productivity——Evidence in textile and apparel industry from China. PLoS ONE 17(11):e0277259
119. Zonatti WF, Baruque-Ramos J, Duleba W (2016) Brazilian scope of management and recycling of textile wastes. In: Natural fibres: advances in science and technology towards industrial applications: from science to market. Springer Netherlands, pp 429–439

Sustainability and Circular Economy in the Textile Industry

Purvi Zaveri, Priya Patel, Rushika Patel, and Nasreen S. Munshi

Abstract The textile industry, a traditional and skill-driven sector, faces challenges in circularizing products and needs innovative thinking for weaving sustainable approaches in day-to-day practices. In 2023, the global textile market was estimated to be around 31 trillion USD and employed over 300 million people globally. The enormous presence brings a significant waste generation and carbon footprint from production and processing that affects climate change on regional and global levels. Circularizing the textile industry can create huge impacts, but the integration needs to keep a tight balance between being circular and surviving in the market. The current chapter tries to critically evaluate all the stakeholders, manufacturers, consumers, retail market, and supply chain for their role in making the textile a sustainable market. It attempts to address nuances of balance required from the stage of design re-thinking to manufacturing in the textile industry.

Keywords Planetary boundary · Circular economy · Sustainable waste management · Textile value chain · Sustainable consumption pattern (SCP)

P. Zaveri
Biocare Research (I) Pvt. Ltd., 1st Floor, Kanth Complex, Paldi cross Road, Ahmedabad, India
e-mail: dr.purvi.zaveri@gmail.com

P. Patel · N. S. Munshi (✉)
Institute of Science, Nirma University, Sarkhej-Gandhinagar Highway, Ahmedabad, Gujarat 382481, India
e-mail: nasreen.munshi@nirmauni.ac.in; nasreenmunshi1174@gmail.com

P. Patel
e-mail: 18ftphds45@nirmauni.ac.in

R. Patel
Varahi IO, Chicago, USA
e-mail: rushika@varahi.io

P. Singh (ed.), *Dye Pollution from Textile Industry*, SDGs and Textiles,
https://doi.org/10.1007/978-981-97-5341-3_16

1 Introduction

Nine planetary boundaries were defined in 2009 by renowned scientists, based on processes that represented ecosystem stability and resilience for earth (Fig. 1). These boundaries are indicators of cumulative resources, water present on earth and trends of utilization of them in Anthropocene age. As mentioned by the Stockholm Resilience Center, out of nine boundaries, six have already been crossed through utilization in 2023 [34]. To assist the path for a sustainable future and respect the planetary boundaries, the United Nations has given 17 sustainable development goals and based on them multiple action plans are being laid down. With current energy, water utilization and carbon emission patterns of industrialization it is not possible to imagine a greener future without discarding the linearity of production and consumption of consumer goods. It is now inevitable to rethink production, consumption and supply chain economics and rebuild the models. The circular economy (CE) approach provides practical guidance and helps policy makers to define indicators of industrial circularity for various sectors. From the earliest days of hand-spun fibers to the modern period of sophisticated equipment and high-tech fabrics, the textile industry has grown into a complex ecosystem. The textile industry serves the production of materials needed from apparel and household goods to technical and industrial textiles, providing employment to many. Over the past two decades, there has been a huge rise in global apparel consumption [37]. The global textile industry's growing export has reached 31 trillion USD in 2022, contributing to 3% of global GDP [41].

The increase in garment consumption has resulted in detrimental effects on the environment, including increased energy use, material handling, and material circulation [8]. From agricultural procurement to the production procedures used in the textile industry are notorious for using a lot of resources, using plenty of water, and releasing hazardous chemicals. Next to the oil sector, textile manufacturing is the second most polluting industry in the world in terms of greenhouse gas emissions. By

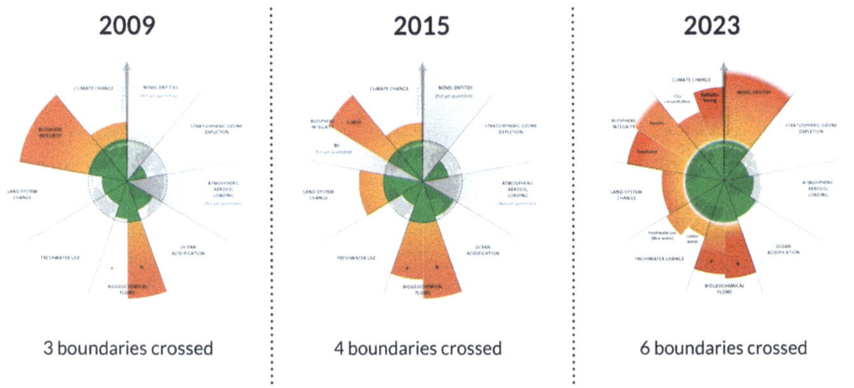

Fig. 1 Planetary boundaries over a period of time (Azote for Stockholm Resilience Centre, based on analysis by Richardson et al. [34])

2050, it is anticipated that the fashion sector would have used up a sizable amount of the global carbon budget [9]. In the past few decades, increased global awareness, involvement of UN like organizations and national leadership have encouraged responsible manufacturing, which is now becoming popular. To make the manufacturer accountable, the transparent reporting process towards sustainable textile, several standards like SBTi and SABS have been adopted by textile giants. These measures not only push the industry towards greener practices but also help level the playing field for businesses committed to sustainability.

Although sustainability and the circular economy are admirable aims, the textile sector faces several significant obstacles in its adoption. The industry's complicated and fragmented supply chain, cost of transition and varied scale of business are the hurdles just to name a few. Furthermore, customers who are used to fast fashion and low costs could be reluctant to spend more for sustainable goods [6]. Unfortunately, in the race to gain customers, companies' marketing strategies have misunderstood the objectives of sustainability resulting in greenwashing. This chapter aims to introduce readers to nuances of textile circularity by questioning critical aspects of manufacturing process and addressing approaches for their circularity. It also addresses the aspect of shared responsibility and policy perspective to implement the net zero related transitions in any textile firm.

2 Textile Industries: Planetary Boundary and Environmental Impacts

When an industry is established in a region, it brings value to the state/province/region/village in a variety of ways. Growth by any such unit is reflected in several parameters like, (i) macroeconomic influence (ii) technological advancements (iii) demographic influence (iv) governmental influences and (v) social influences on nearby population. Global impacts on nations' GDP and global economy are also foreseeable markers. In recent years, innovative fibers like polyester have given a boost to the manufacturing capacity with added features. About 60% of the world's fiber produced is used by the fashion business, with the remaining 40% going to various industries like interior design, industrial textiles, and sanitary textiles. Reduced import/export taxes, cheap labor, pandemic and laxer regulations have contributed to the rise of fast fashion. Driving the global shift from developed countries to the developing nations, the latter are leading the sector with China dominating the market [31]. The sector's contribution to GDP is reflected by the employment of 300 million people, largely women [36].

Enabling people around the establishment and providing employment in developing countries, textile growth doesn't come without any environmental cost. Table 1 indicates planetary boundaries and proxies adopted from [13], and probable negatively contributing factors from the perspectives of textile industries.

Table 1 Planetary boundary, proxy and probable negatively impacting textile waste/process/component

Planetary boundary	Proxy (variables/tools/matrices of measurements)	Probable negatively impacting textile waste/process/component
Atmospheric aerosol loading	Emissions of SO_x, NO_x, and PM (including black carbon) measured (measured in terms of kg/year) Emissions of organic carbon and NMVOC if relevant (measured in terms of kg/year)	Combustor flue gases, VOCs and free gaseous chlorine [29]
Biogeochemical flows	Loss of phosphorous (measured in terms of kg/year), loss of nitrogen (measured in terms of kg/year)	Use of pesticides, fertilizers and eutrophication by effluents
Biosphere integrity	Degradation of land	Leaching of effluent, accidental spillage and disposal in water bodies [26]
	Waste generation	Solid, liquid, and hazardous waste through various streams of manufacturing [4]
	Invasive species	Use of hazardous chemicals, dyes, acids and metal-based processes in the manufacturing [4]
Climate change	Emissions of CO_2eq	1.2 billion tons of CO_2 equivalent (10% of the global GHG) [10]
Freshwater use	Consumption of fresh water	Textile manufacturing is water intensive and used water is discharged as wastewater [18]
Land-system change	Land utilized, damaged and restored	Discharged effluents adversely affect the ecosystem and disturb biodiversity [26]
Novel entities	Use of plastic (measured in terms of kg/year), use of chemicals (measured in terms of kg/year)	Manufacturing of plastic/non-woven type of fabric; packaging of manufactured textile, apparel and footwear (TAF) [41]

(continued)

Table 1 (continued)

Planetary boundary	Proxy (variables/tools/matrices of measurements)	Probable negatively impacting textile waste/ process/component
Ocean acidification	Emissions of CO_2, CO, and CH_4 (measured in terms of tons/ year)	Direct/indirect discharge of effluents, carbon emissions
Stratospheric ozone depletion	Emissions of CFCs, halons, HCFC, and R-40 (measured in terms of tons/year)	Emission of particulate matter, primary and secondary petrochemical-based compounds (chlorinated compounds)

As evident from Table 1, due to the extensive supply networks and energy-intensive manufacturing processes, the apparel and footwear industries contribute 8–10% of the world's carbon emissions. The fast fashion is adding fuel to the problem with negative societal effects, encouraging unsustainable spending, resulting in low-quality products that are unfit for resale or recycling [31].

Up to 20% of industrial wastewater contamination is caused by the dyeing and finishing of textiles [14]. To propose the reduction of waste and introduce circularity in the textile, it is important to understand the flow of materials in the value chain. There are several steps involved in the production of textiles, including the processing of raw fiber, weaving, dying, finishing, and distribution [25] (Fig. 2).

Each step of the textile value chain leads to production of either liquid or solid waste. The next section describes the possibilities of circularizing the production loop and minimizing the waste generation.

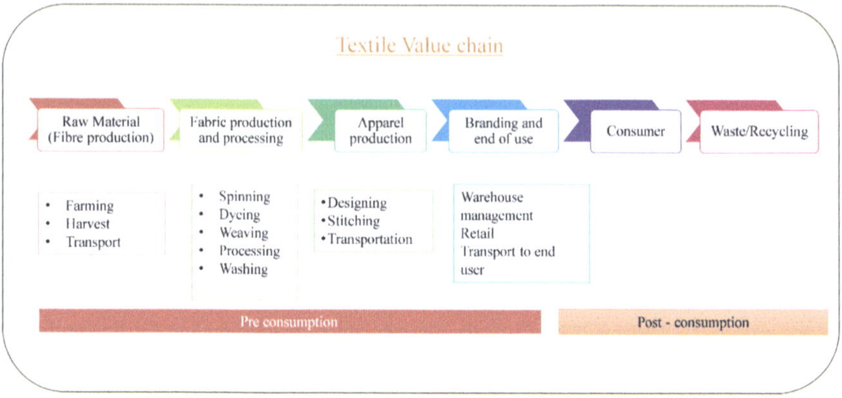

Fig. 2 Textile value chain and its components

3 Circularity for Textile

According to UNEP report, circularity for textile is defined as "an approach where materials are kept at their highest possible value as they move and are retained as long as possible within the textile value chain". As we look at the value chain it is clear that, two types of waste, (i) Pre-consumer textile waste is produced at the industrial level while manufacturing the products whereas (ii) post-consumer textile waste is produced by the consumers after its usage is over (Fig. 2) [23]. The Fiber waste, Yarn waste, Fabric waste as well as overproduction and dead stock are categorized as pre-consumer waste. Whereas reclaimable (reusable fabrics) and non-reclaimable (damaged fabrics) are the two categories of post-consumer waste [5]. Currently linear model is available for managing these wastes where lastly the product is either incinerated or landfilled, ultimately creating environmental pollution by emitting greenhouse gases like CH_4 and black carbon [10]. Low-income countries such as China, India, Turkey, South Korea, Bangladesh, Pakistan, and Vietnam are involved in the pre-consumer processing whereas economically stable countries like the U.S. usually import textile products from them and thus are considered post-consumers [36]. Failure of the linear model to manage textile waste demands the development of circularity in the textile industry which can uplift economic opportunities, control environmental pollution, and even decrease undesirable exploitation of natural resources [33]. According to UNEP, "Every second, the equivalent of a garbage truck full of clothes is thrown away around the world, amounting to an estimated $460 billion in total value" indicating the intensity of the need for circularization of the textile economy [39]. 'Reduce, Reuse, Recycle' (3R's) is the common phenomenon accepted universally for sustainable management of any type of waste [9]. Open-loop (waste is recycled and used in making another product) and closed-loop (recycled waste is used in manufacturing a similar product) recycling systems are the two types of recycling systems [35]. If we categorize the type of input material used in the textile industry, they boil down to fibers, water, dyes and chemicals, energy, fuel and electricity. It is important to consider maximizing the circularity to make the industry sustainable.

3.1 Water Circularity in Textile

Water consumption in producing one unit of fabric depends upon the quality and shade of fabrics, types of chemicals and dyes used while processing as well as the processing conditions [16]. Overall, wet processing stage during fabric synthesis consumes higher amount of water in comparison to fabric production, spinning and weaving stages [16]. Different researchers have compared the water consumption profiles differently. As per one of the reviews, textile industry consumes about 200 L of potable water for making one kg of product [3]. USEPA has stated that 36,000 L of water is consumed while manufacturing 20,000 lbs/day of fabric. Based on types of the fabric, Pattnaik et al. [32] reported usage of 200–250 m^3 of water/ton for cotton clothes manufacturing and man-made fabrics, 20 m^3 of water/ton for nylon and polyester and 150 m^3 of water/ton for viscose rayon in India. Working on the type of apparel, Chen et al. [9] reported that making one cotton t-shirt consumes about 2720 L of water (equivalent to the amount of water intake by one adult in three days), whereas manufacturing of one jeans demands about 7000–10,000 L of water [15]. Annual water consumption by the textile industry has raised from 37 to 86 million Olympic-sized swimming pools [9]. The textile industry not only consumes 20–215 trillion L of water while manufacturing and processing the products but also generates ample amount of wastewater containing varieties of biodegradable and non-biodegradable chemicals such as dyes, chromes, NaOH, starch, acid, other organic and inorganic chemicals as well as pathogens [3, 24, 38]. The major contamination of the wastewater is contributed by improper dyeing system, around 10–50% of the total dye used for fabric dyeing is released directly into the water bodies as it remains unbound to fabrics [38]. Every year 8000 tons of dyes and 6900 tons of chemicals are discharged into the natural water reservoir [16]. Complex and stable aromatic dyes are very difficult to decompose and also affect the photosynthetic activity of water bodies where it is discharged, because it forms a layer over the water hindering sunlight penetration [3, 38]. Apart from dyes and chemicals discharged during the manufacturing process, another major source of pollutants are detergent traces, dyes, and chemicals released while laundering the fabrics. The effluent generated from sizing and de-sizing unit of wet processing system of fabrics preparation is acidic in nature (pH between 4 and 5) and has high BOD ranging between 300 and 450 ppm [3, 24]. Not only wastewater but also plastic fibers of >5 mm size termed as microplastic fibers, are generated while laundering and hence are also a matter of concern [5, 9, 36]. Textile industries are found to be responsible for dumping 190×10^9 tons of microplastics in oceans globally [19].

The only way to bring circularity in water consumption and reduce freshwater usage is to treat the effluent generated efficiently and recirculate it into the process. Different types of biological methods involving application of bacteria, fungi and algae; chemical methods like oxidation; combination of biological and chemical as well as physical methods like adsorption, filtration, coagulation, and ion exchange are known for treating textile industry wastewater. Biological methods being cost effective, eco-friendly and leading to complete mineralization are widely accepted over other available methods [3, 18, 38]. In Europe and the USA, wastewater treatment plants remove the microplastics from the wastewater and retain it into the sludge which is again disposed into the natural environment, creating pollution problems and making it suitable for recirculation [36].

Textile wastewater can be resourceful if important raw materials can be extracted from it [32]. Seeing the current scenario of magnitude of water consumption and wastewater generation in textile industry, recycling of textile wastewater becomes extremely important [18, 38]. Along with the production of reusable water, recycling system also generates ample amount of sludge, so researchers need to account for this factor while establishing recycling or recirculation of the wastewater. Approximately,

70–80 million tons of sludge is generated by textile industries in India [32], which poses a critical issue in handling sludge disposal.

Water recovery/recirculation systems are extensions of the wastewater treatment facility in the industrial setups. If the effluent has nutrients to be recovered, then the primary treatment is designed such that the chemical of interest is precipitated and is recovered. The biological treatment phase has the possibility of recovering nutrients and/or producing electricity. In the last stage of the treatment, the treated water can be recirculated using reverse osmosis membrane. For the sludge generated procedures like, thermal recovering(/recovery) system such as pyrolysis, incineration and waste autoclave are used where waste feedstock is heated at high temperature for recovering useful by-product. For example, textile sludge is heated at around 800 °C temperature for 2 h and converted into sludge ash which can be used as concrete. Such treatment facility demands high expenditure for constructing treatment plants, various machineries, chemicals as well as manpower, and thus capital investment becomes the major hindrance for its adoption [38]. Most of the small-scale textile industries consume freshwater during manufacturing and does not have any wastewater treatment facility and thus release it directly into the natural water bodies. In Germany, about 11% of textile industries (14 out of 127) release wastewater either into natural water system or into municipal treatment system [40]. However, in Europe, the textile industries have treatment plants where they recover 374 m^3/day of water out of 500 m^3/day of wastewater treated. Despite of existence of various textile wastewater treatment plants, very few of them have efficiently reused the water in industry or for agricultural purposes. Thus, there arises a need for development of efficient circular economy-based strategy for managing the consumption of fresh water in textile industry as well as that for managing textile industry wastewater and sludge. Textile industries should also focus on using less harmful materials especially chemicals while manufacturing any product which will not only generate lesser toxic water but also would make its treatment easy [32].

3.2 Solid Waste and Circularity in Textile

In the last 15 years, the fraction of manufacturing clothes has doubled but the number of times the garments are worn is reduced by 36% [5]. Annually worth $ 180 million clothes are landfilled worldwide. Textile solid waste ending up in the municipal solid waste is an unsustainable solution to tackle the waste [20]. Every year, 92×10^6 t of fabric waste is produced, out of which 14% is recycled while rest is landfilled [19]. Ellen Macarthur Foundation has revealed that 150×10^6 tons of garment waste will be landfilled worldwide by 2050 [19]. The numbers are alarming and the lack of awareness about problems associated with used clothes at the post-consumer level is the biggest issue. Nearby 57% of the waste generated by the U.S. is either landfilled or quarter part of it is incinerated whereas only 8% collected via donation or any other related activity is recycled [10]. Approximately 20% of used garments at thrift stores are sold. The remaining are provided to sorting agencies who bifurcate them depending upon their type/quality and either resale them again in markets or dispose them off [36]. It is estimated that bringing the concept of circular economy into the fashion industry will save $191 billion by 2030 [9]. Sorting is the major stage for recycling any garment. It is tedious to sort garments as they are blends of different fibers, yarns, or fabrics [9].

To track the product life cycle, Radio Frequency Identification Tags (RFIDs) were formulated on the basis of quality and type of garments that can be sensed by infrared sensors. This allows tracking of clothes thus ultimately helping in sorting the garments. A solvent separation strategy was formed to sort material on the basis of the melting temperature. After sorting, the next step is recycling. A piece of complete fabric contains 98% of fiber, out of which only 12% is recycled and the remaining is landfilled [9]. Obtained recycled fiber is of poor quality and thus only 1% of it is used in manufacturing cleaning cloths, carpets, mats, rugs, and other related materials. The blending of recycled fiber with new fiber can improve its quality where biodegradable polymers can prove to be a good option [9, 36]. The fashion industry is innovating technologies to blend their waste material from the pre-consumer stage and prepare garments to reduce their footprints.

On an individual level, brands like H&M have initiated the concept of take-back programs wherein they encourage customers to bring their old clothes to reduce the load on the landfills. The re-selling and/or recycling or renting garments can be a viable option for local consumer market too and will play a crucial role in managing textile waste. For the upbringing of the responsible citizens of the planet earth, introducing a curriculum related to stitching and reusing old clothes such as DIY and crafting for making new useful products from old clothes may help increase the life span of a cloth in one's life. Industry, manufacturer and government share the responsibility of the waste generated, however, for the post-consumer waste management, awareness programs and intervention can help improve the system [36].

3.3 Policy Framework

Formulation of strict rules and regulations specifically based on circular economy are crucial for sustainable management of garment waste as well as textile wastewater. In 2007, France had implemented extended producer responsibility (EPR) policy, according to which manufacturers are responsible for managing textiles, linen and shoe waste [7]. Nation Textile Policy 2000 (NTxP—2000) was formulated by Indian government whose aim is to address the issues related to productivity, investment, technology and environmentally sound infrastructure development [27] (Ministry of Textiles, India). USEPA is the major agency involved in Policy framework and does not have any specific board for managing garment waste and formulation, involvement of such organization can help us build a better framework for sustainable management of textile waste [36]. The absence of collaboration between manufacturers, retailers as well as government to manage garment waste also needs consideration [36]. To promote the circular economy in India a collaborative project "Approaches for Circular Textiles and Apparel Industry in India" was initiated under BMZ's public–private partnership program (DeveloPPP.de) by Deutsche Gesellschaft für Internationale Zusammenarbeit (GIZ) GmbH and Aditya Birla Fashion and Retail Ltd (ABFRL). The implementation project partner of the aforementioned scheme is the Center for Environment Education (CEE). Birla Cellulose, Reliance Industries, PurFI, and Kishco are some Indian companies promoting circularity in textiles [5]. The government of India, Ministry of Textiles, Udyog Bhawan, New Delhi had circulated a document recently in June 2023 inviting proposals for mapping the chain of textile waste in India [5]. In addition, initiatives like European Clothing Action Plan had been launched to significantly improve the lifecycle of cloths and sustainability in textile in European Union [28].

Global frameworks like ZDHC (Zero discharge of hazardous chemicals) and The Strategic Approach to International Chemicals Management (SAICM) have recently recognized chemicals as an emerging policy issue and textile as an important concerned sector [39].

4 Sustainability in the Textile Industry

Circularity is the practical approach to attain sustainability in the textile industry. Shifting from linear model to circular model is a must for achieving sustainability in textile industry [22]. Successful implementation of circular economy in textile can be made possible if numerous stakeholders and world's population unite together for achieving it.

4.1 Stakeholders

Manufacturers, designers, brand owners and retailers are the various stakeholders essential for supplying garments [39]. It is necessary for all the stakeholders to work in a coordinated manner as each step in the textile supply chain (SC) is dependent on each other and thus variation in one step disturbs the next one [22]. Amongst all the stakeholders, manufacturers play a crucial role as the circularity of products mainly depends upon the various preliminary decisions taken regarding the type and quality of garments [12]. It is been noted that circular economy from local governmental perspectives have not been explored fully [11].

India has taken various initiatives to achieve circularity in textile industry as it is the sixth largest country for exporting Textiles and Apparel (T&A), which shares around 4% of global trade. For the Indian scenario, Fashion Innovation and Sustainable Design for Circularity by NIFT (National Institute of Fashion Technology), Innovation Center for Natural Fiber by NID (National Institute of Design), and Sustainable

Development in Textile Value Chain by SVPITM (Sardar Vallabhbhai Patel International School of Textiles and Management), Coimbatore are some programs initiated by Indian academic organizations for promoting circular economy in textiles. Sasmira's Institute of Design and Textiles (SIDT) and The Academy for Sustainability (TAS) have also launched various certificate and diploma courses related to sustainability in fashion. IIT-Delhi via its incubation center is also promoting fashion circularity [5]. Such programs can prove to be a boom to the world's second largest exporter and help in developing a sustainable textile culture. Designers should communicate with the garment recycling industry and should consider points highlighted by the industrial team while designing the garment. Currently, the major loophole is the existence of a gap between the fashion designers and the recycling team [36]. Inviting more collaborations on government, policy makers, consumer, designer and waste management workers fronts.

4.2 Consumer Participation

Consumer participation plays a vital role in achieving sustainability in the textile industry. By changing consumption pattern and adopting circular models, the textile industry can significantly reduce waste and emissions across national borders [39]. Additionally, introducing circular apparel where textile waste is recycled to produce new clothes can create additional value for consumers through communication opportunities and enhanced shopping experiences. However, there are obstacles to overcome when it comes to recycling apparel. A structured textile waste management system adopted globally will help control resource management and minimize environmental impact throughout the supply chain. The increasing importance of circular fashion has made consumers more conscious about sustainability in export markets like Turkey, leading to changes throughout the entire value chain driven by technological advancements [21]. One such example is the use of digital transformation in the textile and clothing industries, which enables better tracking and tracing of materials, reducing waste and optimizing production processes.

4.3 Scientific/Engineering Innovation for Textile Sustainability

Sustainable cotton manufacturing and cotton production in water scarcity areas have been helping to reduce the water footprint of the textile industry. Hazardous chemicals have been replaced by green chemistry in manufacturing, where digitalization, thermal control, reuse of water and wastewater recovery are there to name a few. Enzyme application for textile processing, use of plasma, ultrasound and microwave, and development of new dying and printing technologies are reported by Nayak et al.

[30]. It is beyond the scope of this work to enlist details about all the technological advancements, however, it is important to notice that technological advancements can bring the energy and water consumption drastically down to initiate circularity.

5 Current Scenario and Future

A successful transition to net zero in textiles is yet to happen. Companies like Nike, H&M and other giants are adopting the new reporting standards. With the help of technologies, water usage, energy consumption and use of hazardous waste have been reduced. However, the destination is still too far. Continuous monitoring, government subsidization for transitioning towards new technologies as well as public participation would pave the path to sustainable textile via circularization of waste generated.

References

1. https://texmin.nic.in/tender/request-proposal-hiring-consulting-agency-mapping-textile-waste-value-chain-india
2. https://www.dsir.gov.in/sites/default/files/2019-11/2_5.pdf
3. Adane T, Adugna AT, Alemayehu E (2021) Textile industry effluent treatment techniques. J Chem 2021:1–14. https://doi.org/10.1155/2021/5314404
4. Ashfaq A, Khatoon A (2014) Waste management of textiles: a solution to the environmental pollution. Int J Curr Microbiol Appl Sci 3(7). http://www.ijcmas.com
5. Baseline Assessment Report of India (2023) https://www.ceeindia.org/pdf_files/Baseline%20Assessment%20Report%20(Digital%20Version).pdf
6. Bhandari N, Garza-Reyes JA, Rocha-Lona L, Kumar A, Naz F, Joshi R (2022) Barriers to sustainable sourcing in the apparel and fashion luxury industry. Sustain Prod Consum 31:220–235. https://doi.org/10.1016/J.SPC.2022.02.007
7. Bukhari MA, Carrasco-Gallego R, Ponce-Cueto E (2018) Developing a national programme for textiles and clothing recovery. Waste Manag. Res 36(4):321–331
8. Chae Y, Hinestroza J (2020) Building circular economy for smart textiles, smart clothing, and future wearables. Mater Circ Econ 2:2. https://doi.org/10.1007/s42824-020-00002-2
9. Chen X, Memon HA, Wang Y, Marriam I, Tebyetekerwa M (2021) Circular Economy and sustainability of the clothing and textile industry. Materi Circ Econ 3:1–9. https://doi.org/10.1007/s42824-021-00026-2
10. Chopra SS, Liang D, Kaur G, Lens C, Lin CSK (2022) Sustainable process design for circular fashion: advances in sustainable chemistry for textile waste valorisation. Curr Opin Green Sustain Chem 100747. https://doi.org/10.1016/j.cogsc.2022.100747
11. Dagilienė L, Varaniūtė V, Bruneckienė J (2021) Local governments' perspective on implementing the circular economy: a framework for future solutions. J Clean Prod 310. https://doi.org/10.1016/j.jclepro.2021.127340
12. Degenstein LM, McQueen RH, Krogman NT, McNeill LS (2023) Integrating product stewardship into the clothing and textile industry: perspectives of New Zealand stakeholders. Sustainability 15(5):4250. https://doi.org/10.3390/su15054250
13. Erlandsson J, Bergmark P, Höjer M (2023) Establishing the planetary boundaries framework in the sustainability reporting of ICT companies—a proposal for proxy indicators. J Environ Manag 329. https://doi.org/10.1016/j.jenvman.2022.117032

14. European Environment Protection Agency (2023) https://www.epa.gov/
15. Gmurek M, Bilińska L (2023) The role of water recycling in building a circular economy in the textile industry. In: Water in circular economy. Springer International Publishing, Cham, pp 91–102. https://doi.org/10.1007/978-3-031-18165-8_7
16. Guyer GT, Nadeem K, Dizge N (2016) Recycling of pad-batch washing textile wastewater through advanced oxidation processes and its reusability assessment for Turkish textile industry. J Clean Prod 139:488–494. https://doi.org/10.1016/j.jclepro.2016.08.009
17. H&M Sustainability Report (2023) https://hmgroup.com/investors/annual-and-sustainability-report/
18. Halepoto H, Gong T, Memon H (2022) Current status and research trends of textile wastewater treatments—a bibliometric-based study. Front Environ Sci 10:1042256. https://doi.org/10.3389/fenvs.2022.1042256
19. John EP, Mishra U (2023) A sustainable three-layer circular economic model with controllable waste, emission, and wastewater from the textile and fashion industry. J Clean Prod 388:135642. https://doi.org/10.1016/j.jclepro.2022.135642
20. Juanga-Labayen JP, Labayen IV, Yuan Q (2022) A review on textile recycling practices and challenges. Textiles 2(1):174–188. https://doi.org/10.3390/textiles2010010
21. Karell E, Niinimäki K (2019, April 1) Addressing the dialogue between design, sorting and recycling in a circular economy. https://doi.org/10.1080/14606925.2019.1595413
22. Kazancoglu I, Sagnak M, Kumar Mangla S, Kazancoglu Y (2021) Circular economy and the policy: a framework for improving the corporate environmental management in supply chains. Bus Strateg Environ 30(1):590–608. https://doi.org/10.1002/bse.2641
23. Koszewska M (2018) Circular economy—challenges for the textile and clothing industry. Autex Res J 18(4):337–347. https://doi.org/10.1515/aut-2018-0023
24. Kumar PS, Saravanan A (2017) Sustainable wastewater treatments in textile sector. In: Sustainable fibres and textiles. Woodhead Publishing, pp 323–346). https://doi.org/10.1016/B978-0-08-102041-8.00011-1
25. Leal Filho W, Perry P, Heim H, Dinis MAP, Moda H, Ebhuoma E, Paço A (2022) An overview of the contribution of the textiles sector to climate change. Front Environ Sci (Frontiers Media S.A.) 10. https://doi.org/10.3389/fenvs.2022.973102
26. Markandeya, Mohan D, Shukla SP (2022) Hazardous consequences of textile mill effluents on soil and their remediation approaches. Clean Eng Technol (Elsevier Ltd) 7. https://doi.org/10.1016/j.clet.2022.100434
27. Ministry of Textile India. https://texmin.nic.in/sites/default/files/policy_2000.pdf
28. Moorhouse D, Moorhouse D (2017) Sustainable design: circular economy in fashion and textiles. Des J 20(Supp 1):S1948–S1959. https://doi.org/10.1080/14606925.2017.1352713
29. Müezzinoglu A (1998) Air pollutant emission potentials of cotton textile manufacturing industry. J Clean Prod 6
30. Nayak R, Singh A, Panwar T, Padhye R (2019) A review of recent trends in sustainable fashion and textile production. Workforce 13:11–12
31. Niinimäki K, Peters G, Dahlbo H, Perry P, Rissanen T, Gwilt A (2020) The environmental price of fast fashion. Nat Rev Earth Environ 1:189–200. https://doi.org/10.1038/s43017-020-0039-9. https://www.nature.com/articles/s43017-020-0039-9
32. Pattnaik P, Dangayach GS, Bhardwaj AK (2018) A review on the sustainability of textile industries wastewater with and without treatment methodologies. Rev Environ Health 33(2):163–203. https://doi.org/10.1515/reveh-2018-0013
33. Riba JR, Cantero R, Canals T, Puig R (2020) Circular economy of post-consumer textile waste: classification through infrared spectroscopy. J Clean Prod 272:123011. https://doi.org/10.1016/j.jclepro.2020.123011
34. Richardson K, Steffen W, Lucht W, Bendtsen J, Cornell SE, Donges JF, Drüke M, Fetzer I, Bala G, von Bloh W, Feulner G, Fiedler S, Gerten D, Gleeson T, Hofmann M, Huiskamp W, Kummu M, Mohan C, Nogués-Bravo D et al (2023) Earth beyond six of nine planetary boundaries. https://www.science.org

35. Schmutz M, Som C (2022) Identifying the potential for circularity of industrial textile waste generated within Swiss companies. Resour Conserv Recycl 182:106132. https://doi.org/10.1016/j.resconrec.2021.106132
36. Schumacher KA, Forster AL (2022) Textiles in a circular economy: an assessment of the current landscape, challenges, and opportunities in the United States. Front Sustain 3. https://doi.org/10.3389/frsus.2022.1038323
37. Shirvanimoghaddam K, Motamed B, Ramakrishna S, Naebe M (2020) Death by waste: fashion and textile circular economy case. Sci Total Environ 718:137317. https://doi.org/10.1016/j.scitotenv.2020.137317
38. Singh RP, Singh PK, Gupta R, Singh RL (2019) Treatment and recycling of wastewater from textile industry. In: Advances in biological treatment of industrial waste water and their recycling for a sustainable future, pp 225–266. https://doi.org/10.1007/978-981-13-1468-1_8
39. United Nations Environment Programme (2023) Sustainability and circularity in the textile value chain—a global roadmap, Paris
40. Vajnhandl S, Valh JV (2014) The status of water reuse in European textile sector. J Environ Manag 141:29–35. https://doi.org/10.1016/j.jenvman.2014.03.014
41. WTO (2023) https://www.wto.org/english/res_e/publications_e/anrep23_e.htm

Mapping of Emerging Biotechnological Advancement in Textile and Apparel: A Sustainable Approach Towards Circular Economy and Life Cycle Assessment

Rd Sabina and Nazneen Hussain

Abstract The explosion of fast fashion trends amongst the consumer has led to a massive expansion of the textile industry. Excessive complexities generated by textile waste have become a global challenge and thus highlight the need for a scientific, sustainable, eco-friendly, and a cost approach to minimize its toxicological impact on environmental health. The current chapter delivers a wide range of strategic innovations highlighting the following key objectives: (i) to provide an overview of the detrimental changes caused by the wastes generated from textile industries along with the most feasible and sustainable recycling approaches (ii) to provide an overview of current developments utilizing combined biotechnology cutting-edge approaches for strategic management of textile effluents. Studies on integrated nano-based models, genetic engineering infused microbial mediated technologies, and artificial intelligence-based remediation approaches are critically reviewed and emphasized in this chapter (iii) to comprehensively discuss the life cycle analysis (LCA) models with a boundary of the cradle-to-gate system in response to global environmental impact, its challenges on global warming, abiotic depletion, acidification and eutrophication potential (iv) to provide a global perspective on the identification of green innovation and circular economy practices with incorporation of critical design thinking in possible development of new products via innovative and circular solutions. Thus, this chapter delivers a snapshot of the prevailing landscape on the principal hindrances, conceivable opportunities, and untapped economic potential of textile waste management to construct a sustainable environment for the future.

R. Sabina
Department of Microbiology, School of Life Science, Assam Don Bosco University, Kamarkuchi, Assam 782402, India

N. Hussain (✉)
Department of Biosciences, School of Life Science, Assam Don Bosco University, Kamarkuchi, Assam 782402, India
e-mail: nazneen.hussain@dbuniversity.ac.in; hnazneen28@gmail.com

P. Singh (ed.), *Dye Pollution from Textile Industry*, SDGs and Textiles,
https://doi.org/10.1007/978-981-97-5341-3_17

Graphical Abstract

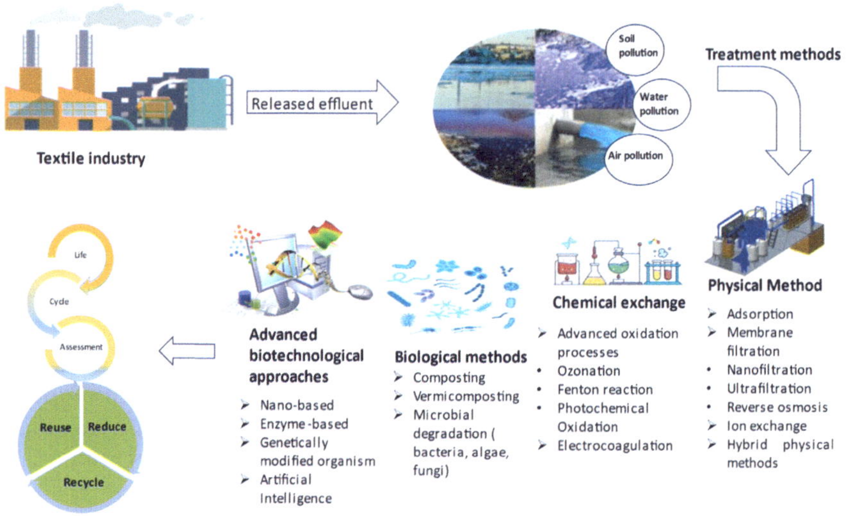

Keywords Textile waste · Eco-efficiency · Sustainable innovations · Life cycle assessments · Circular economy

1 Introduction

The textile sector makes up economic activity for fabricating fibers, yarns, and other materials to manufacture or derive products like clothes. Regarding material consumption, textiles rank fourth in the EU, behind food, housing, and transportation, fifth highest in greenhouse gas emissions, fourth in water consumption, and second highest in land usage [43]. Additionally, this industry relies heavily on non-renewable resources, with a total consumption of 98 million tons annually, and it is predicted that by 2050, it will have used 160 million tons of fossil fuel feedstock (Redesigning Fashion's Future 2018). New technologies like physiochemical, chemical, and biological, methods have revolutionized textile production. Yet, with the rapid economic and population growth, the amount of wastewater released is hazardous to both humans and the environment. Removing dyes and other pollutants from actual or synthetic wastewater is typically assessed using biological, physico-chemical, and physical treatments. Some of the chemical methods used for removing dyes include coagulation, Ion exchange, oxidation, ozonation and adsorption, and other similar processes, besides reverse osmosis, nanofiltration, and ultrafiltration fall under the physical approaches [52]. Biological degradation methods utilize a range of microbes such as bacterial strains, fungal strains, and algal strains, and different enzymes [151]. The release of pollutants like non-biodegradable dyes, from industry is regarded as a key basis of effluent. Economic expansion is currently threatened by water contamination from effluent that the textile sector releases. Hazardous contaminants like dyes, chromium, NaOH, starch, acid, and others are found in the effluent discharged from the fabric industry, etc. [52]. Different industries use dyes to color various material objects. These are sophisticated organic substances that the cosmetic, paper, textile, rubber, printing, and leather industries discharge into waterways as pollutants. Usually, dyes may be grouped into two: synthetic and natural, reliant on where they are produced. In the industrial setting, synthetic dyes are typically produced using a chemical process, but natural materials like flowers, leaves, tree bark, and other elements are used to create natural dyes [138]. Additionally, dyes may be divided into classes grounded on how well the dyes dissolve in water, such as reactive dyes (anionic), basic dyes (cationic), diffuse dyes (non-ionic), acid dyes, and direct [85]. All dyes aside from non-ionic dyes are typically soluble in aqueous media and based on these they may be divided into dissoluble and indissoluble. The indissoluble dyes may be grouped into sulfur, vat, disperse, and solvent dyes including azo dyes [126]. Textile dyeing processes consume a vast amount of water, leading to the production of a significant amount of wastewater. Over 10,000 kinds of fabric dyes have been marketed globally, with a projected yearly production of 8,105 metric tons. Coloring, printing, and functional finishing are the main steps that produce the most wastewater and for operative conduct classification of textile water is important. The percentages of Chemical oxygen demand (COD), Total dissolved solids (TDS), Total suspended solids (TSS), color, pH, biochemical oxygen demand (BOD), Total Kjeldahl nitrogen (TKN), total phosphate (TP), conductivity, total oxygen demand (TOC), metals, Cl, grease, surfactants, alkalinity, rigidity, volatile suspended solid (VSS), sulfide, total solids, turbidity, dissolved organic carbon (DOC), total carbon (TC), absorbable organic halogen (AOX), are the criteria for textile wastewater characterization. A specific chemical group in the dye molecule is targeted during the assessment of the toxicities of textile dyes, and their generalized aspects are then researched. Textile effluent contains toxic chemicals derived from the degradation of dye which are mutagenic and carcinogenic, known to cause cancer. Azo dyes may have carcinogenic and mutagenic properties, as well as toxic effects which may be brought on directly by the dye molecules themselves or by the creation of aryl amine derivatives. The result that occurs due to the reduction process of the azo dye bonds. The byproducts of the breakdown of synthetic dyes and other derivatives in wastewater may be harmful to human health and could result in severe conditions like mucous membranes, hemorrhage skin ulcers, and nausea. The color released by the textile industries has high turbidity that prevents sunlight and oxygen from penetrating the area, these sources are necessary for any living thing to survive, so this barrier inevitably annihilates aquatic life. The textile effluent treatment method is shown in Fig. 1. These harmful toxic contaminants should be eliminated from textile wastewater to reduce contamination once effluent is combined with river or sea or recycled meant for agricultural and manufacturing processes. Therefore, it's vital to indulge in the treatment of dye effluent before its discharge to reduce toxicity, and contamination, and to protect the environment. In this review, several degradation technologies such as biodegradation (bacteria, fungi, algae, earthworms), chemical

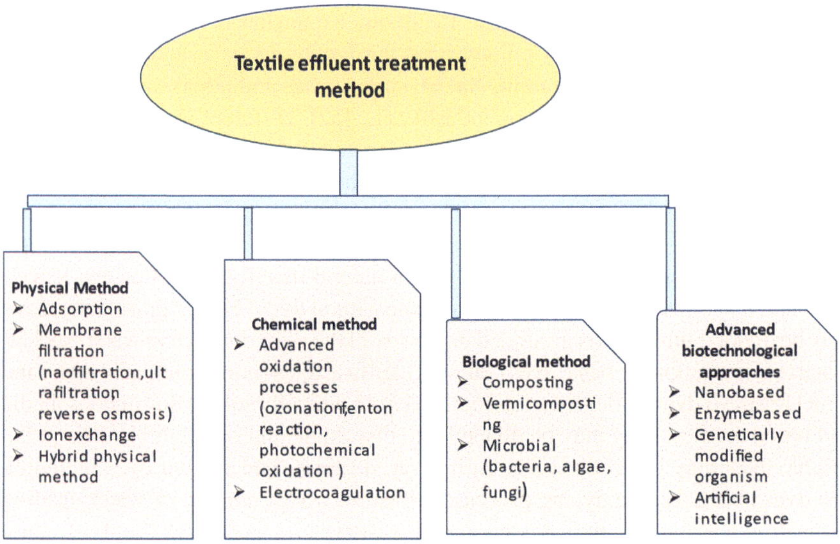

Fig. 1 Textile effluent treatment method

approaches (ion exchange, electrocoagulation, oxidation, ozonation, etc.), and as well as physical approaches are studied. Establishing complicated processes using physical processes is both expensive and inefficient. Therefore data-driven models require time and money to build, but they cost less time and money than physical-based models. As a result, employing artificial intelligence (AI) technologies is also a promising method to enhance the accuracy and operation of wastewater treatment plants (WWTPs). Additionally, degradation using nano-based technology and enzyme base are also simultaneously presented in this review.

2 Textile Effluent and Its Effect on the Environment

Textile industries are known to emit significant quantities of hazardous waste that contain various chemicals, finishing agents, and dyes. The presence of lead, cadmium, mercury, copper, cobalt, arsenic, and nickel in the effluent poses a significant risk due to their harmful nature. Textile wastewater is commonly characterized by high temperature and alkalinity. It is typically loaded with colloidal substances that contribute to the opacity and unpleasant color and odor of the wastewater. The disturbance of oxygen transmission at the air–water boundary is detrimental to the quality of water, which is crucial for the survival of marine life [18]. The perilous waste of the fabric industry transmits like Cu, Cr, As, and Zn which are heavy metals along with total dissolved solids, Chemical oxygen demand (COD), biological oxygen demand (BOD), chemicals, color, odor, which intensifies rate of contamination and is hazardous towards humanity. Untreated wastewater from the textile industry clogs sewer pipes and contaminates drinking water. The water contaminated by textile industries, characterized by bright colors, is a breeding ground for many serious illnesses, including rashes, skin ulcers, hemorrhage, acute inflammation, and nausea. Hint number of metals present in cotton waste blocks light and disturbs the photosynthetic process. The existence of dyes raises the BOD for water, lowering the re-oxygenation work this impacts the development of organisms that rely on photosynthesis for growth. Textile wastewater hinders the self-purification of water by limiting the transmission of oxygen air–water articulation. When effluent is utilized for irrigation of agricultural lands, it can clog the pores as well as solidify the soil surface, which reduces soil fertility and restricts root growth. Various colorants are related to common heavy metals like chromium, zinc, cobalt, nickel, lead, manganese, and copper, which when released into water can harm the environment and human health. Compounds with azo-nitro groups in biotic colorants can lead to the creation of arylamines through incomplete decay. When persistent substances infiltrate the food chain, they accumulate and harm humans and animals. Indestructible biological complexes remain in the environment and then interrupt nutrient cycles [102]. Metals inevitably infiltrate wastewater from numerous sources, such as metallic components like pumps, valves, and pipes, as well as chemicals like acids, bases, pigments, electrolytes, dyes, pesticides, herbicides, and agents. The origin of heavy metals are synthetic dyes.

Aside from heavy metals, organic pigments, dyes, alkalis, acid, bleaching, salt, and finishing agents exhibit a huge undesirable influence on the well-being of living creatures. There exists a link between the biology and physical characteristics of fish and water thereby impacting one another directly or indirectly. Excess total suspended solids (TSS) in water systems can have a major negative impact on the biological, chemical, and physical qualities. Soni et al. [118] observed the deadly properties of fabric dye effluent on *Gambusia affinis* a freshwater fish, and discovered significant alterations in the structure and variety of red blood cells. Aquatic animals and fish are additionally impacted by toxicity because of effluent dyes, high values of (BOD), and (COD). Several azo dyes are cytotoxic even if they are not cleaved into aromatic amines. Many researchers have observed that adding textile effluents to the soil, even temporarily, leads to a decrease in water-soluble nutrients such as organic matter, salts, Mg, Na, P, Ca, NH_4-N, and K content when compared to regular water-irrigated soil. When soil health is affected owing to the occurrence of organic compounds in high concentrations, plant ecosystems, animal ecosystems, and humans will be affected adversely [158].

Most processes produce ambient pollutants like carbon monoxide, carbon dioxide, nitrogen gases, and sulfur dioxide, which are produced by thermal packs, boilers, and diesel generators, resulting in air pollution. Aside from that, they also spread dust into the air that contains combinations of organic debris, viruses, fungus, pesticides, and other poisons during cleaning and cotton processing. The air pollution produced during the phase of manufacturing causes serious health hazards, resulting in occupational ailments. According to survey data done in regions with 639 people, 67% of the villagers reported asthma, bronchitis, cough, chronic obstructive pulmonary disease (COPD), and flu, 56% reported respiratory issues, and 77% reported having a fever [136]. Chemicals that volatilize into the air pollute the environment, certain pollutants enter the system and end up as pollutants in groundwater or surface water, and water contaminants have the ability to volatilize into the atmosphere or seep into the ground.

3 Permissible Limit of Contaminants

In different countries where the textile industry constitutes a substantial economic activity, there are particular regulations pertaining to the wastewater that is produced. The standard limits of different regions are shown in Table 1. By utilizing the best control technology available, using sustainable practices to safeguard the well-being of the public, and reducing pollutant indicators to meet final discharge standards, the regulations' main goal should be to promote industry sustainability [26]. Biological oxygen demand (BOD) levels are typically maintained between 20 and 150 mg/L, Chemical demand oxygen demand (COD) levels are typically kept at range from 8 to 250 mg/L, and pH value 6–9 units, heavy metals may be present owing to the coloring practice of fabrics [101]. The permissible limits for textile wastewater in India are regulated by the Central Pollution Control Board (CPCB) under the Water (Prevention and Control of Pollution) Act, 1974, and the Environment (Protection) Act, 1986. The permissible limit varies within countries.

4 Textile Waste Treatment

Textile waste is widely known for offering significant challenges to conventional treatment methods due to its poor biodegradability index. Fibers are the major type of solid waste produced which is used for making fabric and related goods. Additionally, Yarn manufacturing, production of fibers and their extraction, textile assembling and fabrication, packing, supply and shipping, consumption, reprocessing, and disposal all play a role in the production of solid waste [73]. The demand for textile items is increasing, as is the pattern of waste generation, as a result of seasonal variations and the improvement in living conditions around the world. Textile solid waste materials can be grouped into production, pre-consumer, and post-consumer wastes. "Production wastes" are from manufacturing, while "pre-consumer wastes" are unsold or damaged items. The majority of wastes come from after-use and are called post-consumer wastes. Regarding sustainable consumption, industries such as clothing and textiles are widely regarded as wasteful by users [71]. Through enormous load of solid waste creation and its successive dumping problem internationally, the textile sector must reconsider its greening strategy. The negative environmental impacts of textile solid waste products could be considerably mitigated by extending their life cycle through sustainable management [71]. Because of these environmental

Table 1 Standard Limit (Zero discharge of hazardous chemicals program literature review leaders in advancing environmental responsibility iii, 2015)

Parameters		pH	BOD	COD	TSS	Cr total	Cr VI	Fats and oils	Phenols	Sulfides	Cyanide	Cu
Unit			mg/L	mg/L	mg/L	mg/L	mg/L	mg/L	mg/L	mg/L	mg/L	mg/L
Limit/ standard (country)	India	6.5–8.5	30	250	100	2	0.1	10	1	2	0.2	3
	Mexico	6–9	250	150	60	NA	NA	15	NA	NA	1	4
	Taiwan	6.0–9.0	30	100	30	2	0.5	NA	NA	1	1	3
	Turkey	6.0–9.0	NA	200	100	1	0.5	NA	0.5	0.1	0.5	3
	China	6.0–9.0	20	80–100	50	1.5	0.5	NA	0.5	1	0.5	0.5
	Sri Lanka	6.0–8.5	60	250	50	2	0.5	10	1	2	0.2	3
	Thailand	6.0–9.0	60	400	50	NA	0.25	NA	1	1	0.2	2
	Bangladesh	6.0–9.0	50	200	150	0.5	0.1	10	1	1	0.1	0.5
	Brazil	5.0–9.0	NA	NA	NA	0.5	NA	~50	0.5	1	1	1
	Italy	6.0–9.0	40	160	80	2	0.2	NA	0.5	1	0.5	0.1
Parameters		Cd	Pb	Hg	Zn	As	Ni	Mn	Total nitrogen	Temperature	Color	References
Unit		mg/L	mg/L	mg/L	mg/L	mg/L	mg/L	mg/L	mg/L	°C	Pt–Co units	
Limit/ standard (country)	India	2	0.1	0.01	5	NA	3	2	50	40	400	IS: 2490 Tolerance limits for industrial effluents discharged into inland surface waters (Bureau of Indian Standards, 1981
	Mexico	NA	0.2	0.01	10	NA	2	NA	25	35	NA	Mexican Official Standard NOM-001-Semarnat-2021 (Secretariat of Environment and Natural Resources)
	Taiwan	0.03	1	0.01	5	0.5	1	NA	NA	35	550	Taiwan Effluent Standards (Taiwan Environmental Protection Administration, 2003)
	Turkey	0.1	2	0.05	5	NA	NA	NA	NA	NA	NA	Regulation for Water Pollution Control, No. 25687, 12/31/2004 (US Library of Congress, 2004)

(continued)

Table 1 (continued)

Parameters		Cd	Pb	Hg	Zn	As	Ni	Mn	Total nitrogen	Temperature	Color	References
Unit		mg/L	mg/L	mg/L	mg/L	mg/L	mg/L	mg/L	mg/L	°C	Pt–Co units	
	China	0.1	1	0.05	2	0.5	1	NA	15	NA	NA	GB 4287-2012 + XG1-2015 Discharge Standards of Water Pollutants for Dyeing and Finishing of Textile Industry (China Ministry of Environmental Protection, 2015)
	Sri Lanka	0.1	0.1	0	5	0.2	3	NA	NA	40	NA	National Environmental (Protection and Quality) Regulations, No. 1 2008 (Sri Lanka Central Environmental Authority, 2008)
	Thailand	0.03	0.2	0	5	0.25	1	NA	NA	40	NA	Ministerial Notification No. 2 (B.E. 2539, 1996) Issued in Accordance with the Factory Act (B.E. 2535, 1992), Industrial Effluent Standard (Thailand Ministry of Natural Resources and Environment, Pollution Control Department, 1996)
	Bangladesh	0.5	0.1	0.01	5	0.2	1	5	50	40	NA	The Environmental Conservation Rules (Environment Law Alliance Worldwide, 1997)
	Brazil	0.2	0.5	0.01	5	0.5	2	1	NA	40	NA	The National Environmental Council (CONAMA) Resolution 430 (CONAMA, 2011)
	Italy	0.02	0.2	0.01	0.5	0.5	NA	NA	NA	30	NA	Legislative Decree No. 152, Code on the Environment (Food and Agriculture Organization of the United Nations, 2006)

concerns, the proper management and minimization of textile related waste products is a crucial prerequisite. Textile wastewater containing suspended and dissolved particles and dyes must be filtered before being released into water bodies. Furthermore, several approaches have been created to accomplish water treatment productively and economically. The treatment methods can be either chemical, physical, or biological, which can be further classified into different approaches.

4.1 Physical Treatment

Physical methods are additionally useful for textile waste management. To be sustainable, the textile industry must focus on lowering consumption and implementing completely waste-free practices through eco-designing to increase the value of the supply chain. This involves prolonging the life cycle of generated wastes by valorizing them rather than just eliminating them. Interestingly, certain efforts have recently been made by contemporary textile businesses to include sustainability factors. The sustainability initiatives are primarily focused on employing biodegradable materials, reducing textile wastes, and embracing open-minded trade without affecting employees, nevertheless, sometimes people overlook the problem of managing the solid waste materials produced during the production and consumption of textiles and apparel. Composting is an environmentally friendly way to cut waste for textile solid wastes made of biodegradable natural fibers, but it is ineffective for textiles made of synthetic materials [79]. The mixture of dyes, organic, and inorganic pollutants and heavy metals, dyes, and inorganic and organic contaminants are present in the form of sludge. According to Avelar et al. [17] to produce energy, the sludge can be allowed to air dry using processes like densification and combustion, which not only produces more environmentally friendly energy but also lessens the damaging properties of their contaminants to the environment. Bulky, low-density, and polydisperse cotton-based wastes after being densified can be used to create biofuels like briquettes, this procedure ends in the biodegradation of the residues and a reduction in their ability to induce spontaneous combustion [91]. Textile solid waste materials, especially sludge, contain precipitated dyes, heavy metals, and precipitated salts. Pollutants increase sludge's biochemical and chemical oxygen requirement, rendering it harmful to living species if it enters food chains. The current methods for disposing of solid textile remnants involve dumping and burning. This process can lead to the creation of leachate, which contaminates the soil as well as groundwater resources while also restricting utilization of additional land by taking up a proportion of areas [122]. Large amounts of toxic sludge are produced by dye effluents, which contain refractory heavy metal ions, chemicals, surfactants, and solvents, which cannot be degraded easily [29]. It is challenging to treat the dye content in textile solid waste products from effluent treatment plants. Certain researchers used inexpensive adsorbent substances like polymeric resins, fly ash, bentonite clay, and peat for treating textile wastewater. Additionally, a variety of biotic materials, particularly wheat residue, processed ginger wastewater, potato plant waste, charcoal of nutshell, and date stones, to facilitate the decolorization of textile effluent were also investigated [145]. Yet, the uses for these adsorbents seem limited given that there are numerous challenges, including regeneration plus/or dumping, sludge output, and the expensive nature of the adsorbent. Adsorbents must therefore only be used in methods with low pollution levels, if they can be made cheaply or quickly regenerate [52]. The following subsection outlines some of the physical approaches that could be followed to remediate effluent.

4.1.1 Adsorption

The process of adsorption relies on porous adsorbents to increase surface area for the efficient removal of textile dye molecules [144]. Activated carbon, silica gel, zeolites, and alumina are a few of the frequently used adsorbents for removing dye present in wastewater. This idea of adsorption relies on adsorbents, which frequently feature porous forms that enhance the amount of surface area exposed, which is necessarily intended for quick as well as effective adsorption of dye [2]. The method of adsorption is presented in Fig. 2. The advantages of the adsorption process include high efficacy, rapid turnaround times, and the capacity to reuse adsorbents for the removal of dye from the effluent [6]. Using Mus-C as an adsorbent, a significant dye elimination percentage of roughly 99.98% for Methylene blue (MB) and 99.89% for crystal violet (CV), respectively, was accomplished by maintaining the optimum conditions for a dye concentration of 70 mg/L, adsorbent dosage at 1.5 g/L, and pH at 6 [162]. The highest MG adsorption efficiency onto Ag-MOF was discovered to be 809.71 mg/g. The mesoporous Ag-MOF adsorption material offered a simple and effective way to handle water filtration and industrial wastewater management [13]. On an industrial level, the most commonly used adsorbent is activated carbon. In addition to eliminating dye, activated carbon is known to eradicate hazardous impurities such as heavy metals [63]. Large amounts can boost elimination rates, yet regeneration or repurposing of the carbon produces a significant drop in performance, rendering the elimination of dye performance undetermined and reliant on massive carbon dosages [170].

4.1.2 Membrane Filtration

The use of membrane filtration represents one of the most advanced methods available for eliminating color, salinity levels, and COD from effluent. In this approach, effluent runs over a membrane containing minute pores, causing larger solutes to be trapped which forms a filter cake layer. The filtration method is shown in Fig. 3. The size of membrane pores is one of the properties that can be used to characterize membranes [170]. Membrane performance is commonly defined by penetration fluxes and rejection. It has several distinctive characteristics that are incomparable to other techniques, such as resistance to microbial attack, chemical toxicity, and temperature [144].

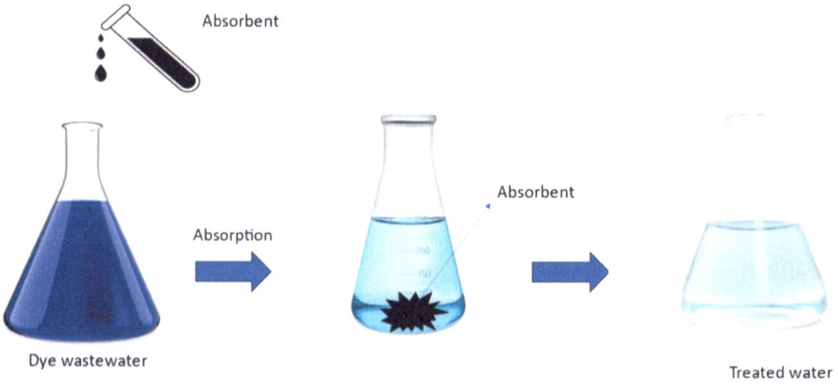

Fig. 2 Adsorption process [10]

Fig. 3 Membrane filtration [10]

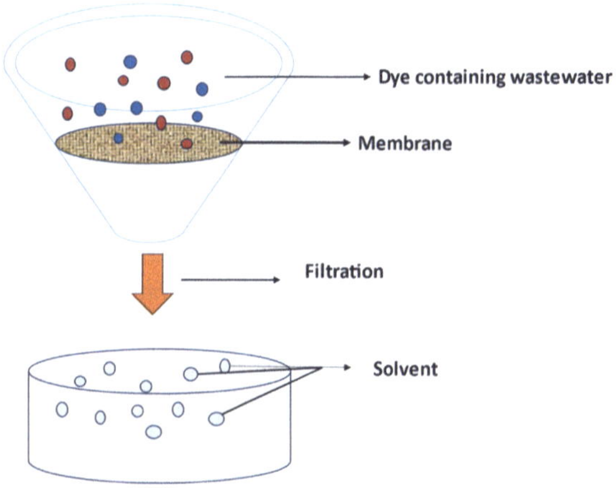

Nanofiltration

A widely used technique in the treatment of wastewater is nanofiltration membranes, which will usually have a measurement of 0.5–0.2 nm in diameter, which has recently been employed to clean wastewater that contains dyes [20]. Nano filtering offers various advantages over other conventional methods of separation, including lower energy use and slight or zero harm to the material being treated [18]. The primary function of NF is the selective removal of ions and organic compounds, and it is used in the desalination of seawater. According to Abid et al. [1] acidic red, reactive blacks, and reactive blue dye were removed with an efficiency rate of 97.2%, 99.58%, and 99.9% respectively using nanofiltration. Hassani et al. [57] reported the clearance rates of 97%, 95.67, and 93.77%, for blue, black, and red dye correspondingly which can be achieved through the nanofiltration method of treatment. The cost-effectiveness and environmental friendliness of polyvinyl chloride and nanocellulose@titanium (PVC-NC@TALCM) nanocomposite membrane was also highlighted by the researchers as a potential remedy for the 98.6% removal of methylene blue (MB) from industrial effluent [42]. Consequently, dye molecules in wastewater solutions may be removed by nanofiltration technology using the principles of size and electrostatic repulsion.

Ultrafiltration

Ultrafiltration is a type of membrane separation technology, that could eliminate particulate components from solutions. Ultrafiltration (UF) membranes can also be used for removing pigments that are organic in textile effluent, with membrane widths ranging from 0.1 to 0.001 m [10]. Ultrafiltration has been studied for its efficacy in discoloring reactive dyes, with a removal effectiveness of up to 90%, and requires less pressure than nanofiltration and reverse osmosis [18]. The findings showed that ultrafiltration (UF) membranes with a range of molecular weight cut-offs (MWCOs) ranging from 6050 to 17,530 Da could achieve 97.47% dye recovery and 99.84% desalination efficiency [65]. Removal of reactive black 5 at the rate of 73.2% for 500 mg/L has also been reported using ultrafiltration. Total dissolved solids (TDS), color, turbidity, conductivity, and removal percentages employing direct nanofiltration were 50–72%, 90–95%, 90–92%, and 45–70%, respectively. Ultrafiltration in conjunction with nanofiltration can improve all of these percent removal rates in textile wastewater treatment [90]. Despite being cheaper and exerting a smaller amount of pressure as compared to nanofiltration, the low rate of separation in ultrafiltration is produced by the great pore size of the membrane.

Reverse Osmosis

Reverse osmosis, also known as RO, is a membrane-based filtration technique that is extensively employed in the industrial sector to eliminate dyes from effluent and generate high-quality water. Reverse osmosis can be used to eliminate all mineral salts, chemical intermediates, and hydrolyzed sensitive dyes. Fouling of RO membranes is a significant challenge for reliable membrane functioning [47]. Ebenezer the removal of acid blue dye with an efficiency rate of more than 99% by Ebrahim [41]. The capacity to produce separation and concentration without state change or heat energy is one of the benefits of RO technology [93].

4.1.3 Ion Exchange

The ion exchangers are either liquid or solid solutions that can release an equal amount of additional ions into the aqueous mixture while concurrently absorbing negatively or positively charged ions from an aqueous electrolyte. The process of this method

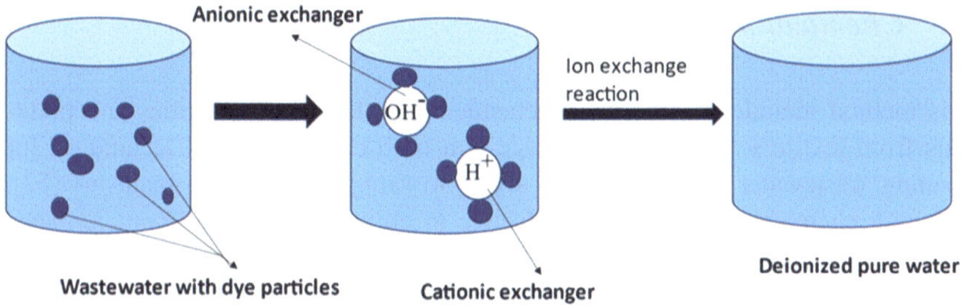

Fig. 4 Ion exchange [10]

is shown in Fig. 4. At a dye concentration of 10–2 M, Acid Orange 10 is said to be 97% removed from wastewater using Amberlite IRA 400 anion-exchange resin [107]. Additionally, Naphthol green B dye can be removed using Purolite MB400 ion exchange resin with an absorption efficiency rate of 98.30% [100]. This method has several benefits, including the removal of dissolved dyes, solvent recovery following usage, and no adsorbent loss during regeneration. An obvious drawback is the high price [170]. For dispersed dyes the ion-exchange approach is ineffective, and organic solvents are costly.

4.1.4 Hybrid Physical Methods

Combining methods is one of the emerging innovations that can produce a good outcome. Coagulation-flocculation-nanofiltration method achieved a nearly 100% removal rate for four of the investigated dyes (Procion Yellow H-EXL, Procion Crimson H-EXL, and Procion Navy H-EXL, Procion Blue H-EXL) and 90% for Reactive black 5 (RB5); offering superior results than either technique alone. Moreover, this combination permits a decrease in the quantity of resins needed for the procedure. Great separation performance against toxic anionic dyes (methyl orange, 99%), cationic dyes (methylene blue, MB), and dye/salt combinations from textile dyeing wastewater is provided by CS/MIL-100 (Fe) based nanofiltration membranes [46]. Yadav et al. [117] reported the removal efficacy of 70% of the TOC and 94% for MB dye in 94.23 min of operation at optimal initial parameters using a catalyst-nanofiltration membrane. Post-reverse osmosis-carbon (RO900) as an adsorbent can be utilized to eliminate MB and MO dyes with an efficiency rate of >99% [143]. According to Hamdan et al. [53], the direct red (DR80) filtration test performance on the membrane Polysulfone-chitosan ultrafiltration (PSF-CS) with chitosan content of 0.7 wt% had a maximum rejection of 99.78% of dye. This combination of processes can ensure a multi-stage treatment approach that targets different types of contaminants at different stages.

4.2 Chemical Methods

This method includes using various chemicals to eliminate impurities and pollutants from textile wastewater. The basic aim is to create practical techniques for cleaning wastewater from the textile sector to safeguard aquatic organisms [52]. Although chemical treatments are effective in eliminating effluent contaminants, they pose the risk of producing secondary contamination due to chemical abuse [170]. The key drawbacks of commercially using chemical techniques for color removal from textile effluent are the necessity to use large amounts of chemicals, the need for high electrical energy, and the high cost of chemicals [84]. Chemical methods present beyond complications because of the byproducts and hazardous metabolites created throughout the elimination procedure.

4.2.1 Advanced Oxidation Processes

The idea revolves around the actual generation of strong oxidizing agents called hydroxyl radicals (OH·), which are used for treating dye-containing wastewater [132]. A few examples of advanced oxidation methods are Fenton, electrochemical, and ozonation oxidation processes which can remove dye under unfavorable environments and without producing sludge [10]. Hydrogen peroxide (H_2O_2), is the main oxidizing substance that requires UV light, to activate it. Chemical bleaching methods will differ based on the method used for the activation of H_2O_2 [170].

Ozonation

An oxidative treatment method enhances solids' biodegradability and sludge hydrolysis. It is used to investigate trace organic chemicals (TrOCs). Maximum results can be achieved when used in combination with Biological activated carbon (BAC) filtration. The ozonation process is shown in Fig. 5. Ozone (O_3) is a potent oxidant due to its extraordinarily high redox capacity of 2.08 V. It interacts with organic molecules more frequently in their ionized and dispersed forms, producing immediate and deliberate oxidation. The reaction proceeds in two phases: the immediate interaction with molecular ozone (O_3) that has dissolved in the water and the indirect collision with radical species (OH, HO_2) that are produced when ozone degrades in the water [170]. According to Wijannarong et al. [175] at reaction periods of 6 h, ozonation can remove color in wastewater by more than 90%. Turhan et al. [169] reported a reduction rate of 64.96% of COD of basic dyes and total removal of Methylene blue can be achieved under basic circumstances like (pH 12), within 12 min. However, the main drawback and difficulty with ozone use is the potential for hazardous byproducts to form, starting with the biodegradable materials found in wastewater.

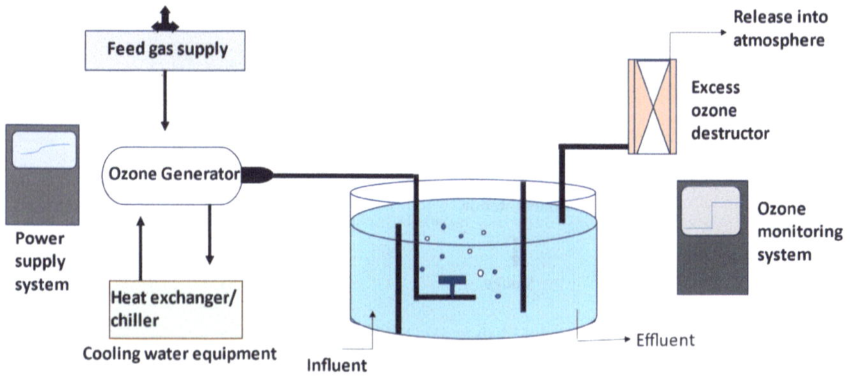

Fig. 5 Ozonation

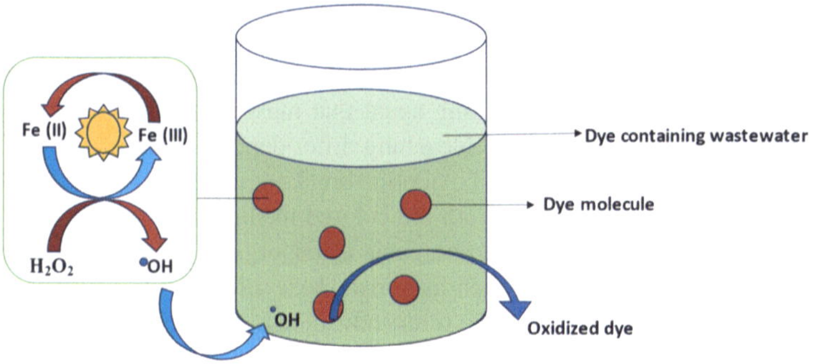

Fig. 6 Fenton reaction [10]

Fenton Reaction

This process utilizes Fenton's reagent like the combined hydrogen peroxide and iron catalyst to remove dye from the effluent. The combined chemical is more powerful than the hydrogen peroxide. For treating both insoluble and soluble dye Fenton's reagent is preferred which is shown in Fig. 6. However, dyes, like disperse dyes and vat dyes, have a high aversion to these reagents [170]. Fenton's reagent can effectively remove dyes, including, indanthrene blue GCD, Helizarin blue BGT, Remazol brilliant blue B, and Irgalan blue FGL [88]. The two main drawbacks of this process are the production of sludge and dye molecule flocculation. The contaminants from the sludge are incinerated which can further affect the environment.

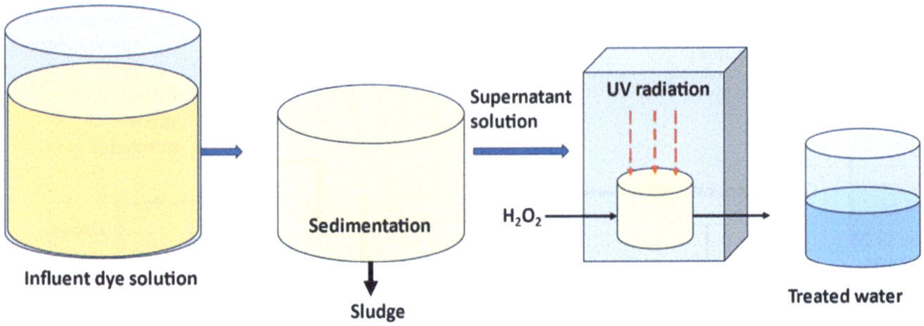

Fig. 7 Photochemical oxidation (Mukhlish et al., 2020)

Photochemical Oxidation

Using UV light in the presence of H_2O, this technique converts dye molecules into CO_2 and H_2O as shown in Fig. 7. UV light intensity, dye structure, pH, and composition of solution are all factors that influence dye removal rate. Hydrogen peroxide (H_2O_2) is the most prevalent oxidizing agent that must be activated by the action of UV light and means of chemical bleaching differ depending on the activation of H_2O_2 [170]. The process of chemical oxidation breaks down the aromatic rings in the dye molecules, removing the dye from the colored media through oxidation [144]. A wide range of residues, including inorganic acids, organic acids, halides, metals, and organic aldehydes can form depending on the base substance as well as the depth of the reaction. This treatment of dye offers advantages, such as a large decrease in odor and no accumulation of sludge. Nevertheless, there are several drawbacks to this approach, including higher material and energy costs.

4.2.2 Electrocoagulation

Electrocoagulation (EC) is a method that removes pollutants from wastewater by supplying an electric current through sacrificial electrodes. Some variables need to be considered to determine the energy consumption for aluminum (Al) electrodes, including the distance between electrodes (1–2 cm), current (0.03–0.09 A), electrolytic concentration (1–3 g/L), electrolysis time (20–60 min) and pH (3–7.5). These variables also affect the percentage removal efficiency of color, chemical oxygen demand (COD), turbidity, and other variables [40]. Aluminum iron anodes act as a catalyst and coagulants in this process, while the cathode emits hydrogen gas as shown in Fig. 8. Wastewater is poured over two metal electrodes in the electrocoagulation process. The hydrogen gas helps the flocculated species ascend into the atmosphere. On the spot, coagulant particles form [170]. Electrocoagulation has the following benefits such as wastewater treatment may be done in natural settings, is easy, has the ability to handle a high volume, no chemical materials are required has a quick reaction time, is cost-effective, creates less sludge, and is environmentally

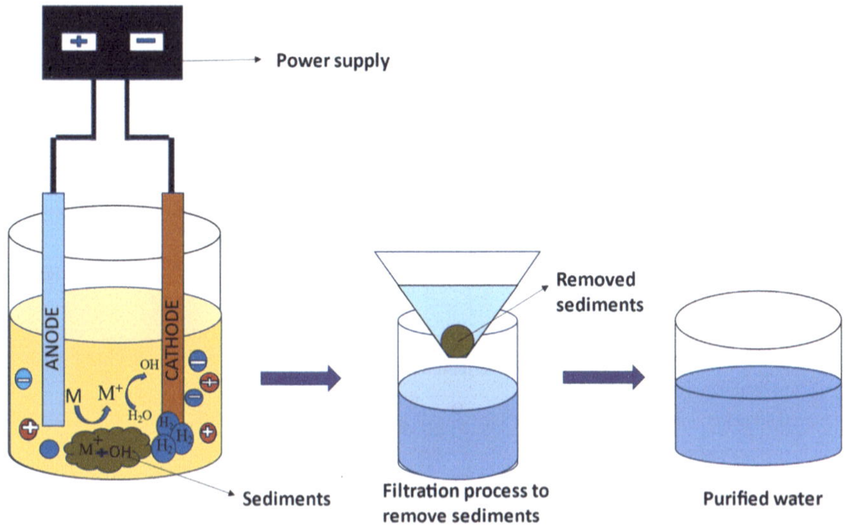

Fig. 8 Electrocoagulation [10]

advantageous [144]. This method consists of three steps. Firstly, sacrificial anode electrodes are oxidated and coagulants are generated, secondly, pollutants are destabilized and finally the formation of flocs takes place. With proper consideration of the said condition, 99.6% of the dye can be eliminated successfully. The importance of the material forms of anodic and the voltage applied, have a considerable influence on the amount of color and organic load [170]. The key challenges of electrocoagulation are the cost of power, sludge generation is quite high and there is a possibility of chlorinated organic pollution and heavy metal contamination by indirect oxidation [86].

4.3 *Biological Treatment*

The biological procedure (also known as bioremediation) produces less sludge and is more ecologically friendly than conventional treatments. Bioremediation is a critical study field in environmental sciences. The biological method for soil remediation is given in Fig. 9. They are simpler to use, as they utilize smaller amounts of chemical reagents, produce fewer sludge, more economic, possess characteristics of energy efficiency, safer for the environment, and the byproducts produced are non-toxic [12]. Physical and chemical treatments are typical methods to remove undesirable effluent elements. These approaches have downsides contributing to high costs and production of hazardous intermediates. Thus, to subsidize the prevalence of toxic additives in physicochemical methodologies, biological treatment is considered to

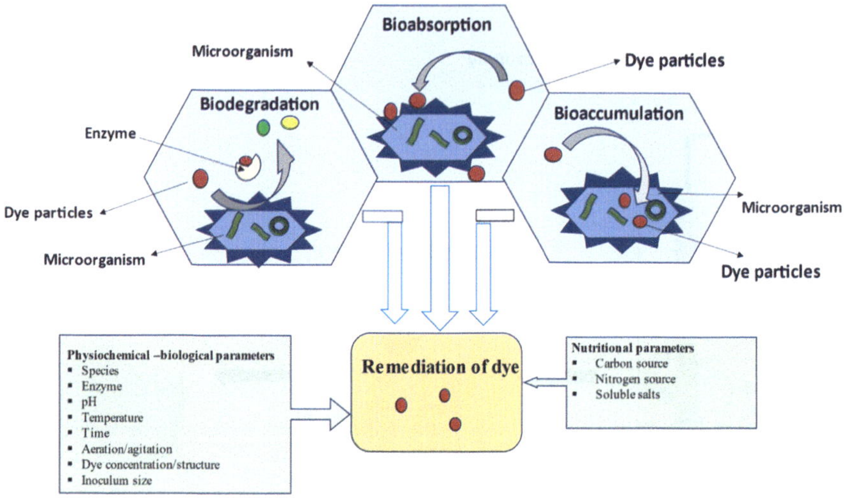

Fig. 9 Biological method (Sudarshan et al. 2022)

be an effective substitute for textile waste. In addition, biological techniques are cost-effective in developing nations and lead to significant dye mineralization. A biological approach's primary benefit is the use of microorganisms with a high capacity for biodegradation, either alone or in groups [96]. The two basic processes in the elimination of dye are adsorption and degradation. These take place either under aerobic conditions releasing products like biomass, CO_2, and H_2O, or by anaerobic conditions and releasing products like methane. The process of anaerobic digestion produces primarily methane, while aerobic treatment yields biomass, CO_2, and H_2O. Removing the link (i.e., the chromophoric group) causes synthetic dyes to decompose into a less hazardous form of inorganic chemical, which then aids in the removal of color [31].

4.3.1 Composting

Composting is an aerobic digestion method. The process of composting is a potential approach for handling waste from numerous outlets. Composting minimizes waste volume and size, making it easier to dispose of. The procedure is used to handle waste produced by industrial processes in addition to waste from homes. Through this aerobic process, hazardous waste is converted into advantageous biofertilizers [159]. Subsequently, those biofertilizers are applied to improve soil fertility and control plant diseases. The process of composting generates harmful gases such as sulfur oxides, Nitrogen oxides, and CO_2. Composting gases are distinguished by having high flow rates, lower pollutant concentrations, and low VOC levels [87]. Biofiltration is frequently applied to reduce emissions generated by composting. The three most popular composting techniques are in-vessel composting, aerated static piles (ASP), and windrows. An outdoor composting method called windrow composting uses mechanical aeration [9]. This approach is appropriate for waste that emits little odor, but it requires an extensive area of land and a longer period for mature composting. For composting to occur using the aerated static pile (ASP) method, oxygen must be forced into the compost pile [111]. This approach is often employed for MSW since it requires a smaller surface area and takes a shorter time to compost. In the in-vessel method, the process is carried out inside multiple vessels and containers. Before composting, sludge can be blended with other waste materials such as wood chips, straw, or green waste to generate a pasteurized product. Composting may transform approximately 20–30% of volatile solids into carbon dioxide. Composting involves natural mesophilic and thermophilic aerobic decomposition inside a largely static system that is aerated naturally and hence has low energy consumption. While direct composting of textile sludge provides plants with both nutrients and toxic substances, it is advantageous for some plants and detrimental to others [50]. Composting is an economical way of managing organic waste, yet the breakdown of organic waste emits greenhouse gases (GHGs) [87]. As a result, there ought to be a mechanism for capturing GHGs and using them to make chemicals and other useful items.

4.3.2 Vermicomposting

The three most common and potentially dangerous contaminants found in soil are pesticides, polycyclic aromatic hydrocarbons (PAHs), and potentially toxic elements (PTEs) [164]. PTEs contain micronutrients like copper (Cu), nickel (Ni), and zinc (Zn), which are dangerous if their concentrations surpass an acceptable limit, as well as heavy metals and metalloids like lead (Pb), mercury (Hg), cadmium (Cd), chromium (Cr), and arsenic (As), which can be highly toxic to the environment even at low levels [165]. Earthworms can absorb and convert toxic elements in the soil to harmless substances, making the soil free from toxins. Rodriguez introduced the idea of vermiremediation, describing how it uses biotic and abiotic interactions, the life cycle of the organism, burrowing, and feeding habits to change, degrade, and remove pollutants from the soil and the environment [49]. The earthworms serve a significant part in organic matter management by dispersing components throughout the soil strata and altering the structure of pores through excavating and soil-downing actions [15]. Industrial wastewater can be treated with vermicompost, which is known to improve soil fertility physically, chemically, and biologically. Vermicomposting refers to a biotechnological technique that employs earthworms to transform organic matter into a humus-like substance known as vermicompost [23]. Due to its ease of laboratory reproduction, quick rate of reproduction, and chemical sensitivity, *Eisenia fetida*, an epigeic earthworm, is one of the most frequently used earthworm species in PTE vermiremediation studies and utilized as the standard test organism in terrestrial environmental toxicology [123]. Of all the elements, cadmium (Cd) has been studied the most, the mean reduction in percentage was 55.38% [62], 92.86, and 91.27% [128]. At the same time the highest level of reduction (100%), in the pharmaceutical sector's waste was achieved after adding 50–75% cow dung

[62]. Earthworms on vermibeds utilizing textile effluent from the cotton industry decreased the amount of cadmium by 91.27% [128]. The gradual mineralization and degradation of organic materials through vermicomposting may accumulate and enhance the PTE concentrations aside from *E. fetida's* capacity to take up metals from the environment [103].

Although contamination levels can vary, *Lumbricus terrestris* has demonstrated exceptional efficiency, with average reduction rates consistently exceeding 80% for all examined elements aside from Zn (69%), moreover the compound that adapts itself best to vermiremediation with *L. terrestris* is Pb, conferring to the authors analysis of the reduction percentages brought about by the presence of earthworms in every soil type examined [164]. A reduction rate of 70.96% was obtained with an initial level of 59.03 mg/kg Cd when *E. eugeniae* was applied to cotton textile effluent to evaluate its vermiremediative potential. In addition, it was found that the most efficient removal rates of Zn, Cr, and Cu occurred when cotton sludge/effluent was mixed with cow dung in a 1:1 or 2:1 ratio [128]. The process of converting textile industrial sludge (TIS) into vermicompost comprised of *Eudrilus eugeniae* and cow dung amended resulted in a highly mature, stable, and nutrient-rich compost that had the highest levels of micronutrients (Ca^{2+}, Mg^{2+}, and Na^{+}) and NPK values [38]. Leather tanneries, where chromium is frequently used as a tanning agent, are the primary source of chromium pollution in industry. *E. eugeniae* reduced PTE by 63.95% in pure matrix, but with 50% CD, it reduced by 89.45%. On the other hand, aerobic composting without earthworms had a low or negative reduction, showing the importance of *E. eugeniae* [164]. The *Perionyx excavatus* earthworm can effectively remediate harmful elements in paper mill wastewater sludge, and the best results were seen when mixed with 50% cow dung, resulting in reduced levels of Pb, Cr, Cu, and Cd, and improved earthworm health, Pb had the largest decline (56%), then Cr (46%), Cu (42%), and Cd (27%) [180]. Research indicates that although numerous species are solely efficient in eliminating a particular kind of pollution, *E. fetida*, *Amynthas robustus*, and the consortia of *Drawida modesta* and *Lampito mauritii* have effectively carried out vermiremediation to eliminate a range of pollutants. Researchers studied the *Allopophora caliginosa's* ability to bioremediate crude oil and showed that TPH concentrations decreased depending on time and doses with a reduction of 54.9% at concentrations of 45.2 mg kg^{-1} and 71% at concentrations of 62.4 mg kg^{-1} respectively thereby proving *Allopophora caliginosa*, which has a strong ability to remove these kinds of pollutants, with the highest reduction observed at the highest TPHs concentration [118]. Earthworms, specifically *E. fetida*, were found to be effective in removing TPHs from contaminated substrates, with a degradation rate of 60.81%. *A. morrisi* also showed some effectiveness at 46.74%, but *E. fetida* was more resistant and had higher survival rates while the combination of both species resulted in a degradation rate of 45.2%. *E. fetida* and shown to promote the degradation of crude oil in soils by working with other advantageous microbes [121]. Two types of earthworms, *Drawida modesta* and *Lampito mauritii* when used to clean up soil contaminated with PAHs and TPHs, at a dose of 7.5 mL, the earthworms reduced the levels of PAHs and TPHs by 68.6 and 34.3% [133]. Earthworms have a symbiotic relationship with bacteria and fungi to break down toxins and neutralize microplastic effects [66]. Earthworms can absorb heavy metals and resist toxicity, making them useful for evaluating contamination, soil quality monitoring, and vermiremediation.

4.3.3 Microbial Degradation

For the biological treatment of textile dye wastewater, bacteria, algae, yeast, fungus, and enzyme-based systems are all viable options for changing the molecules of dye into non-toxic compounds. Biosorption is the breakdown or buildup of substances by microbial biomass. When treating dyes and aromatic amines, enzymes like peroxidase, azoreductase, laccase, and exoenzymes are very helpful. Of these, laccase is most effective at breaking down a variety of aromatic amines [21]. Pollutant degradation is more commonly accomplished by active consortia of organisms rather than a single species at work that are well acclimatized to the present environment. Mordant Black 17 dye degradation was prominently noted by employing *Moraxella osloensis* and *Citrobacter freundii* [60]. The azoreductase gene from *Bacillus laterosporus* RRK1 is transferred to *Escherichia coli* DH5a to create genetically modified microorganisms like *Escherichia coli* SS125. It has been reported that plasmid pAZRSS125 degrades Remazol red dye [28, 171]. Considering significant gains in the development of biological treatments, there are still certain limits because the majority of studies (80%) are conducted with simulated or synthetic effluents, as well as nutritional sources [27]. As reported by Chen et al. [30] when 3.00 g/L of beef extract, 2 g of peptone, and 5 g of glucose per liter were incorporated degradation by specific microbial communities can reach an efficiency rate of 71.3 and 96.3% in synthetic effluent of 600 mg/L concentration. Even though decolorization and chemical oxygen demand (COD) are reduced admirably, these results are typically achieved for dye concentrations of no more than 200 mg L^{-1} [27]. It should be understood that hazardous chemicals, pH, temperature, waste load, and oxygen deficiency all affect how well microorganisms degrade materials.

Degradation by Bacteria

The process of bacterial strains decolorizing dyes starts with the anaerobic reduction or breakage of azo bonds, which is catalyzed by azoreductase. This is followed by the aerobic or anaerobic degradation of the resulting aromatic amines by a mixed bacterial community. The list of dye-degrading bacteria is shown in Table 2. Anaerobic, aerobic, or a combination of the two phases can be used for bacterial degradation. It can work on a variety of dyes, as it is economical, environmentally friendly, and produces relatively little sludge [155]. According to Patel et al. [126], preliminary research, *Pseudomonas mendocina* uses a variety of unknown metabolites to break down dye into phenol, which subsequently enters the ketoadipic acid pathway. John et al. [69] described the use of *Salinivibrio kushneri* HTSP a halophilic bacterium that can degrade Congo red, CBB G-250, and safranin synthetic dyes with approximately

80% decolorization within 48 h [78]. Textile azo dyes (T-blue, yellow GR, and orange 3R) can be removed by *Pseudomonas parafulva* and *Bacillus cereus*, with removal efficiencies for T-blue and yellow GR at 89.21% and orange 3R at 91.69%. Textile dye effluent contains *Bacillus* sp. and *Staphylococcus aureus* that can decolorize and degrade BY HP-2R, BR-RS, and BR RS 01 (BB) dyes. After five days, Bacillus sp. increased the effectiveness of BY dye to 79% while *Staphylococcus aureus* increased the clearance efficiency of BR and BB to 71% and 83%, respectively [8]. *Vibrio* sp. isolated from marine can remove 90.33% of trypan blue dye (direct blue 14) [81]. Dye biodegradation in consortium-based is greater than in pure culture. Consortia are groups of microbiological cultures that work together. Bacterial, fungi, can also occur in consortia. As a result of synergistic metabolic activity, consortia provide a surprisingly efficient breakdown [171]. Karim et al. [74] also reported the elimination of dyes by using pure culture and in consortia. The elimination of dye from diverse textile effluents has been demonstrated to be effective using a bacterial consortium.

Mycoremediation

Mycoremediation is the remediation of textile waste using fungi. Fungal strains have the ability to break down a broad range of pollutants. Both extracellular and intra-enzyme support the activity of fungal metabolism. Enzymes like laccase, manganese peroxidase (MnP), and lignin peroxidase (LiP) aid in the breakdown of dyes. While LiP catalyzes the oxidation of non-phenolic aromatic substances like veratryl alcohol, MnP converts Mn^{2+} to Mn^{3+}, which can oxidize a variety of phenolic compounds [156]. The list of dye-degrading fungi is shown in Table 3. Researchers devised a method to quantify ligninolytic behavior based on the decolorization of a range of were the first to document the decolorization of dye by fungi. Fungi can oxidize dyes, whether they are phenolic or non-phenolic, soluble or insoluble in water. During the solid-state fermentation of wheat straw, *Pleurotus ostreatus* produced an unusual enzyme that was able to decolorize Remazol Brilliant Blue R (RBBR) [168]. *Geotrichum candidum* Dec1, a soil isolate having DSP, glycosylated haem-based peroxidase enzyme that differs from horse-radish peroxidase, MnP, and LiP other peroxidases can decolorize a variety of anthraquinone dyes. Numerous intricate metabolic processes, including the hydroxylation of complex polyaromatic hydrocarbons, dyes, organic matter, wastewater, and steroid chemical compounds, are carried out by fungi [155]. It was discovered that some myco-based dye remediation fungi eliminated up to 100% of the colors from the wastewater. This is due to the enormous number of extracellular enzymes found in fungi. Furthermore, the presence of hyphae and mycelium may have facilitated dye removal [126]. *Piptoporus betulinus* a white rot fungus has the potential to degrade reactive-1 and Direct-1 dye with an efficiency rate of 97.29% and 93.03% respectively. However, employing white rot fungi to remove dyes has several drawbacks, including an extended growth phase, the requirement for nitrogen-restrictive conditions, inconsistent enzyme production, and a large reactor size.

Table 2 List of dye-degrading bacteria

Bacteria	Type of dye	Efficiency	References
B. subtilis (E1), *Exiguobacterium acetylicum* (D1), *Klebsiella terrigena* (R2), *S. aureus* (A22), *P. Pseudo alcaligenes* (A17), and *P. plecoglossicida* (A14)	Whale, Medi blue, fawn, and mixed dye	97.04, 80.61, 94.93 and 81.64%	[126]
Sulfate-reducing bacteria (SRB)	Orange II	95%	
A. hydrophila SK 16	AFY MR	>91%	
Serratia liquifaciens	Azure B	90%	
Bacillus cereus SKB12	RB-5	88.70%	
B. cereus and *P. parafulva*	Azo dyes: T-blue, yellow GR, and orange 3R	91.69 and 89.21%	
B. circulans BWLI 061	MO	99.22%	
Bacterial consortium-Bacillus flexus TS8 (BF), *Proteus mirabilis* PMS (PM), and *P. aeruginosa* NCH (PA)	Indanthrene blue RS	99%	
Citrobacter sp.	CR	93%	
Bacillus subtilis HAU-KK01	CR	92.80%	
Thiosphaera pantotropha	RY	100%	
Klebsiella sp.	Reactive blue 19	90%	
Aeromonas hydrophila SK16	Acid-fast yellow MR	91.25%	
Bacillus sp. and *S. aureus*	BY HP-2R, BS-RS, and B. blue RS 01 (BB)	71–83%	
Alcaligenes faecalis AZ26, *Bacillus cereus* AZ27 and *Bacillus* sp.	NSB-G	90%	
Brevibacillus laterosporus	DB-MR, MO, Blue-2B, Golden yellow (GY) Brilliant blue	100%	
Shewanella oneidensis (MFC)	Acid orange (AO 7)	>98%	
Alcaligenes faecalis and *Rhodococcus erythropolis*	Monoazo dye Acid orange	80–95%	
Neisseria sp., *Vibrio* sp., *Bacillus* sp., *Bacillus* sp. and *Aeromonas* sps.	N Orange FN-R, N Brilliant blue FN-R, N Super black G, BY S8-G, and BR S2-B	60–90%	
Salinivibrio kushneri HTSP	CBB G-250 and CR and safranin	>80%	

(continued)

Table 2 (continued)

Bacteria	Type of dye	Efficiency	References
Bacillus odyssey SUK3, *Morganella morganii* SUK5 and *Proteus* sps. SUK 7	Red HE3B, Reactive blue 59	97–99%	
B. circulans NPP1	MR and other dyes	98%	
P. desmolyticum NCIM 2112	DB 6, Red HEB7, and Green HE4B	70, 90%	
Acinetobacter calcoaceticus	MR, MO	90–98%	
Pseudomonas sp., *Bacillus mycoides*, *Micrococcus* sp.	Red, green, black, and yellow dye	78, 76, 77%	
Vibrio sp.	Trypan blue dye (direct blue 14)	90.33%	[81]
Enterobacter CU2004	Reactive blue 222 (RB222)	≥90%	Lakshmaiah et al. (2023)
Pseudomonas aeruginosa	Azo dye brown 703	71.36%	[163]
Brevundimonas diminuta	Rhodamine-B	90–95%	[148]
Bacillus sp. AS2	Methylene blue (MB), Congo red (CR), and malachite green (MG)	97.01 ± 0.36%, 92.49 ± 0.21%, and 94.23 ± 0.47% respectively	[147]
Bacillus cereus SKB12 and *Enterobacter hormaechei* SKB16	Reactive red 198 (RR 198)	80% by *Bacillus cereus* SKB12 and 85% by *Enterobacter hormaechei* SKB16	[166]
Bacillus megaterium KY848339.1	Acid red 337	91%	[43]
Sterigmatomyces halophilus SSA1 575	RB-5, RB-5	100%	[14]
Anoxybacillus ayderensis SK3-4	DG 6	100%	[33]
Staphylococcus sp. K2204	RBB	100%	
Staphylococcus sp.	Reactive blue	97%	[59]
Consortium of *P. aeruginosa*, *B. pumilis*, *B. thuringiensis*, *Enterococcus faecium* in different combinations	NR, N Black, N Blue dk, N Navy, N Yellow	82–97%	[127]
Acinetobacter (ST16.16/164) and *Klebsiella* (ST16.16/034)	Monoazo dye RO 16 and diazo dye RG-19	More than 80%	[110]

Table 3 List of dye-degrading fungi

Fungi	Dye removed	Efficiency rate in %	References
Schizophyllum commune IBL-06	Brilliant red 80	84.8%	[151]
Pleurotus florida	Malachite green	90%	
Penicillium spp. SA5, *Alternaria* spp. SA4, *Aspergillus tereus* SA3, *Aspergillus flavus* SA2	Orange II (mono-azo)	55.68, 58.66, 44.67 and 48%	
Penicillium spp. SA5, *Aspergillus tereus* SA3, *Alternaria* spp. SA4, *Aspergillus flavus* SA2	Acid red 151 (di-azo)	65, 55, 98.47 and 97.7%	
Phanerochaete chrysosporium	Amido black 10B	98%	[126]
Funalia trogii, *Irpex lacteus*	Remazol brilliant blue	92–98 and 100%	
Trametes polyzona KU-RNW027	RB-B	100%	
P. chrysosporium	Ranocid Fast Blue, Rhodamine-B, Orange-II	83, 91 and 85%	
Pleurotus ostreatus	Acid red 129	77%	
Emmia latemarginata (MAP03, MAP04, and MAP05) and *Mucor circinelloides* (MAP01, MAP02, and MAP06)	Azo, indigo, and anthraquinone dyes (Acetyl yellow G)	~100%	
Cyberlindnera samutprakarnensis S4	Acid red	97%	
Phanerochaete sordida	RR-120	90.6%	
Trichoderma (*T. virens*, *T. viride*)	CR ang MG	81–99%	
Trichoderma tomentosum	AR-3R	99.2%	
A. fumigatus	Direct violet blue	51.38–93.74%	
Trametes hirsuta, *Microporus xanthopus*, and *Ganoderma applanatum*	Turquoise blue textile dye	82.17, 78.50 and 85.84%	
Aspergillus niger, and *Phanerochaete chrysosporium*	BF (81.85%), Nigrosin (77.47%), MG (72.77%)	*P. chrysosporium* maximum with Nigrosin (90.15%) > BF (89.8%), MG (83.25%)	
Aspergillus niger, *A. terreus* and *Rhizopus oligosporus*	Acid Blue 161 e Procion Red MX-5B	58–68%	

(continued)

Table 3 (continued)

Fungi	Dye removed	Efficiency rate in %	References
P. simplicissimum (n-WRF)	Triphenylmethane Dyes: CV, MV, MG, CB	96.1–98.7%	
Diutina rugosa	Indigo	100%	
Asp. Lentulus	VAT Novatic Grey (100%), RR-Red (98.47%), RY (91.55%), and Indanthrane Blue (99.28%) Reactive S	91–100, 55.33%	
Aspergillus foetidus	Azo reactive dyes, Drimarene dyes	>95%	
Aspergillus flavus	Azo textile dye, RR198	84.96%	
Aspergillusoryzaedead cells	RB, MB, and MO	100%	
Trichoderma harzianum	AR and anthraquinone (Basic blue	70 and 51%	
Saccharomyces cerevisiae, Mucor circinelloides	Remazole Blue	100 and 95%	
Fusarium oxysporum	Aniline blue, reactive black 5, orange II, and crystal violet	97.06, 89.86, 91.38, and 86.67%	[167]
Trametes hirsuta D7	Reactive black 5 (RB5), acid blue 113 (AB113), and acid orange 7 (AO7),	RB5 (92%), AB113 (97%), and 30% of AO7	[11]
Aspergillus flavus and *Aspergillus niger*	Congo Red (CR) and Reactive black 5 (RB5)	90 and 96% of CR and RB5 respectively	[51]
Ganoderma lucidum	Direct Red 5B, DR5B	95.16%	[163]
Piptoporus betulinus	Direct flavine 5-G (Direct-1) and reactive red S3B (Reactive-1)	93.03 and 97.29%	[7]

Phycoremediation

The remediation of textile waste using algae is referred to as phycoremediation. Algae additionally comprise a great amount of enzymes and other compounds that aid in the elimination of the color from textile effluent. They are a prospective candidate because photosynthetic algae are frequently seen growing in textile industry effluents. The list of dye-degrading algae is shown in Table 4. According to Jamee and Siddique [64] Azo dyes are broken down into aromatic amines by algae, which then use them as a source of carbon and energy to create simpler organic and inorganic compounds. More than 30 azo dyes have been degraded and decolorized by *Chlorella* sp. and *Oscillatoria* sp. into less complicated aromatic amines. Algae use three different intrinsic mechanisms to remove azo dyes: they use chromophores' metabolism to produce algal biomass, they produce CO_2 and H_2O by converting colored particles into non-colored ones, and they absorb chromophores through their biomass [170]. According to Patel et al. [126], ¦of dye elimination was accomplished up to 99.5% in various experiments of phycoremediation of dyes. Interestingly, the hybrid culture of algae demonstrated the greatest clearance, demonstrating the efficiency of 99.5% in degrading azo dye. Similar to bacteria, algae rely on azoreductase's activated form to break down azo bonds and produce aromatic amines as a byproduct. The type of algae used, how they metabolically process, and the molecular structures of the dyes all affect how quickly dyes degrade.

5 Advanced Biotechnological Approaches Associated with Textile Waste Degradation

5.1 Nano-Based Degradation

The practice of nanotechnology for dye treatment is the most recent promising option. To absorb and remove dyes from textile wastewater, nanomaterials, and nanoparticles are utilized. Nanomaterials may be used to remove organic pollutants from wastewater and purify water. A study found that adsorption by nanomaterials is the most efficient approach to eliminating color contamination from an aqueous media [124]. Green synthetic nanoparticles proved to be more efficient in degrading synthetic dye [56]. Methylene blue adsorption was increased by placing silver nanoparticles on activated carbon, and it has been shown that dyes like Congo red, Eosin Yellow, Bromophenol Blue 2, and Brilliant Blue bind to silver nano-silica powder. Additionally, various metallic nanoparticles have also demonstrated considerable promise in dye treatment [137]. The following outlines some of the nano-based technologies which can be used to degrade textile waste.

5.1.1 Zinc Oxide Nanoparticles

When used for photocatalytic degradation of brilliant green (BG) and indigo carmine (IC) dyes in aqueous solution, zinc oxide nanoparticles (ZnO NPs) with an average size of 33.31 nm showed a degradation rate of 98.07% and 80.8%, respectively, demonstrating that ZnO NPs are capable of degrading hazardous and toxic chemical dyes that persist in the environment [131]. Congo red (CR) dye is removed by zinc oxide nanoparticles (ZnO NPs), which have an efficiency rate of >98% and are produced using an environmentally friendly green process. The pH range for these

Table 4 List of dye-degrading Algae

Algae	Dye removed	Rate of efficiency	References
Chlorella vulgaris	Basic fuchsin	91.20%	[126]
Oscillatoria rubescens	Basic cationic	85.80%	
Nostoc linckia, Lyngbya lagerlerimi	Methyl red	81.97 and 35.61%	
Green algae	Indigo, Direct blue, Remazol brilliant blue, and crystal violet	89.3, 79, 75.3 and 72.5%	
Spirogyra sp., *Oscillatoria* sp.	Blue dye and red dye	*Spirogyra* sp. (78.29 and 64.21%), *Oscillatoria* sp. (76.48 and 62.83%)	
Phormidium	Indigo dye	91%	[151]
Synechococcus sp.	Remazol brilliant blue	11.53%	
Lyngbya sp.	Textile dye	73%	
Cosmarium sp.	Malachite green	92.40%	
Green algae	Mono-azo and diazo dyes	68%	
Kluyveromyces maraxianus	Remazol black-B	98%	
Chlorella sp.	Indigo, Direct blue, Remazol brilliant orange, and crystal violet	89.3, 79, 75.30 and 72.50%	
Chlorella vulgaris, *Chlorella pyrenoidosa*	Methylene blue and AO7	83.04 and 84%	[126]
C. vulgaris UMACC 001	Supranol red 3BW (SR 3BW)	41.8–50.0%	
Chlorella vulgaris (immobilized)	SB 3BW, Lanaset Red 2GA and Levafix navy blue EBNA	44.0–49%	
Sphaerocystis schroeteri	Blue dye and Green dye	63.87 and 60.00%	
Anabaena flos-aquae UTCC64, *Phormidium autumnale* UTEX1580, and *Synechococcus* sp. PCC7942	Indigo, RBBR, and sulfur black	80–90%	
Phormidium valderianum, a marine cyanobacterium	Acid red, Acid red 119, and Direct black 155	>98%	
Gloeocapsa pleurocapsoides and *Phormidium ceylanicum*	AR 97 and FF Sky blue dyes	80%	

(continued)

Table 4 (continued)

Algae	Dye removed	Rate of efficiency	References
Achaetomium strumarium	Acid red	99.50%	
Anabaena flos-aquae, Nostoc ellipsosporum, Nostoc linckia, Anabaena variabilis and *Chlorella vulgaris*	Azo dyes	*N. ellipsosporum* (100%), *C. vulgaris* (96.16%), *A. variabilis* (88.71%), *N. linckia* (88.71%) and *A. flos-aquae* (50.81%)	
Arthrospira platensis NIOF17/003	Methylene blue (MB)	93%	
Arthrospira complete dry biomass (ACDB) and *lipid-free biomass*	Ismate violet 2R, IV2R	75.7 and 61.11%	
Oedogonium subplagiostomum	Methyl orange (MO)	97%	
Cyanobacterium phormidium	Indigo	91%	
Lychaete pellucida	Reactive blue 4 (RB4), Reactive red 120 (RR120), Reactive brilliant yellow 3G (RBY3G), and Reactive green 12 (RG12)	95%	[77]
Candida Albicans (NCIM 3665)	Chrysophenine GX	86.25%	[113]
Shewanella algae B29	Reactive red 66	91.04%	Chaieb et al. (2023)

NPs is 2–10 [141]. The ability of zinc oxide (ZnO) nanocomposites to remove colors from synthetic and textile industrial effluents was investigated under optimum conditions of 600 ppm initial dye concentration and 0.9 mg/mL ZnO nanocomposite dose at 30 °C and pH 6, and the results depicted that ZnO nanocomposite can eliminate 99% of dye (azo) from synthetic and textile industrial effluent [140]. According to Al-Arjan (2022), ZnO-NPs removed Ismate Violet 2R (IV2R) dye from synthetic and textile industry effluents with an efficiency rate of 91.5% and 65.6%, respectively.

5.1.2 Cerium Oxide (CeO_2) Nanoparticles

A simple one-step chemical process was used to create high-grade silver (Ag) decorated CeO_2 nanoparticles, and the unique Ag-decorated CeO_2 catalyst was used to decompose the harmful textile contaminant Rose Bengal (RB) dye in vis light irradiation. Murugadoss et al. [115] reported 96% degradation of RB in 3 h by using Ag/CeO_2 via photocatalytic. According to Aoun et al. [16] when the effective methylene blue and crystal violet by CeO_2 photodegradation was investigated under UV and sunlight irradiation which was revealed after 30 min. 100% of the methylene blue was eliminated, while 95% of the crystal violet deteriorated after 120 min of UV light exposure. Ahmad et al. [4] stated that Ag/CeO_2 photocatalytic activity demonstrated enhanced photocatalytic activity of up to 94%, with 60% of the bromophenol blue (BB) degradation coming from CeO_2 nanoparticles.

5.1.3 Silver Nanoparticles (AgNPs)

Solena amplexicaulis plant leaf extract was used to make the AgNPs, and UV spectroscopy was used to characterize them, revealing absorbance and a wavelength of 393 nm. Photocatalytic degradation of Victoria blue or basic blue dye at varied time intervals can be achieved by using green synthesized Ag nanoparticles with solar irradiation. At 72 h, AgNPs efficiently disintegrated the dye by 99% [70]. Investigations into AgNPs synthesized from Myristica fragrans seed shells (agricultural waste) showed improved photocatalytic degradation (>90%) of methyl violet 10B (cationic), rhodamine B (zwitterionic), and Remazol brilliant blue reactive (anionic) under UV light (Thomas and Thalla 2023). FT-IR, UV-visible spectrophotometry, and HR-TEM, analyzes were utilized to characterize the biosynthesized silver nanoparticles (AgNPs) and test the photocatalytic breakdown of methyl violet revealing that in the presence of sunlight, methyl violet can be degraded rapidly (85.56%) within 20 min [68].

5.1.4 Copper Oxide Nanoparticles

Purpureocillium lilacinum produced Copper oxide nanoparticles (CuO NP) reported 57.5 and 63% catalytic activity for navy blue and safranin respectively [54]. When studied under ideal conditions (pH 11.84), with a dye concentration of 7.49 ppm and an irradiation time of 4 h, copper oxide (CuO) nanoparticles biosynthesized by neem extract were able to degrade reactive orange M2R (RO-M2R) with an efficiency rate of 84.04% [134].

5.2 Enzyme-Based Degradation

Enzymes can efficiently degrade a specific contaminant while having little influence on the other contents of the wastewater. Textile wastewater is the most harmful effluent released from the industrial sector. The limitations of using microbes for degrading activities can be overcome through the application of enzymes. Currently, two types of enzymes are frequently employed for textile dye cleanup: laccase and peroxidase. The list of enzymes employed for the degradation of dye is shown in Table 5. These enzymes are effective and commonly utilized by researchers for removing dyes from textile effluent [126]. According to Gacem et al. [44], azoreductases derived from *Streptomyces* sp. can be used for the elimination of methyl red wastewater. *Yarrowia lipolytica* is a yeast that contains an enzyme laccase that can be used for the degradation of safranin, phenol red, bromocresol purple, and bromothymol blue [37]. The degradation of reactive Blue-4 using algal laccase isolated from Spiriluna platenis was reported by Afreen et al. [3]. Othman et al. [125] reported the use of Laccase obtained from the edible mushroom (*Agaricus bisporus*) for the enzymatic breakdown of Acid Blue. Dai et al. [36] reported the use of laccase isolated from novel bacteria (transformed *E. coli* BL21 (DE3) for degradation of Coomassie brilliant blue, acid violet 7, and bromophenol blue (BpB). The breakdown of lignin and dyestuff depends on the extracellular enzymes of white rot and ligninolytic fungi. Extracellular enzymes that break down refractory aromatic color compounds with remarkable efficiency include LiP, MnP, laccase, and Mn-independent versatile peroxidases (VP), which are non-substrate specific. The use of Fungal lignin peroxidase isolated from *Bjerkandera adusta* CX-9 [22] and *Ganoderma lucidum* IBL 05 [150] with the ability to degrade a set of colors has been reported. Microbial enzymes have numerous attractive qualities, including the capacity to change their genetic makeup, accessibility due to the absence of periodic oscillations, faster growth than enzymes generated by plants and animals, and great stability and ease of production [151]. Azoreductases are the most significant bacterial enzymes engaged in degradation activities that can break azo bonds (–N=N–), releasing aromatic amines that are then transformed into CO_2 (Chang and Kuo 2000).

5.3 Genetically Modified Organism (GMO) for Bioremediation

In the realm of bioremediation, genetic engineering has caused a huge revolution. When a gene of interest is added to a microorganism for a specific purpose that is not typically present in the target host, the result is an organism that has undergone genetic modification. These days, synthetic dyes are created to withstand typical degradation processes, which makes them labor- and time-intensive to work with [149]. Genetically modified organisms (GMOs) can enhance dye degradation in certain environmental conditions. GMOs are produced by modifying or transferring genes from one species to another [88]. Functional genes from a variety of bacterial strains, such as *Mycobacterium marinum*, *Pseudomonas putida*, *Bacillus idriensis*, *Ralstonia eutropha*, *Sphingomonas desiccabilis*, and others, were used and inserted into other species to produce genetically modified organisms (GMOs) [149]. According to Sandhya et al. [146], by transferring the azoreductase gene from *Bacillus laterosporus* RRK1 to Escherichia coli DH5a and plasmid pAZR-SS125 for the degradation of Remazol red dye, Escherichia coli SS125 was produced. In the lab, the azoreductase gene was incorporated into the expression vector pGEX4T-1 to create genetically modified *E. coli* JM109 (pGEX-AZR) that decolorizes direct blue 71. The vector

Table 5 List of enzymes employed for degradation of dye

Dyes/substrate	Enzyme	Species/sources	References
Azo dye	Laccase	*Pycnoporus sanguineus*	[151]
Bromophenol red, Bromocresol purple	Manganese peroxidase (MnP), MiP	*Pleurotus ostreatus*	
Bleach plant effluents	Laccase	*Pycnoporus sanguineus*	
2,6-Dimethoxy phenol (DMP)	MnP	*Bjerkandera adusta*	
Phenolic compounds	Manganese peroxidase (MnP)	*P. chrysosporium*	
Non-phenolic aromatic compound	LiP	*Penicillium chrysogenum*	
Solar golden yellow R	MnP, laccase	*Schizophyllum commune*	
Reactive blue 59	Lignin peroxidase	*Streptomyces krainskii*	
Remazol brilliant blue R	Peroxidase and laccase	*Pleurotus ostreatus*	
Azo dye	Lignin peroxidase	*Penicillium ochrochloron*	
Azo dye	Lignin peroxidase, tyrosinase, azoreductase, and riboflavin reductase	*Bacillus* sp.	
Remazol brilliant blue R, Remazol black B, Reactive orange 122 and Reactive red 251	Laccase	*Trametes versicolor*	
Malachite green and methyl red dyes	Laccase	*Kluyveromyces dobzhanskii* DW1 and *Pichia manshurica* DW2	
Bromocresol Purple, Safranin, Bromothymol Blue, and Phenol Red	Laccase	*Yarrowia lipolytica*	
Acid blue	Laccase	*Agaricus bisporus*	

(continued)

Table 5 (continued)

Dyes/substrate	Enzyme	Species/sources	References
Acid violet 7, bromophenol blue (BpB), and Coomassie brilliant blue	Laccase	*E. coli* BL21 (DE3)	
Fluorene and phenanthrene	Manganese peroxidase (MnP)	*Cerrena unicolor* BBP6	
Indigo carmine	Laccase	*Bacillus amyloliquefaciens*	
Basic Orange 2 dye	Azoreductase	*Escherichia coli*	[61]
Methyl red	Azoreductase	*Pseudomonas aeruginosa*	
Direct Red (DR) 28	Azoreductase and laccase	*E. indicum* ES31G, *K. gibsonii* ES43G, *K. cryocrescens* ES45G, *C. lapagei* ES48G, *E. wuhouensis* ES49G, *A. caviae* ES50G, *L. sphaericus* ES51G, *A. haemolyticus* ES52G, *E. soli* ES53G, and *C. aquatic* ES54G	[55]
Reactive Yellow F3R and Joyfix Red RB	Azoreductase and laccase	*L. sphaericus* MTCC 9523	[161]
Congo red (CR) and malachite green (MG)	Reductase, dehydrogenase, kinase, hydrolase, transferase, catalase, and, dioxygenase	*Klebsiella pneumoniae* K2, *Enterobacter* sp. K16b and *Vibrio tritonius* K20	Chaieb et al. (2022)
Indigo carmine and malachite green	Laccase	*Bacillus subtilis* DS	[153]
Malachite green	Laccase	*Bacillus licheniformis* NS2324	[32]
Trypan blue dye (direct blue 14)	Azoreductase and tyrosinase	*Vibrio* sp.	[81]
Reactive Blue 19 (RB19)	Lignin peroxidase	*Bacillus* sp. JF4	Zhang et al. (2023)
Acid Maroon V (100 mg L^{-1})	Azoreductase, NADH-DCIP reductase, laccase, manganese peroxidase, and lignin peroxidase	*Enterobacter dissolvens* AGYP1 and *Pseudomonas aeruginosa* AGYP2	Patel and Gupte (2023)

(continued)

Table 5 (continued)

Dyes/substrate	Enzyme	Species/sources	References
Direct black G, methyl orange, Congo red, direct black 38, acid black 172	Dye-decolorizing peroxidase (DyP)	*Serratia* sp. AXJ-M	[177]
Congo red (CR)	Laccase	*Trametes flavida* WTFP2	[152]
Reactive red-195	Tyrosinase and azoreductase	*Halomonas meridiana* SAIBP-6	[142]

was then produced and transformed in E. coli JM109 under the guidance of a lac operon [67]. An azoreductase gene that has been integrated into Escherichia coli and copied from *Bacillus laterosporus* RRK1 can degrade Remazol red in the presence of 0.8 mg/L of oxygen [5]. *Pantoea ananatis* an endophytic bacterium Sd-1 that produces the Laccase gene (Lac4) that can be inserted in *E. coli*, demonstrated an excellent azo dye bioremediation capacity for Congo Red (89%) and Remazol Brilliant Blue R (35%) [80]. To degrade a variety of textile azo dyes the *Bacillus subtilis* CotA laccase gene can be integrated into the recombinant strains of *Synechococcus elongatus* PCC7942 heterologously [94]. The list of Genetically modified organisms for dye degradation is shown in Table 6. Numerous genetic techniques and methods, such as 16S rDNA sequencing, randomly amplified polymorphic DNA, polymerase chain reaction (PCR), single-stranded conformation polymorphism, and other emerging sequencing technologies, are available to evaluate the expression of the microbial genome.

5.4 Textile Wastewater Treatment Optimization Using Artificial Intelligence

Artificial Intelligence (AI) is the power of computer programs to simulate human intelligence in self-governed learning, judgment, intellectual, and policymaking. Artificial Intelligence (AI) can be used in the wastewater treatment process to reduce complications. AI can be used to predict water treatment performance, design sensors that estimate important components, optimize operating parameters, assess effluent quality, design treatment units, and optimize micropollutants and other emergent pollutants. Artificial Neural Network (ANN), Search Algorithms (SA), and Machine Learning (ML) are three types of Artificial intelligence models that are used frequently for optimization and prediction in wastewater management plants [173]. Artificial Neural Network (ANN) models that are commonly used include Fuzzy Neural Networks (FNNs), Deep Neural Networks (DNNs), Convolutional Neural Networks (CNNs), and Recurrent Neural Networks (RNNs). AI can also be used to predict preventative measures using diagnosis of faults, optimize energy utilization in

Table 6 List of genetically modified organisms for dye degradation

Gene extracted from	Genes extracted	Gene expressed in	Dye	References
Rhodobacter sphaeroides AS1.1737	Azoreductase	*Escherichia coli* JM109	Acid red GR	[171]
Rhodobacter sphaeroides AS1.1737	Azoreductase gene	*E. coli* JM109	Direct blue 71	
K. pneumoniae MGH 78578	AzoK gene	*E. coli* DH5α	Methyl orange	
Bacillus laterosporus RRK1	Azoreductase gene	*E. coli* strain DH5a	Remazol red	
Halomonas elongata	Azoreductase gene	*E. coli* DH5	Methyl red and Remazol black B	
Halomonas sp. strain GT	AzoG gene	*E. coli* DH5α	Azo dye wastewater	
Rhodococcus sp.	Azoreductase	*Escherichia coli* DH5α	Reactive red 22	
Escherichia coli K12	CueO laccase	*Pichia pastoris* GS115	Congo red, malachite green, and Remazol brilliant blue R	Ma et al. (2017)
Ganoderma sp. En3	Laccase rLAC-EN3-1	*P. pastoris*	Malachite green	Zhuo et al. (2015)
Aspergillus sp. TS-A	Laccase, MnP, and LiP	*P. pastoris* GS115	Congo red dye	[95, 98]

treatment methods, and provide guidelines for operation based on performance. Artificial neural networks (ANNs), machine learning, and other soft computing techniques have all been used recently to forecast water quality and related variables. The use of AI in WWTPs can improve management, guarantee higher effluent quality, and subsequently encourage the recycling of wastewater and the recovery of various resources, including electricity [182]. The digital era could improve the management of wastewater by lowering operational expenses, energy, and chemical usage, and facilitating decisions [109], as well as facilitating the supervision and streamlining of WWTPs [105]. By swiftly and automatically making assessments, AI-based strategies can optimize process parameters. For example, depending on input statistics, AI systems can control industrial machinery settings and alert operators [172]. The most frequently utilized method in applications of AI for wastewater management is the artificial neural network (ANN). A non-linear statistical model called artificial neural network (ANN) uses the same learning process as the human brain to improve its performance over time [108]. The maximum of studies has concentrated on the predictive management of both influent and effluent quality factors like CO, temperature, BOD, pH, overall nitrogen, total suspended solids, turbidity, total phosphorus, sludge retention time (SRT), hydraulic retention time (HRT), food to microorganism ratio (F/M), gaseous emission and sludge volume index (SVI) [157]. Yang et al. [178] developed a dynamic model that can forecast chemical oxygen demand (COD) and TN in wastewater. Wang et al. [174] in their work demonstrated that ANN models that can predict the amount and quality of wastewater influent and effluent were trained using historical data to improve the management of wastewater treatment plants and reduce energy and waste consumption. Artificial intelligence tools can model, anticipate, and improve electrochemical methods utilized in the treatment of wastewater and design artificial intelligence models with the ability to estimate the accuracy of eliminating phosphates from effluents employing the process of electrocoagulation [154]. Picos-Benítez et al. [130] implemented an artificial intelligence model to enhance the conduct of dye wastes and decrease the usage of energy, the efficiency of ANN was evaluated using RMSE and MAPE values computed from real and projected values of 10.73% and 8.81%, respectively, yielding an efficacy of 88.8 0.3%, near to the model's prediction of 95.5%. Matheri et al. [108] studied the link between trace metals and chemical oxygen demand, by developing a prediction model based on artificial intelligence. Additionally, the sequencing batch reactors (SBR) treatment process, and total coliforms in effluent were predicted using ANN [82]. Membrane filtration approaches to treating wastewater have been forecasted and improved using artificial intelligence (AI) models. AI-based methods can help to reduce membrane fouling [72], by improving the effectiveness of chemical and physical cleanup procedures as well as enhancing operational conditions [120]. Additionally, the fabrication of the membrane can be optimized by employing artificial intelligence [92]. Models based on artificial intelligence are exceptional at simulating the treatment of wastewater with limited input data. Yet, it is critical to perform additional research to investigate how they operate in real-world processes with huge data sets.

6 Life Cycle Assessment (LCA)

At every stage of their lifecycle, textile and apparel products' environmental impact has been evaluated using lifecycle assessment (LCA), comprising extraction of raw materials, processing fiber, manufacturing of textile and/or textiles and apparel, distribution stage, consumer usage, and to the end-of-life stage [48]. Methodologies like ReCiPe, Eco-indicator 99, USEtox 2.01, and TRACI 2.1 are frequently used in Life cycle assessment [176]. The framework and principles for LCA are set out by The International Organization for Standardization (ISO) 14040 whereas requirements and guidelines for LCA are provided by ISO 14044. The sublines represent the LCA at each stage.

6.1 Extraction of Raw Material

Cultivating and harvesting natural fibers is essential for material extraction, which necessitates the use of chemical pesticides and fertilizers as well as a lot of water. Moazzem et al. [112] reported that one of the most widely used natural fibers in the production of clothing, cotton has a significant environmental impact because it requires a lot of water, land, fertilizers, pesticides, and energy (0.63–0.95 MJ of electricity per kilogram production). La Rosa and Grammatikos [89] noted that one ton of cotton textile requires approximately 1736 m^3 of water. Conventional cotton is one of the most widely used fibers; from 2019 to 2020, consumers spent $38.54 billion on it (Global Cotton Market Report 2021–2027). Moazzem et al. [112] found that, given that raw cotton fiber accounts for 96.53% of the impact, the textile production stage contributes 98.37% of the impact on agricultural land occupation. Besides energy consumption, irrigation leads to water depletion, and spinning cotton yarn can require 3984 kWh of electricity [95]. In summation, Kazan et al. [76] stated that electricity used in cotton yarn is the leading Global warming Potential Impact (GWP). One of the main synthetic fibers, polyester, makes up more than half of all fiber produced annually in the world (Preferred Fiber and Materials Market Report 2023–2030). Gonzalez et al. [48] reported synthetic material polyester as the most commonly used which depends on non-renewable resources like petroleum and energy-intensive methods. Production of polyester requires the removal and refining of crude oil to develop blocks for PET and use by other industries. The transformation of chemical blocks into polyester and textile demands a high energy process, a high temperature of 290 °C for poly-condensation, and the usage of metal oxides and acetates as catalysts this high temperature results in the emission of carbon dioxide [104]. The synthetic fiber economy was estimated to be worth USD 59.95 billion in 2020 and was expected to grow at a compound annual growth rate (CAGR) of 6.6% between 2021 and 2028 (Synthetic Fibers Market Growth Analysis Report 2021–2028). Moazzem et al. [112] reported that the energy used in the fiber-processing phase and textile-processing of a polyester knit jacket has the biggest impact on the climate change potential (CCP). Polyester dyeing and finishing were listed as being the most unsustainably high stages, accompanied by the preparation of yarn and production of fiber. All impact categories were found to be most negatively impacted by the production stage [116]. Silk despite being a luxurious, biodegradable natural fiber, its production has an important undesirable effect on the environment. Using organic or phosphorous fertilizers caused metal emissions into the soil, water, and air, which contributed significantly to these effects and caused the water table to drop [19].

6.2 Manufacturing Stage

This stage involves bleaching, dyeing, finishing, weaving, and knitting, each requiring significant energy. With several steps in the textile dyeing process, water is utilized to wash off the chemicals used in the previous step before the start of the following step. Manufacturing textiles continues to have a large impact on the environment, with the industry increasingly based in developing nations such as Bangladesh, China, India, and Vietnam. In a conventional dyeing and finishing process, one ton of materials can pollute up to 200 tons of water with various kinds of dangerous chemicals while also requiring a significant amount of energy. Volatile pollutants are particularly problematic since they evaporate into the air or get into the body through foods or skin. Certain chemicals are oncogenic, and harm children before being born, whereas others may induce allergic responses in certain individuals. According to Liu et al. [95, 98] Spinning cotton uses around 3984 kWh of energy. Moazzem et al. [112] found that the spinning process posed the greatest threat to acidification, abiotic resource depletion, and climate change. As reported by Zhang et al. [181] the Cotton dyeing stage requires a huge amount of coal, water, dye, steam, and auxiliaries that contribute to the depletion of abiotic. In addition, this stage is also responsible for 20% of water contamination globally as batch dyeing for 1 kg of fabric dye demands around 150 L of water. Wastewater recycling can result in significant savings in terms of chemical, energy, water, and consumption.

6.3 Transportation and Consumer Stage

The inconvenience of the distance between the industry/manufacturer and the location of retailers has led to the use of routes through either sea or air or both for transporting consumer goods (Gonzalez et al. 2023). With the emission of around 6000,000 tons of CO_2, transportation constitutes the second carbon footprint provider [112]. Activities like washing, drying, and ironing come under consumer usage. The temperature and load size determine how much energy is used and how much CO_2 is released during washing. Approximately 45.45% of the total energy is used [181]. The amount of energy consumed depends on the behavior of the user, it varies depending on the country of origin, age, lifestyle, and climate. Eventually, a significant amount of non-renewable energy is used up as electricity, primarily in laundry practices upon consumer usage, rather than during the manufacturing of textiles.

6.4 End-of-Life Stage

Disposal of clothing involves resale, giveaway, donation, landfilling, swapping, and reusing. In the Netherlands around 67% of disposed clothes are incinerated which is an energy-intensive process and highly polluting [48]. To reduce/put an end to such disposal methods, by 2025 the European Union instructed the inhabitants to divide wastes into assigned bins. Clothing can be recycled, remanufactured, or reused after its initial use cycle, this 3Rs principle can help to eliminate unnecessary waste and limit the consumption of non-renewable resources.

6.5 Possible Substitutes for Cotton/Synthetic Material

Organically grown cotton is a viable substitute for regular cotton where chemical herbicides, fertilizers, insecticides, boll openers, stimulators/growth regulators, and defoliants are not used in production. As reported by the Textile Exchange the requirement for organic cotton has gone up to 31% (Organic Cotton Market Report 2022). When compared to conventional cotton, denim made from organic cotton has the minimum negative environmental effects. The overall GWP impact has decreased from 20.11 to 19.48 kg CO_2 eq due to the restricted use of fertilizers and pesticides. In addition, the use of organic cotton has increased the Freshwater Aquatic, Ecotoxicity Potential (FAEP) impact category by 47.6% [139].

The use of jute and other bast fibers, such as hemp, has increased due to their superior mechanical qualities over other naturally occurring fibers. To effectively use plant fibers in various applications, it is essential to comprehend how they interact with other materials [106]. Jute also has an advantage over cotton in terms of the environment because it uses a lot less water to grow has less of an impact on things like acidification and human toxicity and constantly performs worse than cotton in categories like eutrophication, acidification, cumulative energy demand, and depletion of water resources [89].

Compared to other virgin fibers, flax's production processes have one of the least negative environmental effects [114]. When compared to cotton, flax yields are higher while requiring fewer inputs and flax has a significantly lower environmental impact in terms of GHG emissions, energy use, and water use. Flax fiber could greatly reduce the potential for acidification, the use of agricultural land, and the impact on water demand [112]. From this perspective, excellent substitutes for other plant fibers are flax and hemp. Annually hemp and flax yields in the United States range from 1.0 to 5.0 tons per hectare (flax-hemp), which is significantly higher than the 0.8–0.9 tons per hectare produced by conventional cotton [39]. Additionally, flax does have a lower environmental impact during the raw materials abstraction procedure than other fibers like jute, hemp, polyester, and silk.

7 Circular Economy

The circular economy (CE) method strives to extend the life of textile items through reuse and repair while also recycling end-of-life (EoL) materials. Making the switch to a CE is essential for protecting our earth's resources, fostering domestic, sustainable growth, and creating jobs, and the security and economic prosperity of our country [146]. The circular economy is characterized as a transition wherein waste generation is reduced but resources, materials, and products are kept in the economy for a long time so that they can be used to the greatest extent possible. The rapid advancement of the fashion industry, rapidly evolving fashion trends, the use of large-scale production on a global scale, the spread of manufacturing into underdeveloped countries, landfilling of leftovers have all considerably increased environmental challenges [34]. In 2011, there were 79 Gt of materials consumed globally, and by 2060, that number is expected to rise to 167 Gt (OECD, Outlook to 2060). The average number of wears a garment receives before being discarded has fallen by 36% over the past 15 years, meaning that the frequency with which useful clothing is disposed of (before it reaches end-of-life) has increased. In recent years, the idea of a circular economy has gained popularity as a way to address sustainability issues brought on by the conventional economic system, such as waste production, exhaustion of resources, global warming, and material shortages [83]. The main tactics for achieving a circular economy are reduced use of resources, using fewer resources, and closing the resources loop between post-used and production. The 3Rs (Reduce, Reuse, and Recycle concept) are encouraged globally for making a viable resource-cycle culture achieved via efficient resources and material use. Clothing can be recycled, remanufactured, or reused after its initial use cycle. The emergence of retailer-owned second-hand clothes can help address sustainability issues related to international markets for second-hand clothing, provide new possibilities for reuse models, and develop a new customer base (Lüdeke-Freund et al. 2019). By taking on the duty of returning, repairing, and reselling worn apparel, retailers would improve the likelihood of customer returns and resale opportunities while building consumer trust.

Remanufacturing is the creation of new high-quality clothing that is equivalent to new clothing, this procedure entails disassembling discarded clothing, recovering useful component elements, and redesigning and replicating them. It is a much more economical and resourceful method of dealing than reprocessing since it allows for the avoidance of high energy used as well as the use of high amounts of chemicals and methods of treatment. The clothing industry produces an enormous quantity of waste. A UK-based company called Worn Again is taking different activities to recycle textile waste, for example, by creating a biochemical recycling process and collaborating with Hennes & Mauritz (H&M) to reprocess fabric waste and repurpose it in a circular loop. Used garments are sorted manually for recycling which has many disadvantages such as the inability to differentiate complex materials, cost-effective, time-consuming, and non-standardization [135].

7.1 Resources

The material used in the textile industry must be renewable, regenerative, and non-toxic. By using fewer resources to produce goods, resource efficiency aims to cut the resource loops. Three key strategies can be used to increase a fashion product's resource efficiency: using raw materials that are sustainable and renewable, cutting back on resource usage, and minimizing waste. Decreasing land, and water, and the use of chemicals in farming ought to be a top concern for natural fibers like cotton, however, for synthetic fibers like polyester, the energy and use of natural gas consumption should be minimal. Plant-derived recyclable sources created by Kusaga Athletic an Australian-based company can be used in place of common textiles like cotton and polyester. Reusing or recycling waste results in economic savings and less reliance on natural resources. As reported by a UK-based company called Eileen Fisher sells eco-friendly clothing and invites returns so that the items can be redesigned and resold. Water consumption can be reduced by utilizing cutting-edge technologies and sophisticated machinery. Conservation of energy can be achieved by using some measures such as converting from thermal to renewable energy sources and utilizing energy-saving equipment and processes [24].

7.2 Designs

Designers have the authority to choose fabrics, fittings, patterns, colors, values, and designs in the traditional design process sustainability and the circular economy because their choices have a substantial impact on the product's environmental impact [34]. Better design and recycling may lead to the creation of new business models and useful products while also reducing the cost of initial supplies and landfills. Key elements of a durable product are superior fabrics, robust stitching, and colors that hold up to washing and wearing. With trans-seasonal products, we can keep an item in use for longer than just one season of fashion and delay its disposal. By customization method, each customer throughout the customization process, every customer receives personal interest, and by actively engaging in the process, they become co-designers. These environmentally friendly methods enable personalization by providing distinctive designs in each garment and lowering the consumption of fossil fuels, water, and chemical-based resources as well as reducing excess manufacturing and overconsumption while increasing product life and minimizing the generation of waste. Clothing that allows for easy deconstruct can ensure that all useful parts are reused, recycled, or remanufactured. As a result, it is critical to design with disassembly and end-of-life treatment strategies in mind. Wear2™ Microwave technology sewing thread can be utilized to disassemble clothes by exposing them to microwave radiation which dissolves the sewing threat without affecting other parts of the textile. Reducing material complexity by reducing the usage of mixed resources and increasing the utilization of pure resources may enhance the efficiency of reprocessing. For instance, Candiani Denim, an Italian-based company, creates denim keeping durability at the forefront and encouraging recycling at the conclusion, minimizing debris and promoting speedy use [35]. Nevertheless, one promising way to achieve biodegradable fabrics is the creation and usage of bio-inspired polymers in textile manufacture. For example, the Swiss company Freitag created a biodegradable fabric (FABRIC) comprised of hemp, flex, and modal fibers. While incredibly durable, it will wear out with time and compost after a few months, consisting of buttons made of nuts and sewing threads.

7.3 Consumption

Extending product life is the most effective approach to reducing the need for virgin materials as well as the amount of energy used in raw material extraction and processing. Product-service systems (PSS), provide a novel method of extending product life that places a strong emphasis on such as repair, leasing, and rental models [119]. Maintenance and repair procedures can increase the longevity of clothing and minimize the need for new materials. Consumers can also fix their products at the residence. In this case, Nudie Jeans ships repair supplies to people who are unable to reach their local store but still want to mend their clothing. These do-it-yourself (DIY) are becoming more and more popular as they reduce waste by focusing more on reprocessing (Lüdeke-Freund et al. 2019). To keep clothing in use regularly, rental and leasing businesses might create cooperative business models. DressYouCan in Italy is a company where shoes, accessories, and dresses, can be rented by women at a reasonable price for special events by enabling customers, and up-and-coming businesses, and allowing new designers to promote what they create on this site at a fixed, lower-cost, the organization created an innovative model for recycling and reuse [35]. Haverdash is the rental aid that gives women a taste of fashion which charges a set monthly cost and offers free delivery, return shipping, and dry-cleaning services. Peer-to-peer clothing exchange opens up new consumption and supply opportunities, allowing items to be diverted from dumpsters and into alternative value chains [58]. Clothes switching is helpful when consumers want to try something new or don't have sufficient funds to purchase new items. It also helps extend the life of clothing [25]. Increased wear of clothing results in less waste, less pollution, and fewer purchases of new goods.

8 Conclusion and Future Perspective

The results of many authors indicated that dyes and heavy metals are carcinogenic and mutagenic to humans, animals, and aquatic organisms. Industrial contaminants and effluents/sludges pose significant health risks because they can contaminate the soil and other wildlife if they are not properly treated. Treatment of textile effluent is challenging because the water includes hazardous substances that degrade slowly. The impact of dye contamination is covered in this chapter. And for the same the treatment methods using physical, chemical, biological, and nano-based technology are discussed. The effectiveness of microorganisms in degrading matter makes them interesting organisms. These include bacteria that are particularly helpful for degrading the widely used resistant azo dyes. Algae can be frequently utilized for degrading and decolorizing textile effluents. Different colors, colorants, and organic contaminants can all be broken down by different algae species. Similarly, to this, fungi are also vital to the bioremediation process. The white rot fungus is the most significant in the removal of dye, they can break down complicated wastes by means of a variety of enzymes. Enzymes can offer an alternative to traditional effluent treatment methods due to their ability to absorb substances and break them down. The biological catalysts known as enzymes can bind to a variety of colors, including the most resilient azo dyes. Nanotechnology is currently one of the most efficient and promising methods for dye degradation. It has been determined that bioremediation will be more effective and cost-effective at removing dyes from contaminated water sources and environmentally safer than conventional physical, and chemical methods, which need more energy. A systemic approach is necessary for the adoption of sustainable fashion, involving the engagement and commitment from manufacturers, designers, dealers' suppliers, and consumers with an optimistic outlook. The textile manufacturing sector needs to provide renewable resources, and the all-round textile processes must be improved to efficiently avoid any toxicity. AI setup is a critical factor in enhancing adaptation, decision-making, and mitigation capacity. It could bring about a revolution in the way we deal with complex problems besides tackling challenges that were previously insurmountable. With AI technology, we can unlock new insights and make more informed decisions, leading to better outcomes for society as a whole. Additionally, since chemicals and dyes impair both biotic and abiotic elements of our ecosystem, we should avoid utilizing them and choose synthetic or natural dyes and natural raw materials. Future studies on the biological breakdown of dyes used in textiles must prioritize research that is more directly related to "actual" field or ground settings and focus on the fundamental and practical aspects of the topic.

References

1. Abid MF, Zablouk MA, Elfikylustau-Alameer AM (2012) Experimental study of dye removal from industrial wastewater by membrane technologies of reverse osmosis and nanofiltration. Iran J Environ Health Sci Eng 9(1):17. https://doi.org/10.1186/1735-2746-9-17
2. Abu-Nada A, Abdala A, McKay G (2021) Removal of phenols and dyes from aqueous solutions using graphene and graphene composite adsorption: a review. J Environ Chem Eng 9(5):105858. https://doi.org/10.1016/j.jece.2021.105858
3. Afreen S, Shamsi T, Baig M, Ahmad N, Fatima S, Qureshi MI, Hassan M, Fatma T (2017) A novel multicopper oxidase (laccase) from cyanobacteria: Purification, characterization with

potential in the decolorization of anthraquinonic dye. PLOS ONE, 12. https://doi.org/10.1371/journal.pone.0175144
4. Ahmad A, Javed MS, Khan S, Almutairi TM, Mohammed AAA, Luque R (2023) Green synthesized Ag decorated CeO_2 nanoparticles: efficient photocatalysts and potential antibacterial agents. Chemosphere 310:136841. https://doi.org/10.1016/j.chemosphere.2022.136841
5. Ajaz M, Shakeel S, Rehman A (2020) Microbial use for azo dye degradation—a strategy for dye bioremediation. Int Microbiol 23:1–11. https://doi.org/10.1007/s10123-019-00103-2
6. Akpomie KG, Conradie J (2020) Advances in the application of cotton-based adsorbents for heavy metals trapping, surface modifications, and future perspectives. Ecotoxicol Environ Saf 201:110825. https://doi.org/10.1016/j.ecoenv.2020.110825
7. Akram MA, Javaid Asad M, Mahmood T, Shah MB, Nazir S, Khan J, Rashid K, Aslam S, Rawalpindi P (2023) First report on the biodegradation of direct flavine 5-G and reactive red S3B textile dyes by *Piptoporus betulinus*
8. Akter K, Zerin T, Banik A (2020) Biodegradation of textile dyes by bacteria isolated from textile industry effluents. Stamford J Microbiol 9:5–8. https://doi.org/10.3329/sjm.v9i1.45649
9. Al-Rumaihi A, Mckay G, Mackey H, Al-Ansari T (2020) Environmental impact assessment of food waste management using two composting techniques. Sustainability 12:1595. https://doi.org/10.3390/su12041595
10. Al-Tohamy R, Ali SS, Li F, Okasha KM, Mahmoud YA-G, Elsamahy T, Jiao H, Fu Y, Sun J (2022) A critical review on the treatment of dye-containing wastewater: ecotoxicological and health concerns of textile dyes and possible remediation approaches for environmental safety. Ecotoxicol Environ Saf 231:113160. https://doi.org/10.1016/j.ecoenv.2021.113160
11. Alam R, Mahmood RA, Islam S, Ardiati FC, Solihat NN, Alam MB, Lee SH, Yanto DHY, Kim S (2023) Understanding the biodegradation pathways of azo dyes by immobilized white-rot fungus, *Trametes hirsuta* D7, using UPLC-PDA-FTICR MS supported by in silico simulations and toxicity assessment. Chemosphere 313:137505. https://doi.org/10.1016/j.chemosphere.2022.137505
12. Ali SS, Al-Tohamy R, Sun J (2022) Performance of *Meyerozyma caribbica* as a novel manganese peroxidase-producing yeast inhabiting wood-feeding termite gut symbionts for azo dye decolorization and detoxification. Sci Total Environ 806:150665. https://doi.org/10.1016/j.scitotenv.2021.150665
13. Aljohani MM, Al-Qahtani SD, Alshareef M, El-Desouky MG, El-Bindary AA, El-Metwaly NM, El-Bindary MA (2023) Highly efficient adsorption and removal bio-staining dye from industrial wastewater onto mesoporous Ag-MOFs. Process Saf Environ Prot 172:395–407. https://doi.org/10.1016/j.psep.2023.02.036
14. Altohamy R, Sun J, Fareed M, Kenawy E-R, Ali S (2020) Ecofriendly biodegradation of reactive black 5 by newly isolated *Sterigmatomyces halophilus* SSA1575, valued for textile azo dye wastewater processing and detoxification. Sci Rep 10:12370. https://doi.org/10.1038/s41598-020-69304-4
15. Alvarez MF, Poch RM, Olarieta JR, Wiedner K (2023) Charcoal and biological activity in formiguer soils of Catalonia (Spain): application of a micromorphological approach. Soil Tillage Res 234:105810. https://doi.org/10.1016/j.still.2023.105810
16. Aoun N, Boucheloukh H, Belkhalfa H, Harrouche K, Boughrara B, Sehili T (2023) Combined action of cerianite, UV and direct sunlight irradiation for the removal of violet crystal and methylene blue from aqueous solution. React Kinet Mech Catal 136(3):1607–1623. https://doi.org/10.1007/s11144-023-02410-2
17. Avelar NV, Rezende AAP, Carneiro AdCO, Silva CM (2016) Evaluation of briquettes made from textile industry solid waste. Renew Energy 91:417–424. https://doi.org/10.1016/j.renene.2016.01.075
18. Azanaw A, Birlie B, Teshome B, Jemberie M (2022) Textile effluent treatment methods and eco-friendly resolution of textile wastewater. Case Stud Chem Environ Eng 6:100230. https://doi.org/10.1016/j.cscee.2022.100230
19. Barcelos SMBD, Salvador R, da Graça Guedes M, de Francisco AC (2020) Opportunities for improving the environmental profile of silk cocoon production under Brazilian conditions. Sustainability (Switzerland) 12(8). https://doi.org/10.3390/SU12083214

20. Behera M, Nayak J, Banerjee S, Chakrabortty S, Tripathy SK (2021) A review on the treatment of textile industry waste effluents towards the development of efficient mitigation strategy: an integrated system design approach. J Environ Chem Eng 9(4):105277. https://doi.org/10.1016/j.jece.2021.105277
21. Bhatia D, Sharma NR, Singh J, Kanwar RS (2017) Biological methods for textile dye removal from wastewater: a review. Crit Rev Environ Sci Technol 47(19):1836–1876. https://doi.org/10.1080/10643389.2017.1393263
22. Bouacem K, Rekik H, Jaouadi NZ, Zenati B, Kourdali S, El Hattab M, Badis A, Annane R, Bejar S, Hacene H, Bouanane-Darenfed A, Jaouadi B (2018) Purification and characterization of two novel peroxidases from the dye-decolorizing fungus *Bjerkandera adusta* strain CX-9. Int J Biol Macromol 106:636–646. https://doi.org/10.1016/j.ijbiomac.2017.08.061
23. Buam I, Hussain N (2021) Generation and management of agricultural waste-worldwide view. In: Sharma P (ed) Trends in environmental research, vol 8. Akinik Publications, pp 63–89. ISBN: 976-93-91538-79-8
24. Çay A (2018) Energy consumption and energy saving potential in clothing industry. Energy 159:74–85. https://doi.org/10.1016/j.energy.2018.06.128
25. Camacho-Otero J, Pettersen I, Boks C (2019) Consumer engagement in the circular economy: exploring clothes swapping in emerging economies from a social practice perspective. Sustain Dev 28. https://doi.org/10.1002/sd.2002
26. Castillo-Suárez LA, Sierra-Sánchez AG, Linares-Hernández I, Martínez-Miranda V, Teutli-Sequeira EA (2023) A critical review of textile industry wastewater: green technologies for the removal of indigo dyes. Int J Environ Sci Technol. https://doi.org/10.1007/s13762-023-04810-2
27. Ceretta MB, Nercessian D, Wolski EA (2021) Current trends on role of biological treatment in integrated treatment technologies of textile wastewater. In: Frontiers in microbiology, vol 12. Frontiers Media S.A. https://doi.org/10.3389/fmicb.2021.651025
28. Chauhan P, Panthi N, Mazumdar I, Hussain N (2023) Bioaccumulation and detoxification of metals through genetically engineered microorganism. In: Mathuriya AS, Pandit S, Singh NK (eds) Green technologies for industrial waste remediation. Springer Nature Switzerland, pp 147–187. https://doi.org/10.1007/978-3-031-46858-2_8
29. Chen C, Wu H (2018) Lightweight bricks manufactured from ground soil, textile sludge, and coal ash. Environ Technol 39(11):1359–1367. https://doi.org/10.1080/09593330.2017.1329353
30. Chen Y, Zhang L, Feng L, Chen G, Wang Y, Zhai Z, Zhang Q (2019) Exploration of the key functional strains from an azo dye degradation microbial community by DGGE and high-throughput sequencing technology. Environ Sci Pollut Res 26(24):24658–24671. https://doi.org/10.1007/s11356-019-05781-z
31. Chen G, An X, Li H, Lai F, Yuan E, Xia X, Zhang Q (2021) Detoxification of azo dye direct black G by thermophilic *Anoxybacillus* sp. PDR2 and its application potential in bioremediation. Ecotoxicol Environ Saf 214:112084. https://doi.org/10.1016/j.ecoenv.2021.112084
32. Chopra NK, Sondhi S (2022) Biodegradation of malachite degradation by laccase from *Bacillus licheniformis* NS2324. Int J Health Sci 8517–8527. https://doi.org/10.53730/ijhs.v6ns2.7195
33. Choudhary N, Yadav VK, Yadav KK, Almohana AI, Almojil SF, Gnanamoorthy G, Kim D-H, Islam S, Kumar P, Jeon B-H (2021) Application of green synthesized MMT/Ag nanocomposite for removal of methylene blue from aqueous solution. Water 13(22). https://doi.org/10.3390/w13223206
34. Claxton S, Kent A (2020) The management of sustainable fashion design strategies: an analysis of the designer's role. J Clean Prod 268:122112. https://doi.org/10.1016/j.jclepro.2020.122112
35. Colucci M, Vecchi A (2020) Close the loop: evidence on the implementation of the circular economy from the Italian fashion industry. Bus Strat Environ 30. https://doi.org/10.1002/bse.2658

36. Dai S, Yao Q, Yu G, Liu S, Yun J, Xiao X, Deng Z, Li H (2021) Biochemical characterization of a novel bacterial laccase and improvement of its efficiency by directed evolution on dye degradation. Front Microbiol 12. https://doi.org/10.3389/fmicb.2021.633004
37. Darvishi F, Moradi M, Jolivalt C, Madzak C (2018) Laccase production from sucrose by recombinant Yarrowia lipolytica and its application to decolorization of environmental pollutant dyes. Ecotoxicol Environ Safety 165:278–283. https://doi.org/10.1016/j.ecoenv.2018.09.026
38. Dhadse S (2022) Vermiconversion of textile industrial sludge; waste management and nutrients recycling. In: Agricultural waste—new insights, p 3. https://doi.org/10.5772/intechopen.107260
39. Duque Schumacher AG, Pequito S, Pazour J (2020) Industrial hemp fiber: a sustainable and economical alternative to cotton. J Clean Prod 268:122180. https://doi.org/10.1016/j.jclepro.2020.122180
40. Ebba M, Asaithambi P, Alemayehu E (2022) Development of electrocoagulation process for wastewater treatment: optimization by response surface methodology. Heliyon 8(5):e09383. https://doi.org/10.1016/j.heliyon.2022.e09383
41. Ebrahim S (2018) Removal of acid blue dye from industrial wastewater by using reverse osmosis technology
42. Elfiky AAEA, Mubarak MF, Keshawy M, Sayed IETEl, Moghny TA (2023) Novel nanofiltration membrane modified by metal oxide nanocomposite for dyes removal from wastewater. Environ Dev Sustain https://doi.org/10.1007/s10668-023-03444-1
43. Ewida A, El-Sesy M, Abou Zeid A (2019) Complete degradation of azo dye acid red 337 by *Bacillus megaterium KY848339*.1 isolated from textile wastewater. Water Sci 33:154–161. https://doi.org/10.1080/11104929.2019.1688996
44. Gacem A, Modi S, Yadav VK, Islam S, Patel A, Dawane V, Jameel M, Inwati GK, Piplode S, Solanki VS, Basnet A (2022) Recent advances in methods for synthesis of carbon nanotubes and carbon nanocomposite and their emerging applications: a descriptive review. J Nanomater 2022:7238602. https://doi.org/10.1155/2022/7238602
45. Gholami Shirkoohi M, Tyagi RD, Vanrolleghem PA, Drogui P (2022) A comparison of artificial intelligence models for predicting phosphate removal efficiency from wastewater using the electrocoagulation process. Digit Chem Eng 4:100043. https://doi.org/10.1016/j.dche.2022.100043
46. Gnanasekaran G, Sudhakaran MSP, Kulmatova D, Han J, Arthanareeswaran G, Jwa E, Mok YS (2021) Efficient removal of anionic, cationic textile dyes and salt mixture using a novel CS/MIL-100 (Fe) based nanofiltration membrane. Chemosphere 284:131244. https://doi.org/10.1016/j.chemosphere.2021.131244
47. Goh PS, Lau WJ, Othman MHD, Ismail AF (2018) Membrane fouling in desalination and its mitigation strategies. Desalination 425:130–155. https://doi.org/10.1016/j.desal.2017.10.018
48. Gonzalez V, Lou X, Chi T (2022) Sustainable development in the textile and apparel industry: a lifecycle assessment (LCA) approach
49. Goswami L, Das S, Hussain N, Bhattacharya SS (2023) Vermi-remediation of metal(loid)s contaminated surfaces, pp 221–240. https://doi.org/10.1007/978-3-031-04931-6_9
50. Guha A, Kanti Guha A, Rahman O, Das S, Sazzad Hossain M (2015) Characterization and composting of textile sludge sustainable sludge management: waste to energy view project characterization and composting of textile sludge. Resour Environ 5(2):53–58. https://doi.org/10.5923/j.re.20150502.01
51. Gul R, Sharma P, Kumar R, Umar A, Ibrahim AA, Alhamami MAM, Jaswal VS, Kumar M, Dixit A, Baskoutas S (2023) A sustainable approach to the degradation of dyes by fungal species isolated from industrial wastewaters: performance, parametric optimization, kinetics and degradation mechanism. Environ Res 216:114407. https://doi.org/10.1016/j.envres.2022.114407
52. Halepoto H, Gong T, Memon H (2022) Current status and research trends of textile wastewater treatments—a bibliometric-based study. Front Environ Sci 10. https://doi.org/10.3389/fenvs.2022.1042256

53. Hamdan J, Sabri NSM, Hasbullah H, Sokri MNM, Rahman SA, Kadirkhan F (2021) Removal of direct red dye using polysulfone-chitosan ultrafiltration membrane. IOP Conf Ser: Mater Scie Eng 1142(1):012018. https://doi.org/10.1088/1757-899x/1142/1/012018
54. Hammad EN, Salem SS, Zohair MM, Mohamed AA, El-Dougdoug W (2022) *Purpureocillium lilacinum* mediated biosynthesis copper oxide nanoparticles with promising removal of dyes. Biointerface Res Appl Chem 12(2):1397–1404. https://doi.org/10.33263/BRIAC122.13971404
55. Haque MA, Mosharaf MK, Islam MS, Islam MM, Hasan AH (2022) Biofilm-mediated decolorization, degradation and detoxification of synthetic effluent by novel biofilm-producing bacteria isolated from textile dyeing effluent. Environ Pollut 314:120237. https://doi.org/10.1016/j.envpol.2022.120237
56. Hashmi SS, Shah M, Muhammad W, Ahmad A, Ullah MA, Nadeem M, Abbasi BH (2021) Potentials of phyto-fabricated nanoparticles as ecofriendly agents for photocatalytic degradation of toxic dyes and waste water treatment, risk assessment and probable mechanism. J Indian Chem Soc 98(4):100019. https://doi.org/10.1016/j.jics.2021.100019
57. Hassani AH, Mirzayee R, Nasseri S, Borghei M, Gholami M, Torabifar B (2008) Nanofiltration process on dye removal from simulated textile wastewater. Int J Environ Sci Technol 5(3):401–408. https://doi.org/10.1007/BF03326035
58. Henninger CE, Bürklin N, Niinimäki K (2019) The clothes swapping phenomenon—when consumers become suppliers. J Fash Mark Manag: Int J 23. https://doi.org/10.1108/JFMM-04-2018-0057
59. Holkar CR, Arora H, Halder D, Pinjari DV (2018) Biodegradation of reactive blue 19 with simultaneous electricity generation by the newly isolated electrogenic *Klebsiella* sp. C NCIM 5546 bacterium in a microbial fuel cell. Int Biodeterior Biodegrad 133:194–201. https://doi.org/10.1016/j.ibiod.2018.07.011
60. Hussain N, Goswami L, Bhattacharya SS (2023) Microbial detoxification of contaminated land. In: Pandey VC (ed) Bio-inspired land remediation. Environmental contamination remediation and management. Springer, Cham, pp 189–219. https://doi.org/10.1007/978-3-031-04931-6_8
61. Ikram M, Naeem M, Zahoor M, Hanafiah MM, Oyekanmi AA, Ullah R, Farraj DAA, Elshikh MS, Zekker I, Gulfam N (2022) Biological degradation of the azo dye basic orange 2 by *Escherichia coli*: a sustainable and ecofriendly approach for the treatment of textile wastewater. Water 14(19):2063, 2969. https://doi.org/10.3390/w14192969
62. IndraKumar Singh S, Singh WR, Bhat SA, Sohal B, Khanna N, Vig AP, Ameen F, Jones S (2022) Vermiremediation of allopathic pharmaceutical industry sludge amended with cattle dung employing *Eisenia fetida*. Environ Res 214:113766. https://doi.org/10.1016/j.envres.2022.113766
63. Islam A, Teo SH, Taufiq-Yap YH, Ng CH, Vo D-V, Ibrahim ML, Hasan M, Khan M, Alam SM, Awual M (2021) Step towards the sustainable toxic dyes and heavy metals removal and recycling from aqueous solution—a comprehensive review. Resour Conserv Recycl 175:105849. https://doi.org/10.1016/j.resconrec.2021.105849
64. Jamee R, Siddique R (2019) Biodegradation of synthetic dyes of textile effluent by microorganisms: an environmentally and economically sustainable approach. Eur J Microbiol Immunol 9:1–5. https://doi.org/10.1556/1886.2019.00018
65. Jiang M, Ye K, Deng J, Lin J, Ye W, Zhao S, Van der Bruggen B (2018) Conventional ultrafiltration as an effective strategy for dye/salt fractionation in textile wastewater treatment. Environ Sci Technol 52(18):10698–10708. https://doi.org/10.1021/acs.est.8b02984
66. Jiang X, Chen H, Liao Y, Ye Z, Li M, Klobučar G (2019) Ecotoxicity and genotoxicity of polystyrene microplastics on higher plant *Vicia faba*. Environ Pollut 250:831–838. https://doi.org/10.1016/j.envpol.2019.04.055
67. Jin R, Yang H, Zhang A, Wang J, Liu G (2009) Bioaugmentation on decolorization of C.I. direct blue 71 by using genetically engineered strain *Escherichia coli* JM109 (pGEX-AZR). J Hazard Mater 163(2–3):1123–1128. https://doi.org/10.1016/j.jhazmat.2008.07.067

68. Jishma P, Narayanan R, Snigdha S, Thomas R, Radhakrishnan EK (2018) Rapid degradative effect of microbially synthesized silver nanoparticles on textile dye in presence of sunlight. Biocatal Agric Biotechnol 14:410–417. https://doi.org/10.1016/j.bcab.2018.04.007
69. John J, Dineshram R, Hemalatha KR, Dhassiah MP, Gopal D, Kumar A (2020) Bio-decolorization of synthetic dyes by a halophilic bacterium *Salinivibrio* sp. Front Microbiol 11. https://doi.org/10.3389/fmicb.2020.594011
70. Joicy A, Selvamani R, Janani C, Balasubramanian C, Prabhu K, Marimuthu K, Giridharan B, Vijayakumar T, Saravanan KM (2023) Photocatalytic degradation of textile dye using green synthesized nanoparticles. Lett Appl NanoBioScience 12:102. https://doi.org/10.33263/LIANBS124.102
71. Jung HJ, Choi YJ, Oh KW (2020) Influencing factors of Chinese consumers' purchase intention to sustainable apparel products: exploring consumer "Attitude–Behavioral Intention" gap. Sustainability 12(5). https://doi.org/10.3390/su12051770
72. Kamali M, Appels L, Yu X, Aminabhavi TM, Dewil R (2021) Artificial intelligence as a sustainable tool in wastewater treatment using membrane bioreactors. Chem Eng J 417:128070. https://doi.org/10.1016/j.cej.2020.128070
73. Karaosman H, Brun A, Morales-Alonso G (2017) Vogue or vague: sustainability performance appraisal in luxury fashion supply chains. Springer Singapore. https://doi.org/10.1007/978-981-10-2917-2_14
74. Karim Md E, Dhar K, Hossain Md T (2018) Decolorization of Textile Reactive Dyes by Bacterial Monoculture and Consortium Screened from Textile Dyeing Effluent. J Genetic Eng Biotechnol 16(2):375–380. https://doi.org/10.1016/j.jgeb.2018.02.005
75. Karunya A, Rose C, Valli Nachiyar C (2014) Biodegradation of the textile dye mordant black 17 (Calcon) by *Moraxella osloensis* isolated from textile effluent-contaminated site. World J Microbiol Biotechnol 30(3):915–924. https://doi.org/10.1007/s11274-013-1509-8
76. Kazan H, Akgul D, Kerc A (2020) Life cycle assessment of cotton woven shirts and alternative manufacturing techniques. Clean Technol Environ Policy 22(4):849–864. https://doi.org/10.1007/s10098-020-01826-x
77. Khalaf HA, El-Sheekh MM, Makhlof MEM (2023) *Lychaete pellucida* as a novel biosorbent for the biodegradation of hazardous azo dyes. Environ Monit Assess 195(8):929. https://doi.org/10.1007/s10661-023-11518-w
78. Khaled JM, Alyahya SA, Govindan R, Chelliah CK, Maruthupandy M, Alharbi NS, Kadaikunnan S, Issac R, Murugan S, Li W-J (2022) Laccase producing bacteria influenced the high decolorization of textile azo dyes with advanced study. Environ Res 207:112211. https://doi.org/10.1016/j.envres.2021.112211
79. Khalili P, Tshai KY, Kong I (2017) Natural fiber reinforced expandable graphite filled composites: evaluation of the flame retardancy, thermal and mechanical performances. Compos Part A: Appl Sci Manuf 100:194–205. https://doi.org/10.1016/j.compositesa.2017.05.015
80. Khandare R, Kabra A, Awate AV, Govindwar S (2013) Synergistic degradation of diazo dye direct red 5B by *Portulaca grandiflora* and *Pseudomonas putida*. Int J Environ Sci Technol 10. https://doi.org/10.1007/s13762-013-0244-x
81. Khandare SD, Teotia N, Kumar M, Diyora P, Chaudhary DR (2023) Biodegradation and decolorization of trypan blue azo dye by marine bacteria *Vibrio* sp. JM-17. Biocatal Agric Biotechnol 51:102802. https://doi.org/10.1016/j.bcab.2023.102802
82. Khatri N, Khatri KK, Sharma A (2020) Artificial neural network modelling of faecal coliform removal in an intermittent cycle extended aeration system-sequential batch reactor based wastewater treatment plant. J Water Process Eng 37:101477. https://doi.org/10.1016/j.jwpe.2020.101477
83. Ki C-W, Chong S, Ha-Brookshire J (2020) How fashion can achieve sustainable development through a circular economy and stakeholder engagement: a systematic literature review. Corp Soc Responsib Environ Manag 27. https://doi.org/10.1002/csr.1970
84. Kishor R, Purchase D, Saratale GD, Saratale RG, Ferreira LFR, Bilal M, Chandra R, Bharagava RN (2021) Ecotoxicological and health concerns of persistent coloring pollutants of textile industry wastewater and treatment approaches for environmental safety. J Environ Chem Eng 9(2):105012. https://doi.org/10.1016/j.jece.2020.105012

85. Kumar A, Dixit U, Singh K, Gupta SP, Beg MSJ (2021) Structure and properties of dyes and pigments. Dyes Pigm. https://doi.org/10.5772/intechopen.97104
86. Kumar P, Joshiba J, Carolin F, Varshini P, Priyadharshini S, Karthik A, Jothirani R (2019) A critical review on recent developments in the low-cost adsorption of dyes from wastewater. Desalination Water Treat 172:395–416. https://doi.org/10.5004/dwt.2019.24613
87. Kumar A, Thakur AK, Gaurav GK, Klemeš JJ, Sandhwar VK, Pant KK, Kumar R (2023) A critical review on sustainable hazardous waste management strategies: a step towards a circular economy. Environ Sci Pollut Res. https://doi.org/10.1007/s11356-023-29511-8
88. Kumar A, Singh R, Kumar A, Rai A, Singh V, Bhadouria R (2019) Genetically engineered bacteria for the degradation of dye and other organic compounds, pp 331–350. https://doi.org/10.1016/B978-0-12-818095-2.00016-3
89. La Rosa AD, Grammatikos SA (2019) Comparative life cycle assessment of cotton and other natural fibers for textile applications. Fibers 7(12). https://doi.org/10.3390/FIB7120101
90. Lafi R, Gzara L, Lajimi RH, Hafiane A (2018) Treatment of textile wastewater by a hybrid ultrafiltration/electrodialysis process. Chem Eng Proc - Proc Intensificat 132:105–113. https://doi.org/10.1016/j.cep.2018.08.010
91. Li Y, Liu H (2000) High-pressure densification of wood residues to form an upgraded fuel. Biomass Bioenergy, 19(3):177–186. https://doi.org/10.1016/S0961-9534(00)00026-X
92. Li B, Yue R, Shen L, Chen C, Li R, Xu Y, Zhang M, Hong H, Lin H (2022) A novel method integrating response surface method with artificial neural network to optimize membrane fabrication for wastewater treatment. J Clean Prod 376:134236. https://doi.org/10.1016/j.jclepro.2022.134236
93. Liang C, Wei D, Zhang S, Ren Q, Shi J, Liu L (2021) Removal of antibiotic resistance genes from swine wastewater by membrane filtration treatment. Ecotoxicol Environ Safety 210:111885. https://doi.org/10.1016/j.ecoenv.2020.111885
94. Liang Y, Hou J, Liu Y, Luo Y, Tang J, Cheng J, Daroch M (2018) Textile dye decolorizing *Synechococcus* PCC7942 engineered with CotA laccase. Front Bioeng Biotechnol 6. https://doi.org/10.3389/fbioe.2018.00095
95. Liu Y, Huang H, Zhu L, Zhang C, Ren F, Liu Z (2020) Could the recycled yarns substitute for the virgin cotton yarns: a comparative LCA. Int J Life Cycle Assess 25(10):2050–2062. https://doi.org/10.1007/s11367-020-01815-8
96. Liu W, Liu C, Liu L, You Y, Jiang J, Zhou Z, Dong Z (2017) Simultaneous decolorization of sulfonated azo dyes and reduction of hexavalent chromium under high salt condition by a newly isolated salt-tolerant strain *Bacillus circulans* BWL1061. Ecotoxicol Environ Saf 141:9–16. https://doi.org/10.1016/j.ecoenv.2017.03.005
97. Liu W, You Y, Sun D, Wang S, Zhu J, Liu C (2018) Decolorization and detoxification of water-insoluble Sudan dye by *Shewanella putrefaciens* CN32 co-cultured with *Bacillus circulans* BWL1061. Ecotoxicol Environ Saf 166:11–17. https://doi.org/10.1016/j.ecoenv.2018.09.055
98. Liu S, Xu X, Kang Y, Xiao Y, Liu H (2020) Degradation and detoxification of azo dyes with recombinant ligninolytic enzymes from *Aspergillus* sp. with secretory overexpression in *Pichia pastoris*. R Soc Open Sci 7(9):200688. https://doi.org/10.1098/rsos.200688
99. Liu Y-Q, Maulidiany N, Zeng P, Heo S (2021) Decolourization of azo, anthraquinone and triphenylmethane dyes using aerobic granules: acclimatization and long-term stability. Chemosphere 263:128312. https://doi.org/10.1016/j.chemosphere.2020.128312
100. Liugė M (2023). Experimental studies on the removal of textile dye of Naphthol green B from artificially contaminated water using Purolite MB400 ion exchange resin. Penerbit Universiti Kebangsaan Malaysia
101. Madhav S, Ahamad A, Singh P, Mishra PK (2018) A review of textile industry: wet processing, environmental impacts, and effluent treatment methods. Environ Qual Manag 27(3):31–41. https://doi.org/10.1002/tqem.21538
102. Madhav S, Ahamad A, Singh A, Sharma S, Singh P (2019) Water pollutants: sources and impact on the environment and human health. https://doi.org/10.1007/978-981-15-0671-0_4
103. Mago M, Yadav A, Gupta R, Garg VK (2021) Management of banana crop waste biomass using vermicomposting technology. Bioresour Technol 326:124742. https://doi.org/10.1016/j.biortech.2021.124742

104. Mandal S, Dey A (2019) PET chemistry. In: Recycling of polyethylene terephthalate bottles. Elsevier, pp 1–22. https://doi.org/10.1016/b978-0-12-811361-5.00001-8
105. Mannina G, Rebouças TF, Cosenza A, Sànchez-Marrè M, Gibert K (2019) Decision support systems (DSS) for wastewater treatment plants—a review of the state of the art. Bioresour Technol 290:121814. https://doi.org/10.1016/j.biortech.2019.121814
106. Marcuello C, Chabbert B, Berzin F, Bercu NB, Molinari M, Aguié-Béghin V (2023) Influence of surface chemistry of fiber and lignocellulosic materials on adhesion properties with polybutylene succinate at nanoscale. Materials 16(6). https://doi.org/10.3390/ma16062440
107. Marin NM, Pascu LF, Demba A, Nita-Lazar M, Badea IA, Aboul-Enein HY (2019) Removal of the Acid Orange 10 by ion exchange and microbiological methods. Int J Environ Sci Technol 16(10):6357–6366. https://doi.org/10.1007/s13762-018-2164-2
108. Matheri AN, Ntuli F, Ngila JC, Seodigeng T, Zvinowanda C (2021) Performance prediction of trace metals and cod in wastewater treatment using artificial neural network. Comput Chem Eng 149:107308. https://doi.org/10.1016/j.compchemeng.2021.107308
109. Matheri AN, Mohamed B, Ntuli F, Nabadda E, Ngila JC (2022) Sustainable circularity and intelligent data-driven operations and control of the wastewater treatment plant. Phys Chem Earth, Parts A/B/C 126:103152. https://doi.org/10.1016/j.pce.2022.103152
110. Maurya P, Ali SA, Alharbi R, Yadav K, Alfaisal F, Ahmad A, Ditthakit P, Prasad S, Jung Y-K, Jeon B-H (2021) Impacts of land use change on water quality index in the upper Ganges river near Haridwar, Uttarakhand: a GIS-based analysis. Water 13. https://doi.org/10.3390/w13243572
111. Michel F, O'Neill T, Rynk R, Gilbert J, Smith M, Aber J, Keener H (2022) Forced aeration composting, aerated static pile, and similar methods. In: The composting handbook. Academic Press, Ch 6, pp 197–269. https://doi.org/10.1016/B978-0-323-85602-7.00007-8
112. Moazzem S, Crossin E, Daver F, Wang L (2021) Assessing environmental impact reduction opportunities through life cycle assessment of apparel products. Sustain Prod Consum 28:663–674. https://doi.org/10.1016/j.spc.2021.06.015
113. Mukkera VK, Katuri S, Musham US (2023) Biodegradation of Chrysophenine GX using *Candida albicans* (NCIM 3665). Mater Today: Proc 77:350–357. https://doi.org/10.1016/j.matpr.2022.11.495
114. Munasinghe P, Druckman A, Dissanayake DGK (2021) A systematic review of the life cycle inventory of clothing. J Clean Prod 320:128852. https://doi.org/10.1016/j.jclepro.2021.128852
115. Murugadoss G, Kumar DD, Kumar MR, Venkatesh N, Sakthivel P (2021) Silver decorated CeO2 nanoparticles for rapid photocatalytic degradation of textile rose bengal dye. Scient Reports 11(1):1080. https://doi.org/10.1038/s41598-020-79993-6
116. Muthu SS (2020) Life cycle assessment and product carbon footprint modelling of textile products. In: Muthu SS (ed) Assessing the environmental impact of textiles and the clothing supply chain, 2nd edn. Woodhead Publishing, Ch 7, pp 131–141. https://doi.org/10.1016/B978-0-12-819783-7.00007-7
117. Naresh Yadav D, Anand Kishore K, Saroj D (2021) A study on removal of methylene blue dye by photo catalysis integrated with nanofiltration using statistical and experimental approaches. Environ Technol 42(19):2968–2981. https://doi.org/10.1080/09593330.2020.1720303
118. Nassar SE, Said RM (2021) Bioremediation assessment, hematological, and biochemical responses of the earthworm (*Allolobophora caliginosa*) in soil contaminated with crude oil. Environ Sci Pollut Res 28(39):54565–54574. https://doi.org/10.1007/s11356-021-13889-4
119. Niinimäki K (2018) Sustainable fashion in a circular economy
120. Niu C, Li X, Dai R, Wang Z (2022) Artificial intelligence-incorporated membrane fouling prediction for membrane-based processes in the past 20 years: a critical review. Water Res 216:118299. https://doi.org/10.1016/j.watres.2022.118299
121. Nobili S, Masin CE, Zalazar CS, Lescano MR (2022) Bioremediation of hydrocarbon contaminated soil using local organic materials and earthworms. Environ Pollut 314:120169. https://doi.org/10.1016/j.envpol.2022.120169

122. Nyika J, Dinka M, Onyari E (2022) Effects of landfill leachate on groundwater and its suitability for use. Mater Today: Proc 57. https://doi.org/10.1016/j.matpr.2022.03.239
123. OECD. Test 222: (2016) Earthworms reproduction test (Eisenia Fetida/Eisenia Andrei) Guidel; testing chemical. In: Chemical; organisation for economic cooperation and development
124. Osagie C, Othmani A, Ghosh S, Malloum A, Kashitarash Esfahani Z, Ahmadi S (2021) Dyes adsorption from aqueous media through the nanotechnology: a review. J Mater Res Technol 14:2195–2218. https://doi.org/10.1016/j.jmrt.2021.07.085
125. Othman AM, Elsayed MA, Elshafei AM, Hassan MM (2018) Purification and biochemical characterization of two isolated laccase isoforms from *Agaricus bisporus* CU13 and their potency in dye decolorization. Int J Biol Macromol 113:1142–1148. https://doi.org/10.1016/j.ijbiomac.2018.03.043
126. Patel H, Yadav V, Yadav K, Choudhary N, Kalasariya H, Alam M, Gacem A, Mohammed A, Ibrahium H, Park J-W, Park S, Jeon B-H (2022) A recent and systemic approach towards microbial biodegradation of dyes from textile industries. Water 14:3163. https://doi.org/10.3390/w14193163
127. Patil SM, Rane NR, Bankole PO, Krishnaiah P, Ahn Y, Park Y-K, Yadav KK, Amin MA, Jeon B-H (2022) An assessment of micro- and nanoplastics in the biosphere: a review of detection, monitoring, and remediation technology. Che Eng J 430:132913. https://doi.org/10.1016/j.cej.2021.132913
128. Paul S, Goswami L, Pegu R, Kumar Chatterjee S, Sundar Bhattacharya S (2022) Epigenetic regulations enhance adaptability and valorization efficiency in *Eisenia fetida* and *Eudrilus eugeniae* during vermicomposting of textile sludge: insights on repair mechanisms of metal-induced genetic damage and oxidative stress. Bioresour Technol 345:126493. https://doi.org/10.1016/j.biortech.2021.126493
129. Paździor K, Bilińska L, Ledakowicz S (2019) A review of the existing and emerging technologies in the combination of AOPs and biological processes in industrial textile wastewater treatment. Chem Eng J 376:120597. https://doi.org/10.1016/j.cej.2018.12.057
130. Picos-Benítez AR, Martínez-Vargas BL, Duron-Torres SM, Brillas E, Peralta-Hernández JM (2020) The use of artificial intelligence models in the prediction of optimum operational conditions for the treatment of dye wastewaters with similar structural characteristics. Process Saf Environ Prot 143:36–44. https://doi.org/10.1016/j.psep.2020.06.020
131. Prerna, Agarwal H, Goyal D (2022) Photocatalytic degradation of textile dyes using phycosynthesised ZnO nanoparticles. Inorg Chem Commun 142:109676. https://doi.org/10.1016/j.inoche.2022.109676
132. Rahmani AR, Mousavi-Tashar A, Masoumi Z, Azarian G (2019) Integrated advanced oxidation process, sono-Fenton treatment, for mineralization and volume reduction of activated sludge. Ecotoxicol Environ Saf 168:120–126. https://doi.org/10.1016/j.ecoenv.2018.10.069
133. Rajadurai M, Karmegam N, Kannan S, Yuvaraj A, Thangaraj R (2022) Vermiremediation of engine oil contaminated soil employing indigenous earthworms, *Drawida modesta* and *Lampito mauritii*. J Environ Manag 301:113849. https://doi.org/10.1016/j.jenvman.2021.113849
134. Ramya E, Thirumurugan A, Govindan N, Aravind J, Gobikrishnan S (2023) Photocatalytic degradation of reactive orange M2R using green route synthesized copper oxide nanoparticles and its optimization studies. Springer International Publishing. https://doi.org/10.1007/978-3-031-29597-3_12
135. Riba J-R, Cantero R, Canals T, Puig R (2020) Circular economy of post-consumer textile waste: classification through infrared spectroscopy. J Clean Prod 272:123011. https://doi.org/10.1016/j.jclepro.2020.123011
136. Roy DC, Biswas SK, Sheam MM, Hasan MR, Saha AK, Roy AK, Haque ME, Rahman MM, Tang S-S (2020) Bioremediation of malachite green dye by two bacterial strains isolated from textile effluents. Curr Res Microb Sci 1:37–43. https://doi.org/10.1016/j.crmicr.2020.06.001

137. Roy M, Saha R (2021) 6 - Dyes and their removal technologies from wastewater: a critical review. In Bhattacharyya S, Mondal NK, Platos J, Snášel V, Krömer P (eds) Intelligent Environmental Data Monitoring for Pollution Management, pp 127–160. Academic Press. https://doi.org/10.1016/B978-0-12-819671-7.00006-3
138. Rudakiya DM, Tripathi A, Gupte S, Gupte A (2019) Fungal bioremediation: a step towards cleaner environment. In: Advancing Frontiers in mycology and mycotechnology: basic and applied aspects of fungi. Springer Singapore, pp 229–249. https://doi.org/10.1007/978-981-13-9349-5_9
139. Şener Fidan F, Kızılkaya Aydoğan E, Uzal N (2023) The impact of organic cotton use and consumer habits in the sustainability of jean production using the LCA approach. Environ Sci Pollut Res 30(4):8853–8867. https://doi.org/10.1007/s11356-022-22872-6
140. Sabrin A, Arafat A (2015) Textile dye removal from wastewater effluents using chitosan-ZnO nanocomposite. J Text Sci Eng 05(03). https://doi.org/10.4172/2165-8064.1000200
141. Sachin J, Singh N, Singh R, Shah K, Pramanik BK (2023) Green synthesis of zinc oxide nanoparticles using lychee peel and its application in anti-bacterial properties and CR dye removal from wastewater. Chemosphere 327:138497. https://doi.org/10.1016/j.chemosphere.2023.138497
142. Saha P, Madliya S, Khare A, Subudhi I, Bhaskara Rao KV (2023) Enzymatic biodegradation, kinetic study, and detoxification of reactive red-195 by *Halomonas meridiana* isolated from Marine Sediments of Andaman Sea, India. Environ Technol 44(17):2648–2667. https://doi.org/10.1080/09593330.2022.2038276
143. Saini K, Sahoo A, Kumar J, Kumari A, Pant KK, Bhatnagar A, Bhaskar T (2023) Effective utilization of discarded reverse osmosis post-carbon for adsorption of dyes from wastewater. Environ Res 231:116165. https://doi.org/10.1016/j.envres.2023.116165
144. Samsami S, Mohamadizaniani M, Sarrafzadeh M-H, Rene ER, Firoozbahr M (2020) Recent advances in the treatment of dye-containing wastewater from textile industries: overview and perspectives. Process Saf Environ Prot 143:138–163. https://doi.org/10.1016/j.psep.2020.05.034
145. Samuchiwal S, Gola D, Malik A (2021) Decolourization of textile effluent using native microbial consortium enriched from textile industry effluent. J Hazard Mater 402:123835. https://doi.org/10.1016/j.jhazmat.2020.123835
146. Sandhya S, Sarayu K, Uma B, Swaminathan K (2008) Decolorizing kinetics of a recombinant Escherichia coli SS125 strain harboring azoreductase gene from Bacillus latrosporus RRK1. Bioresour Technol 99(7):2187–2191. https://doi.org/10.1016/j.biortech.2007.05.027
147. Santal AR, Rani R, Kumar A, Sharma JK, Singh NP (2022) Biodegradation and detoxification of textile dyes using a novel bacterium *Bacillus* sp. AS2 for sustainable environmental cleanup. Biocatal Biotransformation 1–15. https://doi.org/10.1080/10242422.2022.2113518
148. Saravanan S, Carolin CF, Kumar PS, Chitra B, Rangasamy G (2022) Biodegradation of textile dye Rhodamine-B by *Brevundimonas diminuta* and screening of their breakdown metabolites. Chemosphere 308:136266. https://doi.org/10.1016/j.chemosphere.2022.136266
149. Saxena G, Kishor R, Saratale G, Bharagava R (2019) Genetically modified organisms (GMOs) and their potential in environmental management: constraints, prospects, and challenges. https://doi.org/10.1007/978-981-13-3426-9_1
150. Shaheen R, Asgher M, Hussain F, Bhatti HN (2017) Immobilized lignin peroxidase from *Ganoderma lucidum* IBL-05 with improved dye decolorization and cytotoxicity reduction properties. Int J Biol Macromol 103:57–64. https://doi.org/10.1016/j.ijbiomac.2017.04.040
151. Sharif A (2020) Microbial degradation of textile industry effluents: a review. Pure Appl Biol 9(4). https://doi.org/10.19045/bspab.2020.90251
152. Sharma B, Tiwari S, Kumar R, Kumar M, Tewari L (2023) Eco-friendly detoxification of hazardous Congo red dye using novel fungal strain *Trametes flavida* WTFP2: deduced enzymatic biomineralization process through combinatorial in-silico and in-vitro studies. J Hazard Mater 455:131503. https://doi.org/10.1016/j.jhazmat.2023.131503
153. Sharma N (2021) Immobilization of laccase from *Bacillus subtilis* DS on chitosan beads and standardization of process for biodegradation of textile dyes. J Chem Technol Biotechnol 1–8

154. Shirkoohi MG, Tyagi RD, Vanrolleghem PA, Drogui P (2022) Artificial intelligence techniques in electrochemical processes for water and wastewater treatment: a review. J Environ Health Sci Eng 20(2):1089–1109. https://doi.org/10.1007/s40201-022-00835-w
155. Singh P, Singh R (2017) Bio-removal of azo dyes: a review. Int J Appl Sci Biotechnol 5:108. https://doi.org/10.3126/ijasbt.v5i2.16881
156. Singh G, Dwivedi SK, Mishra J (2020) Role of fungal enzymes in the removal of azo dyes. In: Arora NK, Mishra J, Mishra V (eds) Microbial enzymes: roles and applications in industries. Springer Singapore, pp 231–257. https://doi.org/10.1007/978-981-15-1710-5_9
157. Singh NK, Yadav M, Singh V, Padhiyar H, Kumar V, Bhatia SK, Show P-L (2023) Artificial intelligence and machine learning-based monitoring and design of biological wastewater treatment systems. Bioresour Technol 369:128486. https://doi.org/10.1016/j.biortech.2022.128486
158. Singh R, Glick BR, Rathore D (2020) Role of textile effluent fertilization with biosurfactant to sustain soil quality and nutrient availability. J Environ Manag 268:110664. https://doi.org/10.1016/j.jenvman.2020.110664
159. Smith MM, Aber JD (2018) Energy recovery from commercial-scale composting as a novel waste management strategy. Appl Energy 211:194–199. https://doi.org/10.1016/j.apenergy.2017.11.006
160. Soni P, Sharma S, Sharma S, Kumar S, Sharma K (2006) A comparative study on the toxic effects of textile dye wastewaters (untreated and treated) on mortality and RBC of a freshwater fish Gambusia affinis (Baird and Gerard). J Environ Biol / Acad Environ Biol India, 27:623–628
161. Srinivasan S, Bankole PO, Sadasivam SK (2022) Biodecolorization and degradation of textile azo dyes using *Lysinibacillus sphaericus* MTCC 9523. Front Environ Sci 10. https://doi.org/10.3389/fenvs.2022.990855
162. Ssouni S, Miyah Y, Benjelloun M, Mejbar F, El-Habacha M, Iaich S, Ait Addi A, Lahrichi A (2023) High-performance of muscovite clay for toxic dyes' removal: adsorption mechanism, response surface approach, regeneration, and phytotoxicity assessment. Case Stud Chem Environ Eng 8:100456. https://doi.org/10.1016/j.cscee.2023.100456
163. Sun S, Liu P, Ullah M (2023) Efficient azo dye biodecolorization system using lignin-co-cultured white-rot fungus. J Fungi 9(1). https://doi.org/10.3390/jof9010091
164. Tagliabue F, Marini E, De Bernardi A, Vischetti C, Casucci C (2023) A systematic review on earthworms in soil bioremediation. Appl Sci 13(18):10239. https://doi.org/10.3390/app131810239
165. Terzano R, Rascio I, Allegretta I, Porfido C, Spagnuolo M, Khanghahi MY, Crecchio C, Sakellariadou F, Gattullo CE (2021) Fire effects on the distribution and bioavailability of potentially toxic elements (PTEs) in agricultural soils. Chemosphere 281:130752. https://doi.org/10.1016/j.chemosphere.2021.130752
166. Thangaraj S, Bankole PO, Sadasivam SK, Kumarvel V (2021) Biodegradation of reactive red 198 by textile effluent adapted microbial strains. Arch Microbiol 204(1):12. https://doi.org/10.1007/s00203-021-02608-9
167. Thoa LTK, Thao TTP, Nguyen-Thi M-L, Chung ND, Ooi CW, Park S-M, Lan TT, Quang HT, Khoo KS, Show PL, Huy ND (2023) Microbial biodegradation of recalcitrant synthetic dyes from textile-enriched wastewater by *Fusarium oxysporum*. Chemosphere 325:138392. https://doi.org/10.1016/j.chemosphere.2023.138392
168. Thozhukattu Valliyaparambil P, Alagapuram Kaliyaperumal K, Gopakumaran N (2019) *Pleurotus ostreatus* laccase decolorization of remazol brilliant violet 5R dye: statistical optimization and toxicity studies on microbes and its kinetics. J Appl Biotechnol Rep 6(3):88–95. https://doi.org/10.29252/JABR.06.03.02
169. Turhan K, Durukan I, Ozturkcan SA, Turgut Z (2012) Decolorization of textile basic dye in aqueous solution by ozone. Dye Pigment 92(3):897–901. https://doi.org/10.1016/j.dyepig.2011.07.012
170. Valli Nachiyar C, Rakshi AD, Sandhya S, Britlin Deva Jebasta N, Nellore J (2023) Developments in treatment technologies of dye-containing effluent: a review. Case Stud Chem Environ Eng 7:100339. https://doi.org/10.1016/j.cscee.2023.100339

171. Varjani S, Rakholiya P, Ng HY, You S, Teixeira JA (2020) Microbial degradation of dyes: an overview. Bioresour Technol 314:123728. https://doi.org/10.1016/j.biortech.2020.123728
172. Viet ND, Jang A (2021) Development of artificial intelligence-based models for the prediction of filtration performance and membrane fouling in an osmotic membrane bioreactor. J Environ Chem Eng 9(4):105337. https://doi.org/10.1016/j.jece.2021.105337
173. Wang Y, Cheng Y, Liu H, Guo Q, Dai C, Zhao M, Liu D (2023) A review on applications of artificial intelligence in wastewater treatment. Sustainability 15(18):13557. https://doi.org/10.3390/su151813557
174. Wang J-H, Zhao X-L, Guo Z-W, Yan P, Gao X, Shen Y, Chen Y-P (2022) A full-view management method based on artificial neural networks for energy and material-savings in wastewater treatment plants. Environ Res 211:113054. https://doi.org/10.1016/j.envres.2022.113054
175. Wijannarong S, Aroonsrimorakot S, Thavipoke P, Kumsopa C, Sangjan S (2013) Removal of reactive dyes from textile dyeing industrial effluent by ozonation process. APCBEE Procedia 5:279–282. https://doi.org/10.1016/j.apcbee.2013.05.048
176. Wu Y, Su D (2020) Review of life cycle impact assessment (LCIA) methods and inventory databases. In: Su D (ed) Sustainable product development: tools, methods and examples. Springer International Publishing, pp 39–55. https://doi.org/10.1007/978-3-030-39149-2_3
177. Xuejiao A, Yi C, Ningjian L, Shulin Z, Liuwei W, Qinghua Z (2023) Green and sustainable approach for decoloration and detoxification of azo dye by bacterial-mediated thermohalotolerant dye-decolorizing peroxidase (DyP). Process Biochem 130:334–346. https://doi.org/10.1016/j.procbio.2023.03.037
178. Yang Y, Kim K-R, Kou R, Li Y, Fu J, Zhao L, Liu H (2022) Prediction of effluent quality in a wastewater treatment plant by dynamic neural network modeling. Process Saf Environ Prot 158:515–524. https://doi.org/10.1016/j.psep.2021.12.034
179. Yin H, Qiu P, Qian Y, Kong Z, Zheng X, Tang Z, Guo H (2019) Textile wastewater treatment for water reuse: a case study. Processes 7:34. https://doi.org/10.3390/pr7010034
180. Yuvaraj A, Karmegam N, Thangaraj R (2018) Vermistabilization of paper mill sludge by an epigeic earthworm *Perionyx excavatus*: mitigation strategies for sustainable environmental management. Ecol Eng 120:187–197. https://doi.org/10.1016/j.ecoleng.2018.06.008
181. Zhang Y, Liu X, Xiao R, Yuan Z (2015) Life cycle assessment of cotton T-shirts in China. Int J Life Cycle Assess 20(7):994–1004. https://doi.org/10.1007/s11367-015-0889-4
182. Zhao L, Dai T, Qiao Z, Sun P, Hao J, Yang Y (2020) Application of artificial intelligence to wastewater treatment: a bibliometric analysis and systematic review of technology, economy, management, and wastewater reuse. Process Saf Environ Prot 133:169–182. https://doi.org/10.1016/j.psep.2019.11.014

Textile Tales: The Environmental Narrative of Fabrics and Fibers

Shreya Sinha, Rahul Sharma, Noor Jahan, and Pinki Rani Agrawal

Abstract Being one of the largest exporters of textiles and apparel in the world, India's textile industries are expected to touch heights with a market size of 250 billion dollars by 2025. Textile industries involve processes like spinning, weaving, dyeing, and finishing products. These industries use a great amount of harmful chemicals and enormous amount of water during the production of a single fabric. It has been estimated that for the production of a single t-shirt, about 2500–3000 L of water is consumed. Various chemicals are used during the manufacturing of textile fabrics and these chemicals get dumped into the environment in one or the other way polluting our environment. Also, the adverse effect of modernization has led to the growth of textile production. Due to this, global production of textile fiber has almost doubled in the last 2 decades, and with this comes a huge amount of fabric waste. The environmental impact of textile industries has become one of the most gut-wrenching issues in today's world. This chapter encompasses the knowledge of textile industries and various environmental challenges faced by living and non-living beings.

Keywords Textile · Harmful · Pollution · Environment

1 Introduction

Clothes and textiles are considered to be one of the fundamental necessities for a human life, alongside food or shelter. With an increase in the population, the textile industry is facing immense pressure to meet the ever-expanding demands of the people worldwide. The adoption of new manufacturing technology to meet

S. Sinha · R. Sharma · N. Jahan (✉)
Central University of Himachal Pradesh, Shahpur, Himachal Pradesh, India
e-mail: noor.jahan@hpcu.ac.in

P. R. Agrawal
Deenbandhu Chhotu Ram University of Science and Technology, Murthal, Sonipat, Haryana, India

P. Singh (ed.), *Dye Pollution from Textile Industry*, SDGs and Textiles,
https://doi.org/10.1007/978-981-97-5341-3_18

expanding customer expectations has elevated the global significance of manufacturing processes. However, its increased growth has resulted in significant environmental concerns and issues [1].

The textile industry, being one of the largest industries, plays a vital role in the economic development of any country by providing employment opportunities to the people [2]. It is considered to be the bedrock of the global economy [3]. For instance, China was the world's largest trader in the textile industry in the patch of 1997–2007 according to the World Trade Organization report in 2008. In 2006, the industry employed around 10 million people in China, accounting for more than 13% of the total workers [4]. Similarly, in India, the apparel industry directly employed 35 million people becoming the country's second-largest employment producer [5]. Key apparel and textile zones in India are shown in Fig. 1 [6]. But beneath the colorful fabrics, and soothing and eye-catching designs, there are lump of serious implications. Worldwide generation of textile fiber has almost doubled in the last two decades, hitting the highest number of 111 million tons in the year 2019 and maintaining growth predictions through 2023 resulting in enormous textile waste. Figure 2 represents an increase of production volume in thousand metric tons of textile fibers from the year 1975 to 2023 [7].

This chapter will delve extensively into the serious environmental issues arising from the textile industries. In spite of the fact that the textile industry is the main pillar of any country's culture and economy, it also adds levels to the environmental pollution due to the hazardous effluents. Hence, the textile industry serves as a critical intersection as we become environmentally conscious and emphasize the need for ecological sustainability.

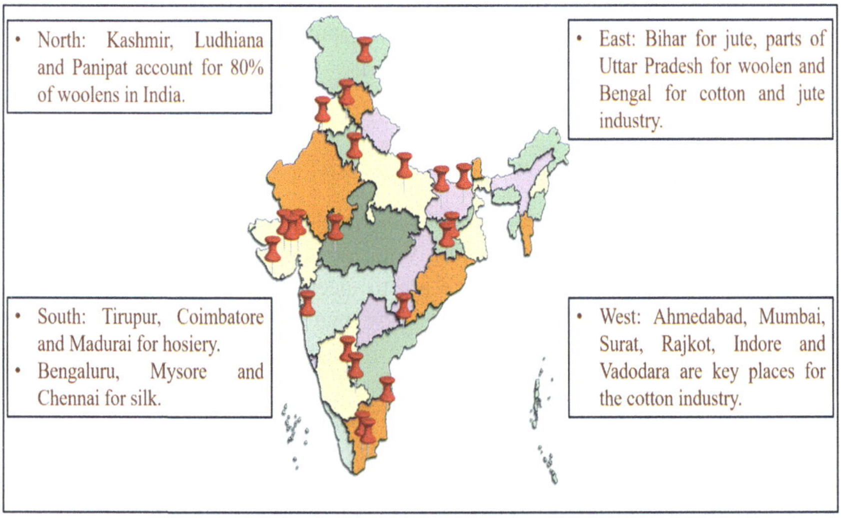

Fig. 1 Key textile and apparel zones in India (redrawn from Ref. [6])

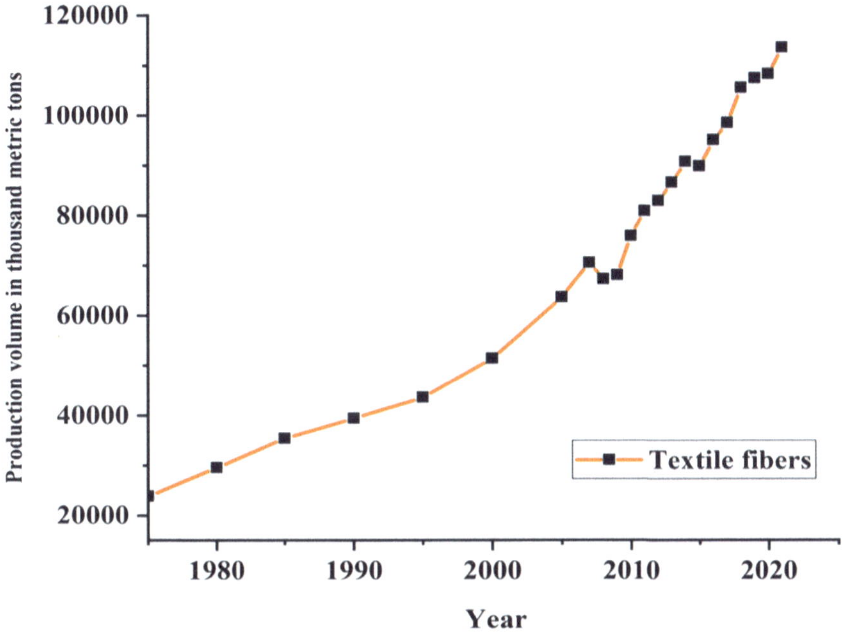

Fig. 2 Year-wise increase of production volume in thousand metric tons of textile fibers (redrawn from Ref. [7])

In this chapter, an attempt has been made to unfold many facets of the environmental challenges stemming from the textile industries. We will also explore the intricate relationship between the textile and the environment, offering insight into the impact of industrial activities on our ecological surroundings. We'll observe how the sector harms the ecosystem and look at the enormous issue of garbage, namely how it's generated and disposed of, which is often in destructive ways that hurt our environment. We will get to know about the growth and evolution of the industry, processes involved in textile production, hurdles it faces, and its environmental impacts such as air quality, release of chemicals and dyes into the water causing pollution, etc.

Furthermore, different regulatory authorities that have been set up nationally and internationally to govern and deal with the issues overlooked by these industries will also be discussed. Finally, we will discuss the emerging innovations, sustainable practices and regulatory frameworks which are promising in the reducing the problematic influence of the textile production.

2 Textile Production Processes

The word textile has been derived from the Latin word textiles and the French word texere which means to weave [8, 9]. Textile is a woven material made from a network of natural and man-made fibers [10].

Textile production processes are broadly divided into two main categories—dry processing and wet processing. Dry processing involves spinning, weaving, and knitting, whereas wet processing involves dyeing and printing, finishing.

2.1 Fiber Preparation

The expansion of the textile industry has resulted in an almost two-fold increase in fiber production, rising from 58 million tons in the year 2000 to 113 million tons in the year 2021. This trend is expected to grow with anticipated growth to 149 million tons by 2030 [11]. Textile fibers represent the delicate, hair-like components found in animal and plant tissues or other substances that are characterized by very small diameter relative to their length. These fibers are elongated material, several hundred times as long as they are wide, and serve as the fundamental building blocks of textile production [12]. Before diving into the steps of textile production, the first and foremost thing involves the cultivation and harvesting of the fiber. Then depending on the types of textiles, these fibers are extracted from their origin. There are two main types of fibers—natural fibers and man-made fibers. On the basis of their source of origin, textile fibers are classified as natural, synthetic, and semi-synthetic fibers [3]. Till 1890, natural fibers were the only fibers available to the textile industry but with the advancement of chemistry, synthetic or man-made fibers entered into the commercial world. Once the fibers are extracted from the source, they are cleaned, sorted, and aligned using machines to form continuous strands. Whether they are natural or synthetic fibers, all affect environment in some way ranging from minor to significant. The extent to which these fibers impact our surroundings depend on the type of fibers used during processing. Manufacturing these fibers require chemical processing and produces pollution. The utilization and disposal of these unsafe chemicals in the process of fiber production raise notable concerns [3].

2.2 Spinning and Yarn Production

This process involves the formation of yarn from raw fibers. After the preparation of the basic material, they are spun into yarn. This method is termed as spinning. Depending on the fiber, there are two types of spinning: dry and wet spinning. Spinning is done by using machines containing steel bobbins pre-loaded with fiber or spinning material commonly referred to as roving. Roving consists of elongated and narrow bundles of loosely twisted fiber. Their primary purpose is to prepare fiber for spinning and it provides more uniform fibers that can be further processed into yarn or thread through spinning [13, 14].

2.3 *Fabric Production*

The subsequent stage in fabric production or manufacturing process is where yarns are transferred from one machine to another in order to produce a continuous piece of cloth. This process is also known as weaving. The warp or lengths require precise fitting between two sets of heddles before initiating weaving. Once the warp is woven, it is taken off the looms and stitched together to make the final fabric [13]. Weaving is the most commonly used technique and has a wide area of applications such as medical packing, textiles, industrial products, seat belts, apparel, etc. [15]. There are four different types of fabrics such as woven fabric, knitted fabric, nonwoven fabric, and braided fabric [13]. Knitting is the second most commonly used process in fabric manufacturing and it involves the interloping of yarns and the word knitted has been taken from the Dutch word 'knutten' which means to knot. In knitted fabric, yarns are interlocked in loops to create a fabric with the help of knitting machines. The bending of yarns in the form of loops provides better stretchability, extensibility, better comfort thereby used in sportswear, undergarments, and casual wear [15, 16]. Whereas nonwoven fabrics are created by bonding fibers together without knitting or weaving.

2.4 *Pretreatment*

Pretreatment involves the treatment of fibers, yarns, or fabrics before undergoing various processing stages such as dyeing, printing, and finishing. Pretreatment plays a crucial role in ensuring that textiles are free from impurities, clean, and ready for subsequent steps. This enables the material to readily accept dyes and pigments during the dyeing process. The specific steps involved in pretreatment may differ based on the textile material and intended end product, but in general, it includes washing, desizing, scouring, bleaching, and mercerization [17].

2.5 *Dyeing and Printing*

In this process, colors are imparted onto the fabric or fiber by treating them with chemical pigments or dyes in order to produce attractive motifs and color effects on the textile. Dyes and chemical pigments are colored due to the presence of chromophore and auxochrome groups [18]. The dyeing process needs water to transfer dyes onto the fabric. The fixing of dyes to fabric, fiber, or yarn should be fast in order to provide reliable and durable fabric against heat, sunlight, washing, and rubbing. Dyed fabric releases around 10–50% of dyestuff to the environment during washing which is the major environmental concern [16]. Every kilogram of yarn requires around sixty liters of water during the dyeing process [1]. Cotton, being the most widely used fiber requires a large amount of water. Printing is a specialized branch of dyeing which is also referred to as 'localized dyeing'. It is the process of decorating fabrics by the application of dyes, pigments, or other materials in the form of patterns [8]. Printing is primarily accomplished using flat or rotary screens and after each session, a thick residual paste finds its way into the wastewater. Dyeing and Printing constitutes the same essential reactions. The only difference is that color is applied in form of solution (aqueous) known as dyebath in dyeing whereas thick paste of color is used in printing [18]. Both of these processes involve a huge amount of water. The discharge of microplastics during the dyeing and washing process pollutes the oceans and seas [19]. Methods adopted during dyeing a textile fabric or fiber decide the amount of pollution an environment will face. Among effective dyeing processes, dyeing in microwave and sonicator and applying compounds using pad-dry cure method are proven to adsorb dyes to a higher extent and ultimately protect the environment [3].

2.6 Finishing

After dyeing and Printing, a thorough inspection of the dyed fabric is done. Any irregular patterns or defective dyed fabric is sent for further correction. Finishing involves scouring, bleaching and mercerization steps. These steps are employed to increase the fabric quality and appearance [3]. To remove dirt, oils, untreated dyes, or residual chemicals during dyeing and printing process, the fabric is washed properly. This washing process involves the usage of scouring agents to enhance the appearance of the finished product. Mercerization is the process which increases luster and strength. Depending on the requirement, wrinkle-resistant, antimicrobial finishing, stain, and water-repellant coatings are performed on the fabric. Throughout the finishing process, regular quality checks are conducted to ensure the quality, color, and texture of the fabric. All the steps used during finishing involve toxic chemicals that are heavily water-based and participate in the environmental pollution [3] (shown in Fig. 3).

Fig. 3 Textile production processes

3 Chemical, Water, and Energy Consumption by Textile Industries

The World Bank estimates that the dyeing and finishing process accounts for 17–20% of industrial wastewater. Numerous chemicals are used in textile manufacturing for tough durability, accumulation in living organisms, and toxicity. Some of these are of volatile nature, get easily soluble in the wastewater and others get absorbed by human skin. Textile effluent contain high biological oxygen demand (BOD), high chemical oxygen demand (COD), total suspended solids, oils, grease, sulfates, heavy metals and other chemicals depending on the type of method applied. Water quality gets reduced with the presence of these chemicals as they react with water changing its minerals, odor, taste, and color. Some of the organic pollutants are biodegradable which consume oxygen in order to degrade, influencing marine life by decreasing water quality. The wastewater of scouring contains very high TDS, BOD, COD, and pH values. Unattached dye when discarded in wastewater contains hazardous contaminants and even though they are present in very minute quantities, they are extremely dangerous due to their resistance to biodegradation. Sodium salts present in dyes pose a serious threat to the environment as they get dissolved in the wastewater resulting in incomplete removal. Formaldehyde is widely used during reactive and direct dyeing causing skin irritations and respiratory problems.

Every stage of wet processing in textile production consumes a huge amount of water from dyeing to finishing. Also, Textiles are washed before moving on to the next stage of textile production. In addition to this, water is consumed during cooling water, boiler, steam drying, and cleaning. In dyeing, 30–50 L of water is consumed per kg of fabric depending on the dye used. About 1/5th of total wastewater is released during the dyeing process. To produce a single cotton shirt, the amount of water used is approximately 2500–3000 L. This water is mostly used during cotton harvesting and the rest of that is consumed during wet textile processing. Nowadays, saving water and reusing it is essential in textile industries. In countries like China and India, businesses don’t get funding assistance until they are sure to solve water consumption and wastewater problems.

Along with chemical and water consumption, textile industries consume enormous amount of energy to operate machines and textile production processes which results in leaving carbon footprints and emission of harmful greenhouse gases. According to a report, the total amount of energy consumed during spinning is 34%, during weaving is 23%, in chemical processing 38% consumption, and the rest of the 5% is used for miscellaneous purposes. Thermal energy is consumed in the majority during heating and drying of water. We are familiar with the fact that fuel consumption is directly proportional to the amount of water consumed. Therefore, a reduction in water consumption will also save energy [20]. Also, thermal and electric energy consumption per meter of textile is 18.8–23 MJ and 0.45–0.55 kWh. Major energy consumption is done during yarn and fabric manufacturing processes which roughly accounts for 75 or 80% of all the energy needed in a textile mill and around 15–20% of energy is used to run the machinery during wet processing [1].

Huge chemical, water, and energy consumption draws our attention toward renewable energy sources and sustainable practices in order to save our planet.

4 Environmental Challenges

Although textiles are an essential part of our lives, their manufacture and disposal can have adverse effects on the ecosystem and the human race. In this section, we look at the dominant issues that the textile industry is causing which proves to be one of the major factors affecting human health and well-being.

4.1 Exposure to Chemicals and Health Risks

The textile production process involves the use of various dyes for coloring the fabric and also the use of pigments that are toxic in nature. Some of these dyes include crystal violet, rhodamine B, methylene blue, Congo red, rose Bengal, methyl orange, and others [21]. These chemicals produce hazardous heavy metals, carcinogens, etc., which affects primarily the health of the employees of the industries that are under the exposure of these chemicals day and night. Lack of safety while working in the environment can result in skin irritation and respiration issues, and deadly diseases like cancer to the workers of these industries. Besides this, even the consumers who wear the finally prepared units of the designed fabrics may sometimes feel the allergy and irritation issues due to the left out or byproducts of these chemicals.

With increased awareness about health and hygiene, the textile industry has set a new height by introducing antimicrobial fabric into the market after a global pandemic outbreak in 2020. A report by Research and Markets estimated that antimicrobial textile market is expected to grow about 4% between the year 2020 and 2025. Having many health benefits along with h, these antimicrobial fabrics also contain hazardous chemicals [22].

4.2 Ecosystem Pollution

During the different stages of fabric production such as pretreatment of fabric, dyeing, and printing, a considerable amount of chemicals like PVA, sodium hydroxide, hydrogen peroxide, etc., are used [9]. These unfiltered fabric coloring waste chemicals have a negative impact on every aspect of the ecosystem including air, water, and also the health of the humans living there. Particulates from the dyes used for textile production increase the density and cloudiness of water resulting in the unpleasant taste and odor of the water [23].

When fine particulates from the spinning, weaving, and finishing processes of textiles are inhaled, they can pose respiratory concerns. Additionally, several chemicals used in textile processes, such as dyes and solvents, release volatile organic compounds that contribute to both indoor and outdoor air pollution. Furthermore, combustion processes in industries, such as boilers used for steam generation, release pollutants like sulfur dioxide (SO_2) and nitrogen oxides into the atmosphere which makes the situation even more severe.

Even these effluents and untreated intermediates can also affect the fertility of the soil as these effluents can imbalance the pH of the soil, changing its acidity or alkaline nature which ultimately affects the ability of the soil to support plant life. This is because some of the effluents can exhibit phytotoxic effects which can damage plant tissues and inhibit their growth. These contaminants can affect the purity of drinking water by reaching the groundwater resources leaching through the soil. Cotton harvesting alone uses 0.5 million tons of pesticides and 9 million tons of fertilizers worldwide. The farmers working in cotton harvesting suffer drastic deterioration in health [19].

Numerous studies have discovered that, in comparison to soil irrigated with ordinary water, adding textile effluents to the soil might cause a temporary reduction in the amount of hydrophilic organic salts like sodium, calcium, magnesium, and potassium [24]. It has been predicted through many studies that when there is high number of various effluents such as textiles present in the soil then there is improper or decreased sprouting in plants [25, 26] (shown in Fig. 4).

4.3 Waste Generation and Disposal Challenges

The fashion industry is among the most unsustainable industries. One major aspect is the sheer volume of textile waste generation. The fast fashion culture and rapid turnover of trends result in a continuous stream of discarded clothing. These garment wastes are often sent to landfills, where they contribute to environmental degradation and slow decomposition rates [30]. When unsold or have reached the end of their usable lives, billions of garments are thrown away, ending up in landfills [31]. According to Ellen MacArthur Foundation, only a fraction of the garment undergoes recycling, with approximately 87% of fibers used to make garments either burned

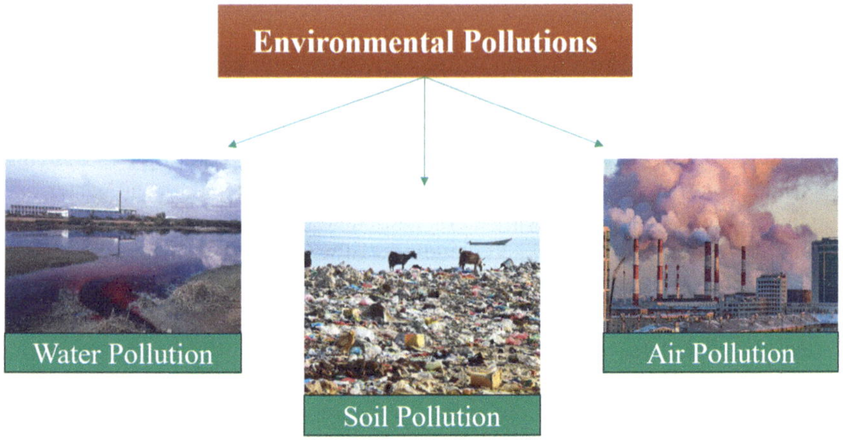

Fig. 4 Environmental pollution—image source for water pollution [27], soil pollution [28] and air pollution [29]

or dumped in a landfill [8, 9]. The fashion industry has a harmful environmental impact, serving as the primary source of micro-plastic pollution and contributing to 10% of the world's carbon dioxide emissions [32]. Textile recycling is one of the main solutions identified to tackle the issue of textile waste, alongside strategies such as waste reduction, garment lifetime extension, and the growth of the second-hand economy [33].

5 Sustainable Practices and Innovation

Fulfilling current needs without sacrificing the standard of living for upcoming generations is generally described as sustainability [1]. Sustainable development is also known as "Socially responsible economic development" which protects the environment for future generations [19]. The emphasis on sustainable practices and innovation has gained much prominence in recent years, with a sudden rise in attention during this decade [34]. Sustainability involves the adoption of eco-friendly practices for minimizing environmental impacts. This involves various key aspects such as renewable resources, waste reduction, recyclability, reduction in chemical usage, certificates, and standards.

5.1 Renewable Resources

Collaborative efforts between government and non-government organizations can play an important role in nurturing the eco-conscious industrial sector. The textile industries need to emphasize the utilization of eco-friendly materials for fabric production. The increased public awareness, government policies, and sustainability introduce the idea of green innovation across a range of sectors with less side effects [35].

Fashion business has seen a sudden peak since the emergence of digital platforms and ever-growing technologies. The consequences of destructive fashion cannot be overlooked by industries. Keeping ecological integrity in the frame, fashion industries are including materials that focus more on ethical and ecological qualities. Many homegrown and sustainable brands have taken their business to digital platforms post COVID [36]. The Sustainable Textiles for Sustainable Development (SusTex) project by the United Nations Climate Change entity promotes sustainable textile production and employs eco-friendly textiles [37].

Customized apparel is becoming more popular among consumers than the mass-produced items. To support sustainable fashion, repairing and renovating are among the newest trends in fashion. Brands are ensuring that the overall experience of the consumers is long-lasting, fulfilling their demands and needs. Some brands are also providing repair facilities for their clothes. Further, clothes that cannot be repaired, are being redesigned or recycled. Along with this, various sustainable innovative products are making their way into eco-friendly textile production. Natural resources like banana, coffee grounds, pineapple, lotus, stinging nettles, etc., are being used to make fibers and turned into sustainable textiles. This proves that there are many alternatives to replace conventional fiber production processes with sustainable products [38].

By adopting these ideas, fashion industries will become more sustainable in the near future. Green fashion is helping to keep our plants clean minimally impacting the environment [39] (Fig. 5).

5.2 *Waste Reduction*

Waste reduction or handling waste is a sustainable innovation focusing on recycling, reusing, or reducing waste [41]. In order to deal with waste, we need to pay attention to the selection of raw materials, and waste management by installing waste treatment plants and increasing general awareness. Understanding and guiding consumers is of great importance as it will increase awareness toward recycling garments and reusing them for multiple purposes [34]. Although the public has become far more aware of the environment since the last few years yet there is a need to take action in order to bring this eco-friendly, greener world into reality.

Fig. 5 Sustainable textile made using **a** lotus [40] **b** coffee grounds [40] **c** banana [40] **d** nettle [40] **e** Pineapple [38]

5.3 Reduction in Chemical Usage, Water, and Energy Consumption

The acceleration of modernization and advancement of technologies have given rise to issues such as environmental pollution, ozone layer depletion, and water and air contamination. To maintain a balance between industrialization and environmental preservation, new methods are being explored with eco-friendly processes. Preventing pollution is a more cost-effective approach than cleaning up afterward. Once the situation slips from our hands, maintaining ecological balance will become extremely difficult [1]. The motivation for less chemical usage has forced industries to look for alternative textile processes. India's largest fabric processing facility BRFL Textiles Private Limited (BTPL) has unveiled an innovative sulfur dyeing method that enables continuous dyeing without needing water [37, 42]. Minimum use of chemicals at each stage of textile production processes will solve half of the problem. Employing natural dyes and digital printing during textile manufacturing will help in lowering water and energy consumption thus reducing pollution.

Fig. 6 GOTS logo [44], ANSI logo [45] and OEKO-TEX logo [46]

5.4 Certifications and Standards

Certifications and Standards are rules, regulations, guidelines, and specific criteria that industries must satisfy to achieve a particular level of quality and safety of the products. The main aim of these standards is to establish uniform practices that must be followed by industries in order to achieve sustainability [43]. There are various sustainable certificates and standards such as GOTS (Global Organic Textile Standard), ANSI (American National Standards Institute) standards, OEKO-TEX, ISO 14000 Environmental Standards, TE (Textile Exchange) standards, etc., which verify and approve the adherence to these standards. The industries that get certified with such standards ensure consistent quality, better products, environmental responsibility, and consumer's trust. These standards ensure that the processing involved in textile production should meet both environmental and toxicological criteria [43] (Fig. 6).

6 Regulatory Framework

As reported by the European Environment Agency, in 2020, each person's textile purchases within the EU resulted in approximately 270 kg of CO_2 emissions. Consequently, textile consumption in the EU led to a total of 121 million tons of greenhouse gas emissions. According to projections from consulting firm McKinsey and the World Economic Forum, the annual production of clothing has at least doubled since 2000, which will substantially elevate waste generation and place an increasingly heavy load on landfills. To put it into perspective, roughly 85% of textiles end up in landfills annually [31].

India generates more than one million tons of textile waste annually. Of this total, approximately 51% is attributed to Indian consumers and falls under the category of post-consumer waste [2, 3].

There are no distinct environmental laws for the textile industry. However, any textile industry must adhere to industry-specific norms while establishing or running an industrial unit. Indian environmental legislation is characterized by its commendable regulations, but unfortunately, enforcement is often lacking. While these policies are designed to address environmental concerns within the Indian context, there is a pressing need to update standards for various types of waste generated. Issues such as effluent discharge from paper manufacturing, the presence of microplastic dyes in personal care products, air pollution, and severe river pollution are all consequences of haphazard industrial expansion and the continued use of outdated technologies [47]. Numerous initiatives have already been undertaken to raise awareness of the impending consequences of dye usage, but there is still much work to be done. Recognizing the environmental impacts associated with dyes [48], it is crucial to establish and uphold regulatory frameworks to mitigate these effects.

In this section, the readers will get to know about the spectrum of international as well as national authorities and laws implemented for textile industries, encouraging them to prioritize the welfare of living entities/beings. It also discusses the hurdles faced in the implementation and various systems which monitor their performance.

The Ministry of Environment and Forest (MoEF), Central Pollution Control Board (CPCB), State Pollution Control Board (SPCB) are some of the Indian regulatory authorities to govern the specific standards that are mandatory to follow by any textile Industry (and other industries also) developing in the country [49]. The Ministry of Environment and Forest is in charge of developing and carrying out national policies and programs that promote environmental protection and sustainable development. It also supports the research work that helps in protecting and conserving the environment, also helps in spreading the awareness among public, schools, etc. On the other hand, CPCB being a central entity, primarily focuses on regulatory and technical aspects of pollution control and monitoring. It plays a critical role in the evaluation and approval of Environmental Impact Assessment reports for particular industries and projects. SPCBs are the regulatory authorities at the state level having the responsibility of establishing and implementing the environmental laws, rules, and regulations in the respective states and union territories. In addition to the contributions of these authorities in the state, they also collaborate with international regulations such as REACH (Registration, Evaluation, Authorization, and Restriction of Chemicals), Basel Convention on the Control of Transboundary Movements of Hazardous Wastes and Their Disposal. United Nations Environment Program (UNEP), Environmental Protection Agency (EPA), United States (US-EPA), European Environment Agency (EEA), and International Union for Conservation of Nature (IUCN) are some of the international organizations that collaborate with the government and other stakeholders to address environmental issues globally, to find the remedies for the challenges faced to protect and sustain a healthy ecosystem. For serving the purpose of the environment, a bunch of laws have been suggested and implemented. Environmental Impact Assessments adopted in 2014 [50], Water Pollution Control Laws, Environmental Protection Laws, and Chemical Safety Regulation such as Global Organic Textile Standard (GOTS) are some of them. These laws not only help in protecting the health and well-being of the human population but also the local ecosystem.

This regulatory structure of the textile industry comes across numerous compliance challenges. One of the major challenges is the cost in the adoption of new technology, instruments, etc., which may result in the financial burden on the industrialists. Effluent treatment plants (ETPs) are largely absent from the textile and dyeing sectors. Even the ones who have ETPs function poorly or don't use them very often. Supervision and administration of the various parameters such as water quality, usage of chemicals may also cause difficulties. Adoption of environmentally conscious and sustainable practices is required to create a healthy and eco-friendly environment adoption of which needs green and renewable materials and processes. Additional obstacles faced by the textile and dyeing industries include a lack of environmental consciousness, limited research, inadequate technology, shortage of skilled personnel, etc. These factors collectively pose significant barriers to the development of eco-friendly effluent management within the textile dyeing sector [48].

7 Future Directions and Conclusion

In conclusion, examining the environmental effects of the textile sector reveals many difficulties as well as some encouraging developments. The manufacturing process, energy and water consumption, and chemical usage all have a significant impact on the environment when it comes to textiles. The issue includes the overuse of resources and water pollution. However, there is also good news. A few smart and environmentally friendly methods of operation are emerging. Innovative technologies and improved materials are being developed by people. The trend toward slow fashion and recycling—that is, extending the life of items and minimizing waste—is becoming more and more popular. Regulations and guidelines are also very important. Companies can become more environmentally friendly by adhering to certain certifications and norms. However, these laws must be properly implemented, and international cooperation is required. What's the lesson, then? The textile sector must adapt its practices. It's not simply the industry's task; everyone has a part to play, from businesses to clients. The emphasis ought to be on utilizing resources more wisely, being receptive to fresh perspectives, and being gentler to the environment. It is evident that decisions made now have an impact on the future as we conclude our examination of the environmental effects of textiles. Through the implementation of sustainable practices, innovative problem-solving, and strict adherence to regulations, we can guide the textile sector toward a more environmentally friendly future. Each step counts on this road. Together, we can create a more sustainable and environmentally responsible narrative for the textile sector.

References

1. Toprak T, Anis P (2017) Textile industry's environmental effects and approaching cleaner production and sustainability: an overview. J Text Eng Fash Technol 2(4). https://doi.org/10.15406/jteft.2017.02.00066
2. Keane J. The role of textile and clothing industries in growth and development strategies
3. Shabbir M, Naim M (2019) Introduction to textiles and the environment. In: Shabbir M (ed) Textiles and clothing. Wiley, pp 1–9. https://doi.org/10.1002/9781119526599.ch1
4. You S, Cheng S, Yan H (2009) The impact of textile industry on China's environment. Int J Fash Des Technol Educ 2(1):33–43. https://doi.org/10.1080/17543260903055141
5. Chavan RB. Indian textile industry-environmental issues
6. Textile industry in India, leading yarn manufacturers in India—IBEF. India Brand Equity Foundation. https://www.ibef.org/industry/textiles. Accessed 03 Oct 2023
7. Chemical and textile fibers production worldwide 2021. Statista. https://www.statista.com/statistics/263154/worldwide-production-volume-of-textile-fibers-since-1975/. Accessed 02 Oct 2023
8. Textile | Description, industry, types, and facts | Britannica. https://www.britannica.com/topic/textile. Accessed 08 Oct 2023
9. Ananthashankar R, Production AG (2013) Characterization and treatment of textile effluents: a critical review. J Chem Eng Process Technol 05(01). https://doi.org/10.4172/2157-7048.1000182
10. Halepoto H, Gong T, Memon H (2022) Current status and research trends of textile wastewater treatments—a bibliometric-based study. Front Environ Sci 10:1042256. https://doi.org/10.3389/fenvs.2022.1042256
11. www.fibre2fashion.com. Global fibre production nearly doubles to 113 mn tonnes in 2021. http://www.fibre2fashion.com/news/textile-news/global-fibre-production-nearly-doubles-to-113-mn-tonnes-in-2021-283471-newsdetails.htm. Accessed 03 Oct 2023
12. Link G, Facebook, Twitter, Pinterest, Email, Apps O. Fiber manufacturing introduction. https://www.dspattextile.com/2021/08/fiber-manufacturing-introduction.html. Accessed 07 Oct 2023
13. Complete 5 textile manufacturing process in details—textile details. https://textiledetails.com/textile-manufacturing-process/. Accessed 07 Oct 2023
14. Roving (2023) Wikipedia
15. Textileblog. Textile manufacturing processes for students and professionals. Textile Blog. https://www.textileblog.com/textile-manufacturing-processes/. Accessed 08 Oct 2023
16. Uddin F (2019) Introductory chapter. In: Textile manufacturing processes. IntechOpen. https://doi.org/10.5772/intechopen.87968
17. Get familiar with your textile production processes | Textile Guide. https://textileguide.chemsec.org/find/get-familiar-with-your-textile-production-processes/. Accessed 08 Oct 2023
18. Babu BR, Parande AK, Raghu S, Kumar TP (2007) Cotton textile processing: waste generation and effluent treatment 11(3)
19. Rathore B (2022) Textile industry 4.0 transformation for sustainable development: prediction in manufacturing & proposed hybrid sustainable practices. Eduzone Int Peer Rev Acad Multidiscip J 11(01):223–241. https://doi.org/10.56614/eiprmj.v11i1.229
20. www.fibre2fashion.com. Energy conservation, textile industries, global energy crisis. http://www.fibre2fashion.com/industry-article/3377/energy-conservation-in-textile-industries-savings. Accessed 17 Oct 2023
21. Dos Santos AB, Cervantes FJ, Van Lier JB (2007) Review paper on current technologies for decolourisation of textile wastewaters: perspectives for anaerobic biotechnology. Bioresour Technol 98(12):2369–2385. https://doi.org/10.1016/j.biortech.2006.11.013
22. www.fibre2fashion.com. Antimicrobial fabrics: the future of textiles. http://www.fibre2fashion.com/industry-article/8710/antimicrobial-fabrics-the-future-of-textiles. Accessed 17 Oct 2023
23. Kant R (2012) Textile dyeing industry an environmental hazard. Nat Sci 04(01):22–26. https://doi.org/10.4236/ns.2012.41004

24. Chhonkar PK, Datta SP, JoshF HC. Impact of industrial effluents on soil health and agriculture—Indian experience: part II-tannery and textile industrial effiuents
25. Ramana S, Biswas AK, Kundu S, Saha JK, Yadava RBR (2002) Effect of distillery effluent on seed germination in some vegetable crops. Bioresour Technol 82(3):273–275. https://doi.org/10.1016/S0960-8524(01)00184-5
26. Nagda GK, Diwan AM, Ghole VS. Seed germination bioassays to assess toxicity of molasses fermentation based bulk drug industry effluent
27. The case for natural dyes. Avani Earthcraft. https://avani-earthcraft.com/blogs/blog/the-case-for-natural-dyes. Accessed 22 Oct 2023
28. Land pollution | Causes, effects and solutions | Britannica. https://www.britannica.com/science/land-pollution. Accessed 22 Oct 2023
29. MD AEB. Does air pollution cause Alzheimer's disease? Harvard Health. https://www.health.harvard.edu/blog/does-air-pollution-cause-alzheimers-disease-2020072320627. Accessed 22 Oct 2023
30. Sandin G, Peters GM (2018) Environmental impact of textile reuse and recycling—a review. J Clean Prod 184:353–365. https://doi.org/10.1016/j.jclepro.2018.02.266
31. These facts show how unsustainable the fashion industry is. World Economic Forum. https://www.weforum.org/agenda/2020/01/fashion-industry-carbon-unsustainable-environment-pollution/. Accessed 26 Oct 2023
32. Kumar R, Verma A, Shome A, Sinha R, Sinha S, Jha PK, Kumar R, Kumar P, Shubham, Das S, Sharma P, Vara Prasad PV (2021) Impacts of plastic pollution on ecosystem services, sustainable development goals, and need to focus on circular economy and policy interventions. Sustainability 13(17):9963. https://doi.org/10.3390/su13179963
33. Leal Filho W, Ellams D, Han S, Tyler D, Boiten VJ, Paço A, Moora H, Balogun A-LA (2019) Review of the socio-economic advantages of textile recycling. J Clean Prod 218:10–20. https://doi.org/10.1016/j.jclepro.2019.01.210
34. Harsanto B, Primiana I, Sarasi V, Satyakti Y (2023) Sustainability innovation in the textile industry: a systematic review. Sustainability 15(2):1549. https://doi.org/10.3390/su15021549
35. Yang Y, Wang Y (2020) Supplier selection for the adoption of green innovation in sustainable supply chain management practices: a case of the chinese textile manufacturing industry. Processes 8(6):717. https://doi.org/10.3390/pr8060717
36. www.fibre2fashion.com. Let's talk sustainable fashion. http://www.fibre2fashion.com/industry-article/8680/let-s-talk-sustainable-fashion. Accessed 17 Oct 2023
37. Textile industry in India, leading yarn manufacturers in India—IBEF. India Brand Equity Foundation. https://www.ibef.org/industry/textiles. Accessed 24 Oct 2023
38. Preuss S. 6 Sustainable textile innovations that will change the fashion industry. FashionUnited. https://fashionunited.com/news/business/6-sustainable-textile-innovations-that-will-change-the-fashion-industry/2017100917734. Accessed 22 Oct 2023
39. www.fibre2fashion.com. 5 Ways to make fashion industry more sustainable in the year 2020. http://www.fibre2fashion.com/industry-article/8520/5-ways-to-make-fashion-industry-more-sustainable-in-the-year-2020. Accessed 17 Oct 2023
40. The 6 sustainable fabric innovations that already shaping the textile industry. TexCovery. https://texcovery.com/resourses/the-6-sustainable-fabric-innovations-that-already-shaping-the-textile-industry. Accessed 24 Oct 2023
41. Hens L, Block C, Cabello-Eras JJ, Sagastume-Gutierez A, Garcia-Lorenzo D, Chamorro C, Herrera Mendoza K, Haeseldonckx D, Vandecasteele C (2018) On the evolution of "Cleaner Production" as a concept and a practice. J Clean Prod 172:3323–3333. https://doi.org/10.1016/j.jclepro.2017.11.082
42. Team E. BTPL sets new benchmark; introduces water-free fabric dyeing method. Apparel Industry | Textile Industry. https://www.apparelviews.com/btpl-sets-new-benchmark-introduces-water-free-fabric-dyeing-method/. Accessed 24 Oct 2023
43. Amutha K (2017) Sustainable practices in textile industry: standards and certificates. In: Muthu SS (ed) Sustainability in the textile industry. Textile Science and Clothing Technology, Springer Singapore, Singapore, pp 79–107. https://doi.org/10.1007/978-981-10-2639-3_5

44. Home—GOTS. https://global-standard.org/. Accessed 19 Oct 2023
45. ANSI logo use. American National Standards Institute—ANSI. https://www.ansi.org/resource-center/logo-use. Accessed 19 Oct 2023
46. OEKO-TEX®—tailor-made solutions for the textile and leather industry. https://www.oeko-tex.com/en/. Accessed 19 Oct 2023
47. Rajput RS, Pandey S, Bhadauria S (2017) Status of water pollution in relation to industrialization in Rajasthan. Rev Environ Health 32(3). https://doi.org/10.1515/reveh-2016-0069
48. Islam M, Mostafa M (2019) Textile dyeing effluents and environment concerns—a review. J Environ Sci Nat Resour 11(1–2):131–144. https://doi.org/10.3329/jesnr.v11i1-2.43380
49. Chungyalpa W. Examination and analysis of the central pollution control board and the state pollution control board—Indian Administrative Arm for Environmental Protection
50. Arabadjieva K (2023) 'Better Regulation' in environmental impact assessment

Social and Environmental Impacts in Textile Production

Syeda Laiba Haider Naqvi, Mahnoor Nadeem, Fizzah Ayub, Abdullah Yasar, Syeda Hamayal Zahra Naqvi, and Rameesha Tanveer

Abstract The chapter discusses the social and environmental impacts of textile production, a significant global industry with significant impacts on labor practices, human rights, gender equality, community well-being, legal frameworks, and sustainability. Textile workers in underdeveloped countries face labor rights abuses, low pay, long hours, dangerous working conditions, and restricted access to healthcare and education. Gender discrimination is a significant issue, particularly for women. Poor work-life balance, financial inequality, and health issues are also prevalent among textile workers. The textile industry has a significant environmental impact, including solid waste production, water pollution, climate change, and air pollution. Water pollution is a major issue due to the industry's substantial water consumption and contamination with chemicals and colors. Air pollution is also a concern due to high noise levels and dust discharge during fiber processing. The textile sector contributes significantly to global greenhouse gas emissions, with over 10% of all emissions coming from this sector. Mitigation measures for the environmental impacts of the textile industry include embracing corporate social responsibility (CSR), promoting community well-being, adopting circular economy principles, exploring advanced production techniques, improving energy efficiency, reducing air pollution, implementing floating treatment wetlands for effluent treatment, and using bioremediation and gamma radiation for solid waste management. These strategies aim to reduce the industry's ecological footprint, improve social and environmental responsibility, and use fewer resources.

Keywords Textile industry · Social impacts · Health issues · Environmental impacts · Air pollution · Wastewater

---

S. L. H. Naqvi (✉) · M. Nadeem · F. Ayub · A. Yasar · S. H. Z. Naqvi
Sustainable Development Study Centre, Government College University, Lahore, Pakistan
e-mail: laibahaider404@gmail.com

A. Yasar
e-mail: yasar.abdullah@gcu.edu.pk

R. Tanveer
Australian Rivers Institute, School of Environment and Science, Griffith University, 10 Southport, West Leederville, Queensland 4222, Australia

P. Singh (ed.), *Dye Pollution from Textile Industry*, SDGs and Textiles,
https://doi.org/10.1007/978-981-97-5341-3_19

1 Introduction

The Social Impact of Textile Production is a multifaceted topic that includes labor practices, human rights, gender equality, community well-being, legal frameworks, and sustainability issues [101]. It is a rising problem in developing nations, where poor salaries, long working hours, hazardous working conditions, and restricted access to healthcare and education are common [122]. Workers' rights are protected by international labor laws and agreements, and international standards such as ISO 14001 and ISO 26000 govern sustainable business practices. Because the sector is notorious for its resource-intensive operations and emissions, sustainability is a critical component [119]. Companies and consumers interested in sustainable textiles can seek help from certifications such as the Global Organic Textile Standard and Fair Trade [93]. Textile firms may enhance their social effect and contribute to a more sustainable future by incorporating these concepts into their operations [1]. This study investigates the influence of the textile industry on labor practices, human rights, gender equality, and community well-being. It investigates the connections between global firms, local producers, and disadvantaged communities, exposing discrepancies in labor conditions, pay, and social welfare. The study also investigates the environmental effects of textile production, such as pollution and climate change, and provides recommendations for sustainable practices (Fig. 1).

The textile industry stands as one of the foremost consumers of two vital resources: water and fuel, which are indispensable for various energy-intensive operations such as generating electric power, and steam, and facilitating transportation. The number

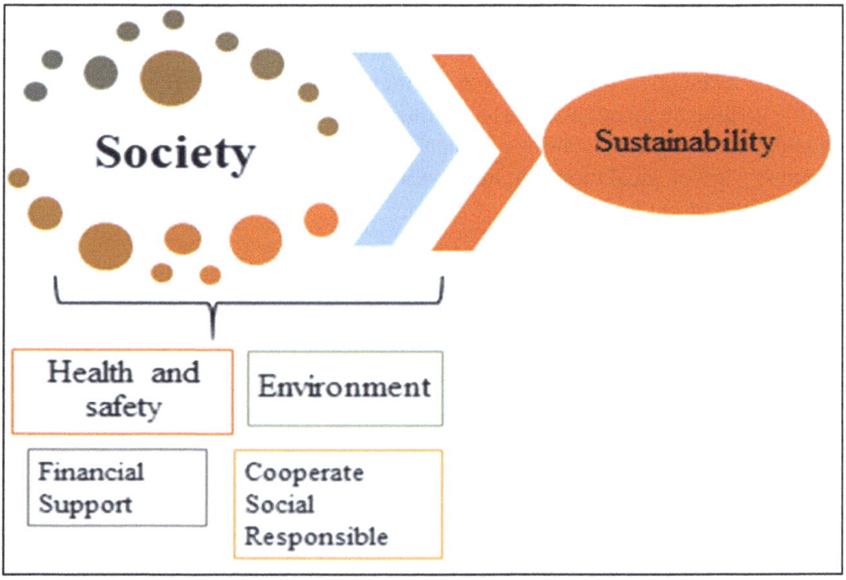

Fig. 1 Indicates the factor effect on society and ultimately leads to sustainability

of textiles consumed per person worldwide annually is around 20 kg, and this figure is increasing. Among other industrial sectors, the textile industry stands out due to its broad and intricate chain of operation, which covers both the forward and backward stages of the production process [35].

This broad business combines a wide range of sub-sectors, starting with the early phases including agriculture, shearing, and the manufacturing of synthetic fibers. The manufacture of finished products designed for customers, covering a diverse range including apparel, home textiles, carpets, and industrial clothing, then follows through intermittent manufacturing processes including weaving, knitting, dyeing, and finishing. The textile business is a vast and complex sector within the global industrial landscape since its reach goes beyond production to include retailing, services, waste management, and life cycle assessment sectors [91, 115].

The two main types of fibers used in the textile business are natural and man-made. Animals and agriculture are the sources of natural fibers, and their production and processing involve significant resource use, including water and energy. According to previous studies [14, 19], this is particularly apparent in the setting of complex manufacturing processes. In addition, the sector is known for using excessive chemicals, especially during the dyeing and finishing processes [128]. Due to its severe effects on the environment, particularly in relation to climate change, namely global warming, the growing worldwide demand for energy has become an issue of utmost significance during the past ten years.

According to the World Health Organization (WHO), over 160,000 people die each year as a result of the negative consequences of climate change, with that figure expected to be nearly fourfold by the year 2020. Increased cases of illnesses including malaria, malnutrition, and diarrhea are a result of climate change, which is also characterized by changes in food production, protracted droughts, and rising temperatures across various regions of the world. Furthermore, the 2003 deadly heatwaves that claimed tens of thousands of lives throughout Europe are evidence that global warming has had disastrous effects, as reported by Mekhilef et al. [78] and Muneer et al. [81]. As reported by Son [114], the impact of climate change extends beyond health and natural catastrophes and is starting to affect people's fashion preferences. This demonstrates how climate change has a profound and wide-ranging influence on several aspects of human life.

It is critical to acknowledge that human actions, especially the burning of fossil fuels and deforestation, have significantly increased the greenhouse gas emissions that cause climate change. It is important to consider as a global issue that exceeds national and regional boundaries. Industrial operations, which produce a variety of pollutants, contribute significantly to environmental pollution. As demonstrated by Bhuyar et al. [22], Farhana et al. [35], and Ho et al. [44], textile production, along with other industrial activities like mining, chemical production, and soap and detergent manufacturing, plays a critical part in adversely affecting the quality of both surface water and groundwater. Therefore, this book chapter addresses the social and environmental impacts of textile production. It is essential to assess the social and environmental impacts of textile production due to the industry's significant global influence.

2 Social Impacts

2.1 *Imbalance Quality of Work Life*

The textile industry research on Quality of Work Life (QWL) emphasizes its relevance in creating organizational success and employee well-being. It emphasizes the importance of motivating insights and increased compensation in improving QWL. QWL is shaped by factors such as attitudes, work environment, job features, stress levels, and career progression, and it is critical for a harmonious workplace [65]. The study done by Indumathy and Indumathy [50] uses a descriptive research design and data analytic methods to evaluate the Quality of Work Life (QWL) of 60 textile industry employees in the Tirupur District. The attitude, work environment, nature of the job, interpersonal interactions, stress levels, career prospects, difficulties, and rewards are important determinants of QWL. Moreover, [104] summarizes that active businesses have strong revenues, cutting-edge technology, and an emphasis on employee-centric corporate responsibility. Midsize businesses have better levels of satisfaction in areas such as working factors mentioned in Fig. 2.

2.1.1 Workplace Safety Issue

Workplace safety is critical in the textile sector, and it is frequently matched with legislative requirements and ISO standards. Compliance with national labor laws and international standards such as ISO 45001 is required, which necessitates extensive safety programs, risk assessments, and the supply of personal protective equipment (PPE) [110]. For environmental management, chemical exposure concerns must be managed in accordance with ISO 14001. Textile mill machinery safety should adhere to ISO 12100 criteria, which include machine guarding, regular maintenance, and personnel training [108]. Ergonomics is critical in the prevention of musculoskeletal ailments, while ventilation and air quality management are critical in the prevention of respiratory disorders [107]. Fire safety necessitates fire-resistant materials and drills, while noise levels necessitate noise-canceling devices and hearing protection. A complete strategy is required to ensure worker safety and ethical corporate practices.

2.2 *Child Labor and Exploitation*

Child labor and exploitation in the textile sector are severe ethical and human rights breaches that continue in low-income nations, according to a study. International labor agreements and national legislation seek to ban and abolish child labor, with a focus on minimum working age, controlled hours, and safe working conditions [73]. ISO 45001 and ISO 26000 urge businesses to identify and address child labor hazards in their supply chains and many other laws mentioned in Table 1. Child

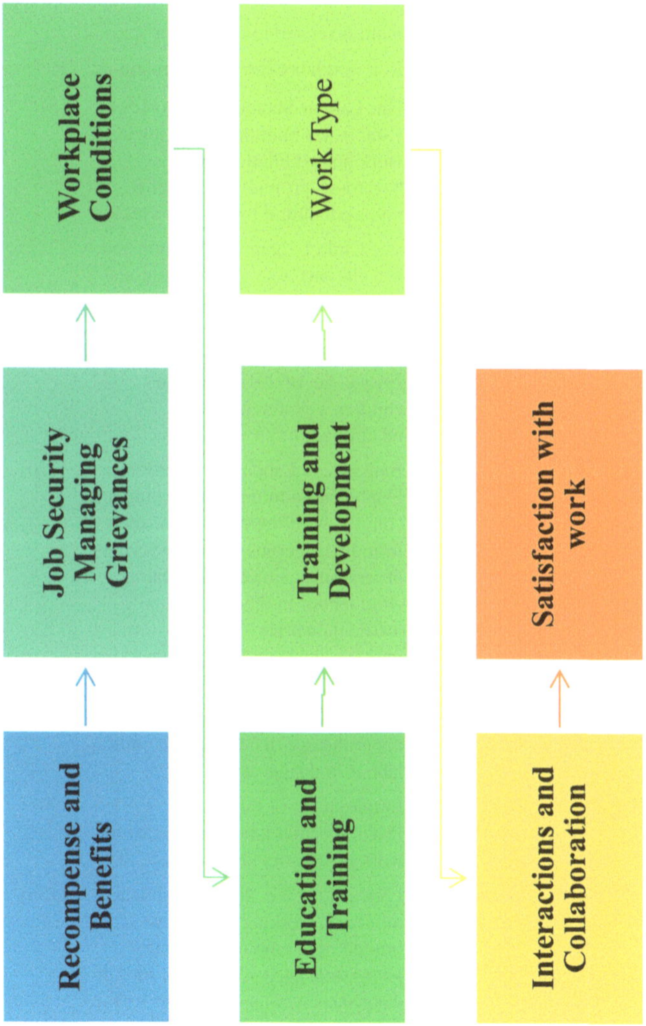

Fig. 2 Factors that impact quality of work life at the industrial scale

labor has a terrible impact on children's physical and psychological well-being, educational chances, and general quality of life, according to research [24]. As a result, governments, companies, and organizations must collaborate to eliminate child labor, defend children's rights, and respect the textile industry's ethical and legal commitments.

Table 1 Laws and regulations of different countries related to child labor

Country	Minimum working age	Relevant labor laws and regulations	References
United States	14 (with exceptions)	Fair Labour Standards Act (FLSA): Establishes minimum working age, work hour limitations, and hazardous professions guidelines for non-agricultural jobs such as textiles	[8]
India	14	The Child Labour (Prohibition and Regulation) Act of 1986 prohibits child labor under the age of 14, classifies dangerous vocations, and requires rehabilitation.—Juvenile Justice Act of 2015: Increases children's protection and rehabilitation	[28]
Bangladesh	14 (with exceptions)	Bangladesh Labour Act of 2006: Establishes a minimum working age, controls working conditions, and forbids dangerous labour.—Several programmes aimed at combating child labour in the apparel and textile industries	[120]
China	16 (with exceptions)	Labour Law and Labour Contract Law: Establish a minimum working age, limit working hours for minors under the age of 15, and ban child labour in dangerous jobs	[118]
Pakistan	14	Employment of minors Act, 1991: Prohibits child labour and hazardous work for minors under the age of 14	[94]
Turkey	15 (with exceptions)	Labour Law No. 4857: Establishes a minimum working age, allows limited light employment for youngsters aged 13–14, and forbids dangerous work.—Initiatives by the government to combat child labour	[138]

2.3 Gender Equality

2.3.1 Pay to Work Ratio

The study looks at how the Basic Laws Act (BLA) is being implemented in Bangladesh's RMG business, demonstrating a lack of education and awareness among 20 female workers [83]. As a result, women's rights are not equally enforced, leading to gender discrimination and maternity leave breaches. The study emphasizes the need to increase awareness and involve stakeholders in order to solve these concerns, but it has drawbacks such as a small sample size and a lack of quantitative or mixed-methods research [79].

2.3.2 Workplace Conditions

Creating safe and inclusive workplaces for women in the textile industry is both a moral imperative and a legal necessity in many countries. Equal treatment, harassment protection, and safe working conditions are all required by labor laws and international conventions. True inclusivity, on the other hand, necessitates a cultural shift inside organizations that fosters diversity, enforces accountability, and promotes women's success in all parts of their careers. Businesses must recognize that creating a gender-inclusive workplace is not a foregone conclusion. According to the research done by Boekhorst [23], Knowledge sharing helps to mitigate workplace prejudice, which is greatly impacted by laissez-faire leadership. High citizenship behavior is possible even in the absence of good leadership, particularly among 18- to 30-year-olds. Beneito and Gómez [17] demonstrates that, despite a higher mortality rate among working-age women, women's contributions to household income may have positively benefited female life expectancy by reducing relative mortality rates of female babies and girls. The typical emphasis on men's work in historical explanations of industrialization is challenged by this.

2.3.3 Influence on Female Leadership Role

The textile sector is an important stakeholder in improving gender equality, with governmental legislation and company efforts aimed at removing gender disparities [58]. To encourage women's participation in leadership roles, industrialized nations such as the United States and Europe have enacted anti-discrimination legislation and gender diversity policies [31, 51]. Companies must provide gender diversity data and set objectives for female representation on boards and in senior positions [39]. In developing nations such as Bangladesh and India, legal frameworks to address gender disparities are emerging, such as the Female Workers' Representation and Participation in Workplaces Act, 2018. The gender gap in leadership, on the other hand, is a global issue that needs ongoing lobbying and policy execution.

2.4 *Work-Life Balance Issues*

Khare and Kapoor [57] reported by using data from 133 workers in the National Capital Region, the study examines the difficulties female textile workers in India have in establishing work-life balance, focusing on elements that affect satisfaction and productivity. Wahyuni and Rahmasari [131] examine elements like asset ownership and working conditions, the study looks at how 3691 Indonesian garment industry employees combine their work and personal lives. The findings demonstrate that health conditions greatly enhance work-life balance, having a beneficial effect on employee satisfaction and retention. Valarmathi and Bhalakarishnan [127] examines the effects of QWL on Coimbatore's textile industry workers, emphasizing fair compensation, safety, career advancement, and work-life balance while addressing issues related to work-life balance. The difficult working conditions in the textile sector, which include long hours, monotonous jobs, and exposure to environmental elements like noise and heat, can cause increased stress and physical exhaustion, which affects both male and female employees' ability to maintain a healthy work-life balance [129]. Long hours and strict schedules might make it difficult for female employees to balance domestic and childcare responsibilities. These pressures have been emphasized by scientific studies, along with the effects they have on work-life balance [133].

2.5 *Wages and Income Disparities*

Textile workers in the formal sector get consistent pay, but regional inequities and employment instability continue, resulting in long-term poverty and the potential impact on future generations [20]. Hussain [48] examines that ineffective minimum wage structures are revealed as a result of international value chains, partisan politics, and government reluctance in the study, which looks at wages, trade union bargaining strength, and workers' survival in Bangladesh's ready-made garment industry. Ahmed et al., [4] look into the variables affecting worker performance in Bangladesh's textile and apparel sectors. Employee satisfaction is influenced by a variety of factors, including the likelihood of advancement, appreciation for one's efforts, wage increases, and supportive peer relationships. Rehman and Rashidi's [97] study examines the unequal distribution of wealth among people or households; referred to as income inequality. In Pakistan, the top 20% of the population controls 50% of the national revenue. This study examines the Primary Enrollment Rate, Completion Rate, and Adult Literacy Rate in order to estimate income inequality in the textile and other industrial sectors of Pakistan. According to the study, which looked at the relationship between labor productivity and salaries in Romania's manufacturing sector from 2008 to 2016, high labor productivity has a favorable effect on wages, but long-term discrepancies between them might result in falling labor shares and social inequality [43].

2.6 Poor Health and Well-Being Conditions

Noise exposure, sensitivity, and irritation each explained 0.09, 0.19, and 0.06 of the variation in job stress, and 0.09, 0.12, and 0.05 of the variation in job satisfaction, respectively. Noise sensitivity had the greatest impact on increasing occupational stress and job satisfaction [1]. The study by Yuan et al. [137], which involved 350 garment industry workers in Bangladesh, discovered a substantial positive association between work-related stress and health risk, with females expressing more concerns. To lessen stress and health concerns, it suggests treating serious illnesses and diseases. Kabir et al. [53] found that Bangladeshi ready-made garments (RMG) industry workers experience both physical and mental health problems, with female workers being more susceptible and those who work in outlying areas suffering from headaches, stress, anxiety, and despair. For workers, occupational health is essential, especially in sectors like textiles. The purpose of services is to preserve psychological, mental, and physical stability. Concerns about health and safety are influenced by things like chemical exposure, cotton dust, organic dust, and noise exposure [10]. In order to ascertain the prevalence of work-related musculoskeletal diseases (wMSDs) and lower their risk, Das et al. [29] examined and organized risk factors for 63 sewing, ironing, and packing workers. According to the study, lower back discomfort, knee pain, wrist pain, shoulder pain, and neck pain are typical musculoskeletal complaints. Awkward posture, uncomfortable movements, and extended standing all enhance worker danger while wearing protective clothing [55]. The Bayesian Best–Worst Method (BWM) and VIKOR are used to give a comprehensive occupational risk assessment (ORA) in this research. The method assesses electricity risks and associated hazards in a textile manufacturing facility. In order to show the approach's applicability and contribute to increasing workplace safety [5, 98].

2.7 Cultural and Indigenous Impact

The role of cultural businesses in job creation, revenue production, poverty reduction, and equitable development is emphasized [9]. It focuses on the textile sector in Adanwomase, Ghana, highlighting six major cultural values and emphasizing the need for governmental interventions and fostering an environment favorable to entrepreneurship and economic success mentioned in Fig. 3.

The research emphasizes the considerable effect of Irish cultural history on textile design approaches, highlighting the significance of traditional crafts such as weaving and embroidery in Irish identity. Beckette et al. [15] demonstrates a revival of old Irish textile processes mixing with rising design concepts driven by global influences and industrialization pressures, demonstrating the effect of Irish history on contemporary designers. The relevance of environmental orientation (EO) in the Chinese textile and apparel sector is highlighted in this study by Zhang et al. [141] and implies that a strong EO is essential for establishing Green Dynamic Capabilities

Fig. 3 Six major cultural values impact due to textile production (self)

and Organisational Green Culture (OGC), which will lead to long-term sustainability, cost savings, and operational efficiency. In a turbulent corporate climate, cultivating a green culture promotes long-term viability.

2.8 *Human Trafficking*

Human trafficking in the textile business is a serious problem that involves forced labor, horrible working conditions, long hours, and little pay, impacting vulnerable people all over the world [7]. Another study interviewed Spanish labor trafficking survivors in order to better understand their experiences, viewpoints, and liberation processes. Its goal was to take a survivor-centered approach and make suggestions for institutional reform [34]. According to Yoder et al. [136], non-citizen victims experience more coercive and misleading control measures in human trafficking cases across 52 nations than citizens. The study investigates how the COVID-19 epidemic has affected human trafficking, sexual exploitation, labor abuse, and victim care. Given the worldwide fragility of migrant reality, it emphasizes the necessity for an inclusive socio-sanitary strategy [76].

3 Environmental Impacts

3.1 *Water Pollution*

The textile sector uses a lot more water than other industries. The proportion of water used in different procedures is 3.2%, including desizing, bleaching, bio-polishing, mercerizing, scouring, dyeing, printing, and other finishing operations. Water contamination results from mixing water with various colors, chemicals, or other auxiliary substances after it has been used in various procedures. The textile industry has recently seen a tremendous rise in the usage of synthetic colors, fibers, and finishes [106] (Table 2).

The textile industry utilizes a wide range of chemicals as well as auxiliary materials to give textile fibers the necessary quality. Unfortunately, the wastewater produced by this sector has a noticeably high alkaline nature and a high concentration of many contaminants which include biological oxygen demand (BOD), total dissolved solids (TDS), alkalinity, and chemical oxygen demand (COD) [82]. This effluent can cause serious environmental problems if not appropriately treated before disposal.

Additionally, the textile sector contributes to air pollution and water contamination. Dust and lint are generated during the processing of fibers, both before and during the spinning and weaving operations, which can have a negative impact on the working conditions in textile facilities. These particles of dust and lint not only degrade the workplace as a whole, but also put the workers' health at risk by raising

Table 2 Different textile processes, their pollutants, and the quality of effluent [80]

Process	Possible pollutants	Nature of effluent
Desizing	Starch, CMC, glucose, PVA, fats, resins, and waxes do not exert a high BOD	CMC and PVA, small volume, High BOD: (30%–50% of total)
Scouring	Caustic Soda, waxes and grease, soda ash, sodium silicate and fragments of cloth	High BOD: (30% of total), small volume, dark color, strongly alkaline
Bleaching	Hypochlorite, caustic soda, Chlorine, hydrogen peroxide, acids	Strongly alkaline, small volume, Low BOD: (5% of total)
Mercerization	Caustic soda	Strongly alkaline, small volume, low BOD: (less than 1% of total)
Dyeing	Dyestuff and reducing agents such as acetic acid, sulfides, and soap	Fairly high BOD: (6% of total), large volume, strongly colored
Printing	Starch, dyes, China clay, gums oil, mordant, acid and metallic salts	Minute volume, fairly high BOD, oily appearance
Finishing	Traces of starch, salts, tallow, special finishes, etc.	Less alkaline, very small volume, low BOD

the possibility of respiratory illnesses. Workers exposed to dust from cotton, flax, and hemp can get byssinosis, a chronic lung condition. Additionally, the numerous processes used in the textile sector frequently generate noise levels that can approach 90 dB(A), which adds to noise pollution [113]. These issues with the environment and public health serve as a reminder of the necessity of implementing sensible policies that would lessen the textile industry's environmental impact.

Water-related concerns are widespread worldwide, but China, in particular, struggles with serious water pollution problems, with 32% of its water sources suffering from severe contamination [27, 72]. In this sense, the textile sector is significant to the nation's economy. Viscose fabrics are particularly popular with consumers because of their outstanding comfort, gentleness toward the skin, and breathability [135]. Viscose fiber is made in several steps, using raw materials like cotton linters, wood, or bagasse treated with alkali to produce alkali cellulose. This is followed by a reaction with sodium cellulose sulfonate to produce carbon disulfide, which finally leads to a spinning solution inside a weak alkali solution. The regenerated fiber is subsequently polished by drawing and cutting procedures, molded in a plasticizing solution, and ultimately dried and packaged as a completed product. It is then spun and wove to form a gray fabric that is then colored and processed to create textile items. The intricacy and large amount of water used in the textile-producing process define it.

In comparison, it takes 65 tons of freshwater to produce 1 ton of viscose staple fiber, while it takes 72 tons of freshwater to produce 1 ton of cotton pulp. In addition, 100–120 tons of fresh water are used to dye one ton of textiles. Additionally, the wastewater produced by this method contains significant amounts of organic debris, acidic pollutants, alkalis, and hazardous compounds, which significantly negatively impact the aquatic environment [135, 142].

In response to these concerns, Hoekstra et al. [45] presented the idea of the water footprint. This idea is based on the concept of virtual water. The water scarcity footprint has been used to measure water use in various industries, including poultry, cables, cocoa, dairy products, and textiles [11, 42]. It acts as an indication for determining the influence of water use on regional water resources. Conversely, the water degradation footprint measures the effects of wastewater contaminants on the ecosystem. This footprint may be divided into three categories: water eutrophication, water acidification, and water ecotoxicity. Each category is stated regarding environmental effect equivalents, such as PO_4^{-3}(or P), SO_2, and effluent, respectively [142, 61]. While Linhares and de Amorim [67] presented results for water eutrophication and water ecotoxicity footprints using P equivalent and 1,4-DB equivalent, respectively, researchers like Huang et al. [46] and Bai et al. [11] calculated the water eutrophication footprint for crops and poultry, respectively.

The water footprint of viscose textile manufacture was carefully examined by Zhu et al. [142]. The analysis found that this industry's water shortage footprint was 60.511 m^3 H_2O eq/ton, with the manufacturing of viscose fiber accounting for 85.71% of it. Additionally, it was determined that the water eutrophication footprint was 12.439 kg $PO_4{}^{3-}$ eq/ton, with COD and BOD_5 accounting for most of the impact (84.37% of the overall impact). According to calculations, the water's acidity and alkalinity footprints are 81.453 kg SO_2 eq/ton and 55.675 kg OH- eq/ton, respectively. The main cause of these was the addition of H_2SO_4 and NaOH during the spinning process. The presence of Zn^{2+} in spinning wastewater is the primary cause of the high water ecotoxicity footprint for manufacturing viscose textiles, which was measured at 3828.169 km^3 H_2O eq/ton. Additionally, the spinning stage carried the greatest environmental impact, followed by pulping and dyeing operations, according to life cycle assessment (LCA) polygon analysis [135].

3.2 Climate Change and Air Pollution

The increasing emissions of greenhouse gases, the resulting global warming, and the complex responses of climatic systems to this rise in temperature are the main causes of climate change [75]. The textile industry significantly contributes to global greenhouse gas emissions due to its intricate supply chain, which includes transferring textiles and clothing from production centers that are usually located in developing nations [26]. This industry alone is responsible for over 10% of the world's greenhouse gas emissions, or around 1.2 billion tons of CO_2 equivalent emissions [126].

The constant exploitation of natural resources is required to meet the rapid demand for apparel, which is being fueled by the fast-paced growth of the fast fashion business model. According to estimates, the global carbon budget for the fashion sector might be used up to 25% by 2050 [33]. An intensified situation for both ecological degradation and climate change is created by the rising demand for rapid fashion, which increases resource consumption [18]. Global fiber output roughly quadrupled over the previous ten years, rising from 58 to 109 million tons [123].

The ecologically expanded multi-regional input-output model by Peters et al. [92] examining how fast fashion influences clothing and shoe consumption in relation to climate change shows that there was a 23 percent increase in greenhouse gas emissions between 2000 and 2015. The process of manufacturing and the generation of materials used in the fashion industry both contribute to this growth. With their rapid expansion, synthetic fibers, notably polyester, have taken over cotton as the most popular type of fashion fabric [84]. Nevertheless, as synthetics are made from fossil fuels such as crude oil and petroleum products, their carbon emissions are much greater and considerably impact climate change. For instance, producing one polyester t-shirt is predicted to need 5.5 kg of CO_2 equivalent instead of 2.1 kg for producing a cotton t-shirt [59].

Further, deforestation is fueled by the need to supply the growing production demands of the fashion industry. Wood pulp is largely used in the production of fabrics like rayon, viscose, and modal, which are semi-synthetic fibers made from cellulose that has been purified. According to Leal Filho et al. [64], the rayon sector is anticipated to develop in response to rising demand, thus worsening the danger to the world's forests. The viscose and rayon used in the textile industry come from old-growth forests that are in danger of extinction. Extreme heat events become more intense in temperate parts of North America and Europe due to deforestation which is the second-largest global source of greenhouse gas emissions [117]. While the relationship between deforestation and severe weather patterns in the developing South has yet to be conclusively proved, a comparable situation could develop. In addition to these environmental problems, the fast fashion business model, defined by brief product life cycles, greatly increases textile waste generated after the textiles have been used once. When burned or dumped in landfills, this material releases hazardous emissions that worsen the environment. It is certain that the environmental effects will worsen in the absence of a robust recovery and recycling plan, given the substantial amounts of materials, water, and energy required in the production of new textile items, which consumers mostly discard as waste [36] (Table 3).

Direct and indirect emissions are the two main ways the textile sector releases greenhouse gases (GHGs) into the environment. Notably, 10% of all carbon dioxide emissions are attributed to the textile and apparel industry. Numerous researchers [86–47] have supported this alarming trend.

Our world is seriously threatened by the rising levels of GHGs in our atmosphere, which are mostly caused by nitrous oxide (8%), methane (14%), and carbon dioxide (77%), These greenhouse gases absorb solar infrared radiation, enhancing radiative forcing effect of the Earth. Numerous experts agree that their persistent existence enhances global warming and subsequently changes our climate, causing temperatures to rise. Limiting anthropogenic activities that produce GHG emissions is essential to protect human civilization and the global economy from the disastrous effects of climate change. Everyone must work together to successfully reduce GHG emissions [103, 116].

Table 3 Air emissions during textile manufacturing [77]

Process	Pollutants
Energy production	Particulates, sulfur dioxide (SO_2), nitrous oxides (Nox),
Coating, drying, and curing	Volatile organic components (VOCs)
Cotton handling activities	Particulates
Sizing	Sulfur oxide, carbon monoxide, nitrogen oxides,
Bleaching	Chlorine, chlorine dioxide
Dyeing	H_2S
Printing	Hydrocarbons, ammonia
Chemical storage	Volatile organic components (VOCs)
Wastewater treatment	Volatile organic components, toxic emissions

There is no doubt that carbon dioxide emissions, which are principally responsible for climate change, significantly influence the growth and continuation of human societies. China is responsible for a major portion of worldwide energy consumption and carbon dioxide emissions, accounting for 23.4% of global energy consumption and 23.4% of carbon dioxide emissions [32]. Additionally, current statistics show that energy consumption continuously increases, with an average yearly increase of 2.3%. The simultaneous production of 21.30 billion tons of carbon dioxide from burning fossil fuels makes it a pressing worldwide issue. The globe faces a wide range of environmental problems, from different contaminants to degrading ecosystems. These issues together highlight the urgent need for comprehensive international action to reduce the harmful consequences of climate change.

3.3 Energy

One of the most urgent issues is the greenhouse effect, in addition to the well-known environmental risks such as stratospheric ozone depletion, acid rain, and the imminent threat of catastrophic climate change. Our energy consumption habits are directly responsible for this occurrence [56]. However, it is crucial to extend our viewpoint to consider a wider range of effects that, despite being less obvious, have significant ecological and societal repercussions. These consist of:

(i) Increased Mortality Caused by Hyperthermia and Heatwaves: Studies have revealed the sobering truth that heat-related hyperthermia is responsible for thousands of fatalities each year [35].
(ii) Crop Failures and Global Food Shortages: Crop failures have become more frequent, significantly worsening the world's food shortages [16].
(iii) Disruption of Power Supply: According to Li et al. [66], increasingly frequent and severe climatic events pose a hazard to power infrastructure.

(iv) Increasing Sea Levels and Coastal Submergence: According to Reddy et al. [96], rising sea levels gradually bury low-lying coastal areas, accelerating coastline erosion.
(v) Increased Risk of Flooding and Saltwater Intrusion: According to Bhattachan et al. [21], climate change increases the risk of flooding and saltwater intrusion, which endangers populations and ecosystems.
(vi) Increased Storm Damage: According to Agache et al. [2], more frequent and intense storms cause more environmental harm.
(vii) Prolonged Droughts: According to Du Plessis and Du Plessis [30], prolonged droughts are growing more frequent, placing additional stress on ecosystems and water supplies.
(viii) Increased likelihood of Wildfires: According to Di Virgilio et al. [130], increased temperatures and drier conditions enhance the likelihood of wildfires, posing serious hazards to natural landscapes and populated areas.

Additionally, it is imperative to recognise the substantial environmental impact that industries such as the textile sector. Spinning, weaving, cutting, and sewing operations are well known for their huge consumption of power, which results in a large increase in CO_2 emissions and a raised risk of acidification. A detailed breakdown of the energy consumed by the main sectors of the textile industry can be seen in Table 4.

The textile industry, in particular, has received severe criticism in recent years for its role in environmental pollution. Due to the enormous scale of operations, the high energy and chemical requirements in different industrial processes result in certain gaseous emissions and solid waste released into the environment. As a result, textile mills were among the top water and carbon dioxide polluters. Different colors, including azo dyes (–N = N–), pesticides, detergents, and others can pollute water. These azo dyes have already been banned for extremely damaging operations in recent years (Fig. 4).

When compared to cotton dying, which is one of the most harmful processes in the textile manufacturing process, the aforementioned activities have a significantly smaller negative environmental impact. Cotton is dyed with a variety of chemicals, including bleaching agents, dyes, wetting agents, soap, softeners, and salts, in addition to copious amounts of electricity, water, and steam in order to get the desired color [100]. Furthermore, a lot of effluent is produced during the dyeing process, which is bad for the environment [99]. This wastewater, according to Akarslan and Demiralay [6] and Zhang et al. [140], contaminates continental water bodies, notably in terms of the prevalence of dangerous dyes and various illnesses that affect human health, especially skin-related issues.

In addition to these concerns, the inefficiency of the dyeing process causes the textile industry to lose up to 200,000 tons of dyes annually, which are then discharged into effluents during the dyeing and finishing processes. Ogugbue and Sawidis [89] claim that these fugitive colors often evade conventional wastewater treatment methods, ending up in the environment and posing a threat to it.

Table 4 Energy use in the textile industry's various processes

Name of the process	Form of energy	Detail	References
Fiber preparation	Mainly electricity and fossil fuels for others	Typically, 21% of the energy needed to create fabric is used to process the fiber	[125]
Spinning	Mainly electricity and fossil fuel	Energy usage typically ranges between 11 and 18% of the total energy utilized to create cloth	[74]
Weaving	Mainly electricity and thermal energy	In weaving, energy is often estimated at around 23%. Here, the energy consumption for winding is 0.4 kWh/kg, and for warping, it is 0.1 kWh/kg	[62]
Sizing	Electricity and thermal energy	Thermal energy is usually used at a rate of 5 MJ/kg, and electrical energy at 0.03 kWh/kg	[62]
Desizing	Electricity and thermal energy	In a typical case, 1000 kg of cotton uses 2500–21,000 L of water	[111, 125]
Bleaching	Electricity and thermal energy	Bleaching and drying 8–33 MJ/kg	[60]
Dyeing and finishing	Electricity, water, chemicals, and fossil fuel	Dyeing and finishing 8–18 MJ/kg	[105]
Printing	Electricity, water, chemicals, and fossil fuel	For cotton, 8000–16,000 l of water is used to produce 1000 kg of fabric	[111]

The industrial sector, which includes manufacturing, agriculture, construction, and mining, consumes 51% of the world's total energy, while the commercial sector, which includes businesses, institutes, services, and organizations, consumes 7% of it, according to ISO 50001:2011, a standard governing global energy management [49].

Energy consumption costs are a critical factor impacting total production costs in the textile industry. However, it is crucial to understand that these prices might vary greatly from nation to nation, frequently corresponding with the regional energy pricing system. This difference is particularly evident in areas with high energy costs. For instance, when producing processed yarn in Brazil, energy costs account for just

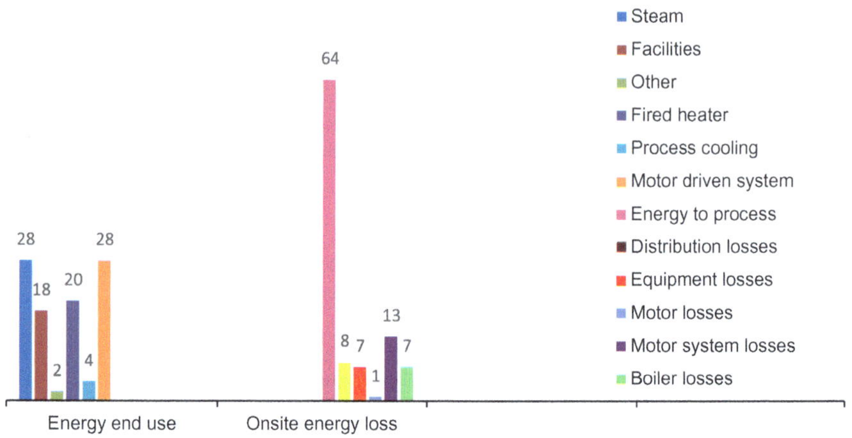

Fig. 4 Standard breakdown of energy usage in the composite textile sector, including thermal and electrical (self)

5% of the overall cost of production, whereas in India, this percentage rises to 12% [41].

In the textile sector, it is crucial to comprehend the complex interactions between consumption, energy prices, and production. According to Ozturk et al. [90], this detailed connection has a distinct linear trend: when manufacturing prices rise, so do the expenditures related to energy usage. Electricity, which is needed for motor drives, lighting, air conditioning, office operations, and steam generation, and other energy sources like coal, oil, LPG, and natural gas, make up the majority of the energy consumed by the textile sector. This steam is crucial as thermal energy in several operations, including dyeing, washing, bleaching/scouring, and other finishing treatments.

According to research studies by Grave et al. [38] and Ozturk et al. [90], the temperature requirements for these applications often span a spectrum extending from low to medium temperatures, roughly lying within the range of 60–200 °C. According to Farhana et al. [35], the structural makeup of the textile industry has a complex relationship with its energy use. The breakdown of average power and thermal energy use in the composite textile sector is shown in Fig. 5. Energy and water resources must be used extensively while manufacturing textile products, with these requirements being spread across the industry's several sub-sectors. Electricity emerges as the main energy source in the spinning industry, using fossil fuels as a backup for things like lights, lighting systems, and compressed air systems. In contrast, the weaving industry relies mostly on electricity, with fossil fuels only necessary for certain operations like humidification, air conditioning, and sizing. The main characteristic of the knitting part is how much power it uses. Water, however, plays a crucial role as a major cost component in the wet processing industry, which stands out as the most voracious in terms of energy and water consumption. The thermal energy released from burning fossil fuels to produce water steam is essential

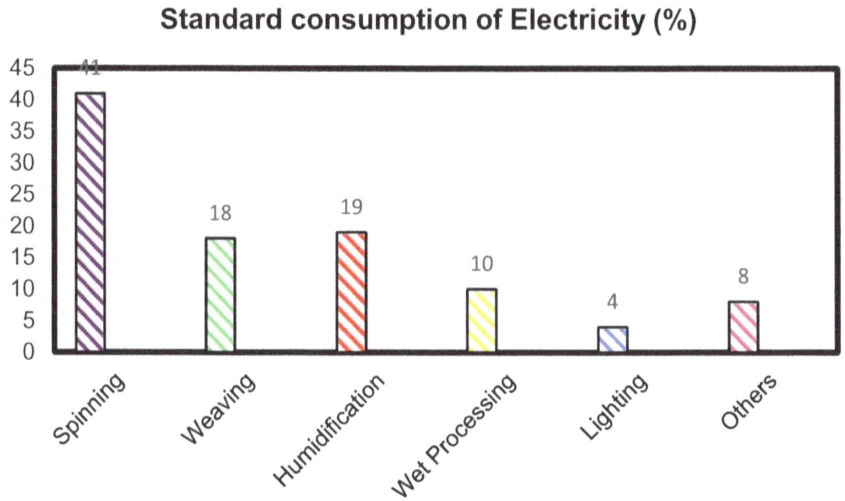

Fig. 5 Breakdown of typical power and thermal energy usage in a composite textile industry (self)

to many operations, including pre-treatment, bleaching/scouring, dyeing, washing, and finishing. Furthermore, while most machinery operations are powered by electricity, research by Hasanbeigi and Price [40], Liserre et al. [68], and Sharma et al. [112] highlights how important it is to conserve thermal energy. According to Shah and Shah [109], the energy used to produce yarn and fabric within the textile sector typically ranges from 40 to 45%, with the energy-intensive wet processing unit accounting for an even greater consumption range of 35–60%. This emphasizes how crucial it is for the textile industry's pursuit of sustainability and efficiency to understand the complex dynamics of energy consumption and cost management.

3.4 Solid Waste

Solid waste materials in the textile industry frequently take the form of fibers that are used to make fabric and other products [52]. Solid waste is produced at several phases of the fiber processing process, from raw materials to completed goods and consumer delivery. A wide range of jobs are covered in these critical steps, such as designing, folding, cutting, preparing for sewing, actually sewing, finishing, ironing, packing, maintaining track of product stocks, shipping, and customer service [52] (Table 5).

The production of yarn, the synthesis of fiber, the assembly of textiles, manufacturing, packing, distribution, transportation, use, recycling, and ultimately disposal are all complex operations in the textile business. All of these processes greatly influence the creation of solid waste [54]. Seasonal variations and an increase in living standards worldwide have increased the demand for textile items, which has increased

Table 5 Solid waste produced throughout the textile manufacturing process [86]

Textile production stage	Solid waste material generated
Shipping	Paperboard, adhesive tape
Packaging	Labels, toner, plastics
Finishing	Paperboard, plastics, labels, trims, leftover fabric, stitching yarn, sticky paper, and thread
Sewing	Stitching yarn, trims, needles, plastic cones, fabric scraps, thread
Cutting	Plastic, metal clips, fabric scraps, plotter pens, paperboard, paper
Folding	Adhesive tape, plastic, paper, paperboard, fabric scrap,
Collection planning	Printer cartridges, packing, defective parts, magazines, paper and paperboards

Table 6 Various solid wastes produced by the value chain of the textile industry [132]

Stage	Solid waste produced
Production	Sludge, selvage, cut-offs and ends of fabric rolls, sweepings, flocks, leftover cloth, yarn, and fiber
Pre-consumer	Things that are damaged and unsold in retailers
Post-consumer	Items that are out-of-date in terms of style, outgrown, damaged, and worn-out

waste creation [139]. As a result, this situation is made worse. Table 6 provides a concise breakdown of the numerous types of solid waste materials produced at various stages of textile processing to provide a thorough understanding. Yalcin-Enis et al. [134] detailed the classification of solid waste products within the textile sector. Production waste, pre-consumer waste, and post-consumer waste are the three main categories into which they were separated.

4 Mitigation Measures

4.1 Social Responsibility and Sustainability

Stakeholder participation, dedicated leadership, and resource accessibility are critical aspects of CSR implementation. Lagnaoui's research [63] emphasizes the significance of corporate social responsibility (CSR) practices in strengthening social, environmental, and economic values, emphasizing the relevance of sustainability in the Finnish textile sector. Today, businesses are pressured to include environmental responsibility and sustainability in their corporate plans. Doing so can set them apart and make them more competitive while putting forward a positive image of the corporation [37]. However, no policies or regulations are known that dictate how businesses can incorporate social and environmental sustainability into the development of new products.

Textile Industries should use more environmentally friendly manufacturing techniques instead of the traditional methods that utilize excessive resources like water and energy [121]. This can be done by combining environmental and economic techniques for process improvement, cost-cutting, and reduced environmental impact. Among the major challenges facing the textile industry is managing production and effluent treatment [121].

4.2 Community Well-Being

When it comes to production and marketing methods, labor and human rights, health, and other factors, the textile sector has a considerable impact on social sustainability and environmental management, as well as safety. By enhancing transparency and traceability, establishing sustainable working conditions, and investing in raw materials from suppliers that adhere to social and environmental norms; significant efforts are also being made to reduce poverty in local supply chains. Regarding social sustainability, the textile sector works harder than ever to establish moral production procedures, pay all workers fairly, and provide secure working conditions [95].

4.3 Circular Economy

Incorporation of renewable energy during the processing and manufacturing of final products from raw materials should be carried out in the most efficient way possible. Applying models based on the circular economy can be beneficial [12]. Between 2000 and 2015, an average global rise in apparel consumption and manufacturing was almost 60%. However, after using each item between seven and eight times on average, the use-time of clothes fell by 36% [70]. The textile industry's global worth of 807 billion US Dollars in 2019 is only increasing. Countries like Bangladesh, Vietnam, and India depend on their textile industry for economic growth; Bangladesh alone generated 80% of their total export revenue from garments in 2020. In terms of social contribution, the textile industry, in particular, has supported the idea of women's empowerment and gender equality [102].

4.4 Latest Method for Production

The traditional methods used in the textile industry, such as coating, dying, and other chemical-intensive treatments that promote features like antibacterial activity and enhanced durability, decrease the biodegradability of natural fibers. Even though natural fibers like wool, flax, silk, and cotton are organic, the above-mentioned treatments directly affect their ability to decompose [69]. The textile sector confronts sustainability issues because of customer preferences for fast-paced fashion and a concentration on quantity over quality, particularly in developing nations [25].

4.5 *Efficient Use of Electricity*

Given the significant amount of power used by the textile sector, it is imperative to evaluate the effectiveness of its lighting systems and, where practical, put electricity-saving measures in place. The industry also depends on a large number of relatively tiny electric motors. Multiple motors individually controlled by a control board and directly connected to certain machine parts are now more common in modern machines. Greater control and efficiency in machine operations are made possible by this method. To cut costs and improve fuel economy, alternative heating techniques, including steam, gas heating, or direct/indirect fired heating, have mainly replaced electric heating. To reduce air pollution, using fuels for industrial operations with good flue gas qualities, high calorific values, and simple to burn is preferable. Notably, Lancastrian or Scotch-type tubular/smoke tube designs of boilers employed in the textile sector have given way to water-tube boilers, improving overall boiler efficiency [13].

4.6 *Reduced Air Pollution*

Numerous actions may be performed to lower emissions into the atmosphere. This involves switching from goods based on organic solvents to those based on water to reduce emissions of volatile organic compounds. Scrubber installation is another efficient method for removing particulate matter from industrial operations. Locating and measuring the sources of air pollution, utilizing safety data sheets to prescreen compounds in order to guarantee their non-toxicity, optimizing boiler operations to reduce nitrous and sulfur oxide emissions, and designing products that do not release hazardous pollutants into the air are other suggested strategies [124].

4.7 *Wastewater Treatment*

The Floating Treatment Wetlands (FTWs) system is a viable method of treating textile effluent. FTWs have demonstrated their efficacy in removing heavy metals, organic materials, organic dyes, pigments, and other contaminants from textile effluents. In order to remove pollutants, these systems combine the operations of bacteria and plants, using their physical impacts and metabolic processes. For vegetation on floating mats, it is recommended to use plant species with deep root systems and significant biomass for maximum performance. Proper plant selection, efficient plant covering management, improved aeration, and the introduction of certain bacterial strains can increase pollution removal effectiveness. The total effectiveness, sustainability, and visual appeal of FTWs are further improved by appropriate installation and maintenance procedures [85].

4.8 *Bioremediation of Sludge*

Solid waste products from the textile industry, particularly sludge, are known for having precipitated colors, salts, heavy metals, etc. These impurities greatly raise the sludge's oxygen requirements, both chemically and biochemically, making it harmful to species when it enters the food chain. An effective way to solve this issue is by bioremediation, which uses microorganisms and their enzymes to degrade dye and heavy metal components in the waste [87]. According to Nyika [88], this biotechnological technique has advantages in terms of cost-effectiveness, environmental friendliness, and efficacy. However, it could take longer to complete. Gamma radiation, recognized for its deep-penetrating abilities, can also be utilized to breakdown color compounds found in textile sludge in addition to bioremediation [71]. After deterioration, the byproducts, such as compost or chemical fertilizers, can be used as soil additions.

5 Conclusion

This chapter deals with textile industries' and production's social and environmental impacts. The social effect of textile manufacturing encompasses labor practices, human rights, gender equality, community well-being, and environmental sustainability. Poor working conditions, low salaries, child labor, gender inequality, and health concerns are all common in developing countries. To tackle these issues, the textile sector should encourage corporate social responsibility (CSR), ecologically friendly production procedures, and fair labor practices. Addressing financial gaps, encouraging work-life balance, and safeguarding cultural heritage are critical. It is also crucial to combat human trafficking and ensure worker safety. A circular economy strategy and environmentally friendly production processes are critical for lowering the environmental effect of the sector. Air pollution, water pollution, and greenhouse gas emissions are just a few of the environmental effects of the textile industry's massive use of energy, water, and chemicals. The industry's actions worsen global issues such as climate change, water shortages, and deforestation. It is crucial to lessen its environmental impact, including improvements in sustainable materials, more effective procedures, and ethical waste management. It is essential to address the environmental effects of the textile sector if we are to reduce climate change and safeguard the planet's health for future generations. Moreover, energy consumption practices, particularly in energy-intensive industries like textiles, are a major contributor to the escalation of environmental problems, including the greenhouse effect, climate change, and solid waste production. Due to its heavy reliance on electricity, the textile sector greatly increases greenhouse gas emissions, negatively impacting the environment and society. These effects include escalating death rates brought on by heatwaves, crop failures, power outages, coastal inundation, increased floods, storm damage, protracted droughts, and increasing wildfire risks. Additionally, the textile sector produces a variety of materials along its value chain, making it a noteworthy producer of solid waste. Mitigation measures are crucial to overcome these difficulties. The textile industry may minimize emissions by adopting water-based products, energy-efficient processes, and heating techniques. Additionally, implementing technologies like bioremediation for solid waste management and Floating Treatment Wetlands (FTWs) for effluent treatment might assist in reducing negative environmental effects.

References

1. Abbasi M, Yazdanirad S, Habibi P, Arabi S, Madvari RF, Mehri A, Poursadeghiyan M, Ebrahimi MH, Ghaljahi M (2019) Relationship among noise exposure, sensitivity, and noise annoyance with job satisfaction and job stress in a textile industry. Noise Vib Worldw 50(6):195–201. https://doi.org/10.1177/0957456519853812
2. Agache I, Sampath V, Aguilera J, Akdis CA, Akdis M, Barry M, Bouagnon A, Chinthrajah S, Collins W, Dulitzki C, Erny B (2022) Climate change and global health: a call to more research and more action. Allergy 77(5):1389–1407
3. Ahlström LH, Eskilsson CS, Björklund E (2005) Determination of banned azo dyes in consumer goods. TrAC, Trends Anal Chem 24(1):49–56
4. Ahmed SU, Mahtab N, Islam MN, Abdullah M (2017) Impact of working capital management on profitability: a study on textile companies of Bangladesh. J Bus Fin Aff 6(4)
5. Ak MF, Yücesan M, Gül M (2021) Occupational health, safety and environmental risk assessment in textile production industry through a Bayesian BWM-VIKOR approach. Stoch Env Res Risk Assess 36(2):629–642. https://doi.org/10.1007/s00477-021-02069-y
6. Akarslan F, Demiralay H (2015) Effects of textile materials harmful to human health. Acta Physica Polonica A 128(2B)
7. Andrevski H, Lyneham S (2014) Experiences of exploitation and human trafficking among a sample of Indonesian migrant domestic workers. Trends Issues Crime Crim Justice (471):1. http://aic.gov.au/media_library/publications/tandi_pdf/tandi471.pdf
8. Andrias K (2019) An American approach to social democracy: the forgotten promise of the fair labor standards act. Yale Law J 128(3):616. https://digitalcommons.law.yale.edu/cgi/viewcontent.cgi?article=9318&context=ylj
9. Asibey MO, Agyeman KO, Yeboah V (2017) The impact of cultural values on the development of the cultural industry: case of the Kente textile industry in Adanwomase of the Kwabre East District, Ghana. J Human Values 23(3):200–217. https://doi.org/10.1177/0971685817713282
10. Babel S, Tiwari M (2014) Occupational health hazards in textiles industry. Asian J Home Sci 9(1):267–271. https://www.cabdirect.org/abstracts/20143332865.html
11. Bai X, Ren X, Khanna NZ, Zhou N, Hu M (2018) Comprehensive water footprint assessment of the dairy industry chain based on ISO 14046: a case study in China. Resour Conserv Recycl 132:369–375

12. Balanay RM, Halog A (2019) Tools for circular economy. Elsevier eBooks, pp 49–75. https://doi.org/10.1016/b978-0-08-102630-4.00003-0
13. Bantelay DT, Getie MZ, Bruk BD, Workalemayehu RN (2020) Analytical modeling of specific energy consumption and cost share in comprehensive textile industry: case study of Ethiopia. J Optimiz Ind Eng 13(2):47–56
14. Baydar G, Ciliz N, Mammadov A (2015) Life cycle assessment of cotton textile products in Turkey. Resour Conserv Recycl 104:213–223
15. Beckette JJ, Yeats RW (2023) Exploring the irish cultural heritage: a comprehensive examination of its pervasive influence on contemporary and historical textile design techniques. Af J Emerg Issues 5(5):15–28
16. Bedasa Y, Bedemo A (2023) The effect of climate change on food insecurity in the Horn of Africa. GeoJournal 88(2):1829–1839
17. Beneito P, Gómez JJG (2021) Gender gaps in wages and mortality rates during industrialization: the case of Alcoy, Spain, 1860–1914. Fem Econ 28(1):114–141. https://doi.org/10.1080/13545701.2021.1983190
18. Berg A, Karl-Hendrik M (2021) Fashion on climate—how the fashion industry can urgently act to reduce its greenhouse gas emissions. McKinsey and Company and Global Fashion Agenda
19. Beton A, Dias D, Farrant L, Gibon T, Le Guern Y, Desaxce M, Perwueltz A, Boufateh I, Wolf O, Kougoulis J, Cordella M (2014) Environmental improvement potential of textiles (IMPRO-Textiles). European Commission 20
20. Bezuidenhout A, Khunou G, Mosoetsa S, Sutherland K, Thoburn J (2007) Globalisation and poverty: impacts on households of employment and restructuring in the textiles industry of South Africa. J Int Dev 19(5):545–565. https://doi.org/10.1002/jid.1308
21. Bhattachan A, Jurjonas MD, Morris PR, Taillie PJ, Smart LS, Emanuel RE, Seekamp EL (2019) Linking residential saltwater intrusion risk perceptions to physical exposure of climate change impacts in rural coastal communities of North Carolina. Nat Hazards 97:1277–1295
22. Bhuyar P, Trejo M, Dussadee N, Unpaprom Y, Ramaraj R, Whangchai K (2021) Microalgae cultivation in wastewater effluent from tilapia culture pond for enhanced bioethanol production. Water Sci Technol 84(10–11):2686–2694
23. Boekhorst JA (2014) The role of authentic leadership in fostering workplace inclusion: a social information processing perspective. Hum Resour Manag 54(2):241–264. https://doi.org/10.1002/hrm.21669
24. Brandenburg M, Bizuneh B, Teklemedhin TB, Woubou AM (2022) Sustainability in Ethiopian textile and apparel supply chains. Springer eBooks, pp 195–215. https://doi.org/10.1007/978-3-030-78791-2_9
25. Bubicz ME, Barbosa-Póvoa AP, Carvalho A (2021) Social sustainability management in the apparel supply chains. J Clean Prod 280124214. https://doi.org/10.1016/j.jclepro.2020.124214
26. Chen X, Memon HA, Wang Y, Marriam I, Tebyetekerwa M (2021) Circular Economy and sustainability of the clothing and textile Industry. Mater Circ Econ 3:1–9
27. Dang W (2020) How culture shapes environmental public participation: case studies of China, the Netherlands, and Italy. J Chinese Govern 5(3):390–412
28. Das KS (2022) Child labour and its determinants in India. Child Youth Serv Rev 138:106523. https://doi.org/10.1016/j.childyouth.2022.106523
29. Das S, Moorthy MK, Shanmugaraja K (2023) Analysis of musculoskeletal disorder risk in cotton garment industry workers. J Nat Fibers 20(1). https://doi.org/10.1080/15440478.2022.2162182
30. Du Plessis A, du Plessis A (2019) Current and future water scarcity and stress. In: Water as an inescapable risk: current global water availability, quality and risks with a specific focus on South Africa, pp 13–25
31. Dunlop R, Scheepers CB (2022) The influence of female agentic and communal leadership on work engagement: vigour, dedication and absorption. Manag Res Rev 46(3):437–466. https://doi.org/10.1108/mrr-11-2021-0796

32. Edenhofer O (ed) (2015) Climate change 2014: mitigation of climate change, vol 3. Cambridge University Press
33. Ellenmacarthurfoundation.org (2023) Fashion and the circular economy. https://ellenmacarthurfoundation.org/fashion-and-the-circular-economy-deep-dive. Accessed 26 Sep 2023
34. Estiarte CV (2023) Human trafficking for labor exploitation: the survivors' perspective. Int Rev Victimol 02697580231 1679. https://doi.org/10.1177/02697580231167907
35. Farhana K, Kadirgama K, Mahamude ASF, Mica MT (2022) Energy consumption, environmental impact, and implementation of renewable energy resources in global textile industries: an overview towards circularity and sustainability. Mater Circ Econ 4(1):15
36. Geneva environment network.org (2023) Environmental sustainability in the fashion industry. https://www.genevaenvironmentnetwork.org/resources/updates/sustainable-fashion/. Accessed 26 Sep 2023
37. Gonçalves A, Silva CM (2021) Looking for sustainability scoring in apparel: a review on environmental footprint, social impacts and transparency. Energies 14(11):3032. https://doi.org/10.3390/en14113032
38. Grave K, Hazrat M, Boeve S, Blücher FV, Bourgault C, Breitschopf B, Friedrichsen N, Arens M, Aydemir A, Pudlik M, Duscha V (2015) Electricity costs of energy intensive industries
39. Hamplová E, Janeček V, Lefley F (2022) Board gender diversity and women in leadership positions—are quotas the solution? Corp Commun Int J 27(4):742–759. https://doi.org/10.1108/ccij-02-2022-0022
40. Hasanbeigi A, Price L (2015) A technical review of emerging technologies for energy and water efficiency and pollution reduction in the textile industry. J Clean Prod 95:30–44
41. Hasanbeigi A, Price L (2015) A technical review of emerging technologies for energy and water efficiency and pollution reduction in the textile industry. J Clean Prod 95:30–44
42. He WW, Li Y, Wang XP, Wang LL (2018) Calculation and assessment of benchmark water footprint of silk products. Adv Textile Technol 26(2):41–45
43. Herman E (2020) Labor productivity and wages in the Romanian manufacturing sector. Procedia Manuf 46313–46321. https://doi.org/10.1016/j.promfg.2020.03.046
44. Ho YC, Show KY, Guo XX, Norli I, Abbas FA, Morad N (2012) Industrial discharge and their effect to the environment. In: Industrial waste, pp 1–33
45. Hoekstra AY, Chapagain AK, Zhang G (2015) Water footprints and sustainable water allocation. Sustainability 8(1):20
46. Huang J, Ridoutt BG, Zhang H, Xu C, Chen F (2014) Water footprint of cereals and vegetables for the Beijing market: comparison between local and imported supplies. J Ind Ecol 18(1):40–48
47. Huang B, Zhao J, Geng Y, Tian Y, Jiang P (2017) Energy-related GHG emissions of the textile industry in China. Resour Conserv Recycl 119:69–77
48. Hussain MS (2023) Bargaining or begging? The politics behind Bangladesh's readymade garment workers' unwinnable demands for a living minimum wage
49. ISO (2011) ISO Launches ISO 50001 energy management standard, Document Ref: 1434, p 1. https://www.iso.org
50. Indumathy R, Indumathy R (2012) A study on quality of work life among workers with special reference to textile industry in Tirupur district—A textile hub—ProQuest. J Contemp Manag Res Prev Publ Details, Tiruchirappalli, pp 678–690. https://www.proquest.com/openview/79ad6acb60afbf0c8168f8069b95d59b/1?pq-origsite=gscholarandcbl=2039847
51. Islam A, Hack-Polay D, Rahman M, Jantan AH, Mas FD, Kordowicz M (2023) Gender and leadership in public higher education in South Asia: examining the individual, socio-cultural and organizational barriers to female inclusion. Stud High Educ 48(8):1197–1215. https://doi.org/10.1080/03075079.2023.2187771
52. Islam MS, Islam JM (2021) Sources and fates of textile solid wastes and their sustainable management. In: Handbook of solid waste management: sustainability through circular economy. Springer Singapore, Singapore, pp 1–21
53. Kabir H, Bhullar N, Islam MS, Usher K, Haque E, Maple M (2023) Prevalence and risk factors of physical and psychological health among the readymade garment (RMG) workers in Bangladesh. Int J Occup Saf Ergon 1–32.https://doi.org/10.1080/10803548.2023.2260168

54. Karaosman H, Brun A, Morales-Alonso G (2017) Vogue or vague: sustainability performance appraisal in luxury fashion supply chains. In: Sustainable management of luxury, pp 301–330
55. Karuppiah K, Sankaranarayanan B, Ali SM, Kabir G (2020) Role of ergonomic factors affecting production of leather garment-based SMEs of India: implications for social sustainability. Symmetry 12(9):1414. https://doi.org/10.3390/sym12091414
56. Khan MA, Khan MZ, Zaman K, Naz L (2014) Global estimates of energy consumption and greenhouse gas emissions. Renew Sustain Energy Rev 29:336–344
57. Khare AK, Kapoor A (2019) A study on work life balance among women labor in garment units of NCR, India. Int J Manang 10(1). https://doi.org/10.34218/ijm.10.1.2019/004
58. Khushk AA, Zhang Z, Hui Y (2022) Role of female leadership in corporate innovation: a systematic literature review. Gender Manag Int J 38(3):287–304. https://doi.org/10.1108/gm-01-2022-0028
59. Kirchain R, Olivetti E, Miller TR, Greene S (2015) Sustainable apparel materials. In: Materials systems laboratory. Massachusetts Institute of Technology, Cambridge, pp 1–34
60. Kocabaş AM (2008) Improvements in energy and water consumption performances of a textile mill after bat applications. Master's thesis, Middle East Technical University
61. Kounina A, Margni M, Bayart JB, Boulay AM, Berger M, Bulle C, Frischknecht R, Koehler A, Milà i Canals L, Motoshita M, Núñez M (2013) Review of methods addressing freshwater use in life cycle inventory and impact assessment. Int J Life Cycle Assess 18:707–721
62. Koç E, Çinçik E (2010) Analysis of energy consumption in woven fabric production. Fibres Text East Eur 18(2):79
63. Lagnaoui T (2023) CSR and business sustainability in the finnish textile industry: a path to a sustainable future. https://urn.fi/URN:NBN:fi:amk-202304205729
64. Leal Filho W, Perry P, Heim H, Dinis MAP, Moda H, Ebhuoma E, Paço A (2022) An overview of the contribution of the textiles sector to climate change. Front Environ Sci 10:1419
65. Leitão J, Pereira D, Gonçalves Â (2019) Quality of work life and organizational performance: workers' feelings of contributing, or not, to the organization's productivity. Int J Environ Res Public Health 16(20):3803. https://doi.org/10.3390/ijerph16203803
66. Li Q, Punzo G, Robson C, Arbabi H, Mayfield M (2022) A novel approach to climate resilience of infrastructure networks. arXiv preprint. arXiv:2211.10132
67. Linhares T, de Amorim MTP (2017) LCA of textile dyeing with Acacia Dealbata tree bark: a case study research. Procedia Eng 200:365–369
68. Liserre M, Sauter T, Hung JY (2010) Future energy systems: Integrating renewable energy sources into the smart power grid through industrial electronics. IEEE Ind Electron Mag 4(1):18–37
69. Liu J, Liang J, Ding J, Zhang G, Zeng X, Yang Q, Zhu B, Gao W (2021) Microfiber pollution: an ongoing major environmental issue related to the sustainable development of textile and clothing industry. Environ Dev Sustain 23(8):11240–11256. https://doi.org/10.1007/s10668-020-01173-3
70. Luján-Ornelas C, Güereca LP, Franco-García ML, Heldeweg MA (2020) A life cycle thinking approach to analyse sustainability in the textile industry: a literature review. Sustainability 12(23):10193. https://doi.org/10.3390/su122310193
71. MM Islam J, Bashar M, Rahman MA, Ahmed R, Parvin F, Abdulla MA, R Bhuiyan M, A Khan M (2014) Application of ionizing radiation for sustainable textile effluent treatment plant. Int Multilingual Acad J 1(1)
72. Ma X, Ye L, Qi C, Yang D, Shen X, Hong J (2018) Life cycle assessment and water footprint evaluation of crude steel production: a case study in China. J Environ Manag 224:10–18
73. Maconachie R, Howard N, Bock R (2021) Re-thinking 'harm' in relation to children's work: a 'situated', multi-disciplinary perspective. Oxf Dev Stud 50(3):259–271. https://doi.org/10.1080/13600818.2021.2004393
74. Martínez CIP (2010) Energy use and energy efficiency development in the German and Colombian textile industries. Energy Sustain Dev 14(2):94–103
75. Masson-Delmotte V, Zhai P, Pirani A, Connors SL, Péan C, Berger S, Caud N, Chen Y, Goldfarb L, Gomis MI, Huang M (2021) Climate change 2021: the physical science basis.

In: Contribution of working group I to the sixth assessment report of the intergovernmental panel on climate change, 2
76. Medeiros AA, Da Penha Costa Vasconcellos M (2023) Covid-19 as a multifaceted crisis and its implications on human trafficking or other forms of human exploitation. Saude E Sociedade 32(1). https://doi.org/10.1590/s0104-12902023220497en
77. Meenaxi T, Sudha B (2013) Air pollution in textile industry. Asian J Environ Sci 8(1):64–66
78. Mekhilef S, Saidur R, Safari A (2011) A review on solar energy use in industries. Renew Sustain Energy Rev 15(4):1777–1790
79. Memon RA, Lohana K, Naqvi IB (2021) Gender pay gap in textile industrial cluster Kotri, Sindh Pakistan. South Asian J Soc Sci Human 2(2):86–100. https://doi.org/10.48165/sajssh.2021.2208
80. Mia R, Selim MD, Shamim AM, Chowdhury M, Sultana S, Armin M, Hossain M, Akter R, Dey S, Naznin H (2019) Review on various types of pollution problem in textile dyeing and printing industries of Bangladesh and recommandation for mitigation. J Text Eng Fashion Technol 5(4):220–226
81. Muneer T, Maubleu S, Asif M (2006) Prospects of solar water heating for textile industry in Pakistan. Renew Sustain Energy Rev 10(1):1–23
82. Naqvi SLH, Ayub F, Yasar A, Tabinda AB, Nawaz H, Tanveer R (2023) Pollution status monitoring and indices development for evaluating sustainable environmental management practices (SEMP) in Quaid-e-Azam Industrial Estate, Pakistan. J Clean Prod 405:136944
83. Neunsinger S (2019) Translocal activism and the implementation of equal remuneration for men and women: the case of the South African Textile Industry, 1980–1987. Int Rev Soc Hist 64(1):37–72. https://doi.org/10.1017/s0020859019000166
84. Niinimäki K, Peters G, Dahlbo H, Perry P, Rissanen T, Gwilt A (2020) The environmental price of fast fashion. Nat Rev Earth Environ 1(4):189–200
85. Ntagia E, Amulya K, Le SL, Le LT, Tran CS, Ha TL, Bui XT, Lens PN (2023) Pilot and full scale applications of floating treatment wetlands for treating diffuse pollution. Sci Total Environ 165595
86. Nyika J, Dinka M (2022) Sustainable management of textile solid waste materials: the progress and prospects. Mater Today: Proc 62:3320–3324
87. Nyika JM (2021a) Tolerance of microorganisms to heavy metals. In: Recent advancements in bioremediation of metal contaminants. IGI Global, pp 19–35
88. Nyika JM (2021b) The use of microorganism-derived enzymes for bioremediation of soil pollutants. In: Recent advancements in bioremediation of metal contaminants. IGI Global, pp 54–71
89. Ogugbue CJ, Sawidis T (2011) Bioremediation and detoxification of synthetic wastewater containing triarylmethane dyes by Aeromonas hydrophila isolated from industrial effluent. Biotechnol Res Int
90. Ozturk E, Karaboyacı M, Yetis U, Yigit NO, Kitis M (2015) Evaluation of integrated pollution prevention control in a textile fiber production and dyeing mill. J Clean Prod 88:116–124
91. Patel M, Sahu A, Rajak R (2022) Solid waste management in textile industry. In: Handbook of solid waste management: sustainability through circular economy. Springer Nature Singapore, Singapore, pp 1225–1256
92. Peters G, Svanström M, Roos S, Sandin G, Zamani B (2015) Carbon footprints in the textile industry. In: Handbook of life cycle assessment (LCA) of textiles and clothing. Woodhead Publishing, pp 3–30
93. Plakantonaki S, Kiskira K, Zacharopoulos N, Chronis I, Coelho F, Togiani A, Kalkanis K, Priniotakis G (2023) A review of sustainability standards and ecolabeling in the textile industry. Sustainability 15(15):11589. https://doi.org/10.3390/su151511589
94. Ponum M, Khan S, Hasan O, Mahmood MT, Abbas A, Iftikhar M, Arshad R (2020) Stunting diagnostic and awareness: impact assessment study of sociodemographic factors of stunting among school-going children of Pakistan. BMC Pediatr 20(1). https://doi.org/10.1186/s12887-020-02139-0

95. Rathore B (2022) Textile industry 4.0 transformation for sustainable development: prediction in manufacturing and proposed hybrid sustainable practices. UGC Newly Added J (IJNMS) 11(1):223–241. https://doi.org/10.56614/eiprmj.v11i1.229
96. Reddy CC, Khilji IA, Gupta A, Bhuyar P, Mahmood S, AL-Japairai KAS, Chua GK (2021) Valorization of keratin waste biomass and its potential applications. J Water Process Eng 40:101707
97. Rehman S, Rashidi MZ (2017) The relationship between education and income inequality (the case of Pakistan). Int J Account Econ Stud. https://doi.org/10.14419/ijaes.v5i1.6505
98. Rikhotso O, Morodi TJ, Masekameni DM (2022) Health risk management cost items imposed by occupational health and safety regulations: a South African perspective. Saf Sci 150105707. https://doi.org/10.1016/j.ssci.2022.105707
99. Roos S, Peters GM (2015) Three methods for strategic product toxicity assessment—the case of the cotton T-shirt. Int J Life Cycle Assess 20:903–912
100. Roos S, Posner S, Jönsson C, Peters GM (2015) Is unbleached cotton better than bleached? Exploring the limits of life-cycle assessment in the textile sector. Cloth Text Res J 33(4):231–247
101. Ross CB (2022) The social impact of textile production and a fabric's effect on you. https://www.the-sustainable-fashion-collective.com/ethical-fabrics-and-why-we-need-to-use-them/the-social-impact-of-textile-production-and-a-fabrics-effect-on-you
102. Saha K, Dey PK, Papagiannaki E (2021) Implementing circular economy in the textile and clothing industry. Bus Strateg Environ 30(4):1497–1530. https://doi.org/10.1002/bse.2670
103. Saheb Koussa D, Koussa M (2016) GHGs (greenhouse gases) emission and economic analysis of a GCRES (grid-connected renewable energy system) in the arid region, Algeria. Energy 102:216–230
104. Sattar S (2018) Relation of job related factors with different dimensions of quality of work life. World J Public Health 3(1):16. https://doi.org/10.11648/j.wjph.20180301.13
105. Saxena S, Raja ASM, Arputharaj A (2017) Challenges in sustainable wet processing of textiles. In: Textiles and clothing sustainability: sustainable textile chemical processes, pp 43–79
106. Senthil Kumar P, Gunasundari E (2018) Sustainable wet processing—an alternative source for detoxifying supply chain in textiles. In: Detox fashion: sustainable chemistry and wet processing, pp 37–60
107. Serinken M, Türkçüer İ, Dağlı B, Karcıoğlu Ö, Zencir M, Uyanık E (2012) Work-related injuries in textile industry workers in Turkey. Turk J Trauma Emerg Surg 18(1):31–36. https://doi.org/10.5505/tjtes.2012.54376
108. Shabani F, Alimohammadi I, Abolghasemi J, Dehdari T, Ghasemi R (2020) The study of effect of educational intervention on noise annoyance among workers in a textile industry. Appl Acoust 170107515.https://doi.org/10.1016/j.apacoust.2020.107515
109. Shah DS, Shah J (2013) Unconventional techniques for energy conservation in textile wet Processing: Formatex
110. Shah NA, Abbas F, Abbas Y, Haider SA, Khan Q, Asghar N, Noor S, Abbas SN, Ali N, Ali A (2015) Assessment of the workplace conditions and health and safety situation in chemical and textile industries of Pakistan. Sci J Public Health 3(6):857. https://doi.org/10.11648/j.sjph.20150306.20
111. Shaikh MA (2009) Water conservation in textile industry. Pakistan Text J 58(11):48–51
112. Sharma AK, Sharma C, Mullick SC, Kandpal TC (2017) GHG mitigation potential of solar industrial process heating in producing cotton based textiles in India. J Clean Prod 145:74–84
113. Sivaram NM, Gopal PM, Barik D (2019) Toxic waste from textile industries. In: Energy from toxic organic waste for heat and power generation. Woodhead Publishing, pp 43–54
114. Son MY (2012) The impact of climate change on the fashion industry in Korea. Ritsumeikan J Asia Pacific Stud 31:56–65
115. Stenton M, Houghton JA, Kapsali V, Blackburn RS (2021) The potential for regenerated protein fibres within a circular economy: lessons from the past can inform sustainable innovation in the textiles industry. Sustainability 13(4):2328

116. Stocker T (ed) (2014) Climate change 2013: the physical science basis: Working Group I contribution to the fifth assessment report of the intergovernmental panel on climate change. Cambridge University Press
117. Stoy PC (2018) Deforestation intensifies hot days. Nat Clim Change 8(5):366–368
118. Sun P, Chen JY, Rani U (2021) From flexible labour to 'Sticky Labour': a tracking study of workers in the food-delivery platform economy of China. Work Employ Soc 37(2):412–431. https://doi.org/10.1177/09500170211021570
119. Sutherland JW, Richter J, Hutchins MJ, Dornfeld D, Dzombak R, Mangold J, Robinson S, Hauschild MZ, Bonou A, Schönsleben P, Friemann F (2016) The role of manufacturing in affecting the social dimension of sustainability. CIRP Ann 65(2):689–712. https://doi.org/10.1016/j.cirp.2016.05.003
120. Syed RF, Mahmud KT (2022) Factors influencing work-satisfaction of global garments supply chain workers in Bangladesh. Int Rev Econ 69(4):507–524. https://doi.org/10.1007/s12232-022-00403-6
121. Tayyab M, Jemai J, Lim H, Sarkar B (2020) A sustainable development framework for a cleaner multi-item multi-stage textile production system with a process improvement initiative. J Clean Prod 246119055. https://doi.org/10.1016/j.jclepro.2019.119055
122. Telaketju (2023) Social impacts of textile industry from the perspective of novel business models. https://telaketju.turkuamk.fi/en/in-english-en/social-impacts-of-textile-industry-from-the-perspective-of-novel-business-models/
123. Textile Exchange (2021) Preferred fiber and materials market report 2021. https://textileexchange.org/wp-content/uploads/2021/08/TextileExchange_Preferred-Fiber-and-Materials-Market-Report_2021.pdf
124. Tiwari M, Babel S (2013) Study of health and nutritional status of the workers working in knitting industry located at Kanpur, India. J Appl Nat Sci 5(2):416–419
125. Uddin F (2014) Energy management and energy crisis in textile finishing. Am J Energy Res 2(3):53–59
126. Unfccc.int (2020) New report sets out how fashion industry can transition to low carbon cotton and polyester | UNFCCC. https://unfccc.int/news/new-report-sets-out-how-fashion-industry-can-transition-to-low-carbon-cotton-and-polyester. Accessed 26 Sep 2023
127. Valarmathi A, Bhalakarishnan H (2013) A study on quality of work life in textile sector in and around Coimbatore District. J Bus Manag Soc Sci Res (JBM&SSR) 2(2)
128. Verma AK, Dash RR, Bhunia P (2012) A review on chemical coagulation/flocculation technologies for removal of colour from textile wastewaters. J Environ Manag 93(1):154–168
129. Veronica GAS (2023) Impact of flexible work arrangements on work-life integration of middle-level employees in the apparel Industry, Sri Lanka. http://ir.lib.ruh.ac.lk/xmlui/handle/iruor/14846
130. Di Virgilio G, Evans JP, Blake SA, Armstrong M, Dowdy AJ, Sharples J, McRae R (2019) Climate change increases the potential for extreme wildfires. Geophys Res Lett 46(14):8517–8526
131. Wahyuni S, Rahmasari D (2022) Investigating the antecedents and outcomes of work-life balance: evidence from garment industries in Indonesia. DIJB (Diponegoro Int J Bus) 5(1):1–11. https://doi.org/10.14710/dijb.5.1.2022.1-11
132. Wang Y (2010) Fiber and textile waste utilization. Waste Biomass Valoriz 1:135–143
133. Weerakkody SS (2023) Social success of the women workers in the garment industry in Sri Lanka: a case of women workers in Katunayake free trade zone. Sri Lanka J Soc Sci Human 3(2):195–202. https://doi.org/10.4038/sljssh.v3i2.111
134. Yalcin-Enis I, Kucukali-Ozturk M, Sezgin H (2019) Risks and management of textile waste. In: Nanoscience and biotechnology for environmental applications, pp 29–53
135. Yavas A (2013) Color and fastness properties of printed with reactive dye viscose fabrics and fixed with radio frequency energy. Industria textilă 64(2):98–105
136. Yoder BR, Entress R, Huang X, Sadiq A, Miller J, Concha M (2023) Exploring human traffickers' means of control: what does citizenship got to do with it? Victims and Offenders, pp 1–21. https://doi.org/10.1080/15564886.2023.2233958

137. Yuan D, Gazi MdAI, Rahman MdA, Dhar BK, Rahaman MdA (2022) Occupational stress and health risk of employees working in the garments sector of Bangladesh: an empirical study. Front Public Health 10. https://doi.org/10.3389/fpubh.2022.938248
138. Yılmaz B, Yılmaz S (2023) The problem of effective enforcement of labour laws in Türkiye: "Is there any word left unspoken under the sun? Springer eBooks, pp 577–597. https://doi.org/10.1007/978-3-031-32822-0_29
139. Zamani B (2014) Towards understanding sustainable textile waste management: environmental impacts and social indicators. Chalmers Tekniska Hogskola (Sweden)
140. Zhang Y, Liu X, Xiao R, Yuan Z (2015) Life cycle assessment of cotton T-shirts in China. Int J Life Cycle Assess 20:994–1004
141. Zhang X, Yang H-L, Kumar N, Bhutto MH, Wang K, Hu T (2023) Assessing Chinese textile and apparel industry Business sustainability: the role of organization green culture, green dynamic capabilities, and green Innovation in relation to environmental orientation and business sustainability. Sustainability 15(11):8588. https://doi.org/10.3390/su15118588
142. Zhu J, Yang Y, Li Y, Xu P, Wang L (2020) Water footprint calculation and assessment of viscose textile. Industria Textila 71(1):33–40

The manufacturer's authorised representative in the EU is Springer Nature Customer Service Centre GmbH, Europaplatz 3, 69115 Heidelberg, Germany. If you have any concerns regarding our products, please contact ProductSafety@springernature.com

Printed and bound by CPI Group (UK) Ltd, Croydon, CR0 4YY
01/10/2026
02225859-0018